The Rise of Reptiles

320 Million Years of Evolution

Hans-Dieter Sues

Johns Hopkins University Press   Baltimore

© 2019 Johns Hopkins University Press
All rights reserved. Published 2019
Printed in Canada on acid-free paper
9 8 7 6 5 4 3 2 1

Johns Hopkins University Press
2715 North Charles Street
Baltimore, Maryland 21218-4363
www.press.jhu.edu

Library of Congress Cataloging-in-Publication Data

Names: Sues, Hans-Dieter, 1956– author.
Title: The Rise of Reptiles: 320 Million Years of Evolution /
    Hans-Dieter Sues.
Description: Baltimore : Johns Hopkins University Press, 2019. |
    Includes bibliographical references and index.
Identifiers: LCCN 2018039554 | ISBN 9781421428673 (hardcover :
    alk. paper) | ISBN 1421428679 (hardcover : alk. paper) |
    ISBN 9781421428680 (electronic) | ISBN 1421428687 (electronic)
Subjects: LCSH: Reptiles—Evolution.
Classification: LCC QL645.3 .S84 2019 | DDC 597.9—dc23
LC record available at https://lccn.loc.gov/2018039554

A catalog record for this book is available from the British Library.

Special discounts are available for bulk purchases of this book. For more information, please contact Special Sales at 410-516-6936 or specialsales@press.jhu.edu.

Johns Hopkins University Press uses environmentally friendly book materials, including recycled text paper that is composed of at least 30 percent post-consumer waste, whenever possible.

Contents

Preface

Reptiles are one of the most diverse groups of present-day vertebrates, with more than 10,000 described species of crocodylians, lepidosaurs, and turtles. If one adds birds (which are a clade of derived dinosaurs), this number more than doubles. Reptiles also have a rich and varied fossil record spanning more than 300 million years. They include the largest land animals of all time, and they repeatedly and successfully invaded the sea and took flight. An extraordinary wealth of new discoveries of extant and extinct reptiles in recent decades has revolutionized our understanding of the evolutionary history of these animals, but there exists no current overview. Carroll (1988) provided the last major English-language review of this subject. Tatarinov (2006, 2009) attempted a traditional synthesis of the evolutionary history of reptiles, which, in his usage, also included synapsids excluding mammaliaforms. Finally, Benton (2014) published an excellent introduction to the evolution of reptiles in his textbook for beginning students. Over the years several colleagues and students have asked me about a more current and detailed exploration of this topic. The present work reviews the evolutionary history based on our current knowledge. I hope that it will be of use to advanced students of vertebrate zoology and paleontology.

Wherever possible, reptilian clades are phylogenetically defined and diagnosed in terms of unambiguous shared derived features. Many taxonomic names have different meanings for different researchers, and thus phylogenetic definitions of clades are critically important.

The illustrations include numerous color images of actual fossils to introduce readers to the material basis for the study of extinct reptiles. In addition, numerous photographs of present-day reptiles introduce paleontologists to the diversity of these animals. For each major group, I have selected a phylogenetic framework or, where no comprehensive analysis was available, I have combined elements of compatible hypotheses. I have tried to cite competing hypotheses for the (inter)relationships of particular reptilian clades to alert readers to different points of view.

Anatomical terms follow the standard nomenclature used in comparative anatomy. Stratigraphic terms follow the recommendations of the International Stratigraphic Commission, and the absolute dates in Fig. 1.2 are based on the figures published by the commission in 2018. Wherever possible, the periods, epochs, and ages for taxa are listed. For the Permian Period, both the traditional and current tripartite divisions are listed.

Chapters 1 and 2 are introductory in nature. Chapters 3 through 12 review the diversity and evolution of the various reptilian clades. Chapter 13 presents a synopsis of the evolutionary history of reptiles. Chapter 14 discusses the uncertain future of reptiles in the context of the current global biodiversity crisis. A glossary provides the definitions for numerous terms used in this volume. Finally, an extensive reference list invites readers to explore aspects of the evolutionary history of reptiles more deeply. It focuses on primary studies published in recent decades because most of these publications include phylogenetic analyses of the taxa under consideration. This does not reflect the common attitude today that the older literature is "outdated." This body of work remains an indispensable source of primary data and citations.

Much to the dismay of many researchers, classification and nomenclature will always remain in flux as new studies on reptilian diversity and evolution continuously provide new data and insights. The traditional Linnaean system suggested that the relationships among reptiles were far better resolved than is the case. It also created the misleading impression that categories such as "family" were somehow comparable actual biological entities. Thus, this system is not used in this book. I still refer to genera (as most genera of extinct reptiles comprise only a single species), but otherwise the higher-level taxonomy in this book has dispensed with categories such as family, order, and class. Like personal names, names of taxa are treated as singulars and without articles.

Although this work is concerned primarily with the fossil record, it makes frequent references to molecular-based phylogenetic analyses. The latter have become the principal tool for assessing the interrelationships among extant organisms. Morphology and molecules sometimes yield strikingly different phylogenetic hypotheses, but such conflicts lead to reconsideration of both kinds of data sets.

No single researcher can be familiar with all aspects of reptilian diversity and evolution I have tried to cover in this book. Over the years I have learned a great deal from friends through discussions and/or joint research. I particularly thank Sasha Averianov, Bob Carroll, Jim Clark, Nick Fraser, Sean Modesto, Ryosuke Motani, Sterling Nesbitt, Paul Olsen, Kevin Padian, Adam Pritchard, Robert Reisz, Olivier Rieppel, Rainer Schoch, and Xiaochun Wu.

I owe a great debt to friends and colleagues who generously provided images for use in this book: Jérémy Anquetin, Sebastián Apesteguía, Sasha Averianov, the late Don Baird, Chris Bell, Mike Benton, Don Brinkman, Chris Brochu, the late Alan Charig, Jim Clark, Adam Clause, Ross Damiani, François Escuillié and the late Jean-Claude Rage, David Evans, Mike Everhart, Nick Fraser, Greg Funston and Phil Currie, Heinz Furrer, Zulma Gasparini, Stephen Godfrey, Lance Grande, Harry Greene, Romain Houssineau, Tom Jorstad, David Krause, Pete Larson, Alexandra Laube, Stephan Lautenschlager, Jun Liu, the late Junchang Lü, Tyler Lyson, Jessie Maisano and Digimorph, Heinrich Mallison, Thomas Martens, Dave Martill, Gerald Mayr, Ryosuke Motani, Sterling Nesbitt, Bill Parker, Oliver Rauhut, Robert Reisz and Diane Scott, Olivier Rieppel, the late Pamela Robinson, Rodolfo Salas-Gismondi, Torsten Scheyer, Rainer Schoch, Cesar Schultz, Paul Sereno and Carol Abraczinkas, Adam Smith, Krister Smith, Juliana Sterli, Helmut Tischlinger, Mike Triebold, Andre Veldmeijer and Erno Endenburg, Laurie Vitt, Mark Witton, Wolfgang Wüster, Pavel Zuber, and George Zug. Several museums and institutions kindly granted permission to reproduce photographs of specimens housed in their collections. Mike Ellison, Scott Hartman, Jeff Martz, and Paddy Ryan licensed use of images created by them.

Special thanks are due to Zoe Kulik and Stuart Sumida who prompted me to embark on this project. At Johns Hopkins University Press, Vincent Burke invited me to write this book and, together with Tiffany Gasbarrini, patiently saw this book through to publication. Susan Campbell provided superb copy-editing. I also thank John Hoey and his team for their care during copy production.

Last but not least, I thank Liz Sues for tolerating and even encouraging my obsession with reptiles past and present and for editorial comments on several chapters.

Outline Classification

This book uses a phylogeny-based classification of Reptilia. Relative ranking of clades is indicated by indenting. Readers should consult Chapters 2 through 12 for further details.

REPTILIA
    PARAREPTILIA
        Mesosauridae
        Unnamed clade
            Millerettidae
            Ankyramorpha
                Lanthanosuchoidea
                Unnamed clade
                    Bolosauria
                    Procolophonia
                        Pareiasauromorpha
                        Procolophonoidea

    EUREPTILIA
        Captorhinidae
        Romeriida
            Paleothyris
            Diapsida
                Araeoscelidia
                Neodiapsida
                    Drepanosauromorpha
                    Weigeltisauridae
                    Testudinata
                      Testudines
                        Pleurodira
                        Cryptodira
                  "Mesozoic Marine Clade"
                    Ichthyosauromorpha
                      Hupehsuchia
                      Ichthyosauriformes
                        Ichthyopterygia
                          Ichthyosauria
                  Thalattosauriformes

Sauropterygia
Placodontiformes
Placodontia
Eosauropterygia
Eusauropterygia
Plesiosauria

Sauria*

(* denotes separate listing of the taxonomic categories within this clade below)

Phylogenetic classification of Sauria.

SAURIA
Choristodera
Lepidosauromorpha
Lepidosauria
Rhynchocephalia
Squamata
Iguania
Polyglyphanodontia
Unnamed clade
Mosasauria
Scleroglossa
Gekkota
Autarchoglossa
Scincomorpha
Lacertiformes
Amphisbaenia
Anguimorpha
Ophidia
Serpentes
Archosauromorpha
Crocopoda
Archosauriformes
Archosauria
Pseudosuchia
Suchia
Loricata
Crocodylomorpha
Crocodyliformes
Mesoeucrocodylia
Notosuchia
Neosuchia
Eusuchia
Crocodylia

Avemetatarsalia
  Aphanosauria
  Ornithodira
    Pterosauria
    Dinosauromorpha
      Dinosauriformes
        Dinosauria
          Saurischia
            Theropoda
              Tetanurae
                Avialae
                  Aves
            Sauropodomorpha
            Sauropoda
          Ornithischia
            Genasauria
              Thyreophora
                Stegosauria
                Ankylosauria
              Neornithischia
                Marginocephalia
                  Ceratopsia
                  Pachycephalosauria
                Ornithopoda

The Rise of Reptiles

320 Million Years of Evolution

1 Introduction

A lifetime of fascination with, and many years of study of, reptiles past and present has led me to writing this book. For several years I have reviewed all I have learned in the course of my own research and surveyed the vast, widely scattered literature on the subject. What follows is an attempt to present an overview of the evolutionary history of reptiles, from their modest beginnings to the diversity of species with which we share this planet today.

There are more than 10,000 described present-day species of reptiles excluding birds. This impressive, and still growing, number represents just three major lineages—turtles, lepidosaurs (lizards, snakes, and the tuatara), and crocodylians (Fig. 1.1). Along with birds (which, in phylogenetic terms, are a clade of feathered theropod dinosaurs), they are the survivors of a once far more varied group of vertebrates.

For much of the past 300 million years or so, reptiles have been the most numerous and diverse land-dwelling vertebrates on Earth. They repeatedly and successfully conquered the air and adapted to life in the sea. The Mesozoic Era, ranging from 251 to 66 million years ago (Fig. 1.2), represented the acme of reptilian abundance and diversity and thus has long been known as the Age of Reptiles (Mantell 1831). One group of reptiles, dinosaurs, included the largest land-dwelling animals of all time and another, pterosaurs, included the largest flying animals that ever existed. Only after the extinction of dinosaurs other than birds at the end of the Cretaceous Period, some 66 million years ago, did the major lineages of extant mammals (including our own) diversify. Even today, however, reptiles including birds (with a combined total exceeding 20,000 described species) substantially outnumber mammals (about 6,500 species; Burgin et al. 2018). Calling the Cenozoic Era the Age of Mammals merely reflects the fact that we, a species of mammal, are writing the history of life.

Many years ago, Alfred Sherwood Romer, the leading vertebrate paleontologist of the day, asserted that "the general pattern of reptilian evolution has become clear in most regards" (Romer 1971:103). Fortunately for later generations of researchers, this claim proved far off the mark. In recent decades, countless new reptilian fossils have been discovered across the globe, both in previously unexplored regions and in well-known locations. Combined with new approaches to the analysis of the interrelationships of organisms and to the interpretation of fossil remains, these discoveries have revolutionized our understanding of the reptilian Tree of Life.

Every year, driven by concerns about rapidly vanishing natural habitats around the world, exploration efforts have led to the discovery of scores of new species of present-day reptiles. Molecular methods—most recently, the sequencing of entire genomes—provide critical new insights into the diversity and interrelationships of

Figure 1.1. Representatives of the four major clades of present-day reptiles (excluding birds). **A**, Crocodylia: spectacled caiman (*Caiman crocodilus*); **B**, Rhynchocephalia: tuatara (*Sphenodon punctatus*); **C**, Squamata: Boyd's forest dragon (*Lophosaurus boydii*); **D**, Testudines: Adanson's turtle (*Pelusios adansonii*). A, courtesy of Laurie Vitt; B, courtesy of Paddy Ryan; C-D, courtesy of Division of Amphibians and Reptiles, National Museum of Natural History.

extant species. This unprecedented wealth of new data has clarified many issues but has also generated many more questions. Thus, this book can provide only a "snapshot" of our current understanding of the evolutionary history of reptiles.

Fossils

Fossils are one of the principal sources of information for the study of organic evolution. Although impressive in its own right, the diversity of present-day multicellular organisms represents only a tiny fraction of all living beings that have ever existed on Earth—by some estimates as little as 1 percent (Nee and May 1999). Without fossils, our understanding of the diversity and history of life would be extremely limited. However, even the most

exquisitely preserved fossil retains only a small amount of the total biological information the original lifeform once did. Thus, fossils must be interpreted in the context of what we know or can learn about present-day organisms.

Fossils are the remains or traces of ancient life. "Fossil" derives from the Latin verb *fodere*, to dig. In the case of reptiles, fossils are typically bones and teeth, which are already fully mineralized in the living animal and thus more likely to survive the death and dissolution of their owner's body. Only in exceptional circumstances have soft tissues been preserved as impressions or become rapidly mineralized soon after an animal's death. Sometimes trackways and other traces of life activity (such as eggs and burrows) can be confidently attributed to

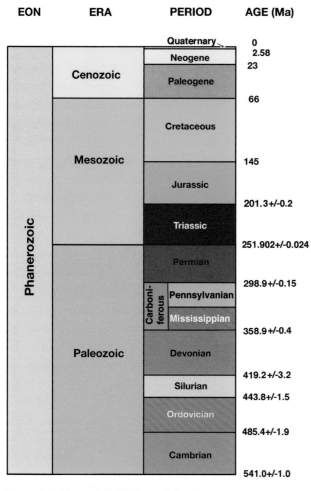

EON	ERA	PERIOD	AGE (Ma)

Figure 1.2. The major divisions of the Phanerozoic Eon. Radiometric ages as of 2018. Nomenclature and color scheme are based on the standards of the International Commission on Stratigraphy.

actions with the surrounding sediment take place. Occasionally, this leads to rapid precipitation of minerals such as calcium carbonate around the remains, encasing and protecting them from further destruction.

Fossils have established that there were once untold numbers of animals and plants unlike anything alive today. These extinct life-forms frequently show characters or combinations of features not found in any present-day species. Sometimes, fossils reveal that similar characters evolved independently in different groups of organisms. Fossils also help establish a time frame for evolutionary trees by providing minimum ages for the divergence of present-day lineages, and they show us that many groups of organisms were once much more diverse than today in terms of number of body plans and species and had different geographic distributions. Finally, fossils provide important data concerning the countless changes in climate and the ever-changing distributions of land and sea throughout Earth's long history and the effects of these changes on evolutionary events.

Reasonably complete skeletons of extinct reptiles are not common. The spectacular skeletal mounts of dinosaurs and other large extinct reptiles on display in natural history museums represent examples of exceptional preservation or reconstructions based on (often multiple) fragmentary specimens. Even in places where reptilian fossils are abundant, the remains are commonly isolated bones and teeth. Yet, to the trained eye, such remains reveal a wealth of information about the reptiles to which they originally belonged. Bones and teeth record changes both subtle and gross during the life of an animal. Sophisticated analytical methods for studying the microstructure and chemical makeup of skeletal elements continue to provide new information concerning the growth, diet, and health of their former owners.

In a few extraordinary occurrences, known as conservation fossil *Lagerstätten,* particular depositional conditions have led to the preservation of complete, articulated reptilian skeletons, frequently retaining even traces of soft tissues. The fossils of the Paleogene (Eocene) strata of the Messel Pit in Germany provide a particularly spectacular example, with mammals preserving hair and gut contents and birds preserving feathers (some even retaining traces of their original color) and gut contents (K. T. Smith et al. 2018). Such occurrences offer unmatched insights into the

particular groups of extinct reptiles and provide important additional insights into their biology.

The likelihood that an individual organism will become a fossil is vanishingly small. As soon as an animal has died, most of its bodily substance is quickly recycled. Decay and dissolution of the soft tissues start immediately, nourishing legions of microorganisms. Scavengers both large and small will feed on the remains and often scatter what is left. Only if an organism is rapidly buried after death is there any chance that its hard parts will be preserved. Even then, the chemical conditions in the sediment have to be just right to keep those hard parts intact, or they too will eventually dissolve and vanish. As the soft tissues decompose, complex chemical inter-

evolutionary history of many groups of animals and plants and occasionally preserve life-forms for which there is little or no fossil evidence anywhere else.

The quest for fossils in the field requires patience and luck. First, locations of potentially fossil-bearing sedimentary rocks are identified through study of the geological literature and maps. Then the real effort begins—looking for and prospecting exposures of rocks. Some regions of the world have vast open terrain with little or no vegetation and rock exposures everywhere. These are ideal hunting grounds for fossils. By contrast, searching for fossils in densely settled or vegetated areas like eastern North America is much more challenging. There, outcrops of fossil-bearing rocks are confined to quarries, road cuts, construction sites, riverbanks, and sea cliffs.

When visiting a potential fossil locality, collectors first scan the rock face for any bones or teeth already exposed by wind and rain. Occasionally such pieces will lead to more complete skeletal remains still buried in the rock. At other sites, collectors must split rock to find fossils embedded in it. When smaller bones and teeth occur in loose sediments, such as sands and clays, screening dry or wet sediment through sets of sieves with a range of mesh sizes is the most effective method used to recover these remains.

Larger fossils are usually exposed to the degree necessary to determine their extent and position in the rock. They are then encased in protective jackets of plaster-soaked bandages for safe transport back to a laboratory, where preparation—the actual excavation of the bones and teeth—takes place. Standard tools for this work include a kind of miniaturized jackhammer (airscribe), chisels, dental picks, and sharpened rods of steel or tungsten carbide mounted in pin vises. Detailed preparation work is done with the aid of microscopes. Fossil bones are commonly fragile and riddled with fractures, requiring consolidation as well as repair of cracks. Skeletal remains embedded in certain kinds of rock, especially limestone, can often be completely extracted through repeated immersion of the rock in greatly diluted acetic or formic acid. Limestone, composed primarily of calcium carbonate, readily dissolves in these acids, exposing bones and teeth, which are made up of a different calcium mineral, hydroxyapatite, and thus do not dissolve as readily. This technique requires skill and patience but can yield exquisitely preserved specimens. Occasionally, bones have largely weathered away or are soft and crumbly in hard rock. In such cases, any remaining bony substance is carefully removed, leaving a sharp impression in the rock matrix. This impression is then cast with a flexible molding compound (such as silicon rubber) to create a copy of the shape and surface texture of the original fossil.

Darwin (1859:310-311) noted despairingly, "I look at the natural geological record, as a history of the world imperfectly kept, and written in a changing dialect. . . . Only here and there a short chapter has been preserved: and of each page, only here and there a few lines." Although the fossil record has grown immeasurably since Darwin penned these lines, parts of it remain tantalizingly incomplete. Certain intervals of Earth history apparently left little if any fossil record. However, paleontologists have not yet fully explored many regions of the globe, and even historical localities continue to yield important new discoveries.

Molecules

Biological molecules form the other major source of information for reconstructing the evolutionary history of organisms. All living beings are made up of proteins. Proteins, in turn, are composed of chains of smaller molecules known as amino acids. There are 20 common types of amino acids that can be assembled in any order to form a tremendous variety of proteins. The instructions for assembling proteins are encoded in a highly complex molecule, deoxyribonucleic acid (DNA). DNA is a long molecule with a "backbone" composed of alternating phosphate and sugar subunits. Attached to each sugar subunit is one of just four kinds of bases (nucleotides): adenine, cytosine, guanine, and thymine. A set of three bases (triplet) represents a single amino acid in a protein. A chain of such triplets makes up a gene. Over time, bases are changed, inserted, or lost, creating genetic mutations. Such mutations can be harmful by disabling a protein. However, many, perhaps most, mutations appear to do little if any harm, presumably because the genetic code includes much redundancy, and many regions of it do not code for proteins.

Species diverge from common ancestors through changes in their DNA. By comparing the structure of the

DNA molecules between different species, biologists can track changes in the nucleotide sequences that have occurred since the divergence of those species from their most recent common ancestor. Assuming that such changes took place at fairly constant rates over long spans of time, researchers can estimate the amount of time it took for the differences to develop and compare the results to the known fossil record. For example, humans and chimpanzees share some 96 percent of their DNA. Calculations for the time of the divergence of the human and great-ape lineages have estimated dates between six and seven million years, which fits well with the geological ages for the oldest known fossils of human-like primates from Africa. DNA sequencing has become the primary tool for reconstructing the interrelationships of present-day (and a few geologically young) organisms.

Much has been made of the fact that molecular and morphological studies often yield strikingly different results. In such situations, potential issues affecting both sets of data must be carefully examined. Researchers frequently assume that molecular data are inherently superior to morphological information for reconstructing phylogenetic relationships. In most analyses of sequence data, nucleotide sites (or amino-acid residues) are treated as characters. In a large data set, such data can be heterogeneous. Another issue, known as long-branch attraction, emerges when evolutionary divergence between lineages under study occurred far back in time. If some lineages have much longer temporal ranges than others, it is possible that these lineages evolved the same nucleotide at the same site because, after all, there are only four possible nucleotides. Surprisingly, adding taxa to the analysis will compound this problem further rather than resolve it (Felsenstein 1978).

Molecular studies on present-day amniote tetrapods sample only a few lineages in the Tree of Life—mammals, turtles, lepidosaurs, crocodylians, and birds. These groups all diversified within the last 100 million years but are connected to each other by lineages that ranged over hundreds of millions of years (Z. Wang et al. 2013).

Philippe et al. (2005) noted that molecular methods for reconstructing phylogenies assume to at least some extent that rates of molecular evolution are broadly comparable across lineages. In the case of reptiles, a number of studies have already challenged this assumption. Cro-codylians and turtles have the slowest rates of molecular change reported among tetrapods, and snakes have some of the fastest (Hugall et al. 2007). Averaging rates of change across lineages in many molecular studies pulls slowly evolving groups toward one another and pushes rapidly evolving taxa toward the base of the phylogenetic tree. Thus, much of the deep history of evolutionary lineages can probably not be discovered using molecular methods.

Ideally, the comprehensive study of the evolutionary history of any group of organisms should draw on as many sources of information as possible: molecules, morphology, behavior, and ecology (total evidence approach; Eernisse and Kluge 1993).

Establishing the Age of Fossils

Establishing the absolute ages of fossils is essential for establishing time frames for the evolutionary diversification of organisms. In most cases, however, it is impossible to date an individual fossil directly.

Only organic remains younger than about 60,000 years can be dated directly by using a radioactive isotope of carbon, carbon-14. Living animals and plants continuously take up this relatively uncommon isotope from the environment, but this uptake ceases when organisms die. Carbon-14 has a half-life of $5,730 \pm 30$ years, which represents the period of time during which half of the original amount of carbon-14 has changed into nitrogen-14. The amount of this isotope remaining can be measured and provides a fairly accurate estimate of the age of a specimen.

By contrast, most of the fossils discussed in this book are much older—tens to hundreds of millions of years (Fig. 1.2). Although these fossils cannot be dated directly, igneous rocks interlayered with the sedimentary deposits containing the fossils can be dated using isotopes of various chemical elements with much longer half-lives than carbon-14. A widely used system involves two isotopes of the noble gas argon, argon-40 and argon-39. Early attempts to determine the absolute ages of such rocks using isotopes were often fraught with large measurement errors, and reported ages varied considerably, but advances in methodology and analytical instrumentation in recent years have made age determinations increasingly more consistent and precise.

For most extinct species, determination of geological age relies on methods more indirect than radiometric dating. Fossils occur in sedimentary rocks for which the age can often be assessed by the presence of so-called index fossils. Strata deposited at different locations often yield fossil remains of the same species. If such taxa were short-lived (in terms of geological time, existing for only hundreds of thousands or a few million years), one can determine that their host rocks were laid down within that particular time interval. The shorter the geological range of a particular species, the more precisely the sedimentary deposits containing its remains can be dated. In addition to being geologically short-lived, index fossils should ideally be common and widespread.

Pollen and spores have proven particularly useful as index fossils for dating continental sedimentary rocks. Plants produce them in enormous quantities. Wind or water often disperses these tiny propagules over great distances; some even end up in marine deposits. Furthermore, many characteristic types of pollen and spores came from plants that were apparently short-lived (in terms of geological time) and widely distributed. This combination of attributes makes them particularly valuable index fossils.

Another increasingly important tool for correlating rocks on a global level draws on changes in Earth's magnetic field over time. At the present time, the North Magnetic Pole is located close to our planet's north rotational pole—a condition known as normal polarity. For reasons still not fully understood, Earth's magnetic field changes its polarity at irregular intervals, with the North Magnetic Pole moving close to the south rotational pole (reversed polarity). Such reversals have occurred many times during the last 600 million years. They can be traced in the rock record because certain minerals are easily magnetized. One such mineral, an iron oxide known as magnetite, is common in various kinds of rock, including basalts. As molten lava cools down to form basalt, it passes through a threshold temperature known as the Curie point, the temperature at which magnetite and certain other minerals take up and lock in magnetization from Earth's field. In sedimentary rocks, minute particles of magnetic minerals become aligned with Earth's magnetic field when the sediments are deposited. Researchers can measure this preserved magnetization and reconstruct the succession of reversals of Earth's magnetic field in geological time, establishing what is called the Geomagnetic Polarity Time Scale. Using the absolute ages of rocks for precise calibration, a regional magnetostratigraphic succession can be correlated with this global scale.

Linnaean Classification

Classifying things is a basic human activity, and classifying the natural world is no exception. Classification allows communication about life's diversity and bestows a sense of stewardship. As common (vernacular) names of animals and plants differ considerably even among closely related languages, biologists long ago settled on a universal system for classifying them to facilitate international scientific communication, employing the binominal system introduced by Carl Linnaeus in the eighteenth century (Linnaeus 1758). In this system, organisms are grouped together in categories, known as taxa (singular, taxon), within an explicitly hierarchical classification. The smallest commonly used taxonomic category, the species, bears a unique double name (binomen), much like our own personal names—most people have a surname and at least one given name. In the case of species, the equivalent of the surname is the genus name. A genus comprises two or more closely related species. The species name corresponds to the given name for a person. Each species is based on one particular specimen, which is the holotype and provides the ultimate standard of reference for that species. Linnaeus christened our own species *Homo sapiens*, with *Homo* (Latin for "man") being the genus name and *sapiens* (Latin for "wise") the species name. Later researchers also referred to *Homo* several extinct species that proved closely related to anatomically modern humans. However, the species name *sapiens* is restricted to the latter. Traditionally, the names of genera and species were mostly derived from Greek or Latin words because Latin was universally used for scholarly communication in Linnaeus's day, much as English is today. As individual genus names can be employed only once, new names sourced from other languages have become increasingly more common over time.

Linnaeus and his successors developed a formal hierarchy of nested sets of taxa to order the diversity of life:

genera were grouped into families, families into orders, orders into classes, classes into phyla, and phyla into kingdoms. (Linnaeus himself originally used only class, order, genus, species, and variety.) As our knowledge of the diversity of life grew, however, Linnaean classification continuously required introductions of additional formal categories. In recent years, many researchers have begun to abandon these ranks altogether, retaining only species.

Linnaeus and his early followers classified organisms based primarily on overall similarity. Many groups were recognized based on the absence of features: for example, the vast majority of multicellular animals were lumped together as invertebrates because they lack a vertebral column. To Linnaeus and other early students of the living world, the purpose of classification was to bring order to nature's vast diversity.

Phylogeny and Classification

Charles Darwin revolutionized scientific thinking about the history of life. He argued that species continuously gave rise to other species over time. Following his lead, researchers moved from searching for features with which to classify groups of organisms to exploring evolutionary links among groups. They sought to reconstruct detailed ancestor-descendant sequences and then explained them in terms of adaptations to particular modes of life. This led to the common but incorrect assumption that paleontologists can somehow "read" patterns of descent in the fossil record.

The work of Willi Hennig profoundly changed the way biologists examine how organisms are related to each other. In a book first published in English in 1966, Hennig argued that relationship is more meaningfully defined in terms of common ancestry rather than overall similarity. His fundamental premise is that two taxa are more closely related to one another than either is to a third only if they share a more recent common ancestor. If this is the case, the former two should share derived features (which Hennig termed "synapomorphies") that they acquired from their most recent ancestor and that are still absent in the third. If they do, the two taxa form a monophyletic or natural group, also known as a clade. Once synapomorphies for a particular clade have been identified, they can no longer be used for determining relationships within that clade, and other features need to be identified to sort out the internal structure of that clade. If the same or similar derived features are present in unrelated groups, they are considered the result of independent, convergent evolution; such features are known as homoplasies. Hennig's approach is concerned with establishing the degree of relationship between taxa rather than ancestry. Although species obviously have ancestors, ancestry is scientifically untestable. For taxon A to be the ancestor of taxon B, it must be demonstrated that the former represents the inferred structure of the common ancestor of A and B in every detail. Not only is this unlikely, but then there would be no shared derived features to link the two taxa.

In order to determine whether a particular feature is primitive (plesiomorphic) or derived (apomorphic) for a set of taxa, it is compared with the condition in groups that lie outside the set under consideration. For example, when comparing features between different reptiles, a researcher would look at the corresponding traits in frogs and salamanders. Such groups are known as outgroups.

At first glance, an iguana and an alligator resemble each other much more closely than either resembles a chicken. The lizard and the alligator are usually grouped together because they share many similarities, such as scaly skin, typically sprawling limb posture, and a long tail. However, these features are shared by most reptiles and represent primitive character states (which Hennig termed "plesiomorphies") for that group. Closer examination shows that the alligator and the chicken share a number of derived features (synapomorphies), such as the presence of a muscular compartment of the stomach (gizzard), the complete division of the heart into four chambers, and the extensions of the air space in the region of the middle ear into the surrounding bones of the cranium and mandible. They inherited these shared derived character states from their most recent common ancestor. The iguana lacks these features. Thus, the alligator and the chicken are considered more closely related to each other than either is to the iguana.

When looking at the distribution of derived character states in a group or groups of organisms, researchers often find that more than one possibility can account for the observed distribution of features. Each possibility involves different assumptions about how often the various features were acquired or lost. Following the principle of parsimony—the simplest explanation for the data is the most likely—scientists choose the arrangement that accounts for acquisition of the greatest number of features in the simplest way. The states of characters are coded in numbers. The simplest example of such states is the absence (scored as 0) or presence (scored as 1) of a feature. For data sets with large numbers of taxa and characters, computer algorithms assist in the search for the simplest arrangement.

Hennig united taxa solely on the basis of derived features they share. Taxa are arranged into nested sets. In illustrated diagrams, these sets are joined by lines that illustrate the successive acquisition of features. A branching diagram in which all taxa occupy the end points of branches replaces the traditional family tree. Time is no longer considered a critical factor for hypothesizing relationships. The key advantage of Hennig's approach is that researchers can continuously reassess old and add new characters for groups of taxa and falsify or modify hypotheses of relationships.

Linnaeus and his followers classified organisms based primarily on the presence or absence of particular features. More recently, informed by Hennig's approach, biologists began to define groups of organisms based on their ancestry. Under this approach, known as phylogenetic taxonomy and pioneered by Gauthier and de Queiroz (1990), an organism is assigned to a particular group based on its place on the tree of life. Characters are used to reconstruct this tree, and shared derived features are employed to characterize, or diagnose, particular clades of organisms. In the previously discussed example, the alligator and the chicken are assigned to a clade Archosauria (from Greek *archon*, ruler, and *sauros*, lizard), which does not include lizards.

Clades can be defined in various ways. The first definition is based on the presence of one or more apomorphies. An example is a definition of birds (Aves) as "all dinosaurs with wings and primary flight feathers." However, interpretations of features as primitive or derived frequently change as new phylogenetic analyses draw on a broader range of taxa and new or revised character assessments. Most researchers prefer other kinds of definition, which explicitly specify the basal member of a particular clade. Stem-based definitions of taxa refer to a clade that includes all the descendants of an event when a new group diverges from an ancestral stem. An example is a stem-based definition of the dinosaurian clade Saurischia as "all dinosaurs more closely related to *Tyrannosaurus rex* than to *Triceratops horridus*." Finally, a taxon can be defined using a basal node. In the earlier example of Aves, a node-based definition could be "the most recent common ancestor of *Archaeopteryx lithographica* and *Passer domesticus* (house sparrow) and all extant and extinct descendants of that ancestor."

Wherever possible, clades should be defined by bracketing them with present-day species, which provide much more phylogenetically useful information than even the best-documented extinct taxa. Crocodylia can be defined as the most recent common ancestor of all present-day crocodiles, alligators, and gharials and all extant and extinct descendants of that ancestor. This node-based clade is firmly bracketed by extant species and known as a "crown group." Many Mesozoic crocodile-like reptiles were historically classified as crocodylians based primarily on overall similarities. However, although variously related to crown-group crocodylians, these taxa lack some or even most of the features considered diagnostic for the crown group. They are now excluded from the crocodylian crown group and are referred to as "stem-crocodylians." Some authors use the concept of a total group (denoted by the prefix Pan-, from Greek *pas*, all), which comprises a particular crown group and all organisms more closely related to it than to any other crown group. For example, Pan-Aves comprises birds (Aves) and all taxa more closely related to them than to Pseudosuchia (Pan-Crocodylia), which comprises crocodylians and all taxa more closely related to them than to birds (Gauthier and de Queiroz 2001).

Although increasingly widely used, the phylogenetic approach to biological classification has been much criticized (e.g., Nixon and Carpenter 2000), primarily because of the considerable instability of classifications introduced by competing hypotheses of phylogenetic relationships. However, biological classification has never been stable. It is important to remember that, as in any other field of scientific inquiry, new discoveries continuously test existing hypotheses. A biologically meaningful classification can never be fixed but should reflect the continually changing views of relationships.

Importance of Fossils for Phylogeny

Since the publication of Darwin's magnum opus (Darwin 1859), fossils have been touted as the key evidence for evolutionary change. Given the spectacular advances in molecular biology in recent years, however, many biologists have come to question the utility of fossils for reconstructing the Tree of Life (e.g., Patterson 1981). One example pertinent to this book underscores the enduring significance of fossils for this purpose.

Zoologists have long recognized that, among present-day vertebrates, birds are most closely related to crocodylians. This group is, in turn, most closely related to lepidosaurs (lizards, snakes, and the tuatara), to the exclusion of mammals. Gardiner (1982) challenged this orthodox view. Based on 37 derived features for extant amniotes, he hypothesized that birds and mammals are most closely related to each other and grouped them together as Haemothermia (from Greek *haima*, blood, and *therme*, heat; in reference to their warm-bloodedness). Gardiner united Haemothermia and Crocodylia as Thecodontia and then grouped Thecodontia with turtles as Euamniota.

Gauthier et al. (1988a) tested Gardiner's hypothesis. As a first step, they explored only the phylogenetic relationships among extant amniotes. Gauthier and his colleagues critically assessed the characters compiled by Gardiner (1982) and by a supporting study by Løvtrup (1985). They modified some of these characters and their distribution among present-day tetrapods in light of new data and added new ones. Their revised data set had 109 characters, most of which concern "soft" features that rarely if ever become fossilized. Present-day lungfishes (the closest relatives of tetrapods surviving today) and amphibians served as outgroups for identifying particular character states as primitive or derived. Gauthier and his colleagues avoided using any information from fossils in assessing character states for extant groups in the phylogenetic analysis. The result of their analysis differed from Gardiner's study in recovering the traditionally recognized relationship between Crocodylia and Aves, with Mammalia as the sister group to that clade.

Gauthier et al. (1988a) then compiled a second data set, to which they added 24 tasa of extinct tetrapods as well as extinct representatives of the five major groups of present-day amniotes. This second phylogenetic analysis involved 67 "soft" features as well as 207 "hard" (skeletal) characters that could potentially be preserved in fossils. Many of these "hard" characters had previously been excluded from consideration because they were found only in particular present-day taxa and thus were not helpful for elucidating the relationships between groups. When Gauthier and his colleagues added fossil amniotes to their data matrix, many of these seemingly unique (autapomorphic) traits emerged as more widely shared derived features. Again, present-day lungfishes (dipnoans) and amphibians served as outgroups for identifying "soft" features as primitive or derived. By contrast, "hard" characters could now be compared across a wide range of extinct tetrapods, some of which turned out to be more closely related to amniotes than to either amphibians or dipnoans. In this second analysis, Gauthier and his colleagues recovered the traditional phylogeny of amniotes, with mammals (and stem-mammals) as the sister group to reptiles including birds. Their study beautifully underscores the importance of fossils for reconstructing the evolutionary history of amniotes. Although imperfect, fossils remain a key source of data for working out the evolutionary transformations of characters and thus for testing relationships between taxa.

The Skeleton of Reptiles

Here follows a brief account of features of the reptilian skeleton. For additional details, interested readers should consult Romer's (1956) classic *Osteology of the Reptiles*. Furthermore, the benchmark series *Biology of the Reptilia* (1969–2010) contains a number of detailed reviews dealing with the skeleton of present-day reptiles.

Skull

The skull consists of the cranium and mandible (Fig. 1.3). In the cranium, a set of bones surrounds the brain (braincase) and is typically loosely attached to the elements of the skull roof, palate, and sides. An opening for the nostril (external narial fenestra or naris), the orbit (eye socket), and (in most reptiles) one or two temporal openings behind the orbit perforate either side of the cranium.

From the tip of the snout back, the side of the cranium comprises the following bones: the premaxilla and maxilla, both of which usually bear teeth; the lacrimal and

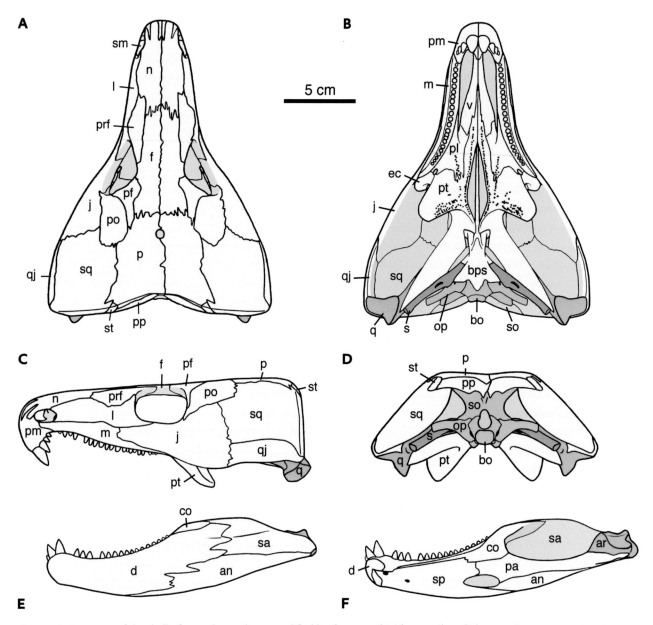

Figure 1.3. Structure of the skull of an early reptile, exemplified by the captorhinid eureptile *Labidosaurus hamatus*. Cranium in **A**, dorsal; **B**, palatal; **C**, lateral; and **D**, occipital views. Lower jaw in **E**, lateral; and **F**, medial (lingual) views. Dermal bones, pale yellow (rendered darker in deeper areas); bones of endochondral origin, orange; bones of the braincase, green; and stapes, pink. Abbreviations: an, angular; ar, articular; bo, basioccipital; bps, basisphenoid + parasphenoid; co, coronoid; d, dentary; ec, ectopterygoid; eo, exoccipital; f, frontal; j, jugal; l, lacrimal; m, maxilla; n, nasal; op, opisthotic; p, parietal; pa, prearticular; pf, postfrontal; pl, palatine; pm, premaxilla; po, postorbital; pp, postparietal; prf, prefrontal; pt, pterygoid; q, quadrate; qj, quadratojugal; sa, surangular; sm, septomaxilla; so, supraoccipital; sp, splenial; sq, squamosal; st, supratemporal; v, vomer. Courtesy of Sean Modesto and Robert Reisz.

prefrontal, which form the anterior margin of the orbit; the postfrontal, postorbital, and jugal, which bound the orbit posteriorly and ventrally; and the quadrate, quadratojugal, and squamosal, which together make up the "cheek" (temporal) region of the cranium behind the orbit. The quadrate forms the jaw, or craniomandibular, joint with the articular bone of the lower jaw. Along with the epipterygoid, a rod-like bone that extends between the palatal complex and the skull roof, the quadrate develops from the embryonic palatoquadrate cartilage.

The skull roof comprises (from front to back) the typically paired nasals, frontals, and parietals. The nasals usually separate the external narial openings, and the frontals extend between the orbits along the midline of the skull. In many reptiles, the parietals enclose between them a foramen, which is related to the pineal gland, a structure of the brain that is sensitive to changes in ambient light and controls circadian rhythms. Basal reptiles retain additional pairs of bones—the postparietals, supratemporals, and tabulars—along the posterior edge of the skull roof, but these elements are mostly or altogether absent in more derived taxa.

Most reptiles have openings in the roof and sidewall of the cranium behind the orbits. These fenestrae, or temporal openings, are related to the development of the adductor jaw musculature (Frazzetta 1968; Werneburg 2012). When these jaw-closing muscles contract, they generate considerable tensile stresses on the bones to which they attach. Thus, muscles tend to attach to thickened areas on bones such as ridges and crests. Development of larger openings with rounded bony edges is also thought to dissipate stresses because forces exerted by the jaw musculature can be distributed along the perimeter of these openings.

On the back of the skull (occiput), the supraoccipital connects the posterior edge of the skull roof to the braincase. Together with the paired opisthotics and exoccipitals, it typically bounds a large median opening in the occiput, the foramen magnum, through which the spinal cord passes from the brain posteriorly. The paroccipital process of the opisthotic extends laterally toward the bones of the temporal region. Together with the prootic, and (when present) the laterosphenoid more anteriorly, the opisthotic forms the sidewall of the braincase. It also houses the semicircular canals of the inner ear. Below the foramen magnum, the basioccipital and exoccipitals form the occipital condyle, which is part of the joint between the head and neck (atlanto-occipital joint). The orbits are separated by a thin interorbital septum, which occasionally is partially or fully ossified.

On the palate, the premaxillae make up the tip of the snout, followed behind by the paired vomers and, further posteriorly, the pterygoids, which extend posterolaterally and contact the quadrates. The palatine and ectopterygoid connect the pterygoid to the maxilla and jugal on either side of the cranium. Along the midline at the back of the palate, the parasphenoid covers, and is typically fused to, the ventral surface of the basisphenoid at the base of the braincase. Anteriorly, it forms a slender (cultriform) process that extends forward in a space bounded by the pterygoids (interpterygoid vacuity). In many reptiles, the palatine, parasphenoid, pterygoid, and/or vomer bear small teeth or denticles.

The mandible comprises two rami (lower jaws or hemimandibles), which are typically connected anteriorly by ligaments to form the mandibular symphysis. In turtles and various other reptiles, the two rami completely fuse at the symphysis. The dentary is the principal tooth-bearing bone of each mandibular ramus. Posteriorly, it contacts the surangular dorsally and the angular ventrally. The lingual (medial) surface of the dentary bears a groove for the Meckelian cartilage and is covered by the splenial anteriorly and the prearticular more posteriorly. The coronoid bone forms a raised point or process for the insertion of adductor jaw muscles and occasionally extends forward between the dentary and the splenial or prearticular. The posterior end of the mandibular ramus is formed by the articular and bears a typically concave surface for contact with the quadrate.

The hyoid apparatus supports the floor of the mouth and the tongue. Parts of this structure are sometimes calcified or ossified and thus can be preserved in fossils.

The ear of most reptiles contains a single rod-like bone, the stapes (columella auris), which transmits airborne sound as vibrations from the eardrum (tympanic membrane) to the inner ear. In some basal reptiles, however, the stapes is a robust element that probably served primarily as a structural brace between the temporal region of the cranium and the braincase (Carroll 1980).

The eyes of most extant and extinct reptiles have a ring of thin bony plates (ossicles) embedded in the sclera around the pupil. This structure is thought to maintain the shape of the eyeball or serve as the site of attachment for muscles for visual accommodation. As these scleral ossicles are delicate, they are usually not preserved in fossils.

Teeth

The free part of a tooth is the crown, and the part that is set in or attached to the jawbone is the root. Not

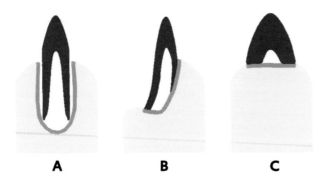

Figure 1.4. Schematic transverse sections through jaws to illustrate the traditionally recognized major types of tooth implantation in amniotes. **A**, thecodont; **B**, pleurodont; and **C**, acrodont. Jaw, beige; tooth, blue; bony lining of the alveolus or attaching the tooth to the jaw, brown.

surprisingly, in view of their varied dietary adaptations, reptiles have a wide range of tooth shapes (Owen 1840–1845; B. Peyer 1968; Edmund 1969; Berkovitz and Shellis 2017). Basal amniotes typically have teeth with simple conical crowns. In more derived reptiles, there are many variations on this pattern. The upper and lower teeth are usually spaced in such a manner that they do not come into contact with each other when the jaws are closed. Various reptilian groups, however, evolved tooth-to-tooth contact with tooth crowns modified for oral processing of food.

Three basic types of tooth implantation are commonly recognized among reptiles (Fig. 1.4). Thecodont implantation refers to the teeth set in typically discrete sockets (alveoli) in the jawbone. Periodontal ligaments anchor and control limited movement of the teeth in the alveoli. Acrodont implantation refers to the firm attachment or fusion of the teeth to the apical ridges of the jawbones. Finally, pleurodont implantation reflects the reduction or absence of the lingual bony wall of the jaw so that the tooth becomes attached only to the labial wall. There are many variations on these three basic types. Proper characterization of the mode of tooth attachment often requires histological thin-sections or high-resolution CT scanning of jaw elements.

Postcranial Skeleton

The vertebral column is divided into (from the skull back) neck (cervical), trunk (dorsal), sacral, and tail (caudal) vertebrae. The cervicals and dorsals are collectively referred to as presacral vertebrae. Each vertebra has a body (centrum) and a neural arch on top of the centrum; these two components surround the spinal cord and are separate or fused together. The neural arch bears a neural spine dorsally and is connected to the arches of adjoining vertebrae by means of articular processes known as zygapophyses. The vertebral centrum develops from tissue surrounding the notochord, a flexible rod in the back of embryonic vertebrates, and the overlying neural tube. It replaces the notochord for the most part. The centra of early reptiles often retain openings for the passage of a persistent notochord.

Unlike in many other tetrapods such as temnospondyl stem-amphibians, the vertebral centra in reptiles are large pleurocentra (Fig. 1.5). If present at all, the intercentra are small, crescent-shaped bones wedged between adjacent pleurocentra. The first two cervical vertebrae are highly modified to facilitate motions between the head and neck. The first cervical or atlas (Fig. 1.6) is composed of six distinct elements: the intercentrum of the atlas, which articulates below the occipital condyle; the pleurocentrum of the atlas; the paired neural arches of the atlas; and the paired proatlas. The atlanto-occipital joint permits rotation of the head on the neck. The second cervical or axis has an intercentrum anteriorly, and its pleurocentrum is fused to the large neural arch. The cervical vertebrae typically bear ribs, which are shorter and less massive than the ribs articulating with the dorsal vertebrae. The dorsal ribs form the ribcage. In early reptiles, each rib head has two points of contact: the capitulum contacts either the intercentrum or, more commonly, a facet (parapophysis) on the anterior edge of the pleurocentrum of the corresponding vertebra, whereas the tuberculum articulates with a facet (diapophysis) on the transverse process of the neural arch. In some taxa, the rib head forms only a single contact with the corresponding vertebra. Reptiles typically have at least two sacral vertebrae, each of which bears short, robust ribs connecting the vertebral column to the pelvic girdle on either side. The tail usually comprises many vertebrae. The centra of the caudal vertebrae bear Y-shaped ventral bones known as hemal arches or chevrons. The two arms of the Y enclose blood vessels and nerves supplying the tail. Many groups of reptiles have rod-like bones known as gastralia that develop in the ventral body wall.

A

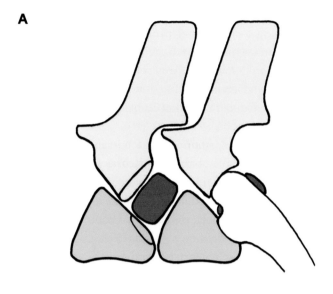

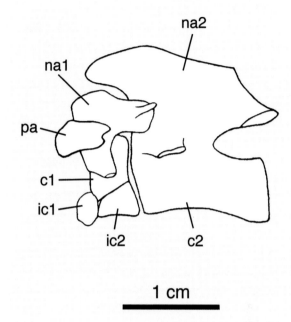

Figure 1.6. Structure of the atlas and axis of the basal diapsid *Petrolacosaurus kansensis* in lateral view. Abbreviations: c1,2, centrum of the axis and atlas, respectively; ic1,2, intercentrum of the atlas and axis, respectively; na1,2, neural arch of the atlas and axis, respectively; pa, proatlas. Modified from Reisz (1981).

B

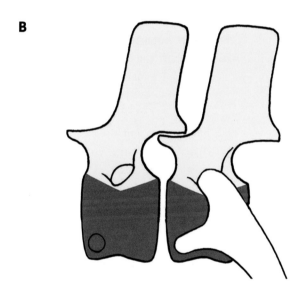

Figure 1.5. Schematic structure of two dorsal vertebrae in lateral view for **A**, a stem-amphibian, and **B**, a derived reptile. Neural arch, yellow; intercentrum, green; pleurocentrum, red; rib-head, white. A, modified from Mickoleit (2004).

Three pairs of major bones form most of each half of the shoulder (pectoral) girdle: the scapula dorsally and the coracoid (metacoracoid: Vickaryous and Hall 2006) and procoracoid ventrally. They often fuse into a single scapulocoracoid. The scapula and coracoid form a compound surface (glenoid) for articulation with the head of the humerus. In early reptiles and their relatives, the glenoid forms a single screw-shaped surface, which restricted the motion of the humerus at the shoulder joint primarily to rotation about its long axis (Jenkins 1971). In more derived groups, the scapula and coracoid each bear glenoid facets, and the humerus has a much greater range of mobility. The cleithra (if present) and clavicles are thin, narrow bones extending along the anterior margin of the scapulocoracoid. The median interclavicle connects the clavicles ventrally. The plate-like sternum is positioned ventromedial to the ribcage. It is either single or paired but is not calcified or ossified in many reptiles.

Three pairs of major elements form the pelvic girdle: the ilium dorsally, the pubis anteroventrally, and the ischium posteroventrally. The ilium is connected to the vertebral column by means of the sacral ribs. The pubis and ischium meet their opposites along a broad symphysis ventromedially. Together with the ilium, they form a prominent articular surface (acetabulum) for contact with the femur.

Each of the four limbs consists of a single proximal bone (propodial) that articulates with two more distal bones (epipodials), which in turn are connected with a series of smaller elements (mesopodials) that make up the wrist (carpus) or ankle (tarsus) and, more distally, the digits (fingers or toes; autopodials). Within each digit, the bones embedded in the palm or sole are referred to as metapodials—metacarpals in the hand (manus) and metatarsals in the foot (pes).

The major bones of the forelimb are the humerus in the upper arm and the radius and ulna in the forearm (antebrachium). The ulna extends lateral to the radius. It bears a large facet for articulation with the humerus and, above this joint surface, occasionally a bony olecranon process for insertion of the triceps brachii muscle, which extends the forearm. Several rows of small bones typically form the wrist (carpus). The proximal row contacts the radius and ulna and comprises the radiale, intermedium, and ulnare. The distal row has five bones, one contacting each digit. Between the proximal and distal rows there is an additional set of bones, the centralia.

The manus primitively has five digits (fingers), each comprising a metacarpal proximally and a series of smaller bones (phalanges) more distally. The phalangeal formula for the manus in basal reptiles is 2-3-4-5-3, but it is often considerably modified in more derived forms. The last phalanx or ungual of each digit typically bears a keratinous claw.

The hind limb comprises the femur in the thigh and the tibia and fibula in the shank. The femur typically bears distinct bony ridges or processes (trochanters) for the attachment of thigh muscles. At the knee joint, it articulates with the tibia, the principal bone of the shank (crus), and the fibula, which is situated lateral to the tibia. The ankle (tarsus) comprises a proximal row composed of the astragalus and calcaneum, an intermediate series of centralia, and a row of distal tarsals. The astragalus originally formed by fusion of the tibiale, intermedium, and one or two centralia (O'Keefe et al. 2006). The pes in early reptiles has five digits (toes) with the same phalangeal formula as the manus, but again, more derived taxa show much variation in phalangeal counts.

2 Amniotes and Reptiles

The meaning of words tends to change over time, and "Reptilia" is no exception. Linnaeus (1758) employed "Reptiles" (from Latin *repere*, to creep) for a subset of limbed members of his "Amphibia" (which included various amphibians and reptiles but also various fishes and the lamprey). Laurenti (1768) used "Reptilia" as a name for all tetrapods other than birds and mammals. De Blainville (1816) first distinguished reptiles from amphibians based on differences in the structure of their skins. Reptilian skin is covered by scales, composed primarily of keratin, that protect the skin against physical damage. Together with proteins and complex lipids in the epidermis, keratin reduces the loss of water through the skin (Alibardi 2003). By contrast, present-day amphibians have moist skin that is rich in glands and much less keratinized than that of reptiles. Gray (1825) also separated amphibians and reptiles, but most zoologists and paleontologists adopted this distinction only later in the nineteenth century.

The most obvious difference between extant amphibians and other tetrapods is the structure of their eggs. Amphibians typically lay eggs in fresh water. The eggs have a clear, jelly-like coating. The developing embryo obtains oxygen for respiration and most of its nourishment from the surrounding water and excretes its waste products into the water. Most amphibians hatch as a larva, which continues to develop in water and eventually undergoes metamorphosis to become an adult.

Extant reptiles, birds, and mammals have a different kind of egg than that of amphibians, the amniotic (or cleidoic) egg (Fig. 2.1). Haeckel (1866) first used the shared possession of this feature to unite these three groups of tetrapods as Amniota (Fig. 2.2). An amniotic egg develops membranes from "outside" the embryo (extraembryonic membranes). One membrane encloses a large sac of yolk, which is connected to the embryo's digestive tract and provides nourishment. Folds of a membrane called the amnion (from Greek *amnos*, lamb; the membrane was noticed in pregnant ewes) meet and form a fluid-filled sac, the amniotic cavity, around the embryo. The sac-like allantois grows from the embryonic hindgut and comes to surround the embryo. It serves as a receptacle for metabolic waste products and aids in respiration. Another extraembryonic membrane, the chorion, envelops almost the entire set of embryonic structures. This package, in turn, is enclosed in an outer shell, which can be either soft and parchment-like or calcified. The shell is porous and connected through the chorion to the allantois, which is well supplied with blood vessels and thus can take up oxygen from and excrete carbon dioxide to the outside environment. The development of the amniotic egg eliminated the need for a water-dwelling larval stage. In reptiles, a miniature version of the adult hatches from the egg and makes its way in life. The acquisition

Figure 2.1. Simplified drawing of an amniotic egg in longitudinal section. Shell, dark gray; chorion, dark blue; amniotic cavity, light blue; embryo, green; yolk sac, yellow; allantois, pink. Modified from Bystrov (1957).

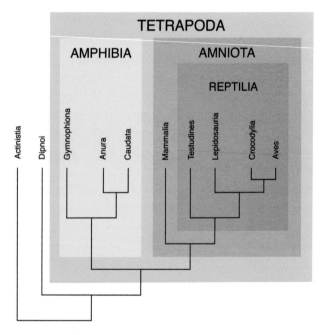

Figure 2.2. Phylogenetic hypothesis of the interrelationships of the major clades of present-day tetrapods, with coelacanths (Actinistia) and lungfishes (Dipnoi) as outgroups.

of the amniotic egg freed its producers from their dependence on water for reproduction.

Packard and Seymour (1997) proposed a scenario for the evolution of the amniotic egg in which the first step involved a reduction of the jelly-like coating of the egg

and its replacement by a yolk sac. This change in the egg's coating would also have increased diffusion of oxygen to the embryo. Simultaneously, the development of a porous membrane around the egg would have strengthened it and further increased its permeability for respiratory gases. These changes would have facilitated an increase in egg size, which, in turn, would have required provision of larger yolk resources for the embryo. In the scenario proposed by Packard and Seymour, the yolk sac was the first extraembryonic membrane to develop. The allantois initially served only for storage of metabolic waste products but subsequently increased in size and assumed a key role in the exchange of respiratory gases.

To date no undisputed Paleozoic amniotic eggs have been recovered. A possible example was reported from the early Permian (Cisuralian) of Texas (Romer and Price 1939) but its identification as a shelled egg remains questionable (Hirsch 1979). The apparent lack of amniotic eggs predating the Mesozoic Era is consistent with the hypothesis that the eggs of early amniotes had parchment-like shells, which would have rarely if ever become fossilized (Oftedal 2002). Parchment-like eggshell is still found today in most lizards and in snakes. It much more readily absorbs water than a mineralized shell (Oftedal 2002), and this attribute may have been important during the initial evolution of the amniotic egg. As the Euramerican region of Pangaea became increasingly drier after the disappearance of the Pennsylvanian tropical forest communities (Sahney et al. 2010), water loss likely became a critical issue for tetrapods. Mineralized eggshells would have limited water loss and increased the eggs' mechanical strength. They would have enabled amniotes to venture from wet habitats into drier settings and to diversify there, unlike other, more water-dependent tetrapods such as stem-amphibians.

The amniotic egg acquires its shell within the female's reproductive tract. As the egg must be fertilized before the shell is formed, the male's sperm must be introduced into the female's reproductive tract. By contrast, the males of most frogs and some salamanders simply shed their sperm on clutches of eggs deposited by the females in water. However, other present-day amphibians internally fertilize their eggs. In most salamanders, for example, the female uses her cloaca to pick up a packet of sperm (spermatophore) deposited by the male. The males of most extant (and presumably extinct) amniotes have special intromittent organs for depositing sperm inside the female's reproductive tract. Turtles, crocodylians, some groups of birds (e.g., ratites), and mammals have a median phallus. Lizards and snakes have paired lateral hemipenes. The tuatara (*Sphenodon punctatus*) lacks an intromittent organ, but Sanger et al. (2015) demonstrated that its embryo has the paired tissue buds from which the phallus develops in other amniotes.

Amniote Origins

It is not difficult to distinguish present-day amniotes from amphibians. However, there has been a long-standing debate whether various Pennsylvanian and Permian tetrapods are early amniotes or even closely related to amniotes (Ruta et al. 2003; Klembara et al. 2014). Traditionally, the origin of reptiles was linked to that of the amniotic egg. However, fossils have since established that the mammal and reptile-bird lineages had already diverged by the time the first amniotes appeared in the fossil record during the Pennsylvanian. Thus, the evolution of the amniotic egg predated this divergence, and it is possible that even some stem-amniotes had already evolved this type of egg.

The group generally considered most closely related to amniotes is Diadectomorpha (Heaton 1980; Ruta et al. 2003; Reisz 2007). Diadectomorphs and amniotes are grouped together as Cotylosauria (from Greek *cotyle*, cup, and *sauros*, lizard; based on an early misinterpretation of the occiput) and share the absence of a distinct intertemporal bone in the skull roof (still present in earlier stem-amniotes) and the shift of the postparietal and tabular bones from the posterior portion of the skull roof onto the occiput. Cotylosauria also shares the fusion of the centra of the first two cervical vertebrae (atlas and axis), the presence of broad, domed ("swollen") neural arches on the dorsal vertebrae, and a sacrum comprising at least two vertebrae. The phylogenetic analysis by Ruta et al. (2003) recovered Amniota plus Diadectomorpha as most closely related to a clade comprising *Westlothiana*, from the Mississippian (Viséan) of Scotland, and Lepospondyli, a diverse but possibly not monophyletic group of mostly small-sized anamniote tetrapods ranging in time from the Mississippian to the late Permian.

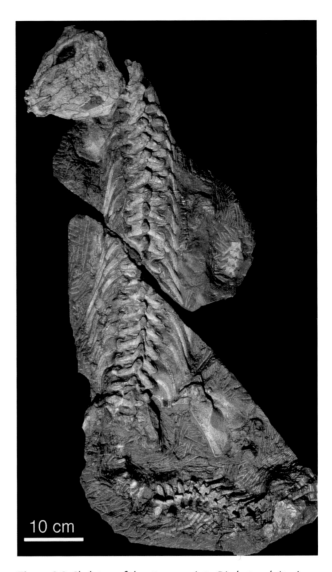

Figure 2.3. Skeleton of the stem-amniote *Diadectes absitus* in dorsal view. Courtesy of T. Martens.

Diadectes, from the Pennsylvanian and early Permian (Cisuralian: Gzhelian-Asselian) of the United States and the early Permian (Cisuralian: Artinskian) of Germany, attained a total length of up to 2.5 m (Olson 1947; Berman et al. 1998; Figs. 2.3, 2.4). Its skull has thick, porous bones, and it is often difficult to delineate individual cranial elements. The parietals are broad transversely. The quadrate is embayed posteriorly. Posteriorly, it contacts a thin plate of bone, which is part of the stapes (Berman et al. 1998). The palatines form a partial secondary bony palate. The distinctive dentition of *Diadectes* consists of forward-projecting (procumbent) incisor-like anterior

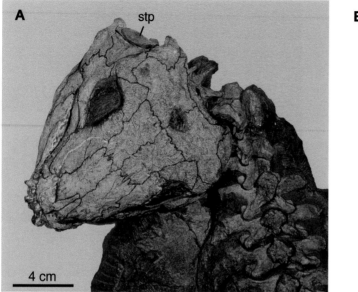

Figure 2.4. A, skull and cervical vertebrae of the stem-amniote *Diadectes absitus* in dorsal view. Abbreviation: stp, ossified stapedial plate. **B**, partial right dentary of *Diadectes sideropelicus* in occlusal view. Note incisor-like anterior and molar-like posterior teeth. A, courtesy of Thomas Martens; B, from Case (1911).

teeth and molar-like posterior teeth with transversely broad crowns (Fig. 2.4). The latter are frequently heavily worn, and scratches on the wear surfaces indicate that the upper and lower teeth partially contacted each other and that the mandible moved fore-and-aft to break down food. The voluminous, barrel-shaped ribcage of *Diadectes* indicates the presence of a capacious gut, which could have accommodated endosymbionts to break down the cellulose in the plant fodder. *Diadectes* and its close relatives were among the earliest known tetrapods capable of feeding on high-fiber plant matter (Hotton et al. 1997; Sues and Reisz 1998).

Another diadectomorph, *Limnoscelis*, from the Pennsylvanian (Gzhelian) of New Mexico and Colorado (Fig. 2.5), has (in plan view) a broadly triangular skull with large teeth at the tip of the snout (Berman et al. 2010). Its slightly recurved tooth crowns have cutting edges close to the tip of the crown, suggesting that *Limnoscelis* was faunivorous. The girdles and limbs are robust. *Limnoscelis* attained a total length of about 2 m.

Some cranial features long considered diagnostic for amniotes (Carroll 1969a) are already present in various stem-amniotes (Ruta et al. 2003; Reisz 2007; Berman et al. 2010). One is the wing-like transverse flange of the pterygoid, which bears teeth in many forms and is related to the differentiation of the pterygoideus jaw muscle. Another is the large and convex occipital condyle that contacts a ring formed by the neural arch and intercentrum of the atlas; the centrum of the atlas contacts that of the axis (Sumida and Lombard 1991). This joint between the head and neck permits a wide range of motion. Finally, amniotes and various stem-amniotes share the possession of well-developed ungual phalanges that are longer than the more proximal phalanges and support keratinous claw sheaths in life.

Amniotes

Amniota comprises the most recent common ancestor of Synapsida (mammals and their close relatives) and Reptilia and all extant and extinct descendants of that ancestor (Reisz 1997, 2007; Fig. 2.6). Early amniotes differ from other Paleozoic tetrapods in a number of skeletal features. On the skull roof, the frontal enters into the dorsal margin of the orbit in basal amniotes, whereas it is excluded from that margin by the prefrontal and postfrontal even in the diadectomorph *Limnoscelis*. The well-ossified occipital condyle is hemispherical. Unlike in *Diadectes*, the posterior portion of the cranium in

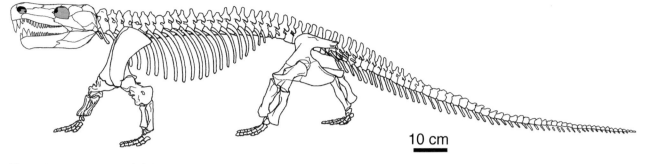

10 cm

Figure 2.5. Reconstructed skeleton of the stem-amniote *Limnoscelis paludis* in lateral view. From Berman et al. (2010).

early amniotes lacks a distinct notch for support of the tympanic membrane. The ankle of amniotes includes a large astragalus, which formed through fusion of at least three smaller bones (tibiale, intermedium, and one or two centralia). These constituent elements are still evident on the astragali in some specimens of various Paleozoic reptiles (Kissel et al. 2002; O'Keefe et al. 2006).

As noted earlier, the skin of amniotes differs from that of extant amphibians. Not only is the epidermis itself more keratinized, but it is typically covered by scales, feathers, or hair, all of which are also composed of keratins. These features protect the skin against physical damage and limit water loss. At the same time, however, this barrier greatly reduces the skin's permeability for oxygen and carbon dioxide. Thus, amniotes rely primarily on their lungs for breathing. Present-day sea snakes and some turtles still are capable of limited cutaneous respiration, and water-dwelling turtles can even employ their cloaca or pharynx to take up oxygen. Extant amphibians, much like air-breathing fishes, push air from their mouth and pharynx into the lungs. By contrast, amniotes draw air into the lungs primarily by expanding the thoracic cavity, typically through movement of their ribs, which are linked by the intercostal muscles (Brainerd and Owerkowicz 2006). In most present-day amniotes, each of the thoracic ribs is made up of multiple segments and articulates with the sternum. This enclosure of the lungs by the ribs is necessary for breathing by means of rib motion (costal aspiration).

The oldest undisputed fossils of amniotes to date have been discovered in stumps of the tree-like lycopsid plant *Sigillaria* from the Pennsylvanian (Bashkirian) of Nova

Scotia (Canada). Initially it was thought that these amniotes, along with other small animals, had become trapped and then entombed inside the hollow stumps. More recent work suggests that these animals probably lived in the stumps or took refuge there from periodic wildfires (Falcon-Lang et al. 2010). The fossils recovered from the lycopsid stump casts include the oldest known reptile, *Hylonomus*, represented by disarticulated skeletons (Carroll 1964). Geologically slightly younger Pennsylvanian (Moscovian) stump casts, also from Nova Scotia, have yielded articulated skeletal remains of the early reptile *Paleothyris* (Carroll 1969a) along with those of two taxa of basal synapsids. Carroll (1964, 1969a,b) interpreted *Hylonomus* and *Paleothyris* as stem-reptiles broadly ancestral to all later amniotes, but recent phylogenetic analyses (Gauthier et al. 1988b; Müller and Reisz 2006) instead found these taxa closely related to diapsid reptiles (Chapter 4).

The absolute age boundaries for the Bashkirian are 323.2 and 315.2 Ma (Davydov et al. 2012). This is consistent with the most recent estimate for the divergence of crown Amniota with a minimum age of 318 Ma (Benton et al. 2015). Based on these data, the book uses an estimated divergence date of 320 Ma for the divergence of Reptilia and Synapsida.

There have been reports of geologically even older skeletal remains referable to amniotes. In the 1990s, *Westlothiana* was widely publicized as the earliest known reptile. However, further preparation of the original find and the discovery of additional specimens have brought this interpretation into question. *Westlothiana* lacks some features such as the transverse flange of the pterygoid that are present in amniotes and their closest relatives.

However, it is certainly related to amniotes (Smithson et al. 1994; Ruta et al. 2003; Clack 2012). *Westlothiana* reached a snout-vent length of 11.5 cm. Although still poorly known, *Casineria*, also from the Mississippian (Viséan) of Scotland, is definitely a stem-amniote (Paton et al. 1999; Clack 2012). Its well-ossified postcranial skeleton has vertebrae with large pleurocentra, slender limbs, and five-fingered manus with curved ungual phalanges. *Casineria* attained an estimated total length of about 15 cm.

Classifying Reptiles

The scientific exploration of extant and extinct reptiles has made extraordinary strides in recent decades. Countless discoveries have uncovered a previously unimagined diversity of these animals spanning some 300 million years. This still-growing wealth of forms has continually defied classification.

Osborn (1903) first distinguished two major groups among amniotes based on the structure of the skull. One group, Synapsida (from Greek *syn*, together with, and *hapsis*, arch; in reference to the bony arch bounding the temporal opening), is characterized by the presence of a single large opening (fenestra) behind the orbit on either side of the cranium. Mammals and their close relatives (which, until quite recently, were classified as reptiles)

have this condition. The other group, Diapsida (from Greek *dyo*, two, and *hapsis*, arch; in reference to the two bony arches bounding the temporal openings), typically has two fenestrae behind the orbit on either side of the cranium. A bony arch formed by the postorbital and squamosal bones separates the upper opening from the lower one, which itself is often bounded by a bony arch below. Crocodylians, lepidosaurs, and many extinct groups of reptiles including stem-birds share this configuration, although it is considerably modified in many forms such as snakes and crown-group birds. Possibly for ease of classification, Osborn also assigned turtles and various other reptiles to Synapsida even though none of these forms have temporal openings. This led Williston (1917) to propose a third group, Anapsida (from Greek *a*, without, and *hapsis*, arch; referring to the absence of bony arches), for reptiles that lack temporal openings. He interpreted the anapsid configuration as the primitive condition for Reptilia. With changing definitions and varying constituents, the Osborn-Williston system of classification has remained in general use ever since (e.g., Romer 1956, 1966; Carroll 1988). However, many reptilian taxa do not fit neatly into this system, repeatedly prompting proposals for additional groupings over the years (e.g., Colbert 1945).

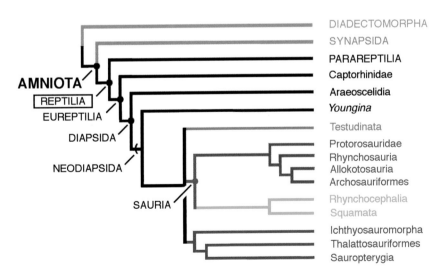

Figure 2.6. Phylogenetic hypothesis of the interrelationships of major clades of Reptilia. Diadectomorpha and Synapsida served as outgroups. Testudinata, brown; Sauria, purple; Lepidosauromorpha, green; Archosauromorpha, red; "Mesozoic marine reptile clade," blue. Dots denote node-based clades and parentheses denote stem-based clades. Topology of the "Mesozoic marine reptile clade" based on Scheyer et al. (2017).

With the adoption of Hennig's methodology for studying the relationships between organisms, the validity of Reptilia was initially questioned. If mammals and birds, historically treated as distinct classes in the Linnaean system, are separated from reptiles, "Reptilia" is reduced to a grade of amniotes that lack the diagnostic features of birds and mammals. Yet the name is historically clearly associated with, and has been in general use for, an assemblage of present-day amniotes comprising turtles, crocodylians, and lizards and their relatives. As Reptilia is a more widely known name, it is given precedence over Huxley's (1869) Sauropsida for the clade comprising reptiles and birds. Gauthier et al. (1988b) defined Reptilia as the clade comprising the most recent common ancestor of extant turtles and saurians (including birds) and all descendants of that ancestor. However, at that time, turtles were still considered most closely related to basal reptiles. The diapsid affinities of turtles are now well established, but their position within Diapsida remains to be resolved. Thus, the definition of Reptilia becomes problematic because, depending on whether turtles are or are not part of Sauria, various taxa traditionally considered reptiles would be excluded from Reptilia. Modesto and Anderson (2004) proposed a definition of Reptilia as the most inclusive clade containing *Lacerta agilis* (sand lizard) and *Crocodylus niloticus* (Nile crocodile) but not *Homo sapiens* (as a representative of Synapsida). This node-based definition is adopted here.

The evolutionary lineages leading to mammals and to birds, respectively, diverged from each other early and can already be distinguished by skeletal features among the oldest known amniotes. The traditional division of Amniota into three "classes"—reptiles, birds, and mammals—has given way to recognition of two major clades: Synapsida (comprising mammals and their close relatives) and Reptilia (comprising reptiles including birds) (Fig. 2.6). The standard textbook statement that "mammals evolved from reptiles" is phylogenetically not justified.

3 Parareptilia

A Group of Their Own

Olson (1947) proposed the name Parareptilia (from Greek *para*, beside, near, and Latin *repere*, crawl) for a group comprising turtles, various late Paleozoic and early Mesozoic reptiles, and various taxa now considered stem-amniotes (e.g., *Diadectes*). He noted that these forms share the presence of a well-developed otic notch, which is absent in other amniotes, which he named Eureptilia (from Greek *eu*, true, and Latin *repere*, crawl). Olson's hypothesis was at odds with then-prevailing ideas concerning classification of reptiles and thus was largely ignored. Laurin and Reisz (1995) formally resurrected the name Parareptilia for an assemblage comprising turtles and all amniotes more closely related to them than to diapsid reptiles. However, starting with Rieppel and deBraga (1996) and supported by all molecular studies, most phylogenetic analyses have hypothesized that turtles are related more closely to diapsids than to parareptiles.

Tsuji and Müller (2009) defined Parareptilia (Fig. 3.1) as the most inclusive clade containing *Milleretta rubidgei* and *Procolophon trigoniceps* but not *Captorhinus aguti*. Parareptiles first appeared in the fossil record in the Pennsylvanian (Modesto et al. 2015) and became diverse during the Permian Period. Most are restricted to this period, but one group, Procolophonoidea, survived the end-Permian extinction event and flourished during the Triassic. Many parareptiles likely occupied ecological roles comparable to those filled by eureptiles during later geological periods. Pareiasauria includes the earliest known large-bodied reptilian herbivores on land. Reassessments of previously described taxa and numerous recent discoveries (e.g., Tsuji et al. 2010; Modesto et al. 2015) hint at a much greater diversity of this clade yet to be discovered.

Parareptilia is characterized by a number of synapomorphies (Tsuji and Müller 2009): absence of caniniform teeth; postorbital region of the cranium short; posterior margin of the skull roof forming a deep median embayment; jugal lacking a subtemporal process; and the absence of a supraglenoid foramen on the scapulocoracoid. The crania of many parareptiles have a small lateral opening in or a distinct ventral emargination of the temporal region. Most have a posterior notch formed by the squamosal and quadratojugal, which is conspicuous in taxa such as *Macroleter* and *Procolophon* and probably supported a tympanic membrane in life. This notch, together with the rather slender stapes, indicates that parareptiles evolved an impedance-matching ear for receiving airborne sound independently from derived eureptiles and from mammals (Müller and Tsuji 2007).

Several lineages of parareptiles evolved cranial and dental features that suggest herbivory (Hotton et al. 1997; Reisz and Sues 2000; Reisz and Fröbisch 2014). This change in diet independently occurred in a number of other amniote clades during

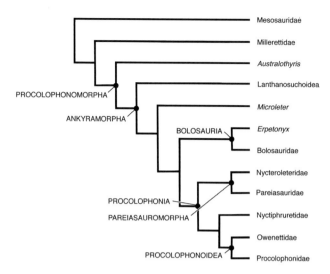

Figure 3.1. Phylogenetic hypothesis of the interrelationships of Parareptilia. Dots denote node-based clades. Based on MacDougall and Reisz (2014).

the Pennsylvanian and Permian. Feeding on plants requires various anatomical and physiological changes. Vertebrates lack the requisite enzymes to break down cellulose, which constitutes the bulk of the available plant material. Thus, herbivores have larger and more capacious guts housing vast numbers of endosymbiotic bacteria and other microorganisms that can convert cellulose into volatile fatty acids, which can then be absorbed by the hosts. The change in gut size is reflected by proportionately longer and/or broader rib cages. The breakdown of cellulose by endosymbionts is most effective if plant fodder is already broken up prior to digestion. This is accomplished either by oral processing or through the use of a muscular gizzard containing hard particles for grinding food (as in many extant birds).

Parareptilia: Mesosauridae

Mesosauridae comprises the last common ancestor of *Mesosaurus*, *Brazilosaurus*, and *Stereosternum* (Fig. 3.2) and all descendants of that ancestor (Laurin and Reisz 1995). It represents the oldest known clade of amniotes adapted to aquatic life. Modesto (2006) first placed Mesosauridae among parareptiles. Mesosaurs are known from the early Permian (Cisuralian: Artinskian) of Brazil, Uruguay, Namibia, and South Africa (Oelofson and Araújo 1987; Modesto 1999, 2006; Piñeiro et al. 2012a,b). This geo-

graphically well-defined distribution in an inland sea on either side of the present-day South Atlantic Ocean historically provided key evidence in support of the hypothesis that Africa and South America were once connected. Laurin and Reisz (1995) listed a number of shared derived features for this clade including the absence of caniniform teeth; the short postorbital region of the skull; the absence of an ectopterygoid; retroarticular process broad transversely and concave dorsally; and a round proximal end of the femur. Mesosaurs attained a total length of up to 1 meter. The greatly elongated, slender jaws bear numerous, needle-like, pointed, and inward curving marginal teeth. The more anterior teeth project forward. The diet of mesosaurs consisted of crustaceans and occasionally small conspecifics (Silva et al. 2017). As the delicate skull is usually badly crushed in known specimens, the presence (Piñeiro et al. 2012b) or absence (Modesto 2006) of a lateral temporal opening remains uncertain.

The neck of mesosaurs is proportionally long. At least *Mesosaurus* and *Stereosternum* have accessory intervertebral articulations with zygosphenes and zygantra (Modesto 1999). Mesosaurs probably used the long laterally flattened tail and the hind limbs for swimming. The forelimbs are shorter than the hind limbs and may have assisted in steering. The structure of the elbow and ankle joints indicates that mesosaurs likely had only limited capacity for moving on land. The ribs in adults of *Mesosaurus* are pachyostotic (with a thick periosteal cortex) and "banana-shaped." The thickening of the ribs and presacral vertebrae probably aided in controlling buoyancy. Pachyostosis is less extensive in other known mesosaurs (Houssaye 2009). Recent discoveries of embryos, including a tiny partial skeleton preserved inside the ribcage of a medium-sized individual, show that mesosaurs were viviparous (Piñeiro et al. 2012a).

Parareptilia: Millerettidae

Millerettidae comprises the most recent common ancestor of *Milleretta* and *Millerosaurus* and all descendants of that ancestor (modified from Laurin and Reisz 1995). It is known only from the middle to late Permian (Guadalupian-Lopingian: Capitanian-Wuchiapingian) of South Africa (Gow 1972, 1997). Diagnostic features include the lateral exposure of the quadrate, the sculpturing

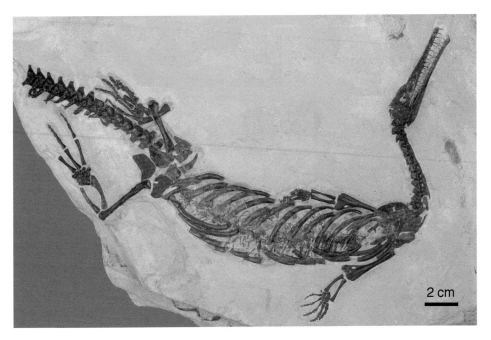

Figure 3.2. Skeleton of the mesosaurid *Stereosternum tumidum*. Courtesy of Robert Reisz and Diane Scott.

of the cranial bones consisting of low domed tuberosities, and the cultriform process of the parasphenoid bearing denticles (Laurin and Reisz 1995). *Milleretta* attained a total length of at least 20 cm and has dorsal ribs with broad rather than rod-like shafts. The conical crowns of the marginal teeth indicate a diet of arthropods and other small animals. Gow (1972) reported the presence of temporal openings in juveniles of *Milleretta*, but these openings were obliterated in older individuals. *Millerosaurus* has a ventrally open lower temporal opening (Carroll 1988). It attained a skull length of about 5 cm. *Eunotosaurus*, from the middle Permian (Guadalupian: Capitanian) of Malawi and South Africa, was considered closely related to Millerettidae (Gow 1997), but several recent studies have reinterpreted it as an early stem-turtle (see Chapter 5).

Parareptilia: Procolophonomorpha

Procolophonomorpha (from Greek *pro*, before, *kolophon*, top, summit, and *morphe*, form) comprises the most recent common ancestor of *Australothyris* and Ankyramorpha and all descendants of that ancestor (Modesto et al. 2009). Diagnostic derived features include the presence of a lower temporal fenestra, a transversely broad occipital condyle, and the distal condyle of the quad-

rate with nearly flat articular surfaces that are shorter anteroposteriorly than wide transversely.

Australothyris, from the middle Permian (Guadalupian: Capitanian) of South Africa, has several diagnostic cranial features such as a contact between the postfrontal and supratemporal, a small interpterygoid vacuity, and the presence of denticles on the ventral surfaces of the basipterygoid processes (Modesto et al. 2009a). It has a lateral temporal fenestra. Modesto et al. (2009a) considered the presence of this opening, along with other derived character states such as the deep posterior emargination of the quadrate, diagnostic for Procolophonomorpha.

Microleter, from the early Permian (Cisuralian: Sakmarian) of Oklahoma, has a large number (32) of slender, slightly recurved maxillary teeth (Tsuji et al. 2010). Its cranial bones bear ornamentation of tiny pits and narrow radiating furrows. The cranium of *Microleter* has a slit-like ventral emargination of the temporal region.

Parareptilia: Procolophonomorpha: Ankyramorpha

Ankyramorpha (from Greek *ankyra*, anchor, and *morphe*, form; in reference to the shape of the interclavicle) comprises the most recent common ancestor of Procolophonia, *Macroleter*, Lanthanosuchidae, *Acleistorhinus*, and all descendants of that ancestor (deBraga and Reisz 1996). This clade is diagnosed by several synapomorphies, notably the presence of a distinctly T-shaped ("anchor-shaped") interclavicle, which has long, slender lateral processes bearing deep, forward-facing grooves for the clavicles.

PARAREPTILIA: PROCOLOPHONOMORPHA: ANKYRAMORPHA: LANTHANOSUCHOIDEA

Lanthanosuchus and its close relatives from the middle Permian (Guadalupian: Capitanian) of Russia have broad, dorsoventrally low crania with pronounced sculpturing and lateral temporal fenestrae (Ivakhnenko 1987; Fig. 3.3). The superficial similarity of this unusual cranial shape to that in temnospondyl stem-amphibians initially led researchers to exclude *Lanthanosuchus* and related forms from amniotes, but more recent studies (e.g., deBraga and Reisz 1996) support parareptilian affinities for these presumably aquatic tetrapods. Lanthanosuchidae is most closely related to *Acleistorhinus*, from the early Permian (Cisuralian: Artinskian) of Oklahoma, which has a rather box-like skull with a lateral temporal opening and with sculpturing composed of shallow round pits (deBraga and Reisz 1996). DeBraga and Reisz (1996) defined a clade Lanthanosuchoidea (from Greek *lanthano*, forget, and *suchos*, crocodile) comprising the most recent common ancestor of Lanthanosuchidae and *Acleistorhinus* and all descendants of that ancestor. Diagnostic features listed by these authors include the presence of a lateral lappet of the frontal inserted between the prefrontal and postfrontal and contributing one third of the orbital margin, the long basicranial articulation, and the absence of basal tubera. Lanthanosuchoidea also includes *Colobomycter*, *Delorhynchus* (Fig. 3.4), and *Feeserpeton*, all from the early Permian (Cisuralian: Sakmarian) of Oklahoma (Reisz et al. 2014). The teeth of these parareptiles have infolding of the dentine (plicidentine). *Colobomycter* has an enormous, tusk-like tooth in each premaxilla and a pair of much enlarged maxillary teeth; the remaining teeth are small. *Feeserpeton* has simple conical teeth, with three enlarged ones in the maxilla and two in the dentary. Some specimens of *Delorhynchus* preserved pieces of chitin still adhering to their palatal teeth, confirming

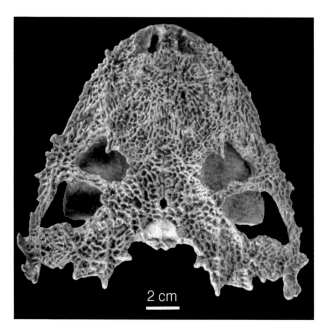

Figure 3.3. Cranium of the lanthanosuchoid *Lanthanosuchus watsoni* in dorsal view. Courtesy of Igor Novikov.

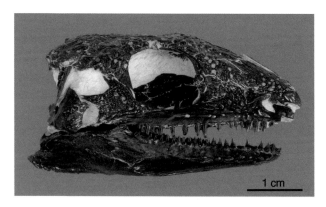

Figure 3.4. Skull of the lanthanosuchoid *Delorhynchus priscus* in lateral view. Courtesy of Robert Reisz and Diane Scott.

that at least this parareptile fed on arthropods (Modesto et al. 2009b).

PARAREPTILIA: PROCOLOPHONOMORPHA: ANKYRAMORPHA: BOLOSAURIA

The closest known relative of Bolosauridae is *Erpetonyx*, from the Pennsylvanian (Gzhelian) of Prince Edward Island, Canada (Modesto et al. 2015; Fig. 3.5). It is distinguished by the presence of 29 presacral vertebrae and lacks the specialized dentition of bolosaurids. Modesto et al. (2015) placed *Erpetonyx* with Bolosauridae in a clade

Figure 3.5. Skeleton of the bolosaurian *Erpetonyx arsenaultorum*. Courtesy of Robert Reisz and Diane Scott.

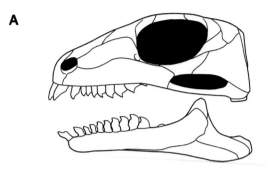

Figure 3.6. A, reconstruction of the skull of the bolosaurid *Belebey vegrandis* in lateral view; **B**, left lower jaw of *Belebey vegrandis* in lateral view. Courtesy of Robert Reisz and Diane Scott.

Bolosauria (from Greek *bolos*, lump, and *sauros*, lizard). Along with Mesosauridae, *Erpetonyx* demonstrates that the initial diversification of parareptiles already occurred during the Pennsylvanian.

Bolosauridae comprises the most recent common ancestor of *Belebey*, *Bolosaurus*, and *Eudibamus* and all descendants of that ancestor. It is known from the early Permian (Cisuralian: Asselian-Sakmarian) of Texas, Oklahoma, France, and Germany and the middle Permian (Guadalupian: Capitanian) of Russia and China (Reisz et al. 2007; Falconnet 2012). The dentition comprises procumbent incisiform teeth in the premaxilla and anterior part of the dentary and thickly enameled and bulbous teeth in the maxilla and the posterior portion of the dentary that occluded with those in the opposing jaw (Fig. 3.6). The maxillary teeth show heavy wear on the lingual surfaces of the crowns, whereas the dentary teeth are heavily worn on the labial sides. Well-developed striations on the wear facets extend more or less parallel to the long axis of the jaw and indicate that the mandible was capable of fore-and-aft motion (Hotton et al. 1997). The lower jaw bears a tall coronoid process for attachment of the adductor jaw muscles (Reisz et al. 2007). Bolosaurs were probably herbivores. The cranium has a narrow lower temporal fenestra bounded ventrally by a slender quadratojugal. *Eudibamus*, from the early Permian (Cisuralian: Artinskian) of Germany, attained a length of about 30 cm (Berman et al. 2000). It has hind limbs that are twice as long as the forelimbs and end in long, slender feet. *Eudibamus* is the oldest known reptile capable of bipedal locomotion, holding its hind limbs close to the body and running on its toes.

PARAREPTILIA: PROCOLOPHONOMORPHA: ANKYRAMORPHA: PROCOLOPHONIA

Following deBraga and Rieppel (1997), the name Procolophonia is used for the grouping Pareiasauromorpha + Procolophonoidea.

PARAREPTILIA: PROCOLOPHONOMORPHA: ANKYRAMORPHA: PAREIASAUROMORPHA

Tsuji (2013) found two synapomorphies for Pareiasauromorpha: expanded epicondyle forming a broad, rectangular, and anteriorly facing flange and astragalus and calcaneum fused to each other, with foramen for the perforating artery.

Nycteroleteridae, including *Macroleter* (which is also known from Oklahoma) and *Nycteroleter* (Ivakhnenko 1987; Tsuji 2006), from the middle Permian (Guadalupian: Capitanian) of Russia, is characterized by a deep embayment in the posterolateral region of the cranium, which is bounded by the quadrate, quadratojugal, and squamosal and probably supported a tympanic membrane (Müller and Tsuji 2007; Fig. 3.7). *Macroleter* also has a small temporal opening between the jugal, quadratojugal, and squamosal (Tsuji 2006). Nycteroleterids lack an

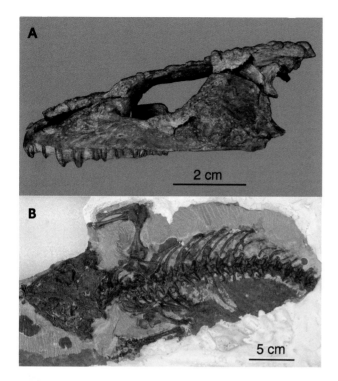

Figure 3.7. **A**, cranium of the nycteroleterid *Macroleter poezicus* in lateral view; **B**, skull and partial postcranial skeleton of *Macroleter poezicus* in dorsal view. A, from Müller and Tsuji (2007)—CC BY 2.5; B, courtesy of Robert Reisz and Diane Scott.

intertemporal (Lee 1995). They have skull lengths ranging from 5 to 10 cm. Tsuji and Müller (2009) and Tsuji (2013) considered nycteroleterids most closely related to pareiasaurs. Tsuji et al. (2012) united Nycteroleteridae with Pareiasauridae in a clade Pareiasauromorpha.

Pareiasauridae (from Greek *pareion*, cheek, and *sauros*, lizard) comprises the most recent common ancestor of *Bradysaurus* and *Anthodon* and all descendants of that ancestor. This clade ranged in time from the middle to late Permian (Guadalupian-Lopingian: Capitanian-Changhsingian) and is known from various regions of Africa and from Brazil, China, Germany, Russia, and Scotland (Ivakhnenko 1987; Lee 1997a,b; Jalil and Janvier 2005; Tsuji 2013). Laurin and Reisz (1995) provide a list of diagnostic features including the exclusion of the frontal from the orbital margin; sculpturing of cranial bones comprising large knobs, pits, and ridges; presence of a distinct ventral boss near the anterior end of the angularr; three to five pairs of sacral ribs; scapula with a prominent acromion; humerus short, robust, and lacking a

distinct shaft; and femur robust, with an only slightly constricted shaft (Lee 1997a,b; Tsuji 2013). Some pareiasaurs attained total lengths of up to 3 m and weights of at least 1,000 kg, making them the oldest known reptilian megaherbivores. The cranium (Fig. 3.8) has a short snout and a transversely broad postorbital region. Large bony bosses are present on the "cheek" region of the cranium, which extends below the tooth rows, and along the ventral margin of the angular. The marginal teeth of pareiasaurs have labiolingually flattened, leaf-shaped crowns with coarsely serrated cutting edges (Lee 1997a) and superficially resemble those of extant plant-eating iguanid lizards. They are closely spaced and form continuous shearing edges. The vertebral column of pareiasaurs comprises only 17 to 20 presacral vertebrae. The massive, barrel-shaped rib cage indicates the presence of a capacious gut for lengthy retention and fermentative digestion of plant fodder (Reisz and Sues 2000). The pelvis has a large ilium and a smaller ischium and pubis, which are aligned behind the ilium in some taxa. The robust limbs of pareiasaurs were held more vertical than in other parareptiles to support the massive body. M. L. Turner et al. (2015) reconstructed *Bunostegos*, from the Permian of Niger, with a more upright forelimb posture than in other pareiasaurs that have more sprawling forelimbs. The broad manus and pes have reduced numbers

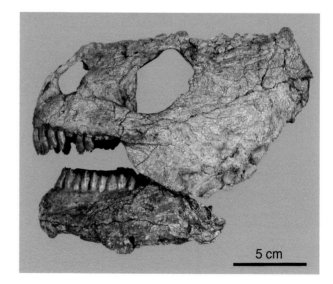

Figure 3.8. Skull of the pareiasaur *Deltavjatkia rossica* in lateral view. Courtesy of Robert Reisz and Diane Scott.

of phalanges. Many pareiasaurs bear extensive dermal armor (Lee 1997a,b; Tsuji 2013). In basal taxa such as *Bradysaurus*, from the middle Permian (Guadalupian: Capitanian) of South Africa, this armor is composed of thick, rounded, and coarsely sculptured bony plates (osteoderms) embedded in the skin along the dorsal midline. In more derived pareiasaurs, such as *Anthodon*, from the late Permian (Lopingian: Wuchiapingian) of South Africa, and *Scutosaurus*, from the late Permian (Lopingian: Changhsingian) of Russia, the osteoderms are arranged in closely spaced transverse rows across the back. Even the limbs bear small conical armor elements.

PARAREPTILIA: PROCOLOPHONOMORPHA: ANKYRAMORPHA: PROCOLOPHONIA: PROCOLOPHONOIDEA

This clade is supported by various derived features including the presence of a narial shelf, the posterior extension of the jugal that contributes to the temporal region, a ventrally open lower temporal opening open ventrally, and three or more bones forming the retroarticular process (MacDougall and Reisz 2014).

Nyctiphruretidae is characterized by the presence of paired surangular foramina (MacDougall and Reisz 2014), and comprises *Nyctiphruretus*, from the middle Permian (Guadalupian: Capitanian) of Russia, and *Abyssomedon*, from the early Permian (Cisuralian: Sakmarian) of Oklahoma. The cranium of *Nyctiphruretus* has a long external

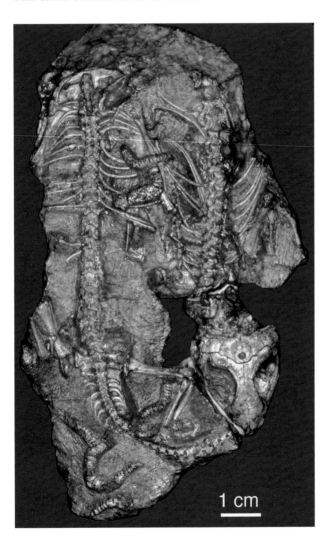

Figure 3.9. Two associated skeletons of the owenettid *"Owenetta" kitchingorum* in dorsal view. The associated worm-like structures represent fossils of millipedes, which presumably scavenged the reptilian remains. Courtesy of Robert Reisz and Diane Scott.

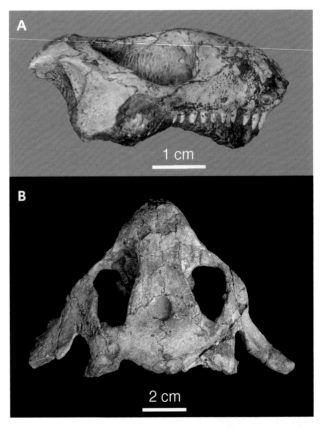

Figure 3.10. Crania of procolophonids. **A**, cranium of the Early Triassic *Procolophon trigoniceps* in lateral view; **B**, cranium of the Middle Triassic *Teratophon spinigenis* in dorsal view. A, courtesy of Robert Reisz and Diane Scott; B, courtesy of Ross Damiani.

narial opening bounded by a large depression (Säilä 2010a). *Nyctiphruretus* attained a total length of about 40 cm.

Procolophonoidea (from Greek *pro*, before, and *kolophon*, top, summit) comprises the most recent common ancestor of Owenettidae and Procolophonidae and all descendants of that ancestor (Lee 1997a). It ranged in time from the late Permian (Lopingian) to the end of the Triassic and attained a worldwide distribution (Ivakhnenko 1979; Cisneros 2008). Diagnostic derived features for this clade include the presence of a large medial process of the prefrontal; the elongation of the frontal; a distinctly domed and large supratemporal; the absence of the postparietal; and the absence of teeth on the transverse flange of the pterygoid (Reisz and Scott 2002). The skull of procolophonoids is proportionately short anteroposteriorly and broad transversely. The posterior margins of the orbits are deeply embayed and probably accommodated part of the adductor jaw musculature (Figs. 3.8, 3.9). The temporal region of the cranium is ventrally emargin-

ated. Furthermore, a few procolophonoids have a lower temporal opening (e.g., *Candelaria*; Cisneros et al. 2004).

Owenettidae ranged in time from the late Permian (Lopingian: Wuchiapingian) to the Middle Triassic (Ladinian) and is known from Brazil, Madagascar, South Africa, and Tanzania. *Owenetta*, from the late Permian (Lopingian: Wuchiapingian) of South Africa, and its relatives have conical, slightly recurved marginal teeth (Reisz and Laurin 1991; Reisz and Scott 2002; Fig. 3.9). Diagnostic apomorphies include the presence of a deep temporal emargination between the jugal and quadratojugal and the separation of the postorbital and parietal by a large postfrontal (Reisz and Scott 2002).

The more diverse Procolophonidae (Figs. 3.10, 3.11) spanned the entire Triassic Period. The largest known procolophonid, *Sclerosaurus*, from the Early to early Middle Triassic (Olenekian-Anisian) of Germany and Switzerland, attained a total length of 50 cm (Sues and Reisz 2008), but most taxa are smaller. Cisneros (2008)

Figure 3.11. Skeleton of the procolophonid *Tichvinskia vjatkensis* in dorsal view, probably originally preserved in a burrow. Courtesy of Igor Novikov.

listed a number of synapomophies for Procolophonidae including the absence of a premaxillary process of the maxilla; maxilla with a lateral depression behind the external naris; three or four conical or incisiform premaxillary teeth; and maxillary and more posterior dentary teeth with labiolingually broad crowns. When unworn, each molariform tooth in the maxilla and dentary typically has a labial and a lingual cusp connected by a sharp crest. Opposing tooth crowns fit between each other like cogs when the jaws closed and could have crushed and sheared food trapped between them. In derived procolophonids such as *Hypsognathus*, from the Late Triassic (Norian-Rhaetian) of Connecticut, New Jersey, Nova Scotia, and Pennsylvania, and *Leptopleuron*, from

the Late Triassic of Scotland, the lower jaw has a tall, sometimes recurved coronoid process, and the jaw joint is placed below the level of the lower tooth row (Sues et al. 2000; Säilä 2010b). These features, along with the structure of the teeth, indicate herbivory for at least the more derived procolophonids (Reisz and Sues 2000). The quadratojugal in many procolophonids is either extended posterolaterally or bears two or more prominent bony spines; these features probably bore keratinous covering in life and may have served as protection against predators or for some other purpose such as digging. Articulated skeletons of owenettids and procolophonids have frequently been found preserved in burrows (Ivakhnenko 1979; Figs. 3.9, 3.11).

4   Basal Eureptilia and Diapsida

Early Evolution of Modern Reptiles

Traditional scenarios of tetrapod evolution depicted reptiles as evolving from "amphibians" (i.e., anamniote tetrapods) and subsequently giving rise to birds and mammals, respectively (Romer 1966; Carroll 1969a,b). The ancestral reptilian stock was thought to comprise small-bodied, superficially lizard-like forms like the Pennsylvanian *Hylonomus*, which have skulls without temporal openings. Reassessments of the interrelationships of amniotes using Hennig's methodology, however, found no support for this view. Instead, they established the existence of two major lineages of amniotes, one leading to eureptiles (including birds) and the other leading to mammals, both of which first appear in the fossil record during the Pennsylvanian (Gauthier et al. 1988a,b; Reisz 1997).

Certain taxa long considered "ancestral reptiles" (e.g., *Limnoscelis*) have been reinterpreted as close relatives of Amniota but not members of the crown group (Heaton 1980; Reisz 2007). Furthermore, the phylogenetic analysis by Müller and Reisz (2006) found *Hylonomus* and various related taxa to be most closely related to diapsid reptiles, based on synapomorphies such as the presence of a suborbital foramen in the palate, the small size of the supratemporal bone, and the long and slender limbs (Heaton and Reisz 1986).

Tsuji and Müller (2009) defined Eureptilia as the most inclusive clade containing *Captorhinus aguti* and *Petrolacosaurus kansensis* but not *Procolophon trigoniceps*. Eureptilia (Fig. 4.1) is distinguished from other amniotes by the absence of a contact between the postorbital and supratemporal on the skull roof, the narrow blade of the ilium, and the distinct ventral constriction of the dorsal vertebral centra.

Eureptilia: "Protorothyrididae" + Captorhinidae

Most Pennsylvanian and Permian eureptiles were long placed in two family-level groups, Protorothyrididae and Captorhinidae (Carroll and Baird 1972; J. Clark and Carroll 1973).

Protorothyrididae (Romeriidae) was used for various taxa of superficially lizard-like reptiles, none of which exceeded 20 to 25 cm in total length (Carroll 1969a,b; Carroll and Baird 1972; J. Clark and Carroll 1973). These forms are known from the Pennsylvanian (Bashkirian) to the early Permian (Cisuralian: Asselian) of North America and Europe. *Hylonomus*, from the Pennsylvanian (Bashkirian) of Joggins, Nova Scotia (Fig. 4.2), is the oldest known reptile. It is documented only by incomplete, disarticulated skeletal remains (Carroll 1964). The supratemporal and tabular bones are small. Anteriorly, the maxilla bears slightly enlarged teeth; the remaining teeth have small conical crowns. The vertebrae have narrow rather than "swollen" neural arches, unlike in other basal reptiles. Müller and Reisz (2006) demonstrated that most "protorothyridids," including *Hylonomus* and *Paleothyris* (Figs. 4.2-4.4), are most closely related to early diapsids. Gauthier et al. (1988b) had already united *Paleothyris* with Diapsida in a clade Romeriida. (The original definition of this clade did not include turtles, which were grouped with Capto-

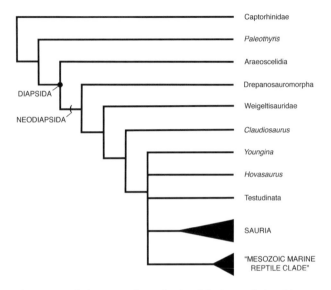

Figure 4.1. Phylogenetic hypothesis of the interrelationships of Eureptilia. Dots denote node-based clades and parentheses denote stem-based clades. Based mainly on Müller and Reisz (2006) and Pritchard and Nesbitt (2017).

rhinidae as "Anapsida.") *Paleothyris* and diapsid reptiles share apomorphies such as the long, slender limbs and the long, narrow manus and pes (Figs. 4.4, 4.5). The fourth digits of both manus and pes are considerably longer than the other digits, as in many diapsid reptiles. Sumida (1997) interpreted this feature as indicating extensive rotation of the hands and feet and push-off involving the elongate digit during locomotion. *Thuringothyris*, from the early Permian (Artinskian) of Germany (Boy and Martens 1991), is now considered a stem-captorhinid, and other "romeriids," including *Romeria*, from the early Permian (Cisuralian: Asselian) of Texas (J. Clark and Carroll 1973), are basal captorhinids (Müller and Reisz 2006). Thus, the name "Protorothyrididae" should no longer be used. Furthermore, the work by Müller and Reisz and the recent reassessment of the phylogenetic position of turtles (Chapter 5) have established that Williston's "Anapsida" is not a natural group.

Captorhinidae comprises the most recent common ancestor of *Euconcordia* and *Moradisaurus* and all descendants of that ancestor. It ranged in time from the Pennsylvanian (Gzhelian) to the late Permian (Lopingian: Wuchiapingian) and was widely distributed, with representatives recorded from the American Southwest,

Figure 4.2. Disarticulated skeletal remains of the eureptile *Hylonomus lyelli*, the oldest reptile known to date. Courtesy of Robert Reisz.

various countries in Europe and Africa, Brazil, China, and India. Müller and Reisz (2005) found four diagnostic synapomorphies for this clade: reduction in the number of maxillary teeth; the anterior position of the parietal foramen; the absence of an ectopterygoid; and the absence of a tabular.

Captorhinids include small-sized basal forms with a single row of marginal teeth in each jaw, as well as often large-bodied derived taxa with multiple rows of teeth in

Figure 4.3. **A**, skeleton of the basal eureptile *Paleothyris acadiana*. No scale provided. **B**, skull of *Paleothyris* in lateral view. A, courtesy of the late Don Baird; B, modified from Carroll (2009).

each maxilla and dentary. The cranial bones have distinctive sculpturing (Fig. 4.6). The neural arches are broad and swollen, with the zygapophyses positioned far laterally. The ribcage is broad. The robust, rather short limbs end in broad hands and feet (Fox and Bowman 1966; Heaton and Reisz 1980; Fig. 4.5). *Captorhinus*, from the early Permian (Cisuralian: Asselian-Sakmarian) of Oklahoma and Texas and possibly Brazil (Cisneros et al. 2015), has either a single row or at most three or four rows of marginal teeth in each maxilla and dentary (Heaton 1979; Kissel et al. 2002). It attained a total length of up to 40 cm. *Labidosaurus*, from the early Permian (Cisuralian: Artinskian) of Texas (Modesto et al. 2007; Fig. 4.7), and *Labidosaurikos*, from the early Permian (Cisuralian: Kungurian) of Oklahoma and Texas (Dodick and Modesto 1995; Fig. 4.8), have additional rows of teeth lingual to the marginal tooth rows in each maxilla and dentary. In captorhinids with multiple tooth rows, new teeth were added on the lingual side of the jaw and, during growth, "drifted" toward the labial side of the jaw because of the addition of new bone lingually and the resorption of bone labially (de Ricqlès and Bolt 1983). The basal captorhinid *Euconcordia*, from the Pennsylvanian (Gzhelian) of Kansas, still replaced its marginal teeth in a manner similar to that in iguanid lizards, with replacement teeth erupting directly lingual to the base of the functional tooth (LeBlanc and Reisz 2015). The pattern of wear on the teeth and the structure of the jaw joint in captorhinids with multiple tooth rows indicate that the lower rows bit between the upper ones and that the mandible was capable of fore-and-aft motion. The largest known captorhinid, *Moradisaurus*, from the Permian of Niger, attained a skull length of more than 40 cm and has at least nine rows of teeth in each maxilla and dentary (de Ricqlès and Taquet 1982).

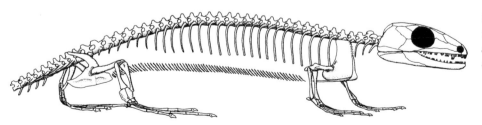

Figure 4.4. Reconstructed skeleton of *Paleothyris acadiana*. Modified from Carroll (2009).

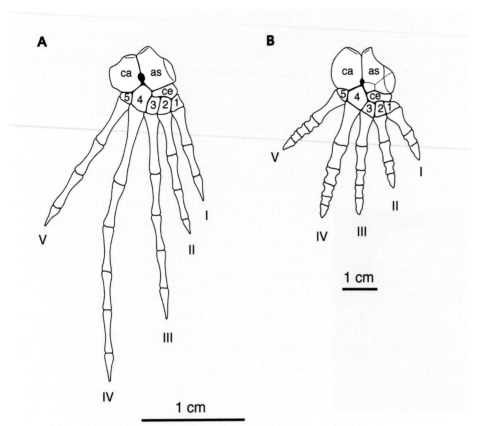

Figure 4.5. Right pes of **A**, *Paleothyris acadiana,* and **B**, *Captorhinus laticeps*. Abbreviations: as, astragalus; ca, calcaneum; ce, centrale. Arabic numerals indicate distal carpals and Roman numerals digits. Note that the astragalus of *Captorhinus* frequently shows sutures (indicated by dotted lines) between the original tarsal elements that became fused during ontogeny. Modified from Heaton and Reisz (1986).

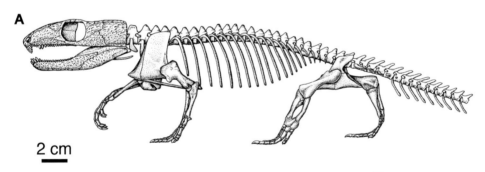

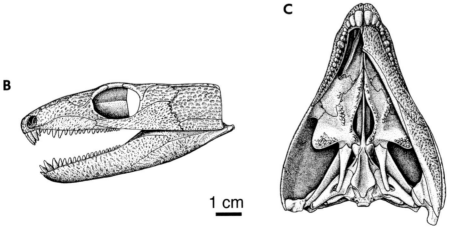

Figure 4.6. A, reconstructed skeleton of the captorhinid *Captorhinus laticeps* in lateral view. (Distal end of tail omitted for layout purposes.) **B-C**, skull of *Captorhinus laticeps* in **B**, lateral, and **C**, palatal views. A, courtesy of Robert Reisz; B-C, from Heaton (1979).

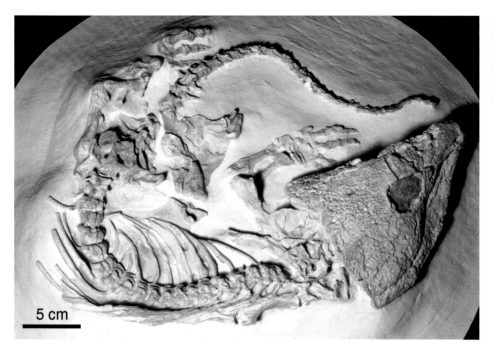

Figure 4.7. Skeleton of the captorhinid *Labidosaurus hamatus* in dorsal view. Courtesy of Department of Paleobiology, National Museum of Natural History.

Figure 4.8. Right maxillary dentition of the captorhinid *Labidosaurikos meachami* in occlusal view. Courtesy of Robert Reisz.

Small-bodied basal captorhinids probably subsisted on arthropods and other small animals, whereas the larger-bodied forms with multiple tooth rows could feed on high-fiber plant matter (Dodick and Modesto 1995; Hotton et al. 1997; Reisz and Sues 2000).

Eureptilia: Diapsida

By far the most diverse group of present-day land-dwelling amniotes, Diapsida includes all extant reptiles and birds. It also has a rich and varied fossil record extending all the way back to the Pennsylvanian and encompassing an amazing variety of reptiles that lived on land, in the sea, and in the air. Extant diapsids belong either to Lepidosauria, comprising the tuatara (*Sphenodon*) and squamates (lizards including snakes), or to Archosauria, comprising crocodylians and birds. Furthermore, recent phylogenetic studies have shown that turtles also form part of Diapsida (Chapter 5).

Diapsida comprises the most recent common ancestor of Araeoscelidia and Sauria and all descendants of that ancestor (Laurin 1991). The key diagnostic feature for this major clade is the presence of upper and lower temporal openings. The upper temporal (supratemporal) fenestra is bounded by the parietal, postorbital (and, if present, postfrontal), and squamosal. The lower temporal (infratemporal) fenestra is framed by the jugal, postorbital, quadratojugal, and squamosal. The development of these openings results in the formation of two bony arches. This temporal configuration is greatly modified in many diapsid groups that lack either one (e.g., many squamates) or both (e.g., snakes) of the bony arches. Some diapsid reptiles have only upper temporal openings, which are bounded laterally by the postorbital and squamosal. This led Gaffney (1980) to argue that the presence of the upper temporal opening, and the arrangement of the bones surrounding it, rather than the presence of two temporal fenestrae should be considered diagnostic for Diapsida. Another synapomorphy for this clade is the presence of a well-developed suborbital opening in the bony palate between the ectopterygoid, maxilla, and palatine.

Diapsida: Araeoscelidia

Araeoscelidia (from Greek *araios*, thin, and *scelis*, leg) comprises the most recent common ancestor of *Araeoscelis* and *Petrolacosaurus* and all descendants of that ancestor (Laurin and Reisz 1995). It differs from more derived diapsid reptiles in the presence of canine-like teeth in the maxilla, the anterior extension of the lacrimal to the external naris, and the dorsoventrally deep posterior (infratemporal) process of the jugal (deBraga and Reisz 1995). As in more basal eureptiles, the quadrate is not emarginated posteriorly and the stapes is robust rather than rod-like, indicating the absence of a tympanum. The manus and pes are more narrow and slender than in stem-amniotes. The fourth digits of the manus and pes are much longer than the other digits. The femur is gently S-shaped and nearly as long as the tibia. Additional synapomorphies for Araeoscelidia include long cervical vertebrae; bony protuberances (mammillary processes) on the neural spines of the posterior cervical and anterior dorsal vertebrae; coracoid with a prominent process for the origin of the coracoid head of the triceps brachii muscle; and the presence of large distal and lateral tubercles on the pubis (Laurin 1991).

Petrolacosaurus, from the Pennsylvanian (Kasimovian) of Kansas, is the oldest known diapsid reptile (Reisz 1981)

and attained a total length of 40 cm (Figs. 4.9, 4.11A). Its skull is proportionately small. The marginal dentition consists of numerous slender teeth. Two of the more anterior teeth in the maxilla have taller crowns than the neighboring teeth. The long cervical vertebrae of *Petrolacosaurus* form a clearly defined neck region. The dorsal vertebrae have swollen neural arches.

Spinoaequalis, from the Pennsylvanian (Kasimovian) of Kansas, differs from *Petrolacosaurus* in lacking a proportionately long neck and the presence of a dorsoventrally deep tail, which is a common feature among water-dwelling reptiles (deBraga and Reisz 1995).

Araeoscelis, from the early Permian (Cisuralian: Asselian-Artinskian) of Texas (Fig. 4.10), is distinguished from *Petrolacosaurus* by a more robust skull with only upper temporal openings (Fig. 4.11B) and slightly bulbous, labiolingually broad teeth in the maxilla and dentary (Vaughn 1955). Its postcranial skeleton closely resembles that of *Petrolacosaurus*. The deep temporal bar and the absence of a lower temporal fenestra likely reflect the development of powerful adductor jaw muscles and specialized dentition (Reisz et al. 1984). *Araeoscelis* probably attained a total length of up to 60 cm.

Diapsida: Neodiapsida

Neodiapsida (from Greek *neos*, new, *dyo*, two, and *hapsis*, arch) comprises all diapsid reptiles more closely related to Sauria (defined below) than to Araeoscelidia (Benton 1985; Gauthier et al. 1988b). Reisz et al. (2011) diagnosed this clade by the absence of caniniform teeth in the maxilla, a slender posterior (infratemporal) process of the jugal, and a lower temporal fenestra that is open ventrally in many taxa. In addition, the parietal forms a distinct ventromedial flange, which increased the cranial area of origin of the adductor jaw musculature.

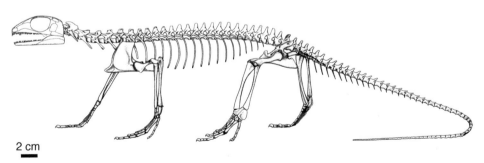

Figure 4.9. Reconstructed skeleton of the araeoscelidian *Petrolacosaurus kansensis* in lateral view. Courtesy of Robert Reisz.

2 cm

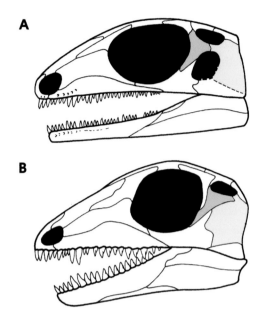

Figure 4.11. Skulls of the araeoscelidians **A**, *Petrolacosaurus kansensis,* and **B**, *Araeoscelis gracilis* in lateral view to show the temporal region. Postorbital, light blue; squamosal, yellow. A, modified from Reisz (1981); B, modified from Reisz et al. (1984).

2 cm

Figure 4.10. Skeleton of the araeoscelidian *Araeoscelis gracilis,* for the most part exposed in ventral view. Courtesy of Robert Reisz and Diane Scott.

The two oldest known representatives of Neodiapsida are *Orovenator* (Reisz et al. 2011; Ford and Benson 2018) and *Lanthanolania* (Modesto and Reisz 2002). The cranium of *Orovenator*, from the early Permian (Cisuralian: Sakmarian) of Oklahoma, has elongate external nares (Reisz et al. 2011). It is uncertain whether its lower temporal opening was open ventrally. *Orovenator* has an estimated skull length of 4.5 cm. Ford and Benson (2018) argued that it is most closely related to Varanopidae, a clade gener-

ally placed among basal synapsids, and that the latter is possibly referable to Reptilia rather than Synapsida. Unlike *Orovenator*, however, varanopids lack upper temporal fenestrae and suborbital openings, and other cranial features support synapsid affinities.

The skull of *Lanthanolania*, from the middle Permian (Guadalupian) of Russia (Modesto and Reisz 2002), has a dorsoventrally low maxilla and a splint-like coronoid bone in the lower jaw. It attained a length of 3 cm. The short posterior process of the slender, crescent-shaped jugal demonstrates that the lower temporal opening was open ventrally.

DIAPSIDA: NEODIAPSIDA: DREPANOSAUROMORPHA

Drepanosauromorpha (from Greek *drepane*, sickle, *sauros*, lizard, and *morphe*, form) is a clade of small-sized reptiles from the Late Triassic with many distinctive skeletal features (Pinna 1984; Renesto et al. 2010; Pritchard et al. 2016; Pritchard and Nesbitt 2017). Renesto et al. (2010) defined it as the least inclusive clade containing *Hypuronector limnaios* and *Megalancosaurus preonensis*. With the exception of *Hypuronector*, from the Late Triassic (Norian) of New Jersey, all drepanosauromorphs share cervical vertebrae with anterodorsally inclined and anteroposteriorly narrow neural spines and with centra that have saddle-shaped articular surfaces. The anterior dorsal vertebrae have highly modified neural spines (Renesto et al. 2010).

Commonly assigned to Archosauromorpha, phylogenetic analyses by Senter (2004), Pritchard et al. (2016), and Pritchard and Nesbitt (2017) placed drepanosaurs outside Sauria. They were initially known only from Italy, but recent discoveries attest to a much greater diversity and wider geographic distribution for these unusual reptiles (Renesto et al. 2010).

The basal depanosauromorph *Hypuronector,* from the Late Triassic (Norian) of New Jersey, has a dorsoventrally deep tail with long hemal bones. Colbert and Olsen (2001) interpreted it as an aquatic form, but Renesto et al. (2010) argued that the lack of flexibility in the vertebral column renders this interpretation unlikely.

The shoulder girdle and especially the forelimb of *Drepanosaurus,* from the Late Triassic (Norian) of Italy (Pinna 1984; Fig. 4.12) and New Mexico (Pritchard et al. 2016), show a suite of unique features. The glenoid faces laterally. The long axis of the crescent-shaped, flattened ulna extends perpendicular to that of the radius. The radius has a flattened proximal end. The greatly elongated ulnare and intermedium are longer than the radius and contact the ulna. Digit II of the manus terminates in an enormous ungual phalanx with a ventrally hooked distal tip. Pritchard et al. (2016) argued that the forelimb of *Drepanosaurus* was capable only of protraction and retraction and could have been used for hook-and-pull digging. The neural spines of the dorsal

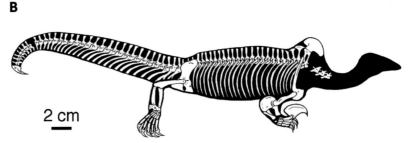

Figure 4.12. A, postcranial skeleton of the drepanosauromorph *Drepanosaurus unguicaudatus;* **B,** reconstructed skeleton of *Drepanosaurus unguicaudatus* in lateral view, based on **A.** A, courtesy of Nick Fraser; B, from Renesto et al. (2010).

vertebrae are tall, with those of the second through fifth dorsals expanded anteroposteriorly. The tail of *Drepanosaurus* has long neural spines and hemal arches. In at least the Italian material, the tail was apparently prehensile and terminates in a claw-like bone distally. *Drepanosaurus* attained a total length of up to 35 cm.

Figure 4.13. Skull and partial postcranial skeleton of the drepanosauromorph *Megalancosaurus preonensis*. Courtesy of Nick Fraser.

Avicranium, from the Late Triassic (Rhaetian) of New Mexico, is noteworthy for the absence of teeth, the anteriorly directed orbits, and a domed skull roof with transversely broad and dorsoventrally tall frontals and parietals (Pritchard and Nesbitt 2017).

Megalancosaurus, from the Late Triassic (Norian) of Italy, has a superficially bird-like skull with huge orbits and a sharply tapered snout (Fig. 4.13). Its forelimb has a somewhat elongated ulnare and intermedium but otherwise lacks the specializations present in *Drepanosaurus*. The long and slender manual digits I, II, and III oppose digits IV and V, resembling the condition in chameleons and suitable for grasping and holding on to narrow supports such as tree branches. This feature indicates arboreal habits for *Megalancosaurus*.

DIAPSIDA: NEODIAPSIDA: WEIGELTISAURIDAE

Weigeltisauridae (named for the German paleontologist Johannes Weigelt and Greek *sauros*, lizard) is a clade of distinctive small-bodied reptiles from the late Permian (Lopingian) of England, Germany, Madagascar, and Rus-

sia (S. E. Evans and Haubold 1987; Schaumberg et al. 2007; Bulanov and Sennikov 2010, 2015). *Coelurosauravus*, from the Changhsingian of Madagascar, and *Weigeltisaurus*, from the Wuchiapingian of England and Germany (Fig. 4.14), attained an estimated snout–vent length of about 20 cm. The parietals and squamosals frame large temporal openings and form a distinct casque at the back of the cranium, superficially resembling that in chameleons. The lateral margin of the parietal and the posterior edge of the squamosal bear toothlike projections. The ventral margin of the temporal region is deeply emarginated. The small quadrate is firmly attached to the squamosal. The premaxilla is long. The cervical and especially the dorsal vertebrae have long centra. A unique feature of the postcranial skeleton is the presence of a set of long, hollow rod-like bones along either flank, which could be folded back and probably supported a gliding membrane in life (Frey et al. 1997; Schaumberg et al. 2007). This differs from the condition in other known gliding reptiles (such as the extant lizards of the genus *Draco*) in which elongated trunk ribs support a gliding membrane. The limbs are long and slender. Bulanov and Sennikov (2010) argued that weigeltisaurids were probably arboreal. Although referable to Neodiapsida, Weigeltisauridae occupies a basal position within that clade (Pritchard and Nesbitt 2017). *Wapitisaurus*, from the Early or Middle Triassic of British Columbia, was originally referred to this group (D. B. Brinkman 1988) but the poor preservation of the only known specimen does not permit a definitive assignment.

DIAPSIDA: NEODIAPSIDA: *YOUNGINA* + "YOUNGINIFORMES"

The phylogenetic position of *Youngina*, from the late Permian (Lopingian: Changhsingian) of South Africa, has long been contentious. Broom (1914) established a group Eosuchia (from Greek *eos*, dawn, and *souchos*, crocodile) for its reception because *Youngina* is much less derived than most other diapsids. Subsequently, many authors came to view Eosuchia as the rootstock from which all other diapsid reptiles evolved (Romer 1956; Carroll 1977, 1988). Any diapsid reptile that could not be readily assigned to Sauria was classified as an eosuchian. Phylogenetic analyses have since established that "Eosuchia"

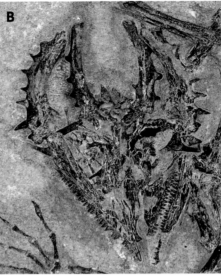

Figure 4.14. A, skeleton of the weigeltisaurid *Weigeltisaurus jaekeli*; **B**, close-up of the skull in dorsal view, showing the casque-like structure formed by the parietals and squamosals. Courtesy of Diane Scott.

comprises mostly unrelated taxa that are united merely by lacking features diagnostic for more derived diapsid clades (Gauthier et al. 1988c; Bickelmann et al. 2009). Thus, this name should no longer be used.

Youngina attained a skull length of up to 5 cm and a total length of up to 40 cm (Gow 1975; Carroll 1977). The cranium (Fig. 4.15) has two temporal openings and a complete lower temporal bar. As in araeoscelidians, the quadrate has a straight posterior margin and the stapes is robust, indicating the lack of a tympanic membrane (Carroll 1977; N. M. Gardner et al. 2010). Although Gow (1975) and Carroll (1977) reported tabular bones, these elements were reconstructed in an anatomically implausible position and their presence is uncertain. The marginal teeth of *Youngina* have conical crowns and subthecodont implantation. The neck is short. As in other basal diapsids, the limbs are slender. A distinctive feature of *Youngina* is the presence of a median row of small osteoderms along the back (with one osteoderm per vertebra; Gow 1975).

Several taxa of diapsid reptiles from the late Permian (Lopingian: Changhsingian) of Madagascar and Tanzania and one from the Early Triassic of Kenya were placed with *Youngina* in a larger grouping Younginiformes

(Currie 1981; Laurin 1991). Gauthier et al. (1988c) assigned this group to Lepidosauromorpha, but Laurin (1991) argued that *Youngina* and its relatives occupied more basal phylogenetic positions among Diapsida. Furthermore, Bickelmann et al. (2009) found no support for the monophyly of Younginiformes.

Unlike *Youngina*, *Acerosodontosaurus* and *Hovasaurus*, both from the late Permian (Lopingian: Changhsingian) of Madagascar, have incomplete lower temporal bars. The structure of their postcranial skeletons and occurrence in nearshore marine deposits suggest an aquatic mode of life (Currie 1981; Bickelmann et al. 2009). *Hovasaurus* attained a length of about 50 cm. Its tail is at least twice as long as the body and has tall neural spines and long, plate-like hemal arches (Fig. 4.16A). It appears well suited for swimming. Most specimens of *Hovasaurus* preserve masses of small pebbles in the abdominal region, which may have served as ballast for buoyancy control (Currie 1981). By contrast, *Kenyasaurus*, from the Early Triassic of Kenya, was a terrestrially adapted reptile (Harris and Carroll 1977). Most of these diapsids have a large sternal plate, which developed from paired smaller elements in juvenile individuals (Currie 1981).

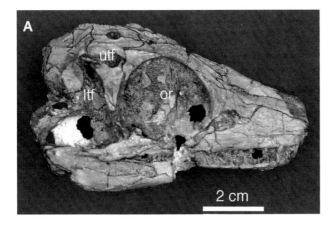

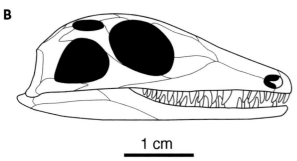

Figure 4.15. A, partial skull of the diapsid *Youngina capensis* in lateral view. Abbreviations: ltf, lower temporal fenestra; or, orbit; utf, upper temporal fenestra. **B**, reconstruction of the skull of *Youngina capensis* in lateral view. A, courtesy of Robert Reisz and Diane Scott; B, simplified from Carroll (1977).

DIAPSIDA: NEODIAPSIDA: *CLAUDIOSAURUS*

Claudiosaurus, from the late Permian (Lopingian: Changhsingian) of Madagascar (Fig. 4.16B), differs from *Acerosodontosaurus* and *Hovasaurus* in the presence of a proportionately long neck (resulting from the relatively posterior position of the shoulder girdle; Carroll 1981). Manual digit III rather than digit IV is the longest, lending the hand a vaguely paddle-like outline. *Claudiosaurus* attained a total length of about 60 cm. Carroll (1981) argued that it shares derived cranial features with Sauropterygia, the most diverse group of Mesozoic marine reptiles (Chapter 6). Like Sauropterygia, *Claudiosaurus* has a deep ventral embayment of the temporal region, without a lower temporal bar. Furthermore, the pterygoid lacks a transverse flange, and the suborbital fenestra and interpterygoid vacuity are small. Unlike *Hovasaurus*, *Claudiosaurus* has a slender tail. Poor ossification of the joint surfaces on its limb bones (Carroll 1981) and some pachyostotic thickening of the limb bones and vertebrae (Buffrénil and Mazin 1989; Houssaye 2009, 2013) suggest that *Claudiosaurus* was primarily aquatic. Phylogenetic analyses (e.g., Rieppel 1993) have not supported Carroll's interpretation of *Claudiosaurus* as a stem-sauropterygian. Sauropterygians share various features such as a bowed quadrate with more derived

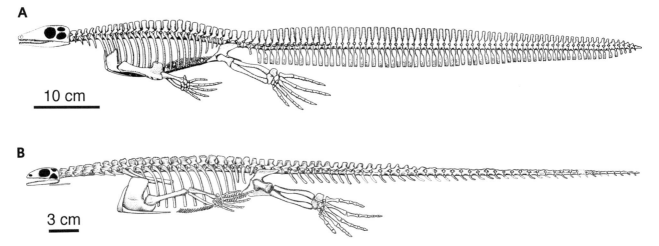

Figure 4.16. Reconstructed skeletons of two late Permian aquatic diapsid reptiles from Madagascar. **A**, *Hovasaurus boulei*. Note gastric ballast in the abdominal region. **B**, *Claudiosaurus germaini*. A, modified from Currie (1981); B, modified from Carroll (1981).

diapsids, but none of these traits is present in *Claudiosaurus*. The phylogenetic analysis by Chen et al. (2014b) found *Claudiosaurus* most closely related to crown-group Diapsida plus all Mesozoic marine reptiles.

DIAPSIDA: NEODIAPSIDA: SAURIA

Most neodiapsid reptiles belong to the clade Sauria (from Greek *sauros*, lizard). Macartney (in Cuvier 1802) first used this name for a group comprising a variety of lizards and crocodiles (*Crocodylus*). Brongniart (1800) had earlier employed the vernacular equivalent "Sauriens" for essentially the same taxonomic content. Starting with Gray (1825), many authors restricted the use of Sauria to lizards (e.g., Camp 1923; Estes 1983). The current phylogenetic concept of Sauria dates back to Gauthier et al. (1988b), who defined Sauria as comprising the most recent common ancestor of birds, crocodylians, squamates, and *Sphenodon* and all descendants of that ancestor. Most recent studies have adopted this definition.

Laurin (1991) provided a list of synapomorphies for Sauria: postparietal small or absent; tabular absent; squamosal confined to the dorsal portion of the temporal region except for a slender process contacting the quadrate; quadrate emarginated posteriorly; stapes slender and lacking a foramen for the stapedial artery; retroarticular process well-developed; shoulder girdle lacking cleithra; lateral centrale small or absent; absence of a distinct fifth distal tarsal; and metatarsal V "hooked" rather than straight.

Ezcurra's (2016) list of saurian synapomorphies includes the shallow posterior emargination of the quadrate and the absence of a cleithrum and added the absence of an anterior process of the quadratojugal and the at most slight extension of the parietals over the interorbital region. The currently known fossil record of Sauria dates back to the late Permian (Lopingian: Wuchiapingian) (Ezcurra et al. 2014).

Although now considered diapsids, turtles and several clades of Mesozoic marine reptiles are discussed in separate chapters of this book. This reflects the current absence of a general consensus regarding the phylogenetic positions of these groups. Although the diapsid relationships of turtles are now firmly established (Chapter 5), it is still not clear whether turtles are part of Sauria. The same obtains for all Mesozoic marine reptiles that are not referable to turtles, squamates, or archosauriforms (Chapter 6).

5 Testudinata

Turtles and Their Stem-Taxa

Present-day turtles comprise only 351 species (as of July 2018; http://www.reptile -database.org), but the fossil record of these reptiles is rich and varied. The clade Testudinata includes crown-group turtles (Testudines; from Latin *testudo*, tortoise, turtle) and various stem-turtles (Joyce et al. 2004; Li et al. 2008; Joyce 2017; Fig. 5.1). Joyce et al. (2004) defined Testudinata as comprising all reptiles that have a complete bony shell homologous to that of the extant turtle *Chelonia mydas* (green turtle). Testudines comprises the most recent common ancestor of *Chelonia mydas* and *Chelus fimbriatus* (matamata) and all descendants of that ancestor (Joyce et al. 2004). The vernacular "turtle" is used here for Testudines and "tortoise" specifically refers to members of Testudinidae.

The possession of a box-like bony shell that incorporates part of the vertebral column and encases the trunk (Figs. 5.2-5.3) readily distinguishes turtles from all other known tetrapods. This shell consists of a dorsal carapace and a ventral plastron, which are usually linked by a bony bridge on either side. It has a large opening for the head, neck, and forelimbs at the anterior end and another one for the hind limbs and the typically short tail posteriorly. In most turtles, large keratinous scutes cover the external surface of the bony shell. These scutes continue to grow throughout life and, in some turtles (e.g., *Chrysemys*), are occasionally shed and regrown. They leave distinct furrows on the underlying bony plates. Although the arrangement of the scutes roughly corresponds to that of the bony plates, their grooves do not coincide with the sutures between the plates. This offset arrangement presumably increases the overall mechanical strength of the shell.

The carapace (Fig. 5.2A) has longitudinal rows of bony plates. Along the midline of the carapace, a row of plates, the neurals, are fused to the underlying dorsal vertebrae. At the anterior end of this row, a single element, the nuchal, is not fused to the underlying dorsal vertebra. The nuchal is considered homologous to the cleithra in the shoulder girdle of more basal reptiles (Lyson et al. 2013). Similarly, at the posterior end of the row of neurals, there are one or two suprapygal plates and a pygal plate, which are also not fused to the underlying vertebrae. Extending laterally on either side of the neurals are long plates, termed costals, that incorporate the ribs. The ribs extend laterally rather than ventrolaterally as in most other amniotes. They remain visible as ridges on the internal surface of the carapace, and the costals are T-shaped in transverse section. The peripheral plates make up most of the outer edge of the carapace. All bony plates are sutured to their neighbors.

In recent decades, embryological studies on extant hard-shelled turtles (Burke 1989; Gilbert et al. 2001; Nagashima et al. 2009) have argued that the carapace developed as

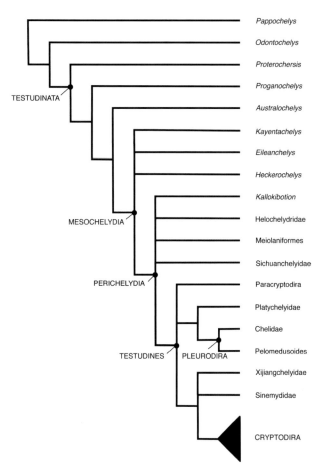

Figure 5.1. Phylogenetic hypothesis of the interrelationships of Pan-Testudines (Testudinata and various stem-taxa). Dots denote node-based clades. Based on Joyce (2007, 2014) and Joyce et al. (2013).

an entirely novel structure, rather than through fusion of dermal bones to the underlying vertebrae and ribs as traditionally assumed (e.g., Zangerl 1969). Early during embryonic development, a structure called the carapacial ridge forms between the front and hind limb buds (Burke 1989). The outgrowth of this ridge leads to the inward folding of the lateral body wall and the deflection of the developing ribs into the ridge. The ribs do not grow ventrally into the lateral body wall but extend laterally in the thick superficial dermis of the carapacial disc. The ribs develop in a fan-shaped pattern, with the anterior ribs ultimately extending anterolaterally over the shoulder girdle and the posterior ribs projecting posterolaterally. As a result, the shoulder girdle ultimately ends up inside the shell of turtles. The ribs and neural spines

of the dorsal vertebrae subsequently become centers for the ossification of the costals and neurals, respectively.

The plastron (Fig. 5.2B) is the smaller, flat ventral portion of the shell. Its anterior portion is made up of paired epiplastra and a median entoplastron, which are considered the homologues of the clavicles and interclavicle in other amniotes, respectively. The more posterior portion is typically made up of three pairs of elements, the hyoplastra, hypoplastra, and xiphiplastra, all of which probably developed from gastralia (Gilbert et al. 2007; Schoch and Sues 2015, 2018a). In some turtles, such as box turtles (*Terrapene*), a transverse hinge divides the plastron into two parts, which can be folded up tightly against the carapace to further protect the head, neck, and limbs.

The structure of the shell is highly variable among turtles, especially among primarily water-dwelling forms. Among extant turtles, Carettochelyidae (pitted-shelled turtles), Dermochelyidae (leatherback turtles), and Trionychidae (softshell turtles) lack keratinous scutes altogether, and the greatly reduced bony shell is covered by tough, leathery skin.

The shoulder girdle of Testudinata has three bony prongs that extend at distinct angles to each other (Fig. 5.3). A dorsal prong (which corresponds to the scapula in other reptiles) is connected to the nuchal and the first dorsal rib by ligaments. The short anteroventral prong (homologous to the acromion process of the scapula) extends medially and contacts the entoplastron. The posterior prong (which most likely corresponds to the coracoid; Vickaryous and Hall 2006) ends freely. The glenoid surface of the shoulder joint permits a considerable range of motion for the forelimb. The pelvis also has a three-rayed structure, with the ilium extending dorsally and connecting to the sacral ribs and the ischium and pubis extending ventrally. The contacts between the pelvic prongs and the shell independently became sutured in the stem-turtle *Proterochersis* and in pleurodiran turtles (Joyce et al. 2013a). The limbs extend horizontally through the openings at the front and back of the turtle shell, respectively. Except in sea turtles, the limbs can be withdrawn into the shell.

Most reptiles breathe by moving their ribs. Because the ribs in turtles are fully integrated into the bony shell, they cannot assist in expanding the thoracic cavity during breathing. Instead the transversus abdominis and

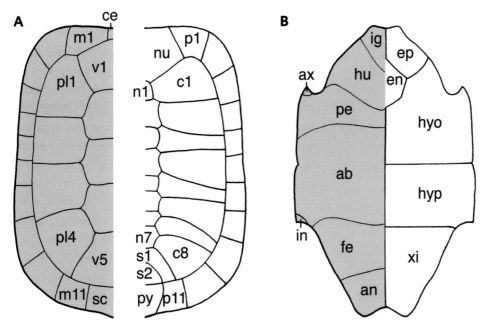

Figure 5.2. Simplified rendering of the keratinous scutes (light brown) and underlying bones (light yellow) forming the **A,** carapace, and **B,** plastron of a tortoise (*Testudo*). Abbreviations: ab, abdominal scute; an, anal scute; ax, axial scute; ce, cervical scute; c1-8, costals; en, entoplastron; ep, epiplastron; fe, femoral scute; hu, humeral scute; hyo, hyoplastron; hyp, hypoplastron; ig, intergular scute; in, inguinal scute; m1-11, marginal scutes; n1-7, neurals; nu, nuchal; pe, pectoral scute; p1-11, peripherals; pl1-4, pleural scutes; py, pygal; s1,2, suprapygals; sc, supracaudal scute; v1-5, vertebral scutes; xi, xiphiplastron.

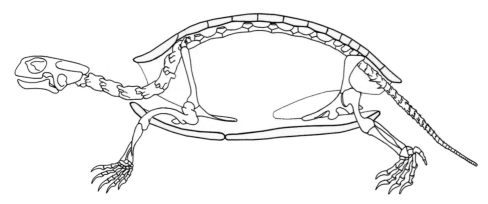

Figure 5.3. Skeleton with longitudinal section through the bony shell of a European pond turtle (*Emys orbicularis*) to show the relationship of the two structures. Simplified from Bojanus (1819-1821).

obliquus abdominis alternate their bilateral activity to produce inhalation and exhalation (Landberg et al. 2003). At rest, the obliquus abdominis curves into the body cavity. When this muscle contracts, it flattens and moves the flank postero- and ventrolaterally. This leads to a decrease in intrapulmonary pressure and inhalation when the glottis is open. The transversus abdominis is situated deep to the obliquus abdominis and cups the posterior half of the lung. Its contraction leads to an increase in intrapulmonary pressure and exhalation when the glottis is open.

Turtles have flexible necks. Most can withdraw their head and neck into the shell. The eight cervical vertebrae have long centra and, at most, greatly reduced ribs. The two principal clades of crown-group turtles differ in their respective mode of neck retraction (Williams 1950;

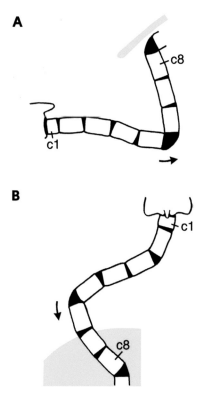

Figure 5.4. Diagrams contrasting the modes of cervical retraction in **A**, cryptodiran, and **B**, pleurodiran turtles. Shell rendered in light brown. Abbreviations: c1, cervical 1; c8, cervical 8. Arrows indicate direction of retraction and black wedges points of maximum flexure. Based on Williams (1950).

Fig. 5.4). In Cryptodira (from Greek *kryptos*, hidden, and *dere*, neck), the neck is vertically withdrawn into a median space within the body cavity so that it appears S-shaped in side view (Fig. 5.4A). In Pleurodira (from Greek *pleura*, side, and *dere*, neck), the neck bends sideways so that the sides of the head and neck lie underneath the anterior edge of the carapace (Fig. 5.4B). In both groups, the eighth cervical vertebra forms a distinct joint with the first dorsal, which is firmly attached to the bony carapace.

In the appendicular skeleton, the humerus has well-developed epicondyles and an ectepicondylar foramen or groove. The femur has a more or less spherical head that is set off from the remainder of the element.

Most turtles lack nasal bones (Fig. 5.5). The external nares are confluent and form a single median opening. The orbits are positioned well forward on the cranium. The septomaxilla, lacrimal bone and duct, ectopterygoid,

and supratemporal are absent. There is usually no parietal (pineal) foramen. The jaws lack teeth and instead bear keratinous sheaths (rhamphothecae). These sheaths are firmly attached to bony projections and ridges on the maxilla and dentary and have smooth or serrated cutting edges. The pterygoids are sutured to the braincase and, together with the palatines and the fused vomers, form a secondary bony palate. Behind this secondary palate the choanae open into the pharynx. The quadrate is deeply notched posteriorly and, together with the squamosal, forms a funnel-shaped recess (cavum tympani) on which the tympanic membrane is suspended. At the bottom of this funnel, the rod-like stapes passes into the recess through a hole or notch. In the mandible, the dentaries are completely fused to one another at the symphysis. Gaffney (1979) has published a detailed review of the cranial structure in turtles.

Although the cranium in Testudinata lacks temporal openings, the back of the skull roof in many taxa is deeply embayed along the dorsal margin of the posttemporal fenestra (an opening between posterior bones of the skull roof and the auditory capsule at the back of the cranium) on either side of a prominent median crest formed by the supraoccipital (Fig. 5.6A). The resulting passage is occupied by the external adductor mandibulae muscle, which originates from the supraoccipital crest and adjoining surfaces on the cranium and passes ventrally to insert on the lower jaw. In many taxa, this passage further increases in size, and the dermal bones in the region of the cranium behind the orbit often become reduced to a narrow temporal arch (Fig. 5.6B) or are lost altogether. These openings in the sidewall of the cranium are not homologous to the temporal fenestrae in other amniotes (Werneburg 2012). In basal amniotes, the external adductor muscle extended more or less vertically from the cranium to the mandible. In turtles, however, a large otic capsule takes up part of the space occupied by the musculature in more basal reptiles. Thus, the external adductor jaw muscle extends backward over the otic capsule. In Testudinata, the prominent tendon of the external adductor muscle (which contains a cartilage [cartilago transiliens]) extends over a pulley-like feature, the trochlear process, as the mandible is brought up against the cranium (Schumacher 1973). The specific configuration of this pulley structure differs between

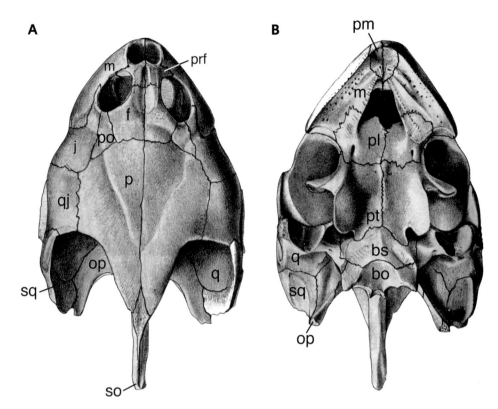

Figure 5.5. Cranium of the South American river turtle (*Podocnemis expansa*) in **A**, dorsal, and **B**, ventral views. From Gray (1855) with sutures added from Gaffney (1979). See Fig. 1.3 for abbreviations.

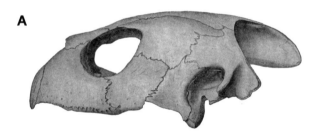

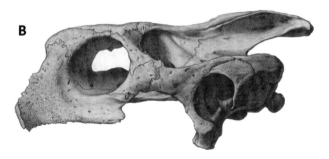

Figure 5.6. Lateral views of crania of two turtles to show **A**, closed sidewall, and **B**, extensive secondary emargination of the sidewall. **A**, Kemp's ridley sea turtle (*Lepidochelys kempii*); **B**, Charles Island giant tortoise (*Chelonoidis nigra*). A, from Hay (1908); B, from Günther (1877).

cryptodiran and pleurodiran turtles (Schumacher 1973; Gaffney 1975). In cryptodires, the trochlear process is positioned on the anterodorsal aspect of the paroccipital process (Fig. 5.7A). In pleurodires, the pterygoid forms a distinct lateral process (Fig. 5.7B,C). Because the former condition is also present in stem-turtles (Joyce 2007; Sterli 2010) it is no longer considered diagnostic for cryptodires.

Phylogenetic Position of Turtles

The phylogenetic position of turtles has long been one of the most contentious issues in the study of amniote evolution. The highly modified body plan of these reptiles makes anatomical comparisons with other groups of reptiles difficult, and molecular and morphological analyses have generated competing hypotheses of relationships.

Ever since Williston (1917), paleontologists have used the absence of temporal openings to unite turtles with various groups of basal reptiles that also lacked such openings, as Anapsida. Olson (1947) linked turtles to

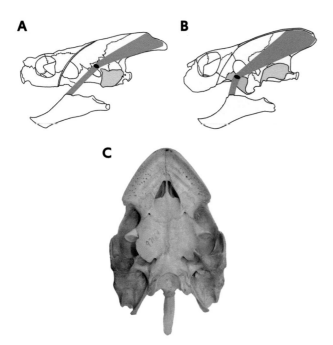

Figure 5.7. A–B, extent of the external adductor muscle (pink) in **A**, a cryptodiran turtle, and **B**, a pleurodiran turtle. Dark ellipse represents the cartilage (cartilago transiliens) in the adductor tendon. **C**, cranium of *Podocnemis expansa* in palatal view with pterygoid trochlea highlighted in green. A–B, cranial outlines redrawn from Gaffney (1979); C, modified from photograph in Cadena (2015)—CC BY 4.0.

diadectid stem-amniotes. Lee (1995, 1997a) related turtles to pareiasaurs, and Reisz and Laurin (1990) linked them to procolophonoids (Chapter 3). Gaffney (1980) and Gauthier et al. (1988b) considered turtles most closely related to captorhinids (Chapter 4). A number of authors placed turtles as the sister group to Diapsida based on morphological and developmental evidence. Others noted that turtles and lepidosaurs share a number of skeletal features, such as the presence of a large opening (thyroid fenestra) between the ischium and the pubis, the posterior embayment of the quadrate, the fusion of the astragalus and calcaneum into a single proximal tarsal, and the presence of a "hooked" fifth metatarsal. Hill (2005) also recovered this relationship in a phylogenetic analysis based on a broad range of amniote taxa. DeBraga and Rieppel (1997) and Rieppel and Reisz (1999) found turtles as the sister taxon to Sauropterygia and, in turn, this pair as the sister group to Lepidosauriformes (Lepidosauria and their closest relatives). Strongly supported by molecular data (e.g., Hedges and Poling

1999), another hypothesis considers turtles close to Archosauria (crocodylians, dinosaurs, and related taxa) but most of the proposed anatomical features cited in support of this hypothesis have not stood up to further scrutiny (Rieppel 2000). Bhullar and Bever (2009) cited the presence of a possible laterosphenoid in *Proganochelys* and the presence of middorsal osteoderms as potential morphological synapomorphies for archosaurs and turtles, but assessment of both features is problematic. Most recently, Bever et al. (2015) and Schoch and Sues (2015, 2018a) affirmed the diapsid affinities of turtles but left their phylogenetic position within Diapsida unresolved.

Analyses of molecular data sets have consistently placed turtles as the sister group to or even nested within Archosauria. An analysis using a large data set for absence or presence of micro RNAs (miRNA) supported a relationship between turtles and archosaurs (Field et al. 2014). This is consistent with other studies that sequenced mitochondrial and nuclear DNA and recovered turtles with archosaurs (e.g., Hedges and Poling 1999; Tzika et al. 2011). Chiari et al. (2012) analyzed DNA sequence data from 248 genes in fourteen amniote taxa and also found significant support for grouping turtles with archosaurs. Similarly, a study of more than 1,100 ultraconserved elements and the flanking DNA sequences in the genomes of a range of reptilian taxa supported this hypothesis (Crawford et al. 2012).

What accounts for these conflicting results? Morphology-based phylogenetic analyses are often affected by uncertainties in establishing the nature of individual characters and by the choice of outgroups. Molecular analyses can be influenced by assumptions concerning rates of nucleotide substitution and by sampling of taxa. Lu et al. (2013) employed a novel approach to the analysis of molecular data. Most analyses of sequence data treat individual nucleotide sites as characters. By contrast, Lu et al. (2013) treated genes as characters, each with multiple states and thus amenable to parsimony analysis. They used a data set comprising 4,584 orthologous genes. ("Orthologous" refers to homologous genes that diverged after a speciation event. Such genes typically have a function similar to that of the ancestral gene from which they evolved.) The analysis surprisingly yielded conflicting results, placing turtles either with archosaurs or with lepidosaurs, or even as the sister group to both. None of these results had particularly robust statistical support. Lu et al. (2013) interpreted this conflict as evidence that turtles, lepidosaurs, and archosaurs may have rapidly diverged from each other and diversified.

Studies by Joyce and his associates (e.g., Joyce et al. 2004; Joyce 2007, 2017) have a developed a comprehensive phylogenetic classification of turtles, which is used in this chapter.

Stem-Turtles

Several taxa have been interpreted as stem-turtles, but they all lack a complete turtle shell. Joyce et al. (2004) united these forms with Testudinata in a clade Pan-Testudines. These early forms have shed much light on the evolution of the turtle body plan.

Eunotosaurus, from the middle Permian (Guadalupian: Capitanian) of Malawi and South Africa, attained a total length of up to 30 cm (Fig. 5.8). It has six short cervical vertebrae, which have bulbous neural spines and bear long, slender ribs. The short, broad trunk comprises only nine vertebrae, which have anteroposteriorly elongated centra and bear expanded ribs that contact each other for much of their length and are distinctly T-shaped in transverse section distally (Lyson et al. 2013). Histological examination of the ribs of *Eunotosaurus* revealed that Sharpey's fibers (collagenous fibers linked to muscle attachment) are restricted to the ventral side of the dorsal ribs, indicating that reorganization of the muscles responsible for respiration had already occurred. *Eunotosaurus* has paired gastralia without lateral and medial elements. It has only a single pair of sacral ribs.

In recent years, several researchers (Lyson et al. 2010, 2013; Bever et al. 2015) have reexamined the skeletal structure and phylogenetic relationships of *Eunotosaurus*, which Watson (1914) first interpreted as a precursor of turtles based on the structure of its trunk region. A phylogenetic analysis by Lyson et al. (2010) found *Eunotosaurus* as the sister taxon of turtles and placed it among parareptiles. However, Bever et al. (2015) reported upper temporal fenestrae in a juvenile skull of *Eunotosaurus* and argued that the large supratemporals covered smaller gaps in the temporal region of adult individuals. However, homologizing the latter with the upper temporal

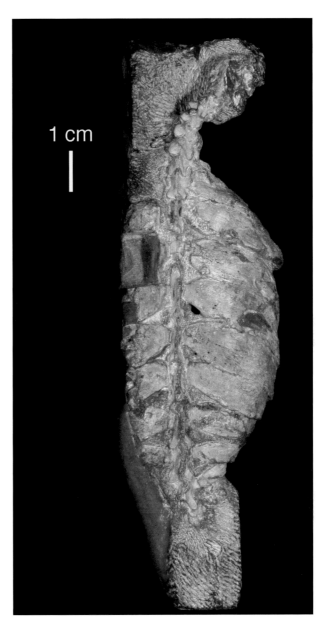

1 cm

Figure 5.8. Skull and partial postcranial skeleton of the possible stem-turtle *Eunotosaurus africanus* in dorsal view. Courtesy of Tyler Lyson.

fenestrae of other diapsids is problematic. The discovery of the clearly diapsid Middle Triassic stem-turtle *Pappochelys* unambiguously demonstrates the diapsid affinities of turtles and their stem-group (Schoch and Sues 2015, 2018a). Thus, the closed temporal region in turtles and most stem-turtles represents an evolutionary reversal rather than the retention of a primitive "anapsid" condition.

Lyson et al. (2013) presented a detailed scenario for the evolution of the turtle body plan, interpreting *Eunotosaurus* as the earliest known stem-turtle. Since then the Middle Triassic stem-turtle *Pappochelys* has provided an additional stage in the sequence between *Eunotosaurus* and the Late Triassic (Carnian) stem-turtle *Odontochelys*. The Late Triassic (Norian) *Proterochersis* is the oldest known member of Testudinata with a fully developed carapace and plastron. This morphological transformation series is consistent with the relative stratigraphic ages of the various taxa under discussion. Lyson et al. (2016) argued that *Eunotosaurus* was a capable digger and that the origin of the turtle shell should be interpreted as an adaptation for this particular mode of life. However, none of the more derived stem-turtles shows obvious specializations for digging.

Pappochelys, from the Middle Triassic (Ladinian) of Germany, has expanded dorsal ribs (Fig. 5.9) and differs from *Eunotosaurus* in the presence of a well-developed ventral cuirass of paired gastralia that are occasionally fused to each other serially (Schoch and Sues 2015, 2018a). Its skull has small, conical marginal teeth in the upper and lower jaws. The temporal region of the cranium of *Pappochelys* has a fully diapsid configuration, with a rounded upper and a ventrally open lower temporal fenestra. Although all known remains were recovered from lacustrine strata, *Pappochelys* has rather compact limb bones and shows no features that suggest a predominantly aquatic mode of life (Klein et al., unpublished data).

Eorhynchochelys, from the Late Triassic (Carnian) of Guizhou (China), resembles *Pappochelys* in the absence of a carapace and plastron (C. Li et al. 2018). It differs from the latter in its much greater size (total length of 1.8 m), the closure of the upper temporal fenestrae, the presence of a beak rather than teeth on the premaxilla and anterior end of the dentary, and the presence of a solid puboischiadic plate. *Eorhynchochelys* differs from other known stem-turtles in the presence of 12 dorsal vertebrae,

Odontochelys, from the same provenance as but stratigraphically slightly younger than *Eorhynchochelys*, has a fully ossified, oval plastron (Fig. 5.10). By contrast, its carapace consists only of at least eight pairs of expanded dorsal ribs and neural plates, which are not fused to the neural spines of the dorsal vertebrae (Li et al. 2008). This combination established that the evolutionary development of the plastron preceded that of the carapace. The scapula is still positioned anterior to the first dorsal rib in *Odontochelys*, indicating that the fan-shaped pattern of rib growth characteristic of derived Testudinata was absent in this stem-turtle (Nagashima et al. 2009). *Odontochelys* has small marginal teeth in the upper and lower jaws, palatal teeth, an unfused basicranial articulation, and lacks a bony wall to the middle ear (Li et al. 2008). It probably lived in river deltas and possibly in nearshore marine settings.

Testudinata

As noted earlier, this clade is diagnosed by the presence of a complete turtle shell (or one derived by modification of such a shell). *Proterochersis,* from the Late Triassic (Norian) of Germany and Poland, and *Proganochelys*, from the Late Triassic (Norian) of Germany, Greenland, and

Figure 5.9. Partially disarticulated postcranial skeleton of the stem-turtle *Pappochelys rosinae*. Arrows point to expanded dorsal ribs.

1 cm

Figure 5.10. Skeleton of the stem-turtle *Odontochelys testacea* in ventral view. Note the fully developed plastron but only expanded dorsal ribs rather than a carapace. Total length c. 40 cm. Courtesy of Tyler Lyson.

Figure 5.11. Carapace of the stem-turtle *Proterochersis robusta* in anterolateral view. Note development of a complete bony shell. Length c. 35 cm. Courtesy of Rainer Schoch.

Thailand, are stem-turtles that already have a fully developed carapace and plastron firmly connected by a bony bridge on either side (Gaffney 1990; Szczygielski and Sulej 2016; Joyce 2017).

Proterochersis, from the Late Triassic (Norian) of Germany and Poland (Joyce et al. 2013a; Szczygielski and Sulej 2016; Joyce 2017), has a strongly domed carapace, which can be up to 40 cm long (Fig. 5.11). It was long considered the earliest known pleurodire because its pelvis

is sutured to its carapace and plastron. However, Joyce et al. (2013a) argued that *Proterochersis* developed this sutural connection independently from pleurodires. Unlike in *Proganochelys*, the first thoracic rib forms a costal. Szczygielski and Sulej (2016) and Joyce (2017) considered *Proterochersis* the most basal member of Testudinata.

Except for a greater number of bony plates in the carapace and plastron, the shell of *Proganochelys* closely resembles those of more derived turtles. The ribs are integrated with the costal plates, and the neural elements are firmly attached to the dorsal vertebrae. The nuchal and peripherals form the margins of the carapace. The carapace of *Proganochelys* (Fig. 5.12) attained a length of about 60 cm. The pectoral and pelvic girdles are positioned inside the shell. Although partially integrated into the plastron, the interclavicle and clavicles remain distinct. *Proganochelys* bears well-developed dermal armor on its neck and tail (including a small tail club). Both manus and pes have only two phalanges per digit (Gaffney 1990). *Proganochelys* has numerous plesiomorphic cranial features, such as the presence of the lacrimal and supratemporal, a canal for the lacrimal duct, an interpterygoid vacuity, and an unfused basicranial articulation. It lacks marginal teeth, but the vomer, palatine, and pterygoid bear small teeth. The middle ear lacks bony

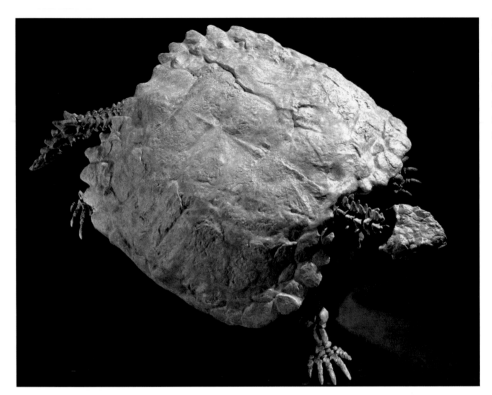

Figure 5.12. Reconstructed skeleton of the stem-turtle *Proganochelys quenstedti* in obliquely dorsal view. Note the osteoderm covers over the neck and tail. Length of shell c. 70 cm. Courtesy of Rainer Schoch.

subdivision into chambers, unlike in crown-group turtles. Gaffney (1990) inferred amphibious habits for *Proganochelys*, but Joyce and Gauthier (2004) argued that it was probably a land-dweller based on the proportions of its forelimb and the reduced phalangeal formula of the manus.

Palaeochersis, from the Late Triassic (Norian) of Argentina (Sterli et al. 2007), and *Australochelys*, from the Early Jurassic (Hettangian-Sinemurian) of South Africa (Gaffney and Kitching 1994), are more derived than *Proganochelys* in the absence of teeth on the palatine and vomer, the partial formation of a cavum tympanicum, the fusion of the basicranial articulation, and the fusion of the lateral end of the paroccipital process to the quadrate and squamosal (Joyce 2007).

Testudinata: Mesochelydia

Joyce (2017) proposed Mesochelydia (from Greek *mesos*, middle, and *chelys*, turtle) for a clade comprising the most recent common ancestor of the Jurassic taxa *Condorchelys antiqua*, *Eileanchelys waldmani*, *Heckerochelys romani*, and *Kayentachelys aprix* and all descendants of that ancestor.

Compared to *Proganochelys*, mesochelydians share a suite of derived features including the absence of a lacrimal bone and lacrimal duct; presence of confluent external nares; fully formed cavum tympani; paired basioccipital tubercles; presence of only 11 pairs of peripherals in the carapace; and pectoral girdle with three strap-like bony processes connected only by minor bony webs (Joyce 2017).

The best-known mesochelydian, *Kayentachelys*, from the Early Jurassic (Pliensbachian) of Arizona, was originally interpreted as the oldest known cryptodire (Gaffney et al. 1987). It shares many apomorphies with more derived turtles such as the absence of the lacrimal and supratemporal, the presence of confluent external nares, a fully developed cavum tympanicum, and the absence of a contact between the nuchal and cleithra (Joyce 2007). If, as Gaffney and Jenkins (2010) argued, *Kayentachelys* were the oldest known cryptodire and *Proterochersis* the earliest pleurodire, all these features would have evolved independently in cryptodires and pleurodires. Thus, it is more parsimonious to reinterpret *Kayentachelys* as a stem-turtle (Joyce 2007, 2017). Based on its dorsoventrally low, rather smooth shell, *Kayentachelys* was probably a predominantly aquatic form (Gaffney et al. 1987). Its shell attained a length of about 20 cm.

Eileanchelys, from the Middle Jurassic (Bathonian) of Scotland, attained a carapace length of at least 20 to 30 cm (Anquetin 2010). Anquetin et al. (2009) interpreted it as an aquatic stem-turtle. *Eileanchelys* has an elongate postorbital region of the cranium and its carapace has several derived features.

Heckerochelys, from the Middle Jurassic (Bathonian) of Russia, is noteworthy for its distinctly aquatic adaptations (Sukhanov 2006; Joyce 2017). Its shell attained a length of 40 to 50 cm and lacks bony connections between the carapace and plastron, both of which have fontanelles. The bones of the shell bear crenulated ornamentation.

TESTUDINATA: PERICHELYDIA

Various Mesozoic and Cenozoic taxa have long been classified as cryptodiran turtles (Gaffney and Meylan 1988), but Joyce (2007), Sterli and de la Fuente (2011, 2013), Anquetin (2012), and Sterli (2015) argued that these forms are not crown-group turtles. Joyce (2017) proposed Perichelydia (from Geek *peri*, near, and *chelys*, turtle) for a clade comprising the most recent common ancestor of *Meiolania platyceps, Helochelydra nopcsai, Sichuanchelys chowi*, and the present-day *Testudo graeca* (Greek tortoise) and all descendants of that ancestor.

Helochelydridae comprises all turtles more closely related to *Helochelydra nopcsai* than to *Meiolania platyceps, Sichuanchelys chowi*, or any extant turtle (Joyce et al. 2016). Diagnostic apomorphies for this clade include the presence of a secondary pair of occipital tubercles formed by the pterygoids, the presence of a triangular fossa formed by the squamosal at the posterior margin of the cranium, and distinct tubercles forming the surface ornamentation of the shell (Joyce 2017). *Helochelydra* is known from the Early Cretaceous (Berriasian-Barremian) of England, France, and Spain and the Late Cretaceous (Cenomanian) of Germany. *Naomichelys*, from the Early Cretaceous (Aptian-Albian) of Maryland, Montana, Nevada, Texas, and Wyoming and the Early to Late Cretaceous (Albian-Cenomanian) of Utah, is a North American representative of this clade (Joyce et al. 2014).

Sichuanchelyidae includes all perichelydians more closely related to *Sichuanchelys chowi* than to *Helochelydra nopcsai, Meiolania platyceps*, or any extant turtle (Joyce 2017). Shared derived features for this clade include the presence of enlarged squamosals that partially or fully cover the anterior region of the neck; the close approximation or contact between the jugal and the quadrate; presence of paired pits on the ventral surface of the basisphenoid; and a nuchal notch delimited by the second peripheral on either side (Joyce 2017). *Sichuanchelys* is known from the Middle to Late Jurassic of Sichuan and Xinjiang (China) and *Mongolochelys* from the Late Cretaceous (Campanian-Maastrichtian) of Mongolia.

Meiolaniformes comprises all turtles more closely related to *Meiolania platyceps* than to Cryptodira and Pleurodira (crown-group Testudines) (Sterli and de la Fuente 2013; Sterli 2015). It is an exclusively Gondwanan clade distinguished by a suite of shared derived features including the exclusion of the frontal from the orbit; prefrontal with large exposure; presence of a ventral crest on the vomer; opisthocoelous anterior caudal vertebrae; external surfaces of dermal cranial bones and shell elements covered by small, scattered pits; and grooves of intermarginal scutes inflected anteriorly along the rim of the carapace (Sterli 2015). The oldest known representative is *Otwayemys*, from the Early Cretaceous (Albian) of Australia.

Meiolaniidae is known from the Paleogene (Paleocene-Eocene) of Argentina, the Paleogene (Eocene) to Quaternary (Holocene) of Australia, and the Quaternary of islands near Australia (Gaffney 1983, 1991, 1996; Sterli 2015; Poropat et al. 2016). Diagnostic features for this clade include the contact of the quadratojugal and squamosal below the cavum tympani; squamosals with large horn-like projections (Fig. 5.13); and caudal vertebrae enclosed in overlapping rings of dermal armor, which form a terminal tail club (Sterli 2015). The manus and pes of *Meiolania* each have only two or fewer phalanges per digit. Gaffney (1991, 1996) interpreted meiolaniids as cryptodires but a suite of plesiomorphic features indicates that they represent a clade of late-surviving stem-turtles closely related to crown-group turtles (Sterli and de la Fuente 2011, 2013; Sterli 2015). Meiolaniids attained total lengths of 2 to 3 m.

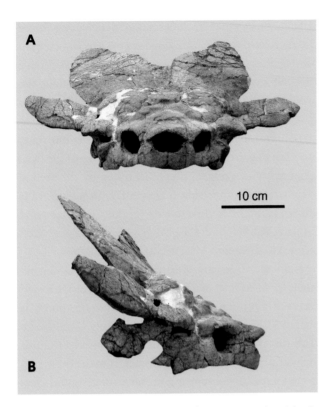

10 cm

Figure 5.13. Cranium of the meiolaniform stem-turtle *Niolamia argentina* in **A**, anterior, and **B**, lateral view. Courtesy of Juliana Sterli.

The shell of the enigmatic perichelydian *Kallokibotion*, from the Late Cretaceous (Maastrichtian) of Romania, reached a length of about 50 cm and has a finely crenulated surface ornamentation. Pérez-García and Codrea (2018) recognized the division of the nuchal plate into two elements as an autapomorphy for this taxon. Gaffney and Meylan (1992) interpreted *Kallokibotion* as a basal cryptodire but later authors considered it a stem-turtle closely related to Testudines. The phylogenetic analyses by Sterli and de la Fuente (2013) and Rabi et al. (2014) found this taxon closely related to Meiolaniidae.

TESTUDINES

Testudines is the crown clade arising from the most recent common ancestor of *Chelonia mydas* (green turtle), *Chelus fimbriatus* (matamata), and all other valid species of turtles listed in Appendix 1 in Joyce et al. (2004). It shares various derived features such as the exclusion of the frontal from the dorsal margin of the orbit, the absence of a contact between the nuchal and the eighth cer-

vical vertebra, and the absence of large cervical ribs (Joyce 2007). Molecular data also support the monophyly of Testudines (Guillon et al. 2012; Crawford et al. 2015). Gaffney (1975) proposed Casichelydia for the clade comprising Cryptodira and Pleurodira but Joyce et al. (2004) argued that the older name Testudines should be used instead. The currently known fossil record of Testudines extends back to the Late Jurassic (Danilov and Parham 2006).

TESTUDINES: PLEURODIRA

Pleurodira is the crown clade arising from the most recent common ancestor of *Chelus fimbriatus*, *Pelomedusa subrufa* (African helmeted turtle), *Podocnemis expansa* (Arrau River turtle), and all other valid species of turtles listed in Appendix 1a, but none of the other valid species listed in Appendix 1b in Joyce et al. (2004). Present-day pleurodirans are restricted to freshwater environments in the Southern Hemisphere (Fig. 5.14) but extinct taxa had a much wider geographic distribution and occupied a variety of habitats. In addition, a number of Mesozoic turtles have been interpreted as stem-pleurodirans. Joyce et al. (2004) and Joyce (2007) united them with the crown-group as Pan-Pleurodira. The latter total clade is characterized by a number of synapomorphies in the postcranial skeleton such as the well-defined articulations between the centra of the cervical vertebrae, the presence of a well-developed anal notch, paired mesoplastra lacking a median contact, and the sutural connection of the pelvis to the shell (Joyce 2007; Cadena and Joyce 2015). Platychelyidae comprises *Platychelys*, from the Late Jurassic (Kimmeridgian-Tithonian) of Germany and Switzerland, and *Notoemys*, from the Late Jurassic (Tithonian) of Argentina, the Late Jurassic (Oxfordian) of Cuba, and the Early Cretaceous (Valanginian) of Colombia. Diagnostic features for this clade include the presence of a central fontanelle in the plastron, a straight anterior margin of the carapace, broad vertebral scutes, and various features of the vertebral column (Cadena and Joyce 2015). Platychelyids are known from nearshore marine deposits. Another group of pan-pleurodirans, Dortokidae, is particularly distinguished by the ornamentation of the carapace with anteroposteriorly elongated pits and ridges. *Eodortoka* is known from the Early Cretaceous (Barremian-Aptian) of Spain

Figure 5.14. Extant representatives of Chelidae. **A**, Murray River turtle (*Emydura macquarii*); **B**, matamata (*Chelus fimbriata*). A, courtesy of Division of Amphibians and Reptiles, National Museum of Natural History. B, courtesy of Laurie Vitt.

(Pérez-García et al. 2014) and *Dortoka* from the Late Cretaceous (Campanian-Maastrichtian) of Spain and possibly France and from the Paleocene of Romania (Cadena and Joyce 2015).

The monophyly of Pleurodira (Fig. 5.1) is well supported by a suite of derived skeletal features (Gaffney and Meylan 1988; Joyce 2007) and by molecular data (e.g., Guillon et al. 2012). The former includes the presence of a wing-like trochlear process on the pterygoid, the absence of a contact between the prefrontal and palatine, the absence of cleithra, and procoelous caudal vertebrae.

TESTUDINES: PLEURODIRA: CHELIDAE

Pan-Chelidae is distinguished by the presence of biconvex fifth and eighth cervical centra and the (plesiomorphic) presence of a cervical scute (Maniel and de la Fuente 2016). Chelidae (snake-necked turtles) is the crown clade arising from the most recent common ancestor of *Chelus fimbriatus*, *Chelodina longicollis* (eastern long-necked turtle), and all valid species of turtle listed in Appendix 2, but none of the other valid species listed in Appendix 1 in Joyce et al. (2004). It differs from stem-chelids in the absence of the quadratojugal and the absence of mesoplastra (except in *Yaminuechelys*, from the Early or Late Cretaceous (Albian or Cenomanian) of Argentina; de la Fuente et al. 2001). Present-day chelids occur in Australasia (Fig. 5.14A) and throughout much of South America. Most taxa such as the matamata (*Chelus fimbriatus*) have aquatic habits (Fig. 5.14B). *Chelus* is first known from the Neogene (Miocene) of Colombia and Venezuela (Ferreira et al. 2016).

TESTUDINES: PLEURODIRA: PELOMEDUSOIDES

Pelomedusoides is the crown clade arising from the most recent common ancestor of *Pelomedusa subrufa*, *Podocnemis expansa*, and all valid species of turtles listed in Appendices 3 and 4, but none of the other valid species listed in Appendix 1 in Joyce et al. (2004). Synapomorphies for this clade are the absence of the nasal, the presence of a median contact between the prefrontals, the absence of the splenial, the absence of cervical scutes, and the presence of anteriorly concave articular contacts between cervical vertebrae 2 and 3 and cervical vertebrae 3 and 4, respectively (Joyce 2007). The molecular-based phylogeny by Crawford et al. (2015) supported recognition of this clade. The oldest known member of Pelomedusoides is *Atolchelys*, from the Early Cretaceous (Barremian) of Brazil (Romano et al. 2014). Pelomedusoides includes the extinct Araripemydidae, Bothremydidae, and Euraxemydidae (Gaffney et al. 2006, 2011). Pelomedusidae and Podocnemididae (Figs. 5.15, 5.16) are also represented by a number of extant taxa. A number of pelomedusoids, especially most bothremydids, were widely distributed in nearshore marine settings in Europe, Africa, and the Americas during the Late Cretaceous and Paleogene. The bothremydid *Puentemys*, from the Paleogene (Paleocene) of Colombia, attained a carapace length of 1.5 m (Cadena et al. 2015; Fig. 5.17). Podocnemididae is the crown clade arising from the most recent common ancestor of

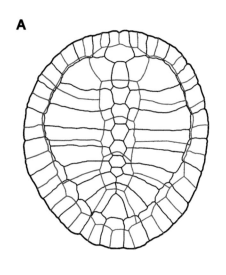

Figure 5.15. Extant representatives of Pelomedusoides. **A**, African helmeted turtle (*Pelomedusa subrufa*); **B**, Geoffrey's side-necked turtle (*Phrynops geoffroanus*). Both courtesy of Division of Amphibians and Reptiles, National Museum of Natural History.

Podocnemis expansa, Peltocephalus dumerilianus (big-headed Amazonian river turtle), and all valid species of turtles listed in Appendix 4, but none of the other valid species listed in Appendix 1 in Joyce et al. (2004). It is first known from the Late Cretaceous (Cenomanian) of Mo-

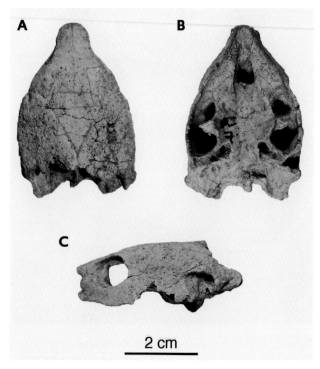

Figure 5.16. Cranium of the podocnemidid *Neochelys arenarum* in **A**, dorsal; **B**, palatal; and **C**, lateral views. From Cadena (2015)—CC BY 4.0.

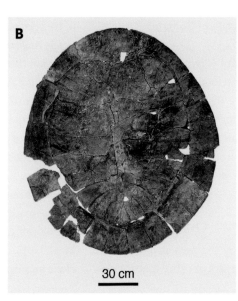

Figure 5.17. A, reconstructed carapace of the bothremydid *Puentemys mushaisaensis* in dorsal view. Red lines represent furrows for the scutes. **B**, carapace as preserved. A, modified from Cadena et al. (2015); B, courtesy of Carlos Jaramillo.

A

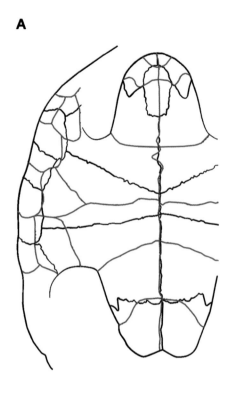

B

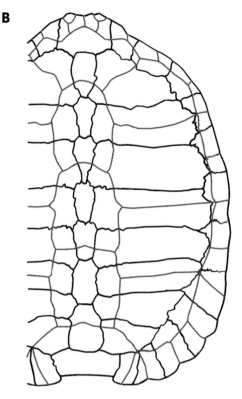

Figure 5.18. **A**, plastron, and **B**, carapace of the baenid *Baena arenosa*. Red lines represent furrows for the scutes. Mid-line length c. 31.5 cm. From H. F. Smith et al. (2017)—CC BY 4.0.

rocco and Brazil (Joyce et al. 2013b) and today occurs in western Madagascar and northern South America. *Stupendemys*, from the Neogene (Pliocene) of Venezuela, is one of the largest known turtles with a carapace length of about 2.3 m (Wood 1976). Pelomedusidae is the crown clade arising from the most recent common ancestor of *Pelomedusa subrufa*, *Pelusios subniger* (East African black mud turtle), and all the other valid species of turtles listed in Appendix 3, but none of the other valid species listed in Appendix 1 in Joyce et al. (2004). It is first definitely recorded from the Miocene of Kenya (Joyce et al. 2013b) and today occurs in freshwater habitats in Sub-Saharan Africa and Yemen as well as on Madagascar and the Seychelles.

TESTUDINES: PAN-CRYPTODIRA

Joyce (2007) recognized a clade Pan-Cryptodira that encompasses Eucryptodira, which comprises Cryptodira and closely related Mesozoic taxa, and Paracryptodira, which includes a variety of Mesozoic and Paleogene turtles (Gaffney 1975). Pan-Cryptodira is characterized by the presence of a contact between the pterygoid and basioccipital, the position of the posterior foramen for the bony canal carrying the internal carotid artery halfway along the suture between the pterygoid and basisphenoid, and the position of the sulcus between vertebrals II and III on the fifth neural. The first feature is problematic because it is also present in *Kallokibotion* and *Meiolania*, both of which were interpreted as cryptodires by Gaffney (1975) but not by Joyce (2007).

TESTUDINES: PAN-CRYPTODIRA: PARACRYPTODIRA

Joyce (2007) listed various shared derived features for Paracryptodira, including a (secondarily) small supraoccipital crest and the reduced exposure of the prefrontal on the skull roof. Pleurosternidae is characterized by the following apomorphies: oblong and anteriorly rounded shape of the skull, which is longer than broad; restricted contact between the pterygoid and basioccipital; and anterior plastral lobe larger than the posterior one (Pérez-García et al. 2015). It includes *Glyptops*, from the Late Jurassic and Early Cretaceous of the western United States, and *Pleurosternon*, from the Late Jurassic (Kimmeridgian-Tithonian) of France and Spain and the Early Cretaceous (Barremian) of England and Germany. The sister taxon of Pleurosternidae is Baenidae, which is the most inclusive clade including *Baena arenosa* (Fig. 5.18) but not *Pleurosternon bullockii* or any extant species of turtle (Joyce and Lyson 2015). This apparently exclusively North American clade ranged in time from the Early Cretaceous (Aptian-Albian) to the Paleogene (Gaffney 1972; Lyson and Joyce 2009; Joyce and Lyson 2015). Diagnostic features for Baenidae are the presence of a contact between the basioccipital and pterygoid (Fig. 5.19), the extension of the axillary and inguinal buttresses to the costals, and the absence of an epiplastral process (Joyce and Lyson 2016). Baenids lived in freshwater and were omnivorous to molluscivorous.

TESTUDINES: PAN-CRYPTODIRA: EUCRYPTODIRA

Joyce (2007) diagnosed Eucryptodira based on the presence of a medial contact between the prefrontals and the posterior position of the posterior foramen for the canal carrying the internal carotid artery at the posterior end of the pterygoid. The phylogenetic positions of basal eucryptodirans such as Sandowniidae, from the Early Cretaceous of England and Texas and the Late Cretaceous to Paleocene of Africa, and various mainly Asian Mesozoic clades (Xinjiangchelyidae, Sinemydidae [Fig. 5.20], and Macrobaenidae) remain problematic (Gaffney and Meylan 1988; D. B. Brinkman and X.-c. Wu 1999; Joyce 2007; Tong and Meylan 2013; Anquetin et al. 2014; Rabi et al. 2014). The phylogenetic analysis by Rabi

et al. (2014) even recovered Xinjiangchelyidae outside Testudines.

The traditionally recognized Eurysternidae, Plesiochelyidae, and Thalassemydidae comprise mostly Jurassic marine turtles from Europe, but assessments of their relationships to each other and other eucryptodirans await a comprehensive review of the many known specimens. Anquetin et al. (2014) noted that these groups possibly form a clade. Some have a fully ossified carapace and a bony bridge on either side between the carapace and plastron (e.g., *Craspedochelys*; Fig. 5.21), whereas others have prominent fontanelles in the carapace and/or plastron and lack bony bridging between the two portions of the shell (e.g., *Eurysternum*; Fig. 5.22).

TESTUDINES: PAN-CRYPTODIRA: EUCRYPTODIRA: CRYPTODIRA

Cryptodira is the crown clade arising from the most recent common ancestor of *Testudo graeca*, *Chelonia mydas*, *Trionyx triunguis* (African softshell turtle), and all other valid species of turtles listed in Appendix 1b, but none of the other valid species listed in Appendix 1a, of Joyce et al. (2004; Fig. 5.23). Joyce (2007) listed four synapomorphies for this clade: the absence of paired pits on the ventral surface of the basisphenoid; the absence of the splenial; the centrum of the eighth cervical vertebra being distinctly shorter than that of the seventh; and the absence of a cleithrum. Crawford et al. (2015) supported Crypto-

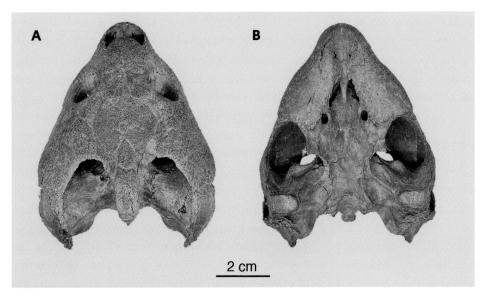

Figure 5.19. Cranium of the baenid *Eubaena cephalica* in **A**, dorsal, and **B**, palatal views. Courtesy of Tyler Lyson.

2 cm

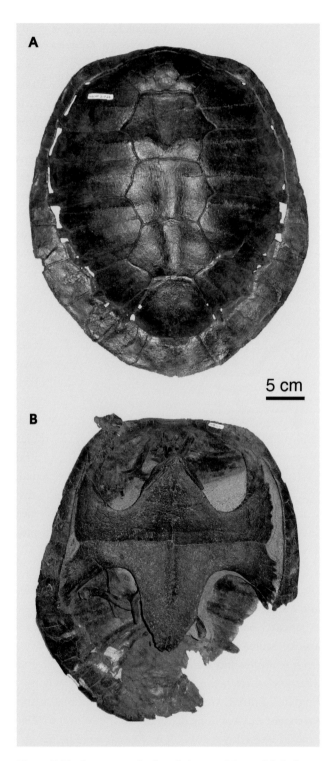

Figure 5.20. A, carapace in dorsal view, and **B**, partial shell with plastron in ventral view of the sinemydid *Judithemys sukhanovi*. Courtesy of Don Brinkman.

5 cm

dira based on molecular data. Most extant and many extinct turtles belong to Cryptodira. Present-day cryptodires occur mainly in the Northern Hemisphere but are also present in Africa and South America. Although the diversity and structure of cryptodirans are well documented, their interrelationships have to be fully resolved.

CRYPTODIRA: PAN-TRIONYCHIA

Pan-Trionychia is united by the presence of extragular scutes and the presence of anteriorly convex articulations between cervical vertebrae 4 through 7 (Joyce 2007). Danilov and Parham (2006) recognized a pan-trionychian clade Adocusia comprising Acocidae, which includes *Adocus*, from the Late Cretaceous to Paleogene (Paleocene) of Montana, New Jersey, and New Mexico (Meylan and Gaffney 1989) and the Paleogene (Eocene) of Inner Mongolia (China), and Nanhsiungchelyidae. The latter includes *Basilemys*, from the Late Cretaceous (Campanian) of Alberta, New Mexico, and Utah and the Late Cretaceous (Maastrichtian) of Alberta, Montana, and Saskatchewan (J. H. Hutchinson et al. 2013); and *Zangerlia*, from the Late Cretaceous (Campanian) of Mongolia and Inner Mongolia (China) (Joyce and Norell 2005). *Basilemys* attained a shell length of almost 1 m and may have resembled extant tortoises in its inferred mode of life (D. B. Brinkman 2005).

Trionychia (from Greek *tri*, three, and *onyx*, claw) is the crown clade arising from the most recent common ancestor of *Trionyx triunguis*, *Carettochelys insculpta*, and all other valid species of turtles listed in Appendix 6, but none of the other valid species listed in Appendix 1, in Joyce et al. (2004). It is characterized by a suite of synapomorphies including the absence of a contact between the prefrontal and palatine; the fusion of the premaxillae and their exclusion from the external naris; the presence of a median contact between the palatines; the absence of a median contact between the pterygoids; the absence of distinct ventral processes on the posterior cervical vertebrae; and flipper-like limbs (Meylan 1987; Joyce 2007). The typically flattened shell lacks bony bridging between the carapace and plastron. At least some carapacial scutes and all plastral scutes are lacking. The entoplastron is boomerang-shaped. The distinctive external sculpturing (Figs. 5.25, 5.26) and characteristic "plywood" microstructure allow identification of even

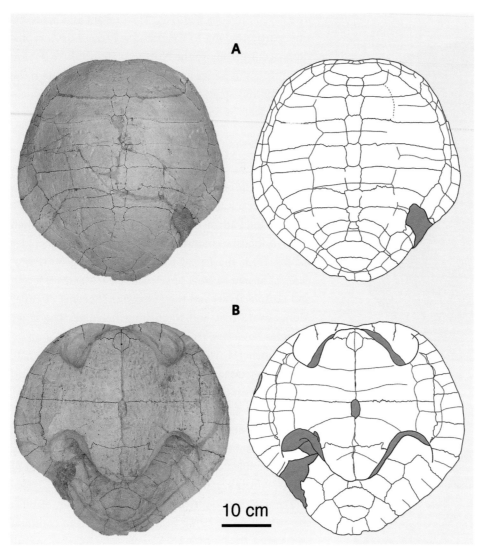

Figure 5.21. Shell of the eucryptodiran *Craspedochelys jaccardi* in **A**, dorsal, and **B**, ventral views. Red lines represent furrows for scutes. From Anquetin et al. (2014)—CC BY 4.0.

10 cm

small fragments of trionychian shells (Holroyd and J. H. Hutchinson 2002; Scheyer et al. 2007).

Joyce et al. (2004) proposed Pan-*Carettochelys* for the total group that includes the extant *Carettochelys insculpta* (pig-nosed turtle; Fig. 5.24A). This clade is characterized by a number of derived features such as the presence of a shallow fossa behind the quadrate, the presence of only 10 peripheral scutes, and the reduction of plastral scutes (Joyce 2014). *Kizylkumemys*, from the Early Cretaceous (Aptian) of Thailand and the Late Cretaceous (Cenomanian) of Uzbekistan, is its oldest known representative (Joyce 2014). Carettochelyidae comprises the most recent common ancestor of *Carettochelys insculpta* and *Anosteira ornata* and all descendants of that ancestor (Joyce et al. 2004). *Anosteira* is known from the Paleogene (Eocene) of

Utah and Wyoming, Guandong and Liaoning (China), and Mongolia (Joyce 2014). Its shell differs from those of other carettochelyids particularly in the presence of carapacial scutes. Carettochelyinae includes *Carettochelys*, from Australia and Papua New Guinea (Fig. 5.24A), and *Allaeochelys*, from the Paleogene (Eocene) of Belgium, England, France, Germany, Myanmar, and Spain, the Paleogene of Guandong (China), and the Neogene (Miocene) of Libya (Joyce 2014). It is characterized by the absence of carapacial and plastral scutes in adult individuals, a transversely broad plastron, and the presence of a deep fossa behind the quadrate (Joyce 2014). *Carettochelys* attains a shell length of up to about 60 cm.

Pan-Trionychidae is diagnosed by various synapomorphies including the exclusion of the fused premaxillae

Figure 5.22. Skeleton of the eucryptodiran *Eurysternum wagleri* in ventral view. Note the reduction of the bony shell. Courtesy of Jérémy Anquetin.

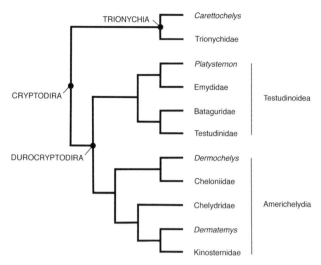

Figure 5.23. Phylogenetic hypothesis of extant cryptodiran turtles. Dots denote node-based clades. Based mainly on Joyce (2007) and Joyce et al. (2013).

Figure 5.24. Extant representatives of Trionychia. **A**, pig-nosed turtle (*Carettochelys insculpta*); **B**, Chinese softshell turtle (*Trionyx sinensis*). A, courtesy of Laurie Vitt; B, courtesy of Division of Amphibians and Reptiles, National Museum of Natural History.

from the external narial aperture; the absence of a contact between the quadratojugal and the maxilla or postorbital; all metaplastic portions of the shell with superficial ornamentation; the absence of peripherals, pygals, and suprapygals; absence of scutes on the shell; entoplastron boomerang-shaped; absence of articulation between the centra of the eighth cervical and first dorsal vertebra; hyperphalangy of the manus and pes; and the presence of three claws on each manus and pes (Vitek and Joyce 2015). *Perochelys*, from the Early Cretaceous (Aptian) of Liaoning (China), is the oldest well-known pan-trionychid and already closely resembles more derived members of this clade in its skeletal structure (L. Li et al. 2015). Pan-Trionychidae has a rich fossil record from the Late Cretaceous to Quaternary of Africa, Asia, Europe, and North America (e.g., J. D. Gardner et al. 1995; Danilov and Vitek 2013; Vitek and Joyce 2015; Georgalis and Joyce 2017) (Figs. 5.25, 5.26). The Late Cretaceous *Aspideretoides* attained a carapace width of almost 50 cm (D. B. Brinkman 2005), and *Drazinderetes*, from the Paleogene (Eocene) of Pakistan, possibly reached a carapace width of 1.1 m (Head et al. 1999). Unfortunately, many extinct taxa of pan-trionychids were based on shell remains that lack clearly diagnostic features, and, furthermore, Meylan (1987) observed high levels of homoplasy even among extant trionychids.

Trionychidae comprises two distinct lineages, represented by the extant *Cyclanorbis* (flapshell turtles) and *Trionyx* (softshell turtles; Fig. 5.24B), respectively (Meylan 1987; Georgalis and Joyce 2017). The oldest known record of *Cyclanorbis* is from the Neogene (Miocene) of Kenya, and *Trionyx* was already widely distributed throughout Europe by the Miocene (e.g., Georgalis and Joyce 2017). Extant trionychids occur in freshwater settings throughout much of Africa, Asia, and North America (Figs. 5.24B, 5.27).

CRYPTODIRA: DUROCRYPTODIRA

Based on molecular data (Parham et al. 2006; Barley et al. 2010; Crawford et al. 2015), cryptodirans other than Pan-Trionychia form a clade that Danilov and Parham (2006) named Durocryptodira (from Latin *durus*, hard, and Cryptodira), in reference to the hard shell shared by these groups. Durocryptodira is the crown clade arising from the most recent common ancestor of *Testudo graeca*, *Kinosternon scorpioides* (scorpion mud turtle), *Chelonia mydas*, and *Chelydra serpentina* (snapping turtle) but excluding *Trionyx triunguis* and *Carettochelys insculpta* (Danilov and Parham 2006). It can be divided into Americhelydia and Testudinoidea (Fig. 5.23).

CRYPTODIRA: DUROCRYPTODIRA: AMERICHELYDIA

Molecular analyses (Parham et al. 2006; Barley et al. 2010; Crawford et al. 2015) found a clade comprising Chelydroidea and Chelonioidea. Joyce et al. (2013b) proposed the name Americhelydia for this clade, which arises from the most recent common ancestor of *Chelonia mydas*, *Chelydra serpentina*, and *Kinosternon scorpioides*.

CRYPTODIRA: DUROCRYPTODIRA: AMERICHELYDIA: CHELYDROIDEA

Knauss et al. (2011) used the name Chelydroidea for the clade including Chelydridae and Kinosternoidea. Mo-

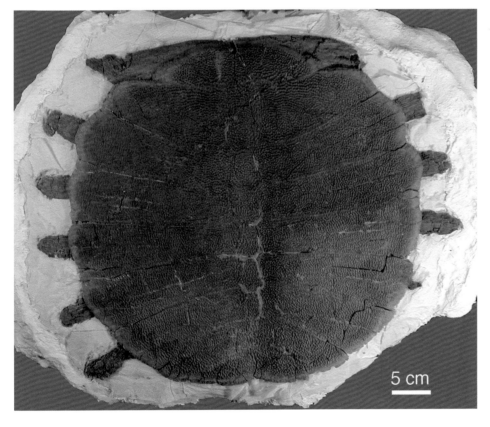

Figure 5.25. Carapace of the trionychid *Axestemys splendida* in dorsal view. Courtesy of Don Brinkman.

Figure 5.26. Carapace of the trionychid *Gilmoremys lancensis* in **A**, dorsal, and **B**, ventral view. Courtesy of Tyler Lyson.

lecular data support recognition of this clade (Bailey et al. 2010). Uniquely among turtles, Chelydridae and Kinosternoidea share the presence of rib-like (costiform) processes on the nuchal that extend along the internal (visceral) side of the carapace and insert into the second (in kinosternoids) or third (in chelydrids) peripherals (but see Joyce et al. 2013b). Knauss et al. (2011) listed additional synapomorphies for Chelydroidea. All basal chelydroids have reduced abdominal scutes that do not contact one another along the midline. The plastron is greatly reduced in size and much thickened along the midline and the bridge.

Pan-Chelydridae is distinguished by derived features such as the exclusion of the frontal from the orbit, the absence of extragular and pectoral scutes, and the presence of three or four contiguous inframarginals (Joyce 2016). The oldest well-known pan-chelydrid is *Protochelydra*, from the Paleogene (Paleocene) of Alaska, Alberta, and North Dakota (Erickson 1973). Chelydridae is the crown clade arising from the most recent common ancestor of *Chelydra serpentina* (Fig. 5.28) and *Macrochelys temminckii* (alligator snapping turtle) but none of the other valid species listed in Appendix 1 in Joyce et al. (2004). The present-day distribution of Chelydridae ranges from Canada and the United States east of the Rocky Mountains south to Colombia. *Chelydra* and *Macrochelys* share the presence of fontanelles in the carapace and plas-

tron of skeletally mature individuals and the absence of bridge sockets on the peripherals (Joyce 2016). *Macrochelys* dates back to the Neogene (Miocene) but *Chelydra* definitely only to the Quaternary (Pleistocene) (Joyce 2016). Chelydrid turtles are freshwater carnivores or molluscivores.

Shared derived features for Pan-Kinosternoidea include the presence of a thickened plastron, the angled shaft of the ilium, and the lack of a distal expansion of the ilium. The oldest known representative of this total group is *Emarginachelys*, from the Late Cretaceous (Maastrichtian) of Montana (Joyce and Bourque 2016). Pan-*Dermatemys* comprises the extant *Dermatemys* (Central American river or tabasco turtle) from Guatemala and Mexico and stem-taxa such as "*Baptemys*," from the Paleogene (Paleocene-Eocene) of New Mexico and Wyoming, and *Hoplochelys*, from the Late Cretaceous (Maastrichtian) of Montana and North Dakota and the Paleogene (Paleocene) of New Mexico (Joyce and Bourque 2016). The diagnostic apomorphy is the presence of a contact between the eighth peripheral and the inguinal buttress. Kinosternidae is the crown clade arising from the last common ancestor of *Staurotypus triporcatus* (Mexican musk turtle), *Kinosternon scorpioides*, and all other valid species of turtles listed in Appendix 7, but none of the other valid species listed in Appendix 1, in Joyce et al. (2004). It shares several derived features such as a

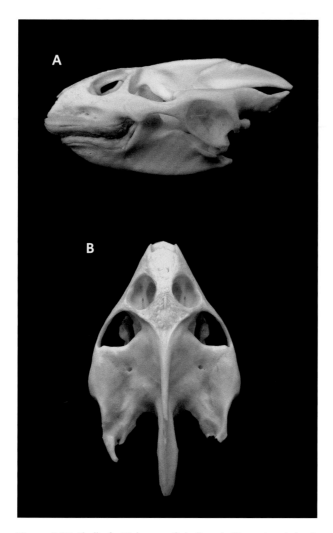

Figure 5.27. Skull of a Malayan softshell turtle (*Dogania subplana*) in **A**, lateral, and **B**, dorsal views. Courtesy of Pavel Zuber.

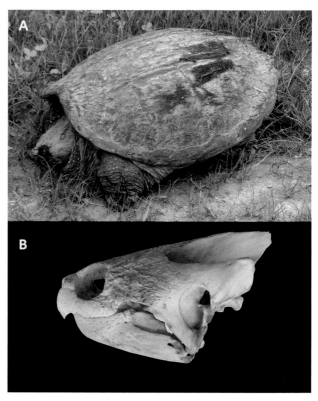

Figure 5.28. A, snapping turtle (*Chelydra serpentina*); **B**, skull of *Chelydra serpentina* in lateral view. A, Wikipedia (photo by D. G. E. Robertson)—CC BY-SA 3.0; B, courtesy of Chris Bell.

contact between the maxilla and quadratojugal, the presence of 10 pairs of peripheral scutes, the absence of abdominal scutes, and grooves for the musk ducts on the anterior peripherals (Joyce and Bourque 2016). The stratigraphically oldest definitive representative of Kinosternidae is *Baltemys*, from the Paleogene (Eocene) of Wyoming (J. H. Hutchinson 1991). *Kinosternon* (mud turtles; Fig. 5.29) has a transverse hinge joint that divides the plastron into a forelobe and hindlobe and facilitates complete enclosure in the shell when the animal is threatened. It dates back to the Neogene (Miocene). Extant kinosternids range from the central and southern United States to Argentina.

CRYPTODIRA: DUROCRYPTODIRA: AMERICHELYDIA: PAN-CHELONIOIDEA

Pan-Chelonioidea is the total clade arising from the most recent common ancestor of *Toxochelys latiremis* and *Chelonia mydas* (Gentry 2017). Gentry (2017) listed several shared derived features for this total group, such as the presence of intergulars, the presence of a platycoelous articulation between cervical vertebrae 6 and 7, and the humerus being longer than the femur. Both the carapace and plastron have large fontanelles in adults, reducing the weight of the shell. The forelimbs in pan-chelonioids are modified as paddles for underwater "flight" (Zangerl 1953). *Toxochelys* is known from the Late Cretaceous (Campanian) of Alabama, Kansas, and South Dakota (Zangerl 1953; Gentry 2017; Fig. 5.30).

Chelonioidea is the crown clade arising from the last common ancestor of *Chelonia mydas*, *Dermochelys coriacea*, and all other valid species listed in Appendix 5, but none

Figure 5.29. Yellow mud turtle (*Kinosternon flavescens*). Courtesy of Laurie Vitt.

of the other valid species listed in Appendix 1, in Joyce et al. (2004). Gentry (2017) listed a suite of synapomorphies diagnostic for this clade, including the coracoid being as long as the humerus, the position of the lateral process of the humerus on the proximal portion of the shaft but distal to the head of that bone, and the presence of a distinct notch between the femoral trochanters.

Pan-Cheloniidae is the total clade arising from the most recent common ancestor of *Ctenochelys* spp., *Puppigerus camperi*, and *Chelonia mydas* (Gentry 2017). Diagnostic features for this clade such as the presence of a well-developed secondary bony palate, the presence of a deep crescentic concavity between the basioccipital tubera, and the reduction in ossification of the carapace so that the peripherals have rib-free medial margins anterior and posterior to the ribs. *Ctenochelys* is known from the Late Cretaceous (Santonian-Campanian) of Alabama (Zangerl 1953; Gentry 2017). Cheloniidae is the crown clade arising from the last common ancestor of *Chelonia mydas* (Figs. 5.31, 5.32), *Caretta caretta* (loggerhead sea turtle), and all other valid species listed in Appendix 5, but none of the other valid species listed in Appendix 1, in Joyce et al. (2004). Synapomorphies are the absence of the prepalatine foramen and the contribution of the vo-

mer to the upper triturating surface (Gentry 2017). The oldest pan-cheloniid is *Puppigerus*, from the Paleogene (Eocene) of Belgium, Denmark, England (Moody 1974), and Uzbekistan.

Pan-*Dermochelys* is the total clade comprising *Dermochelys coriacea* (Fig. 5.33) and various extinct stem-taxa (Wood et al. 1996; Joyce et al. 2013b). *Dermochelys* (which can attain a total length of up to 2.4 m) has the widest geographic distribution of any present-day reptile, ranging from tropical to subarctic seas. The carapace of this taxon and its close relatives comprises a mosaic of many ossicles embedded in leather-like skin (Fig. 5.34). The bony plastron consists of a narrow peripheral ring, the center of which is not ossified. There are no keratinous scutes on either the carapace or plastron. By contrast, the carapace of the pan-dermochelyid *Eosphargis*, from the Paleogene (Eocene) of Belgium, Denmark, and England, lacks the mosaic of ossicles but retains small costal plates (Nielsen 1963). *Mesodermochelys*, from the Late Cretaceous (Maastrichtian) of Japan, is the oldest known pan-dermochelyid (Hirayama and Chitoku 1996; Joyce et al. 2013b).

Bardet et al. (2013) described a large-sized pan-chelonioid, *Ocepechelon*, from the Late Cretaceous

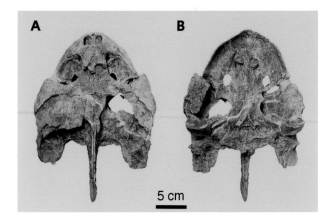

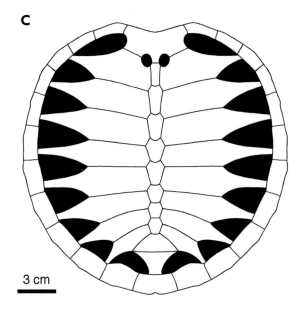

Figure 5.30. The toxochelyid *Toxochelys latiremis*. **A-B**, cranium in **A**, dorsal, and **B**, palatal view; **C**, reconstructed carapace. A-B, courtesy of Tyler Lyson; C, modified from illustrations in Zangerl (1953).

berances. Parham and Pyenson (2010) documented the repeated, independent acquisition of jaw mechanisms suitable for crushing and shearing in various pan-chelonioid lineages.

Protostegidae comprises a variety of Cretaceous marine turtles (Zangerl 1953; Kear and Lee 2006; Gentry 2017). Cadena and Parham (2015) defined it as the most inclusive clade that includes *Protostega gigas* (Fig. 5.36) but no extant turtle nor *Macrobaena mongolica*, Sinemydidae, or Xinjiangchelyidae. Diagnostic features include the T-shaped entoplastron that is not sutured to other elements and the anterior curvature of the radius. Protostegids such as *Archelon*, from the Late Cretaceous (Campanian) of South Dakota, attained total lengths between 3 and 4 m. The phylogenetic position of this clade remains contentious. Kear and Lee (2006), Cadena and Parham (2015), and Gentry (2017) interpreted Protostegidae as stem-dermochelyids. However, Joyce (2007) cautioned that some of the hypothesized synapomorphies could merely be convergent features related to the shared aquatic habits and raised the possibility that protostegids may represent a distinct radiation of marine turtles.

CRYPTODIRA: DUROCRYPTODIRA: TESTUDINOIDEA

Testudinoidea is the crown clade arising from the most recent common ancestor of *Testudo graeca*, *Emys orbicularis*, *Batagur baska*, and all other valid species of turtles listed in Appendices 8-10, but none of the other valid species listed in Appendix 1, in Joyce et al. (2004). It comprises more than half of the known present-day species of turtles (Crawford et al. 2015). Diagnostic derived features of the shell include the contact between the visceral surfaces of the costals and the extensive axillary and inguinal buttresses of the plastron, the presence of an anal notch, and the presence of two rather than four pairs of inframarginal scutes (Joyce 2007). However, the last character state is only present in extant and some extinct testudinoids. Joyce et al. (2013b) noted that all forms older than Eocene retain more than two pairs of inframarginals. In addition, testudinoids share the presence of a biconvex eighth cervical vertebra. *Lindholmemys*, from the Late Cretaceous (Turonian) of Uzbekistan, and *Mongolemys*, from the Late Cretaceous (Maastrichtian) of Mongolia, probably represent

(Maastrichtian) of Morocco, and placed it in an unresolved trichotomy with Protostegidae and Dermochelyidae. The 70-cm-long skull of *Ocepechelon* has an elongate, hemitubular snout with an anteriorly directed opening, which suggests suction-feeding (Fig. 5.35). Subsequently, de Lapparent de Broin et al. (2014) described a second, closely related taxon, *Alienochelys*, from the same deposits, in which upper and lower jaws form a powerful crushing apparatus; the rostral portion of the skull is shaped like a semicircle in plan view. The rough triturating surfaces on the jaws bear short crests and protu-

Figure 5.31. Green turtle (*Chelonia mydas*). Courtesy of Division of Amphibians and Reptiles, National Museum of Natural History.

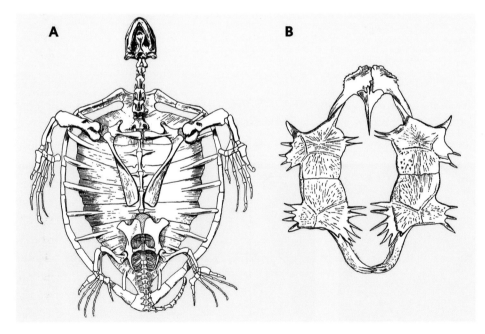

Figure 5.32. The cheloniid *Chelonia mydas*. **A**, skeleton (with plastron removed) in ventral view; **B**, plastron in ventral view. From Reynolds (1913).

stem-testudinoids (Danilov and Sukhanov 2001; Joyce et al. 2013b).

Pan-*Platysternon* is the most inclusive clade containing *Platysternon megacephalum* (big-headed turtle; Fig. 5.37) but not *Testudo graeca* or any of the other valid species listed in Appendix 1 in Joyce et al. (2004). *Platysternon* was traditionally grouped with Chelydridae, but molecular data (Parham et al. 2006; Guillon et al. 2012; Crawford

et al. 2015; Spinks et al. 2016) place it in Testudinoidea as the sister group of Emydidae. It has a proportionately large head that cannot be withdrawn into the shell. *Platysternon* lives in rocky mountain streams in southern China, Laos, Myanmar, Thailand, and Vietnam and feeds on invertebrates including mollusks. *Cardichelyon,* from the Paleogene (Paleocene-Eocene) from Colorado, New Mexico, and Wyoming, possibly represents the earliest

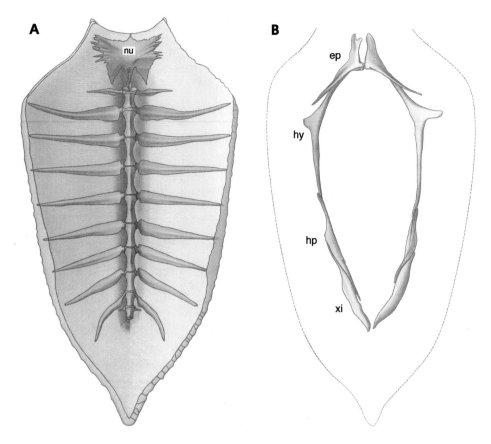

Figure 5.33. **A**, carapace, and **B**, plastron of a leatherback sea turtle (*Dermochelys coriacea*) in ventral view. No scale. Blue denotes cartilages. Abbreviations: ep, epiplastron; hp, hypoplastron; hy, hyoplastron; nu, nuchal; xi, xiphiplastron. From Völker (1913).

Figure 5.34. Partial carapace of the stem-dermochelyid *Psephophorus calvertensis*. Courtesy of Department of Paleobiology, National Museum of Natural History.

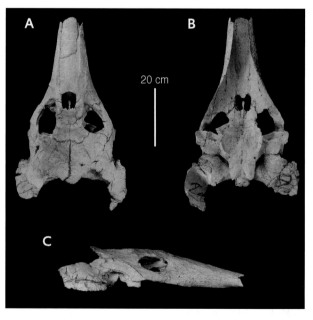

Figure 5.35. Cranium of the pan-chelonioid *Ocepechelon bouyai* in **A**, dorsal; **B**, palatal; and **C**, lateral view. Note tubular snout. From Bardet et al. (2013)—CC BY 2.5.

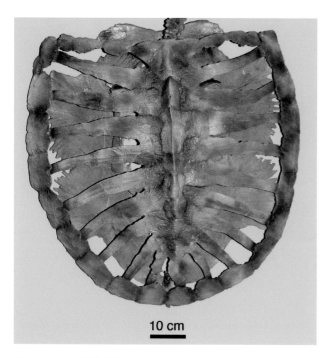

Figure 5.36. Shell of the protostegid *Protostega gigas* in dorsal view. Courtesy of Mike Everhart.

known stem-member of this clade (J. H. Hutchinson 2013; Joyce 2013b).

Emydidae is the crown clade arising from the most recent common ancestor of *Emys orbicularis* (European pond turtle), *Chrysemys picta* (painted turtle), and all other valid species listed in Appendix 10, but none of the species listed in Appendices 1, 8, and 9, in Joyce et al. (2004). Extant representatives of this clade (Fig. 5.38) range widely through most of North and Central America, parts of South America and North Africa, and Europe to the Ural Mountains. *Psilosemys*, from the Paleogene (Eocene) of Wyoming, and *Pseudochrysemys*, from the Paleogene (Paleocene) of Mongolia, are the oldest pan-emydids (J. H. Hutchinson 2013). No modern morphology-based diagnosis is currently available for this clade. The bony bridge connecting the carapace and plastron is often reduced in emydids. *Emys*, *Emydoidea* (which is possibly synonymous with *Emys*; Spinks et al. 2016), and *Terrapene* are distinguished by the presence of a hinge joint in their plastron. *Chrysemys* dates back to the Paleogene (Eocene) (J. H. Hutchinson 1996). The oldest records of *Emydoidea* and *Terrapene* are from the Neogene (Miocene) of Nebraska (Holman 1987). Hervet (2006) proposed Ptycho-

gasteridae for the reception of a number of testudinoid taxa, mostly from the Paleogene (Eocene) to Neogene (Miocene) of Central and Western Europe and traditionally assigned to Emydidae. However, support for Ptychogasteridae is not robust.

Bataguridae (Geoemydidae) is the crown clade arising from the most recent common ancestor of *Batagur baska* (northern river terrapin), *Geoemyda spengleri* (black-breasted leaf turtle), and all other valid species of turtles listed in Appendix 9, but none of the other valid species listed in Appendices 1, 8 or 10, in Joyce et al. (2004). To date no modern morphology-based diagnosis has been published for this clade. Hervet (2004) described a number of mostly Paleogene (Eocene-Oligocene) taxa from Central and Western Europe but compared them only to the extant *Mauremys* (which is first known from the Oligocene of France). The present-day Asian box turtles (*Cuora*) range from southern China to Indonesia and the Philippines. The oldest undisputed record of *Cuora* is from the Neogene (Miocene) of Yunnan (China) (Joyce et al. 2013b). Joyce and Lyson (2010) proposed a clade Palatochelydia for South Asian batagurids, including *Pangshura* (roofed turtles), and characterized it by the presence of a well-developed bony secondary palate. Extant batagurids include exclusively terrestrial to highly aquatic forms and are one of the most species-rich groups of present-day turtles.

Testudinidae is the crown clade arising from the last common ancestor of *Testudo graeca*, *Manouria emys* (Asian forest tortoise), and all other valid species of turtles listed in Appendix 8, but none of the other valid species listed in Appendices 1, 9, or 10, in Joyce et al. (2004). Shared derived features of this clade include the presence of two or fewer phalanges per digit in the manus and pes, the presence of only four digits in the pes, and the coalescence of the femoral trochanters (Auffenberg 1974; Crumly 1985). Joyce et al. (2004) proposed a clade Testuguria if Testudinidae and Bataguridae are sister taxa, which is supported by mitochondrial data (Parham et al. 2006). Le et al. (2006) presented a phylogenetic hypothesis for extant tortoises based on molecular data. Most tortoises have highly domed shells, and all have columnar limbs. All are terrestrial and almost exclusively herbivorous (Figs. 5.39, 5.40). *Hadrianus*, from the Paleogene (Eocene) of Colorado, New Mexico, Texas,

Figure 5.37. Big-headed turtle (*Platysternon megacephalum*). Courtesy of Division of Amphibians and Reptiles, National Museum of Natural History.

Figure 5.38. Extant representatives of Emydidae. **A**, painted turtle (*Chrysemys picta*); **B**, spiny turtle (*Heosemys spinosa*). Courtesy of Division of Amphibians and Reptiles, National Museum of Natural History.

Figure 5.39. Extant representatives of Testudinidae. **A**, pancake tortoise (*Malacochersus tornieri*); **B**, Charles Island giant tortoise (*Chelonoidis nigra*). Courtesy of Division of Amphibians and Reptiles, National Museum of Natural History.

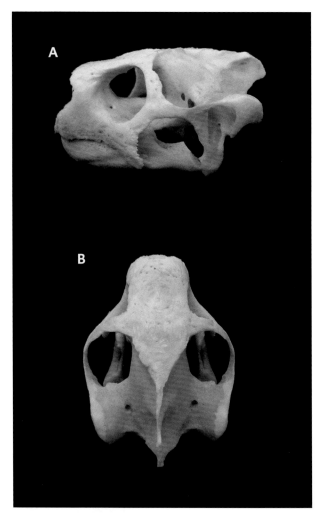

Figure 5.40. Skull of Hermann's tortoise (*Testudo hermanni*) in **A**, lateral, and **B**, dorsal views. Courtesy of Pavel Zuber.

Figure 5.41. Carapace and skull of the testudinid *Stylemys nebrascensis* in dorsal view. Courtesy of Department of Paleobiology, National Museum of Natural History.

Utah, and Wyoming, is either the sister taxon of or referable to Testudinidae (Joyce et al. 2013b). The testudinids *Gopherus, Hesperotestudo,* and *Stylemys* (Fig. 5.41) are common in the Paleogene (Eocene) of Colorado, South Dakota, and Wyoming (J. H. Hutchinson 1996). *Gigantochersina*, from the Paleogene (Eocene) of Egypt, represents the oldest testudinid known from Africa

(Holroyd and Parham 2003). Extant tortoises are widely distributed in southern Europe, the southern United States, Mexico, Central America, and much of Africa, Asia, and South America. They also occur on various oceanic islands including Madagascar, the Galápagos Islands, much of Indonesia, and the Seychelles. *Titanochelon*, from the Neogene (Miocene) to Quaternary (Pleistocene) of Austria, Bulgaria, France, Germany, Greece, Spain, Switzerland, and Turkey, attained a carapace length of up to about 2 m (Pérez-García and Vlachos 2014). Among present-day testudinids, only the Galápagos tortoise (*Chelonoidis nigra*; Fig. 5.39B) reaches comparable carapace size.

6 Sauropterygia, Ichthyosauromorpha, and Related Reptiles

The Early Mesozoic Invasion of the Sea

At the beginning of the Mesozoic Era, several groups of diapsid reptiles adapted to life in the sea. Surprisingly, each apparently evolved different modes of feeding and aquatic locomotion. In each instance, the oldest known representatives indicate that these lineages evolved in lagoonal or nearshore marine settings (Sues 1987a; Storrs 1991; Rieppel 2002). It is possible that the global biotic crisis at the end of the Permian generated new ecological opportunities that facilitated the rapid evolutionary diversification of marine reptiles at the beginning of the Triassic Period.

Only two of the early Mesozoic clades, Sauropterygia and Ichthyosauromorpha, survived the end-Triassic extinction and fully adapted to life in the open sea. Both flourished during the Jurassic and Cretaceous, with sauropterygians persisting until the end of the Cretaceous. At several points during the Mesozoic, various groups of turtles, lepidosaurs, and crocodile-line archosaurs also invaded the sea. Only the marine chelonioid turtles survived the end-Cretaceous extinction and persist to the present day (Chapter 5). Among present-day lepidosaurs, two lineages of elapid snakes independently adapted to life in the ocean during the late Cenozoic (Chapter 7). In addition, *Crocodylus porosus* (saltwater crocodile) and some lizards, especially *Amblyrhynchus cristatus* (Galápagos marine iguana), spend extended periods of time at sea.

Seymour (1982) observed that reptiles readily adapt to an aquatic mode of life because of their low metabolic rates, reliance on heat from external sources (ectothermy), considerable tolerance of anoxia, and ability to use fermentative metabolism for muscle activity. Reptiles can hold their breath for long periods of time and can modify their blood circulation during diving. For *Amblyrhynchus cristatus*, swimming requires only about 25 percent of the metabolic cost of walking on land (Gleeson 1979). Moving on land involves lifting the body off the ground, lowering the center of gravity, and moving on irregular surfaces. Once immersed in water, however, animals essentially become "weightless" (Seymour 1982). The only constraint is that at least most present-day marine reptiles require water temperatures of at least 20°C, unlike marine mammals that can thrive in the cold waters at high latitudes. A notable exception is *Dermochelys coriacea* (leatherback turtle), which can regulate the distribution of its blood flow to raise its body temperature while swimming in cooler water. This large-bodied turtle also has increased body insulation (Bostrom et al. 2010).

Based on the presence of only upper temporal fenestrae, often bounded by dorsoventrally deep temporal bars, Colbert (1945) proposed classifying sauropterygians with ichthyosaurs and an assortment of Permian and Triassic terrestrial reptiles as Euryapsida (from Greek *eurys*, broad, and *hapsis*, arch). Since then, various studies have demonstrated that the euryapsid temporal configuration represents merely a modification of the diapsid condition and independently evolved in several lineages (Carroll 1987; Sues 1987a; Rieppel 1993, 1994). Although considerable progress has been made in recent years in elucidating the evolutionary histories of sauropterygians and ichthyosauromorphs, the relationships of the various major groups of early Mesozoic marine reptiles to each other and to other diapsid reptiles are still far from resolved. DeBraga and Rieppel (1997), and later studies building on their analysis, placed Sauropterygia close to Lepidosauromorpha. The phylogenetic analyses by Neenan et al. (2013) and Scheyer et al. (2017) united Sauropterygia, Ichthyopterygia, and other groups of early Mesozoic marine diapsid reptiles in a clade that is the sister group of Sauria or of Sauria and Testudinata.

A key issue is the phylogenetic assessment of skeletal features that presumably reflect adaptations to an aquatic mode of life. Did these features evolve independently in different lineages related to similar modes of life, or do they, at least to some extent, suggest a shared common ancestry? Certain traits in various groups of aquatic reptiles, such as the posterior shift of the external narial openings from the tip of the snout to closer toward the orbits, superficially appear to be synapomorphies for these taxa, but on closer examination, they often differ in detail and thus possibly evolved independently. Furthermore, many similarities between the various clades of marine reptiles likely reflect a phenomenon that Rieppel (1989c) termed "skeletal paedomorphosis" because these features resemble those in immature individuals of land-dwelling reptiles. Once surrounded by water, the limbs of an animal no longer have to support its body against the pull of gravity while standing or moving. Thus, ossification in the bones of the girdles and limbs slows down during development. Instead of being replaced by bone during the growth of the animal, cartilage (which weighs much less than bone) is retained throughout life. When multi-jointed limbs become flippers, they lose the intricate joint surfaces that facilitate

mobility between the individual sets of limb bones in land-dwelling tetrapods. Finally, certain features, such as the lack of fusion between the neural arches and centra of the vertebrae or the loss of various bones, possibly reflect changes in the developmental processes underlying the formation of the skeleton rather than adaptations to changing functional demands in a particular region of the body. X.-h. Chen et al. (2014b) explored the potential impact of skeletal paedomorphosis on phylogeny reconstruction by undertaking several analyses in which possible aquatic features were coded as either apomorphies or of questionable status. Not surprisingly, these different approaches generated sometimes strikingly different hypotheses of relationships.

Sauropterygia

Owen (1860) first proposed the name Sauropterygia (from Greek *sauros*, lizard, and *pteron*, fin, wing) for a variety of Mesozoic marine reptiles. Phylogenetic studies (e.g., Rieppel 2000b) have since confirmed that Sauropterygia represents a monophyletic group. It was the most species-rich clade of Mesozoic marine reptiles and attained global distribution. Rieppel (2000a) defined Sauropterygia as the clade comprising Placodontia and Eosauropterygia (Fig. 6.1).

Diagnostic cranial features include large premaxillae forming much of the snout anterior to the external nares; anterior teeth in the upper and lower jaws procumbent (Fig. 6.2); absence of a lacrimal, supratemporal, and tabular; lower temporal opening open ventrally; and pterygoid broad posteromedially and lacking teeth on the transverse flange (Rieppel 2000b, 2002).. The most diagnostic postcranial feature of Sauropterygia is the structure of the pectoral girdle in which the clavicles are positioned anteroventral to the interclavicle and contact the medial surfaces of the scapulae, unlike the condition in other amniotes (Rieppel 2000b; Fig. 6.3). The interclavicle has only a short posterior process or lacks one altogether. The humerus is curved and has reduced epicondyles. The bones of the carpus and tarsus are greatly reduced. The sacrum has three or more pairs of sacral ribs that were only loosely connected to the ilium. Typically, the presacral vertebral column exhibits some degree of pachyostosis (Houssaye 2009).

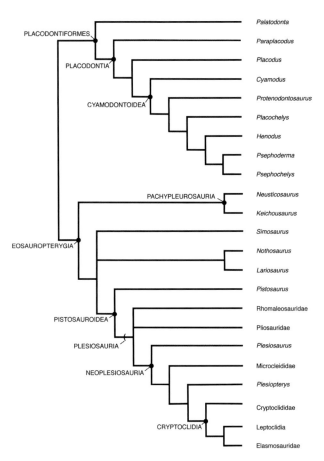

Figure 6.1. Phylogenetic hypothesis of the interrelationships of Sauropterygia. Dots denote node-based clades and parentheses denote stem-based clades. Based mainly on Rieppel (2000a) and Ketchum and Benson (2010). Neenan et al. (2015) hypothesized a clade with *Placodus* and *Paraplacodus*.

Sauropterygia: Placodontiformes

Neenan et al. (2013) reported on the skull of a juvenile specimen of a stem-placodont, *Palatodonta*, from the early Middle Triassic (Anisian) of the Netherlands. *Palatodonta* has a single row of teeth on the large palatine, but these teeth are small and pointed rather than broad and flat. It has an L-shaped jugal and four peg-like, slightly procumbent premaxillary teeth. Neenan et al. (2013) proposed a clade Placodontiformes for *Palatodonta* and Placodontia.

SAUROPTERYGIA: PLACODONTIFORMES: PLACODONTIA

Rieppel (2000a) defined Placodontia as comprising Placodontoidea and Cyamodontoidea. Placodonts are first known from the early Middle Triassic (Anisian) and became extinct near or at the end of the Triassic (Rieppel 2000b,c). They were apparently restricted to the periphery of Tethys, the vast sea that formed a wedge in the eastern margin of the supercontinent Pangaea, and the shallow sea in the adjoining Germanic Basin.

Placodontia (from Greek *plax*, plate, and *odous* (*odon*), tooth) is readily distinguished by the presence of large, thickly enameled teeth in the maxilla and dentary and on the palatine (Fig. 6.4). These teeth closely resemble the crushing dentition in certain extant fishes feeding on hard-shelled prey. Thus, it is not surprising that iso-

A

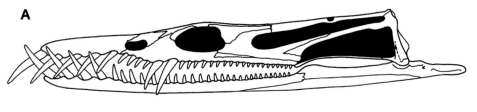

B

Figure 6.2. Skulls of **A**, the eosauropterygian *Nothosaurus jagisteus,* and **B**, the placodont *Placodus gigas* in lateral view, showing the structure of the temporal region. Teeth on the palate of *Placodus* indicated in gray. Modified from illustrations by Rieppel (2000a, 2001).

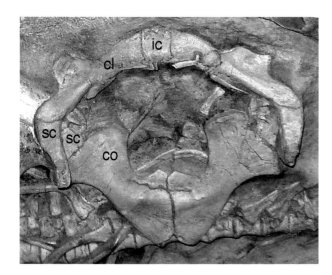

Figure 6.3. Shoulder girdle of *Nothosaurus jagisteus* in dorsal view, showing the contact of the clavicle with the anteromedial surface of the scapula. Abbreviations: cl, clavicle; co, coracoid; ic, interclavicle; sc, scapula.

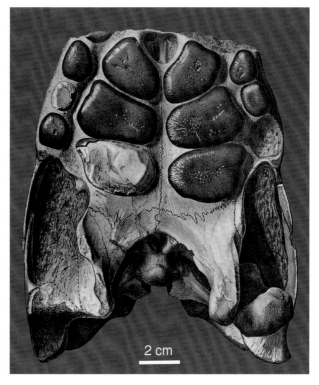

Figure 6.4. Partial cranium of *Placodus gigas* in palatal view to show the crushing teeth on the palate. Modified from Broili (1912).

lated teeth of *Placodus* were first considered those of a fish (Agassiz 1833). Most placodonts presumably subsisted on hard-shelled marine invertebrates such as brachiopods and mollusks. The heavily ossified skeleton and the presence of a well-developed bony carapace in many forms suggest primarily bottom-dwelling habits for these reptiles. The humerus of *Placodus* has a thick cortex (Houssaye 2013).

SAUROPTERYGIA: PLACODONTIFORMES: PLACODONTIA: PLACODONTOIDEA

Placodontoidea comprises *Placodus*, from the Middle Triassic (Anisian-Ladinian) of Central Europe and the Middle Triassic (Anisian) of China, and *Paraplacodus*, from the Middle Triassic (Anisian-Ladinian) of Switzerland (Rieppel 1995, 2000b,c; Neenan et al. 2015). It is diagnosed by the transverse expansion of the palatal teeth and the confluence of the internal nares (Neenan et al. 2015). *Placodus* and *Paraplacodus* have procumbent, chisel-shaped teeth in the premaxilla and anterior portion of the dentary, which were probably used to pick off prey attached to a substrate. The maxilla has robust, somewhat conical teeth, and the palatine bears large, flattened teeth, which have vertical tooth replacement and were probably used for crushing hard-shelled prey (Figs. 6.2B, 6.4). The transversely broad postorbital region of the more or less trian-

gular cranium has only large upper temporal openings and, in most placodonts (except *Paraplacodus*; Rieppel 2000c), dorsoventrally deep, imperforate temporal bars, all of which indicate powerful adductor jaw muscles (Sues 1987b; Rieppel 1995). The lower jaw has a prominent coronoid process for the insertion of these muscles. The braincase is firmly sutured to the palate.

In *Paraplacodus* and *Placodus*, the neck is short and the trunk is long and triangular in transverse section (B. Peyer 1931b; Drevermann 1933; Fig. 6.5). The vertebrae have large transverse processes, tall neural spines, and accessory intervertebral (hyposphene-hypantrum) articulations. The long tail is rounded rather than laterally flattened in transverse section. The proportionately short limbs are not modified as paddles. All placodonts have a set of robust gastralia. A single row of bony nodules extends along the back in *Placodus*, whereas *Paraplacodus* lacks any trace of dermal armor. *Placodus* attained a length of about 2.5 m. It probably used undulation of the tail as the primary means of propulsion, with the limbs employed for steering.

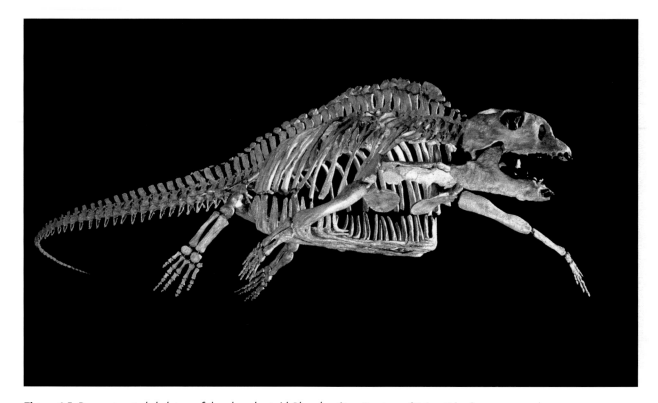

Figure 6.5. Reconstructed skeleton of the placodontoid *Placodus gigas*. Courtesy of Rainer Schoch.

SAUROPTERYGIA: PLACODONTIFORMES: PLACODONTIA: CYAMODONTOIDEA

More derived placodonts are grouped together as Cyamodontoidea (from Greek *cyamos*, bean, and *odous* (*odon*), tooth) (Rieppel 2000b). Diagnostic derived features of this clade include the lateral extension of the palatine contacting the jugal, an anteroposteriorly broad dorsal process of the epipterygoid, and the presence of a persistent palatoquadrate cartilage in life (Neenan et al. 2015). Cyamodontoids also have a large, dorsoventrally flattened carapace composed of many polygonal bony plates (Westphal 1975). Unlike in turtles, the vertebrae and ribs are not integrated into the carapace, although they were probably attached to it by means of connective tissue. In addition to the carapace, some cyamodontoids have a separate, smaller shield of osteoderms covering the pelvic region (Pinna and Nosotti 1989; Scheyer 2010). Cyamodontoidea comprises Cyamodontida and Placochelyida (Rieppel 2000b, 2001). *Cyamodus*, from the Middle Triassic (Anisian-Ladinian) of Germany, Italy, Poland, and Switzerland (Fig. 6.6), has two bulbous teeth in each premaxilla, and the palatine teeth are positioned behind the maxillary teeth (Rieppel 2000b). *Henodus*, from the early Late Triassic (Carnian) of Germany, has a broad, squared-off snout. The premaxillae have an overhanging cutting edge with a palisade of tiny denticles (Rieppel 2001), and the posterior part of each palatine bears a single tooth (Huene 1936). C. Li et al. (2016) interpreted *Henodus* as an aquatic herbivore that scraped algae off the substrate and sucked them into its mouth. Its dermal armor is strikingly turtle-like, with a rather flat carapace, which was covered by keratinous scutes between its dorsolateral edges, and a plastron composed of long, thin osteoderms (Fig. 6.7). *Henodus* attained a length of slightly more than 1 m and was a bottom-dweller in brackish water. The basal placochelyidan *Protenodontosaurus*, from the Late Triassic (Carnian) of Italy, retains premaxillary teeth (Rieppel 2001). More derived placochelyidans such as *Psephoderma,* from the Late Triassic (Norian-Rhaetian) of Europe (Pinna and Nosotti 1989), and *Psephochelys*, from the Late Triassic (Carnian) of Guizhou (China) (C. Li and Rieppel 2002; Neenan et al. 2015), have narrow, edentulous tips of the snout and mandible, which were perhaps used to probe for food on the seafloor. The palatine bears one or two greatly enlarged teeth.

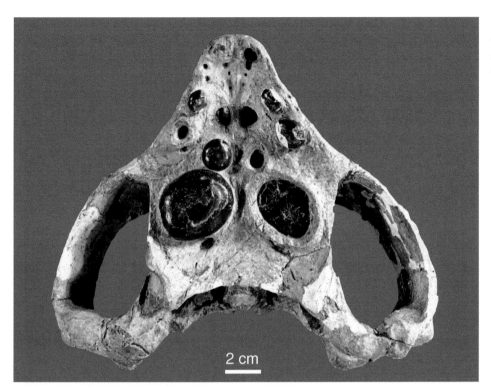

Figure 6.6. Cranium of the cyamodontoid *Cyamodus kuhnschnyderi* in palatal view. Courtesy of Rainer Schoch.

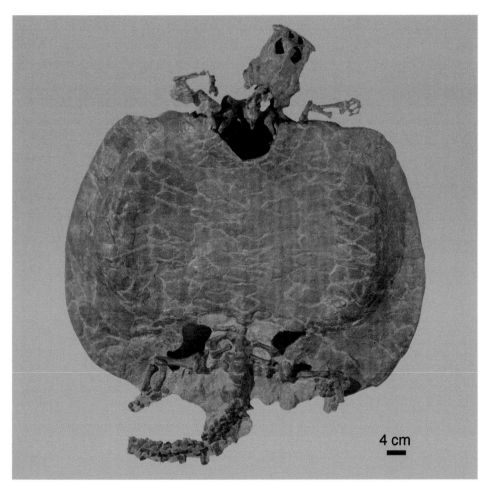

Figure 6.7. Skeleton and armor of the cyamodontoid *Henodus chelyops* in dorsal view. Courtesy of Rainer Schoch.

Sauropterygia: Eosauropterygia

Eosauropterygia (from Greek *eos*, dawn, and Sauropterygia) comprises Pachypleurosauria, Nothosauroidea, and Pistosauroidea (Rieppel 2000b). Rieppel (2000b) listed a number of diagnostic synapomorphies for this grouping: squamosal extending lateral to the quadrate to the ventral margin of the skull; occiput plate-like and lacking major openings; presence of accessory intervertebral (zygosphene-zygantrum) articulations between the neural arches; centra bearing expanded articular facets for contact with the pedicles of the neural arches; blade of the scapula reduced; coracoid distinctly constricted; clavicles broad medially and contacting each other anterior to the interclavicle; and presence of a small internal trochanter on the femur. Many of these features are absent in the most derived clade of eosauropterygians, Plesiosauria (Rieppel 2000b). The oldest known eosauropterygians are from the late Early Triassic (Olenekian) (Jiang et al. 2014).

Traditional classifications distinguished two major groups among eosauropterygians, the exclusively Triassic "nothosaurs" and the mostly Jurassic and Cretaceous plesiosaurs. Whereas the latter do form a clade, "nothosaurs" are a paraphyletic assemblage of taxa, some of which are more closely related to plesiosaurs than others (Sues 1987a; Storrs 1991; Rieppel 1998, 2000a).

SAUROPTERYGIA: EOSAUROPTERYGIA: PACHYPLEUROSAURIA

Pachypleurosauria is the sister group to all other eosauropterygians. It includes *Neusticosaurus*, from the Middle Triassic (Anisian-Ladinian) of Germany, Italy, and Switzerland (Carroll and Gaskill 1985; Sander 1989; Fig. 6.8), and *Keichousaurus*, from the Middle Triassic (Ladinian) of China (Lin and Rieppel 1998; Fig. 6.9). Pachypleurosaurs have a proportionately small skull, relatively long neck, and long tail. They typically attained total lengths between 50 and 60 cm, although one species of *Neusticosaurus* reached a length of up to 1.2 m. The upper temporal opening is considerably smaller than the orbit. As in other eosauropterygians, the lower temporal opening is open ventrally. The quadrate has a posterior notch, which, along with the stapes, indicates that pachypleurosaurs still had a tympanum (Carroll and Gaskill 1985). The vertebrae and especially the ribs are distinctly thickened (pachyostosis), presumably to facilitate buoy-

Figure 6.8. Skeleton of a small individual of the pachypleurosaurian *Neusticosaurus edwardsii* in dorsal view.

Figure 6.9. Skeleton of the pachypleurosaurian *Keichousaurus hui* in dorsal view. No scale provided. Courtesy of Rainer Schoch.

ancy control (Houssaye 2009). The humerus is large but the more distal bones of the forelimb are reduced in size and the articular surfaces for the elbow joint are poorly defined. The humerus shows apparent sexual dimorphism in most known taxa (Sander 1989; Lin and Rieppel 1998). The ilium has a small dorsal process and is only loosely connected to the sacral ribs. The hind limb is shorter than the forelimb in *Neusticosaurus*. Pachypleurosaurs probably used undulating movements of the tail for swimming, with the forelimbs possibly assisting in steering and the hind limbs held close to the body to reduce drag. They probably could no longer move on land. Discoveries of curled-up embryos inside adults of *Keichousaurus* show that the females gave birth to live young (Y.-n. Cheng et al. 2004).

SAUROPTERYGIA: EOSAUROPTERYGIA: EUSAUROPTERYGIA

The more derived eosauropterygians, Eusauropterygia (from Greek *eu*, true, and Sauropterygia), comprise Nothosauroidea and Pistosauroidea (Rieppel 2000b). Diagnostic features for this clade include the separation of the nasals by the premaxillae, which extend to the frontals posteriorly, the presence of a prominent lateral ridge on the surangular, and platycoelous vertebrae (Rieppel 2000b). Unlike in pachypleurosaurs, the upper temporal opening is larger than the orbit in eusauropterygians, which possibly reflects changes in the adductor jaw musculature to facilitate rapid sideward snapping during the pursuit of fast-moving prey (Rieppel 2002). The skull is more flattened dorsoventrally and proportionately longer than in pachypleurosaurs. The tooth rows extend posteriorly below the temporal openings. The maxilla usually bears one or two enlarged, fang-like teeth. The forelimb has a robust humerus and a broad, flattened radius and ulna, which are often widely separated from each other.

SAUROPTERYGIA: EOSAUROPTERYGIA: EUSAUROPTERYGIA: NOTHOSAUROIDEA

Rieppel (2000a) defined Nothosauroidea as the clade comprising *Simosaurus* and Nothosauria. Diagnostic features include the postorbital region of the cranium being distinctly longer than the antorbital region; upper tooth row extending posteriorly to a point ventral to the anterior one-third to half of the upper temporal fenestra;

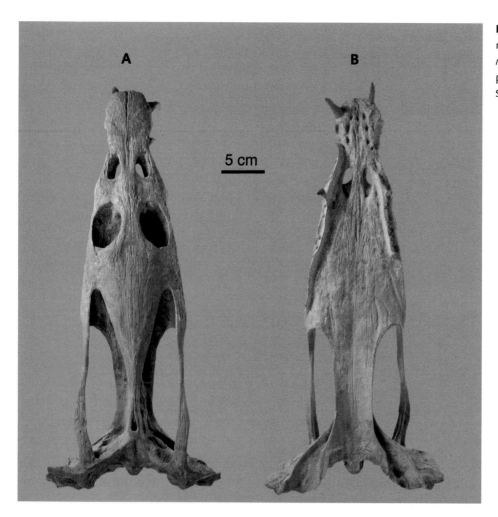

Figure 6.10. Cranium of the nothosaurid *Nothosaurus mirabilis* in **A**, dorsal, and **B**, palatal view. Courtesy of Rainer Schoch.

and iliac blade small, not extending far behind the posterior margin of the acetabulum.

Simosaurus, from the Middle Triassic (Ladinian) of France, Germany, Italy, and the Middle East, attained a length of more than 3 m. Its short and broad snout lacks any constriction and bears teeth with blunt crowns, which indicate feeding on hard-shelled prey (Rieppel 2000a, 2002).

Nothosauria comprises *Germanosaurus* and Nothosauridae (Rieppel 2000a). Shared apomorphies include the distinct constriction of the snout and the dorsoventrally low temporal region of the cranium.

Germanosaurus, from the Middle Triassic (Anisian) of Poland, is distinguished by a relatively short and broad snout and paired parietals. *Nothosaurus*, from the Middle Triassic (Anisian-Ladinian) from Europe, Tunisia, the Middle East, and China (Fig. 6.10), has a long and narrow skull. The procumbent and interdigitating fangs in the premaxilla and the anterior portion of the dentary, as well as the paired maxillary fangs, were probably used to catch and hold on to slippery prey such as fish (Rieppel 2002). The frontal and parietal both are unpaired. The vertebral column of *Nothosaurus* has much taller neural spines and narrower neural arches than those of pachypleurosaurs, indicating that the trunk was less flexible. The ribs show pachyostosis (Houssaye 2009). The forelimbs of *Nothosaurus* and closely related taxa (e.g., *Lariosaurus*; Fig. 6.11) are larger and more robust than the hind limbs. Nothosaurids possibly employed their forelimbs for some form of sculling, an intermediate stage between rowing and underwater flight that would generate thrust by combining drag and lift (Thewissen and M. A. Taylor 2007). *Nothosaurus* attained a skull length of up to 65 cm and a total length of up to 4 m (Rieppel 2001; J. Liu et al. 2014).

Figure 6.11. Skeleton of the nothosaurid *Lariosaurus calcagnii* in ventral view, preserved in association with several skeletons of *Neusticosaurus edwardsi.* Courtesy of Heinz Furrer.

SAUROPTERYGIA: EOSAUROPTERYGIA: EUSAUROPTERYGIA: PISTOSAUROIDEA

Pistosauroidea (from Greek *pistos*, genuine, and *sauros*, lizard) comprises Cymatosauridae and Pistosauria (Rieppel 2000a; Rieppel et al. 2002). It includes a diversity of eusauropterygian taxa that share with plesiosaurs derived cranial features such as a contact between the jugal and squamosal, the presence of a parietal crest, and the presence of an interpterygoid vacuity, unlike the closed palate in more basal eosauropterygians (Ketchum and Benson 2010). The cervical portion of the vertebral column has at least 35 vertebrae. The humerus has a weakly developed deltopectoral crest or lacks this feature altogether.

Rieppel (2002a) defined Cymatosauridae as the clade comprising *Corosaurus* and *Cymatosaurus.* However, statistical support for this grouping is weak (Rieppel et al. 2002). The cranium of *Corosaurus*, from the Early to Middle Triassic (Olenekian-Anisian) of Wyoming, has antorbital and postorbital portions of more or less equal length and a proportionately short snout without a constriction (Storrs 1991; Rieppel 1998). The nasal is unusu-

ally large. *Cymatosaurus*, from the Middle Triassic (Anisian) of Germany and Poland, has a cranium with a constricted snout and distinctly elongated postorbital region (Rieppel 2000b).

Rieppel et al. (2002) defined Pistosauria as the clade including *Augustasaurus*, *Pistosaurus*, and Plesiosauria. *Augustasaurus*, from the Middle Triassic (Anisian) of Nevada (Rieppel et al. 2002), and *Pistosaurus*, from the Middle Triassic (Anisian-Ladinian) of Germany (Sues 1987a), have distinctly elongated, slender snouts terminating in a blunt tip. The nasal is greatly reduced (Sues 1987a; Sato et al. 2014). The frontal enters the anteromedial margin of the upper temporal fenestra. The temporal bar is deep. The squamosal forms a box-like suspensorium as in plesiosaurs. The palate has an interpterygoid vacuity, unlike in more basal eosauropterygians (Rieppel et al. 2002). The pistosaurian *Yunguisaurus*, from the Middle Triassic (Ladinian) of Yunnan (China) (Stato et al. 2014) is noteworthy for its long neck with about 50 cervical vertebrae. Both manus and pes show hyperphalangy. The coracoids of *Pistosaurus* and closely related taxa lack the median constriction present in more basal eosauropterygians and form a broad median symphysis similar to that in plesiosaurs. Basal pistosaurians probably already relied on their limbs for propulsion rather than undulatory movements of the trunk and tail (Sato et al. 2014).

SAUROPTERYGIA: EOSAUROPTERYGIA: EUSAUROPTERYGIA: PISTOSAURIA: PLESIOSAURIA

The derived pistosauroids, Plesiosauria (from Greek *plesios*, near, and *sauros*, lizard) first appeared in the Late Triassic (Rhaetian) and persisted until the end of the Cretaceous Period. They attained global distribution. Plesiosaurs ranged in length from about 2 m to possibly more than 15 m.

Plesiosauria comprises all taxa more closely related to *Plesiosaurus dolichodeirus* and *Pliosaurus brachydeirus* than to *Augustasaurus hagdorni* (a Middle Triassic pistosaurid) (Ketchum and Benson 2010). Members of this clade are distinguished from other eosauropterygians by the possession of four elongate, paddle-like limbs (Fig. 6.12). The humerus and femur are large and robust. The distal limb

bones, including the carpals and tarsals, are polygonal or disk-shaped. The carpals and tarsals are tightly integrated with each other, unlike in earlier eosauropterygians, where they were probably connected by cartilage. The elbow and knee as well as the wrist and ankle joints were no longer mobile, making each limb a single rigid functional unit. Manus and pes show the addition of as many as 17 phalanges in each digit and lack unguals (D. S. Brown 1981; Caldwell 1997b). Skin impressions on excellently preserved Early Jurassic specimens indicate that the limbs formed flippers that projected laterally from the body (J. A. Robinson 1975). Each flipper was rather stiff except near the tip and acted as a hydrofoil, which generated lift during swimming. Related to the changes in limb structure, the pectoral and pelvic girdles of plesiosaurs are greatly expanded horizontally along the ventral surface of the body, providing greatly increased attachment areas for limb muscles (Fig. 6.13). The girdles are separated by up to nine tightly spaced rows of thick gastralia (J. A. Robinson 1975, 1977). The vertebrae lack the

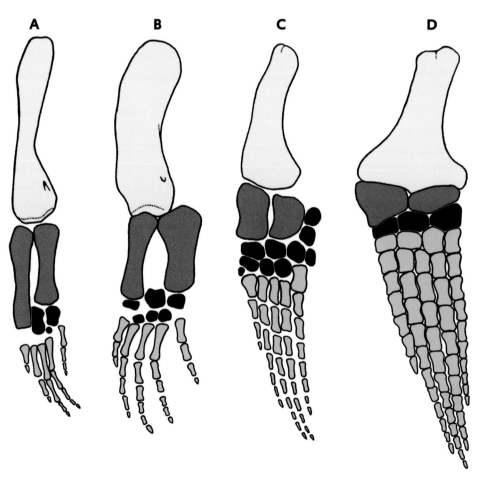

Figure 6.12. Left forelimbs of the sauropterygians, **A**, *Neusticosaurus*; **B**, *Lariosaurus*; **C**, *Plesiosaurus*; and **D**, *Cryptoclidus*. Humerus, yellow; radius and ulna red; carpals black; phalanges, teal. Not to scale. Simplified from Andrews (1910), Carroll and Gaskill (1985), and Carroll (1985).

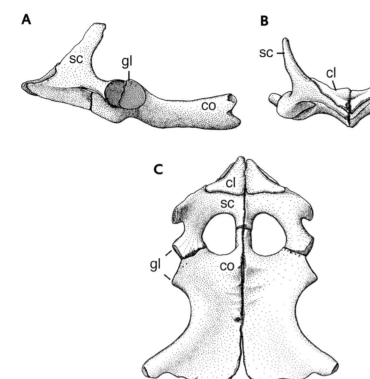

Figure 6.13. Shoulder girdle of the cryptoclidian plesiosaur *Cryptoclidus eurymerus* in **A**, lateral; **B**, anterior; and **C**, dorsal views. Abbreviations: cl, clavicle; co, coracoid; gl, glenoid; sc, scapula. Redrawn by Stephen Godfrey from Andrews (1910).

accessory intervertebral articulations present in more basal eosauropterygians. Overall the trunk is rather rigid and short. The proportionately short tail probably served as a rudder rather than for propulsion. In at least some plesiosaurs, it supported a small caudal fin. Plesiosaurian limb bones have a spongy internal structure and lack a medullary cavity (de Ricqlès and Buffrénil 2001).

The mode of underwater locomotion in plesiosaurs has been a contentious topic because the flipper-like limbs of these reptiles are without close analogues among present-day aquatic tetrapods. Watson (1924) argued that plesiosaurs deployed their flippers for anteroposterior rowing strokes, but later studies rejected this model as inefficient. J. A. Robinson (1975) first reconstructed plesiosaurian locomotion as a form of underwater "flight," similar to that employed by extant penguins and sea turtles. In this mode of swimming, the flippers move mostly in a vertical plane, describing a figure-eight course in side view, while continuously being rotated on their long axes. This motion would have generated thrust in a manner resembling that in the penguins and sea turtles and would have been primarily lift-based (Thewissen and M. A. Taylor 2007). S. Liu et al. (2015) reconstructed plesiosaurs as primarily relying on their forelimbs for un-

derwater flight. In this model, the forelimbs would have generated much of the thrust, whereas the hind limbs served mainly for maneuvering and stabilization. However, simulations by Muscutt et al. (2017) suggested that the hind limbs, working in harmony with the forelimbs, could have generated up to 60 percent more thrust and 40 percent higher efficiency by exploiting the vortices generated by movement of the forelimbs. Plesiosaurs possibly used their heads and necks for controlling direction while swimming.

Krahl et al. (2013) documented an interesting change in the microstructure of the limb bones from *Nothosaurus* to *Pistosaurus* and *Plesiosaurus*. The bones of *Nothosaurus* have primarily lamellar zonal bone tissue, which reflects low growth rates. By contrast, the limb bones of *Pistosaurus* and *Plesiosaurus* have fibrolamellar bone tissue, which indicates high growth rates and high basal metabolic rates. Wintrich et al. (2017) confirmed this observation on the latest Triassic (Rhaetian) plesiosaur, *Rhaeticosaurus*. Fleischle et al. (2018) argued that plesiosaurs and even more basal eosauropterygians were endothermic. Together with an increase in body size, a higher basal metabolic rate was probably critical for the invasion of the open sea and the subsequent evolutionary diversification of plesiosaurs during the Jurassic and Cretaceous. Plesiosaurs were capable of maintaining high body temperatures, estimated to be between $35 \pm 2°C$ and $39 \pm 2°C$ (Bernard et al. 2010).

Consistent with their fully aquatic mode of life, plesiosaurs were viviparous. O'Keefe and Chiappe (2011) reported the discovery of a single rather large embryo within the ribcage of a specimen of a Late Cretaceous polycotylid plesiosaur. They argued that plesiosaurs possibly gave birth to just a single large offspring at a time, much like extant marine mammals.

Pachyostosis has been observed only in a few plesiosaurian taxa, including juveniles of elasmosaurs and pliosaurs, but adult individuals of both groups lack such thickening (Houssaye 2009).

Following Owen (1841), researchers have distinguished two body plans among plesiosaurs, based primarily on the relative proportions of head and neck. O'Keefe (2002) referred to them as "pliosauromorph" and "plesiosauromorph," respectively. Pliosauromorphs have a large head, a relatively short neck, and hind limbs that are larger than the forelimbs. Plesiosauromorphs have a proportionately small head, a sometimes extraordinarily elongated neck, and forelimbs that are typically longer than the hind limbs. Based on these differences in body proportions, Plesiosauria was traditionally divided into Plesiosauroidea and Pliosauroidea. However, phylogenetic analyses have established that the "pliosauromorph" body plan evolved more than once in this clade (O'Keefe 2001b, 2002; Ketchum and Benson 2010; Benson and Druckenmiller 2014). Body proportions in plesiosaurs are quite variable, and the classic body plans just represent end points of a continuum. Based on geometrical analysis of flipper shape and comparisons with bird wings, O'Keefe (2001b) argued that the large-headed, short-necked pliosauromorphs were high maneuverable, fast swimmers, and probably pursuit predators. This is consistent with the large gape and large tooth size in these animals, which, in turn, suggest a preference for larger-bodied, highly mobile prey including other marine reptiles. By contrast, the small-headed, long-necked plesiosauromorphs likely were cruising swimmers that could cover long distances at low to intermediate speeds. The limited gape and slender teeth in these forms suggest that they subsisted on small prey. Noè et al. (2017) ar-

gued that the long neck of plesiosaurs was primarily adapted for ventral bending, with only limited mobility in other directions.

Plesiosauria is diagnosed by various additional cranial and postcranial synapomorphies (Ketchum and Benson 2010): premaxillae partially separating the frontals along the midline; external nares positioned behind the internal ones; splenial participating in the mandibular symphysis; radius with a concave preaxial margin; and ulna with a convex postaxial margin.

Despite several major phylogenetic analyses, the interrelationships of plesiosaurs remain controversial (O'Keefe 2001b; Druckenmiller and Russell 2008; Ketchum and Benson 2010; Benson and Druckenmiller 2014). Revisions of historical finds and numerous discoveries of new taxa in recent years have uncovered a much greater diversity of these marine reptiles than traditionally assumed. This overview follows Ketchum and Benson (2010) and several more recent studies based on their analysis.

PLESIOSAURIA: NEOPLESIOSAURIA

Ketchum and Benson (2010) defined a clade Neoplesiosauria (from Greek neos, new, plesios, near, and sauros, lizard) as comprising the most recent common ancestor of *Plesiosaurus dolichodeirus* and *Pliosaurus brachydeirus* and all descendants of that ancestor.

PLESIOSAURIA: NEOPLESIOSAURIA: PLIOSAUROIDEA

Pliosauroidea (from Greek plion, more, and sauros, lizard) comprises all taxa more closely related to *Pliosaurus brachydeirus* than to *Plesiosaurus dolichodeirus* (Ketchum and Benson 2010). It is diagnosed by the contact of the paroccipital process with both the squamosal and quadrate and the presence of a rounded medial flange formed by the angular and prearticular anterior to the mandibular glenoid fossa for the jaw joint (Ketchum and Benson 2010).

Pliosauroidea comprises Rhomaleosauridae and Pliosauridae. Rhomaleosauridae is distinguished by bowed mandibular rami and a distinctly inclined suspensorium. It includes *Rhomaleosaurus*, from the Early Jurassic (Toarcian) of England (A. S. Smith and Dyke 2008), and *Meyerasaurus*, from the Early Jurassic (Toarcian) of

Germany (A. S. Smith and Vincent 2010; Fig. 6.14). The former reached a total length of about 7 m and the latter about 3.5 m. Both have a proportionately large skull and a relatively long neck with 28 postaxial cervical vertebrae. Benson et al. (2012) referred several additional Hettangian- and Sinemurian-age taxa from England to Rhomaleosauridae, which survived into the Middle Jurassic (Callovian; Sato and Wu 2008). A. S. Smith and Dyke (2008) argued that rhomaleosaurids represented the first group of major predators among plesiosaurs and were subsequently replaced in that ecological role by Pliosauridae, which ranged in time from the Middle Jurassic (Callovian) to the early Late Cretaceous (Benson et al. 2013). Pliosauridae is characterized by a long process of the coronoid bone along the lingual surface of the lower jaw, cervical centra with flat or gently convex ventral surfaces, the asymmetrically flared dorsal end of the ilium, and the convex preaxial margin of the tibia.

Based on skull and dentition, there are two major types of pliosaurs: rather gracile forms with long snouts, long mandibular symphyses, and slender teeth, and large-headed, robust forms with massive teeth. Benson et al. (2013) considered the former predominantly piscivorous, whereas the latter were capable of pursuing and taking large prey. The gracile pliosaurs include *Hauffiosaurus*, from the Early Jurassic (Toarcian) of England and Germany, and *Marmornectes* and *Peloneustes* (Fig. 6.15), both from the Middle Jurassic (Callovian) of England. Wintrich et al. (2017) reported the oldest known pliosaurid, *Rhaeticosaurus*, from the Late Triassic (Rhaetian) of Germany. Benson and Druckenmiller (2014) united *Peloneustes* with *Pliosaurus* and its close relatives in a subclade, Thalassophonea, based on a suite of shared derived features, especially the anteroposteriorly broad dorsal portion of the iliac blade. The robustly built thalassophoneans include *Liopleurodon*, from the Middle Jurassic

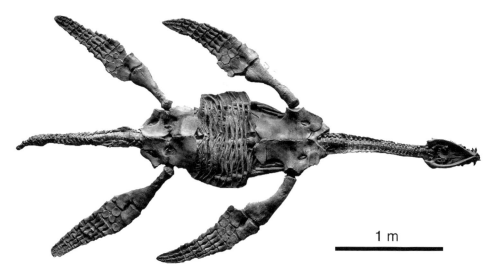

Figure 6.14. Skeleton of the rhomaleosaurid *Meyerasaurus victor* in ventral view. Courtesy of Adam Smith.

1 m

Figure 6.15. Skull of the pliosaurid *Peloneustes philarchus* in lateral view. Courtesy of Adam Smith.

10 cm

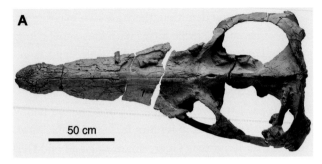

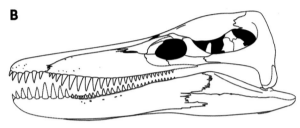

Figure 6.16. A, cranium of the pliosaurid *Pliosaurus kevani* in dorsal view; **B**, reconstruction of the skull in lateral view. From Benson et al. (2013)—CC BY 2.5.

(Callovian) of England and France; *Pliosaurus*, from the Late Jurassic (Kimmeridgian-Tithonian) of England, France, and Svalbard (Fig. 6.16); and *Kronosaurus*, from the Early Cretaceous (Aptian-Albian) of Australia and Colombia. They attained skull lengths of up to 3 m and total lengths of up to 12 m. The robust teeth of *Pliosaurus*, up to 30 cm long, have trihedral crowns with up to three cutting edges and often are heavily worn at their tips (Knutsen 2012; Benson et al. 2013). Gut contents demonstrate that thalassophoneans were active predators capable of taking prey up to half of their own length (Foffa et al. 2014).

Among Thalassophonea, *Brachauchenius*, from the Early Cretaceous (Barremian) of Colombia and the Late Cretaceous (Turonian) of Kansas and Morocco (Benson and Druckenmiller 2014), and closely related taxa such as *Makhaira*, from the Early Cretaceous (Hauterivian) of Russia (Fischer et al. 2015), form a clade Brachaucheniinae. Most appear to have preferred smaller prey items, but *Makhaira* has trihedral teeth with prominent, serrated carinae, which suggest feeding on larger prey (Fischer et al. 2015). *Luskhan*, from the Early Cretaceous (Hauterivian) of Russia, has cranial features, such as a long, slender snout, that resemble those of a clade of large-headed plesiosauroids, Polycotylidae (Fischer et al. 2017).

PLESIOSAURIA: NEOPLESIOSAURIA: PLESIOSAUROIDEA

Plesiosauroidea comprises all taxa more closely related to *Plesiosaurus dolichodeirus* than to *Pliosaurus brachydeirus* (Ketchum and Benson 2010). It includes Plesiosauridae, Microcleididae, Cryptocleididae, Elasmosauridae, and Leptocleidia (Ketchum and Benson 2010; Benson and Druckenmiller 2014). Benson and Druckenmiller (2014) united Elasmosauridae and Leptocleidia as Xenopsaria. In Plesiosauroidea, the supraoccipital is taller than broad and bears a posteromedian ridge.

Plesiosauridae comprises all taxa more closely related to *Plesiosaurus dolichodeirus* than to *Cryptoclidus eurymerus*, *Elasmosaurus platyurus*, *Leptocleidus superstes*, or *Polycotylus latipinnis* (Ketchum and Benson 2010). It is characterized by the participation of the frontal in the margin of the external naris; a strongly inclined suspensorium; cervical centra with gently convex articular surfaces; and anteromedial margin of the coracoid contacting the scapula. Based on recent studies (e.g., Benson et al. 2012), *Plesiosaurus* is known only from the Early Jurassic (Hettangian-Sinemurian) of England (Storrs 1997). Historically, however, this name was indiscriminately applied to plesiosaurian remains of all ages and from around the world.

Microcleididae represents a second group of Early Jurassic (Sinemurian-Toarcian) plesiosauroids from England, France, and Germany. It comprises *Microcleidus homalospondylus* and all taxa more closely related to it than to *Plesiosaurus dolichodeirus*, *Cryptoclidus eurymerus*, *Elasmosaurus platyurus*, *Leptocleidus superstes*, *Pliosaurus brachydeirus*, or *Polycotylus latipinnis* (Benson et al. 2012). Diagnostic features for this clade include widely separated rib facets on the posterior cervical vertebrae, the concave medial surface of the iliac blade, and a prominent anterior flange on the proximal portion of the radius (Benson et al. 2012).

Plesiopterys, from the Early Jurassic (Toarcian) of Germany, has 39 cervical vertebrae (O'Keefe 2004; Fig. 6.17). Benson and Druckenmiller (2014) considered it the sister taxon of Cryptoclididae and Xenopsaria.

Cryptoclidia comprises the most recent common ancestor of *Cryptoclidus eurymerus* and *Polycotylus latipinnis* and all descendants of that ancestor (Ketchum and Benson 2010). Cryptoclididae is defined as the clade

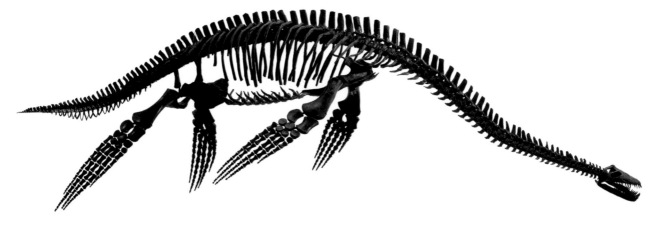

Figure 6.17. Reconstructed skeleton of the plesiosauroid *Plesiopterys wildi* in lateral view. Length 2.38 m. Courtesy of Rainer Schoch.

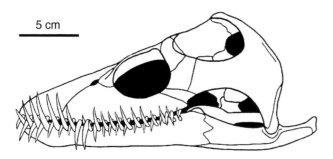

Figure 6.18. Skull of the cryptoclidian *Muraenosaurus leedsi* in lateral view. Modified from M. Evans (1999) and D. S. Brown (1981).

comprising all taxa more closely related to *Cryptoclidus eurymerus* than to *Elasmosaurus platyurus, Leptocleidus superstes, Plesiosaurus dolichodeirus,* or *Polycotylus latipinnis* (Ketchum and Benson 2010). It shares a number of derived features including a narrow and vertical jugal, the moderate ventral embayment of the temporal region, and the presence of an atlantal rib and rib-facet. The ventral portions of the scapulae meet along the midline ventrally, and the anteromedial margin of the coracoid contacts the scapula. *Cryptoclidus* and *Muraenosaurus,* from the Middle Jurassic (Callovian) of England (Andrews 1910; D. S. Brown 1981; Fig. 6.18), have elongated necks (with 44 cervicals in *Muraenosaurus*) and neither apparently exceeded 8 m in total length.

The cryptoclidian clade Xenopsaria (from Greek *xenos,* strange, and *psaras,* angler) comprises all taxa more closely related to *Elasmosaurus platyurus, Polycotylus latipinnis,* and *Leptocleidus superstes* than to *Cryptoclidus eurymerus, Plesiosaurus dolichodeirus, Rhomaleosaurus cramptoni,* or *Pliosaurus brachydeirus* (Benson and Drucken-

miller 2014). Diagnostic synapomorphies for this clade include the smoothly curved suborbital margin of the temporal bar; posteromedial process of the premaxilla constricted by the external naris; raised median ridge in the region of the pineal foramen; and temporal bar not embayed ventrally.

Elasmosauridae comprises all taxa more closely related to *Elasmosaurus platyurus* than to *Cryptoclidus eurymerus, Leptocleidus superstes, Plesiosaurus dolichodeirus,* or *Polycotylus latipinnis* (Ketchum and Benson 2010). Diagnostic synapomorphies for this clade include the massive quadrate and the closure of the pineal foramen (Ketchum and Benson 2010). The cervical vertebrae have nearly flat articular surfaces, and the anterior cervical centra bear a ridge on the lateral surface. In the shoulder girdle, the coracoids form a prominent posterior embayment. The phylogenetic analysis by Benson and Druckenmiller (2014) restricts Elasmosauridae to mostly Late Cretaceous taxa. Some elasmosaurids from the Northern Hemisphere, such as *Albertonectes,* from the Late Cretaceous (Campanian) of Alberta; *Hydrotherosaurus,* from the Late Cretaceous (Maastrichtian) of California (Fig. 6.19); and *Elasmosaurus,* from the Late Cretaceous (Campanian) of Kansas, have extraordinarily elongated necks (Welles 1943, 1952; Kubo et al. 2012). The neck of *Albertonectes* comprises 76 cervical vertebrae and has a length of 7 m in an animal with a total length of 11 m. Noè et al. (2017) argued that the rigid neck was primarily capable of ventral bending for use beneath the body, allowing feeding in the water column and close to or on the seafloor. This is consistent with a report by McHenry et al. (2005) on

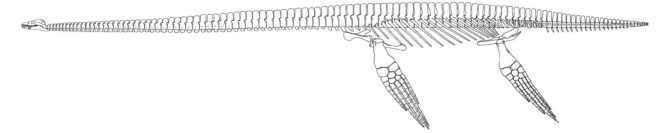

Figure 6.19. Reconstructed skeleton of the elasmosaurid *Hydrotherosaurus alexandrae* in lateral view. Reconstructed length 7.8 m. From Welles (1943).

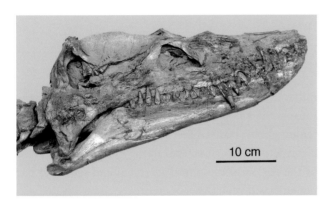

10 cm

Figure 6.20. Skull of the elasmosaurid *Styxosaurus snowii* in lateral view. Courtesy of Mike Everhart.

gut contents from two elasmosaurid skeletons from Australia, which suggest that at least these animals fed close to the seafloor. Both specimens also preserved gastroliths, which may have facilitated the mechanical breakdown of food. The skull of elasmosaurids is proportionately small (only up to about 50 cm long), and the teeth have long, rather slender crowns (Fig. 6.20). The elasmosaurid *Morturneria*, from the Late Cretaceous (Maastrichtian) of Antarctica, and closely related taxa from Argentina, Chile, and New Zealand, have a distinctive cranial structure (O'Keefe et al. 2017). The jaw joint is placed well behind the occiput, and the mandible is long and "hoop-like." Both features indicate a large gape and voluminous mouth cavity. The upper and lower jaws hold interlocking rows of many slender teeth. O'Keefe et al. (2017) interpreted this combination of features as suitable for filter-feeding in a manner not unlike that in baleen whales.

Long considered related to pliosauroids based on the shape and relative size of the skull, Leptocleidia comprises the most recent common ancestor of *Leptocleidus superstes* and *Polycotylus latipinnis* and all descendants of that ancestor (Ketchum and Benson 2010). It can be divided into Polycotylidae and Leptocleididae. Diagnostic cranial features for Leptocleidia include the contact between the maxilla and squamosal, the presence of a prominent dorsomedian ridge on the premaxillae, and the presence of a deep longitudinal trough on the lateral surface of the mandibular ramus. The humerus is shaped like an S, especially in derived polycotylids. Polycotylidae comprises all taxa more closely related to *Polycotylus latipinnis* than to *Cryptoclidus eurymerus*, *Elasmosaurus platyurus*, *Leptocleidus superstes*, *Plesiosaurus dolichodeirus*, *Pliosaurus brachydeirus*, or *Rhomaleosaurus victor* (Ketchum and Benson 2010). *Dolichorhynchops*, from the Late Cretaceous (Turonian) of Kansas, Saskatchewan, and Utah (Fig. 6.21), and its close relatives superficially resemble pliosaurs in the presence of a proportionately large head with a long, narrow snout and a short neck. The dentition consists of slender, widely spaced teeth that are fairly uniform in size. The largest known polycotylids have a skull length of about 1 m. Leptocleididae comprises all taxa more closely related to *Leptocleidus superstes* than to *Cryptoclidus eurymerus*, *Elasmosaurus platyurus*, *Plesiosaurus dolichodeirus*, *Pliosaurus brachydeirus*, *Polycotylus latipinnis*, or *Rhomaleosaurus victor* (Ketchum and Benson 2010). It is noteworthy for occurring in marginal marine and brackish-water to freshwater settings. Leptocleididae includes both long- and short-necked forms (Benson et al. 2013).

Ichthyosauromorpha

Ichthyosauromorpha (from Greek *ichthys*, fish, *sauros*, lizard, and *morphe*, form) is the second most diverse group of Mesozoic marine reptiles (Sander 2000; McGowan and Motani 2003; Motani 2005). It ranged in time from the late Early Triassic (Olenekian) to the early Late

Figure 6.21. Skull of the polycotylid *Dolichorhynchops osborni* in lateral view. Length 53.5 cm. Dark areas restored. Courtesy of Mike Everhart.

Cretaceous (Turonian) and rapidly attained worldwide distribution.

Motani et al. (2015a) defined Ichthyosauromorpha as the clade comprising the most recent common ancestor of *Ichthyosaurus communis* and *Hupehsuchus nanchangensis* and all descendants of that ancestor (Fig. 6.22). Synapomorphies for this clade include the presence of anterior flanges on the humerus and radius; distal width of the ulna equal to or greater than the proximal width; forelimb longer than or almost equal to hind limb; length of the manus at least about three-quarters the combined lengths of the humerus and forearm; fibula extending farther postaxially than the femur; and transverse processes of the neural arches short or absent altogether. Motani et al. (2015a) divided Ichthyosauromorpha into Hupehsuchia and Ichthyosauriformes.

Ichthyosauromorpha: Hupehsuchia

Hupehsuchia (from Hupeh, a version of the name of the Chinese province Hubei, and Greek *souchos*, crocodile) is represented by five genera, all from the Early to Middle Triassic (Olenekian-Anisian) of Hubei (China) (Carroll and Dong 1991; X.-h. Chen et al. 2014a,b, 2015: Scheyer et al. 2014; Fig. 6.23). Attaining estimated total lengths ranging from 40 cm to about 1 m, hupehsuchians are characterized by a long, flattened, and toothless snout and a long, edentulous mandible, which apparently was flexible. Motani et al. (2015b) hypothesized that *Hupehsuchus*, much like extant pelicans and certain whales,

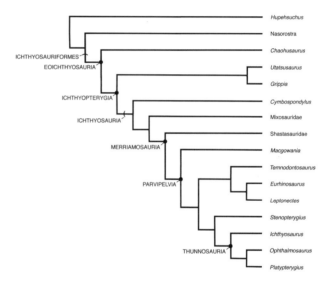

Figure 6.22. Phylogenetic hypothesis of the interrelationships of Ichthyosauromorpha. Dots denote node-based clades and parentheses denote stem-based clades. Based on Motani (1999), McGowan and Motani (2003), and Motani et al. (2015).

was a lunge feeder that used a gular pouch to capture soft-bodied prey. The laterally flattened, spindle-shaped body of hupehsuchians has complex dermal armor above the vertebral column, which comprises more than 30 presacral vertebrae. The dorsal vertebrae of *Hupehsuchus* and *Parahupehsuchus* have a second ossification atop each low neural spine. The top portion of this bipartite structure supports, and often appears to be fused to, an overlying bony armor element. The trunk region of *Parahupehsuchus* is enclosed in a largely rigid bony tube

Figure 6.23. Skeleton of the hupehsuchian *Hupehsuchus nanchangensis* in lateral view. Note the armor elements on top of the neural spines. From X.-h. Chen et al. (2014)—CC BY 2.5.

formed by overlapping ribs and gastralia. The ribs bear posterior flanges, whereas the lateral gastralia elements have anterior flanges. X.-h. Chen et al. (2014a) plausibly interpreted this body tube and the dermal armor along the back as protection against predators. The flipper- or paddle-like limbs of hupehsuchians have one or two additional digits in some specimens. In view of the rather stiff trunk region (which also shows some degree of pachyostosis), hupehsuchians presumably relied primarily on the tail for propulsion, with the limbs serving for steering.

Phylogenetic analyses (e.g., X.-h. Chen et al. 2014b) have consistently found Hupehsuchia as the sister taxon of Ichthyopterygia. The recent discoveries of *Cartorhynchus* and *Sclerocormus* lend additional support to this hypothesis (Motani et al. 2015; Jiang et al. 2016).

Ichthyosauromorpha: Ichthyosauriformes

Motani et al. (2015a) defined Ichthyosauriformes (from Greek *ichthys*, fish, and *sauros*, lizard, and Latin *forma*, form) as comprising all ichthyosauromorphs more closely related to *Ichthyosaurus communis* than to *Hupehsuchus nanchangensis*. Synapomorphies for this clade are the anterior extension of the nasal well beyond the external naris, the large size of the scleral ring (which fills the orbit), the constriction of the snout in dorsal view, and converging digits with little interdigital space.

ICHTHYOSAUROMORPHA: ICHTHYOSAURIFORMES: NASOROSTRA

Jiang et al. (2016) established Nasorostra (from Latin *nasus*, nose, and *rostrum*, snout) for the reception of two

ichthyosauriform taxa from the Early Triassic (Olenekian) of Anhui (China). Diagnostic derived features include the elongate nasals extending to the tip of the snout, the more or less equal lengths of the pre- and postorbital portions of the skull, and the ribcage being deepest near the shoulder.

Cartorhynchus attained an estimated length of only 40 cm (Motani et al. 2015a). The snout is short, unlike in most ichthyopterygians, but the premaxilla is long and the external narial openings are set back from the tip of the snout. Along with the absence of teeth, the large and robust hyoid apparatus suggests possible suction feeding. The vertebral column of *Cartorhynchus* is robust and not elongated as in ichthyopterygians. The ribs are distinctly thickened. The fore fins are unusually large. Motani et al. (2015) argued that they were still capable of bending at the wrists, and this may have allowed seal-like locomotion with bent flippers on land.

Sclerocormus is distinguished from *Cartorhynchus* by its greater body size (length about 1.6 m), the exclusion of the frontal from the dorsal margin of the orbit, and the presence of a basket of robust, flattened gastralia (Jiang et al. 2016).

ICHTHYOSAUROMORPHA: ICHTHYOSAURIFORMES: ICHTHYOPTERYGIA

Ichthyopterygia (from Greek *ichthys*, fish, and *pteron*, fin, wing) comprises all ichthyosauriforms more closely related to *Ichthyosaurus communis* than to Nasorostra (Jiang et al. 2016). The vernacular term "ichthyosaur" is used here for all ichthyopterygians. Ichthyosaurs were the first amniotes to evolve a fish-like body shape with a spindle-shaped (fusiform) body, limbs modified as flippers, and a crescent-shaped tail fin. They were also the first marine tetrapods to attain gigantic body size, with one Late Triassic species attaining a total length of more than 20 m.

Motani (1999) provided a suite of synapomorphies for this clade including the presence of a posterior process of the postfrontal; parietal with a long supratemporal process; ectopterygoid absent; interpterygoid vacuity small or absent; metacarpal I with reduced anterior region of the shaft; metacarpal V without postaxial region of the shaft; digits integrated without any interdigital separation; caudal portion of the vertebral column with "peak"; and anticliny of the caudal neural spines. C. Ji et al. (2015) also found strong support for Ichthyopterygia but provided several different shared derived character states for this clade, including the anterior extension of the nasal beyond the external naris and the absence of a coracoid foramen.

The skull of most ichthyosaurs has a long snout that is offset from the remainder of the cranium, large orbits, and an anteroposteriorly short postorbital region (McGowan and Motani 2003; Fig. 6.24). The large upper temporal opening is bordered below by a deep bony bar. In some Triassic ichthyosaurs, the ventral margin of the temporal bar is distinctly embayed, likely the remnant of a lower temporal opening. This embayment is not present in more derived taxa, in which the large size of the orbits led to considerable modification of the temporal region. The temporal region includes a large supratemporal, which (rather than the squamosal) forms the posterolateral margin of the upper temporal fenestra (Fig. 6.24). *Temnodontosaurus* had the largest eye of any known extant or extinct vertebrate, with a diameter of up to 26.3 cm, but even the smaller-sized *Ophthalmosau-*

rus attained an eye diameter of 25 cm (Motani 2005). Presumably these enormous eyes enabled the animals to see and forage in deeper water where light levels were low. Based on what appear to be pigment bodies (melanosomes) preserved in skin impressions, Lindgren et al. (2014) reconstructed some ichthyosaurs as uniformly dark-colored, without the countershading commonly found among marine vertebrates. This would be consistent with the hypothesis that these animals foraged in dimly lit waters. Dark colors would also have facilitated more rapid warming of these ichthyosaurs when they swam close to the surface of the sea. Based on isotopic data from bones, Bernard et al. (2010) interpreted ichthyosaurs as homeothermic, with high body temperatures, much like plesiosaurs.

The snout of ichthyosaurs is formed primarily by the long premaxillae. The rather small maxilla is usually excluded from the margin of the external naris. In post-Triassic ichthyosaurs, the numerous teeth are typically set in deep longitudinal grooves in the jaws rather than in individual sockets (alveoli) as in most other reptiles. Tooth shape varies considerably among ichthyosaurs. All teeth show distinct infolding of the dentine (plicidentine), as in some other reptiles, stem-amphibians, and basal tetrapods (Schultze 1969). The cranium lacks postparietal and tabular bones, as well as the transverse flange of the pterygoid.

The vertebrae of ichthyosaurs typically have short, disk-shaped centra with deeply concave anterior and posterior articular surfaces (amphicoelous condition) and without protruding facets for articulation with the rib-heads. The long and slender ribs at least partially articulate with the centra. The simple neural arches were apparently connected to the centra by cartilage in life and bear distinct zygapophyses only in basal forms. The

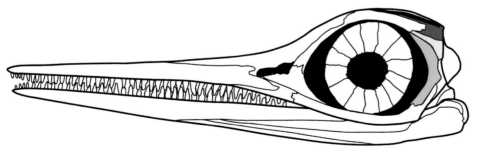

Figure 6.24. Skull of the parvipelvian ichthyosaur *Ophthalmosaurus icenicus* in lateral view. Note the structure of the temporal bar with a supratemporal (purple), squamosal (yellow), and postorbital (light blue). Modified from Andrews (1910) and McGowan and Motani (2003)

neck is short. The length of the trunk is quite variable among ichthyosaurs: some Triassic taxa have more than 60 vertebrae, but most Jurassic and Cretaceous ichthyosaurs have shorter trunk regions with 40 to 50 vertebrae. The sacrum comprises only a single vertebra, and the sacral ribs are not in contact with the pelvic girdle in more derived ichthyosaurs. The tail region is long and slightly "kinked" in Triassic taxa but is shorter and distinctly "bent" downward in post-Triassic ichthyosaurs. This caudal "peak" is marked by a small number of wedge-shaped caudal centra and by a change in the orientation of the neural spines. Exquisitely preserved specimens of ichthyosaurs, with petrified microbial mats tracing the outlines of the bodies, from the Early Jurassic (Toarcian) of Germany show that the distal tail vertebrae supported the lower (ventral) lobe of a crescent-shaped tail fin whereas the upper (dorsal) lobe was entirely formed of soft tissue. The tail fin resembles those of tunas and certain sharks in overall shape and indicates a mode of swimming in which the tail was held at an angle to the direction of motion and beating from side to side, much as in present-day fishes with similar tail fins. The paddle-like limbs of ichthyosaurs were probably used for changing direction and for controlling pitch and roll during swimming. For the Early Jurassic *Stenopterygius*, Motani (2002) calculated cruising speeds of about 5.4 km per hour based on comparisons with extant vertebrates. M. A. Taylor (1987) argued that the tail fin of ichthyosaurs could have produced powerful downward pitching action, in particular when diving after breathing at the water surface. The oldest known ichthyosaurs, such as *Chaohusaurus*, already had a small dorsal tail fin (Motani 2005). In addition, ichthyosaurs had a dorsal fin, which was supported only by soft tissue and may have functioned as a stabilizer during swimming. Houssaye (2009) noted the absence of pachyostosis in ichthyosaurs. Ichthyosaurian limb bones have a spongy internal structure and lack medullary cavities (de Ricqlès and Buffrénil 2010).

Ichthyosaurs have a well-developed shoulder girdle and forelimbs that are larger than the hind limbs in most taxa. The scapula and coracoid are not fused together, and the plate-like coracoids contact each other ventrally along the midline of the body. The forelimbs presumably served as rudders and stabilizers during swimming. The bones of the pelvic girdle are only loosely connected to each other in many forms, and the center of the acetabulum is not ossified. A thyroid fenestra is present between the pubis and ischium in some ichthyosaurs but the two bones are fused in various Jurassic taxa.

The fore- and hind limbs of ichthyosaurs have flattened bones. The elbow and knee as well as the wrist and ankle joints are immobile. Except for the humerus and femur, the limb bones are polygonal or rounded disks (Fig. 6.25). They are often notched along the anterior margin of the fin or, in early forms, along both margins. The carpal bones appear to have been fully integrated into the digits. Ungual phalanges are absent. The soft-tissue flipper extends beyond its bony supports, and its edges are reinforced by radial fibers of connective tissue (Motani 2005).

The fossil record of ichthyosaurs documents the transition from a basal pattern of limb bones still comparable to that in terrestrial reptiles to highly modified patterns in later Mesozoic forms (Caldwell 1997a; Fig. 6.25). In many ichthyosaurs, the fins not only have hyperphalangy, with up to 20 bones per digit, but often also additional digits (hyperdactyly), with up to 10 digits per limb in *Caypullisaurus*, from the Late Jurassic (Tithonian) of Argentina (Motani 2005). However, some taxa show a decrease in the number of primary digits: for example, *Shastasaurus* has only two primary digits.

Ichthyosaurs were no longer capable of venturing onto land. Based on preserved gut contents, ichthyosaurs fed on invertebrates (especially cephalopods), fishes, and occasionally other ichthyosaurs (Böttcher 1989). A few forms have short, robust jaws with blunt crushing teeth and possibly subsisted on hard-shelled invertebrates.

Ichthyosaurs were viviparous. This mode of reproduction is already present in one of the earliest known ichthyosaurs, *Chaohusaurus* (Motani et al. 2014). The newborn still emerged head first in *Chaohusaurus*, presumably as in the terrestrial antecedents of ichthyosauromorphs (Fig. 6.26A). Scores of females of the derived Early Jurassic ichthyosaur *Stenopterygius* contain up to 11 embryos in two slightly offset uteri and some of them were preserved in the process of being expelled from the mother during the latter's decomposition (Böttcher 1990). Unlike in *Chaohusaurus*, but as in extant cetaceans, embryos of *Stenopterygius* developed with their heads point-

A **B** **C** **D**

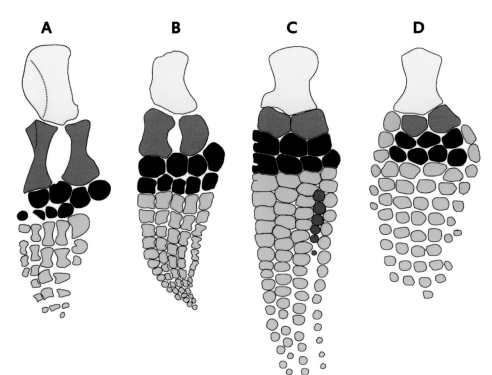

Figure 6.25. Left forelimbs of ichthyopterygians. **A**, *Utatsusaurus hataii*; **B**, *Mixosaurus cornalianus*; **C**, *Stenopterygius* sp.; and **D**, *Ophthalmosaurus icenicus*. Not to scale. Humerus, yellow; radius and ulna, red; carpals, black; phalanges, teal; accessory digits, green or purple. Combined and modified from Sander (2001) and McGowan and Motani (2003).

ing forward in the trunk region of the mother and were born tail first (Fig. 6.26B). Data on bone microstructure indicate that most ichthyosaurs had rapid growth rates (Houssaye 2013).

Motani (1999), Sander (2000), McGowan and Motani (2003), and C. Ji et al. (2015) published detailed reviews of the diversity of Ichthyopterygia that form the basis for the following discussion.

ICHTHYOPTERYGIA: EOICHTHYOSAURIA

Motani (1999) defined a clade Eoichthyosauria (from Greek *eos*, dawn, and Ichthyosauria) as comprising the most recent common ancestor of *Grippia longirostris* and *Ichthyosaurus communis* and all descendants of that ancestor. Diagnostic apomorphies include the manus being longer than the more proximal portion of the forelimb and the presence of bicipital facets for the ribs on at least some cervical vertebrae.

Chaohusaurus and *Utatsusaurus* are the earliest well-documented ichthyosaurs. *Chaohusaurus*, from the late Early Triassic (Olenekian) of Anhui and Yunnan (China), attained a total length of 1.5 m (Fig. 6.27). *Utatsusaurus*, from the late Early Triassic (Olenekian) of Japan and

British Columbia, reached 3 m (Cuthbertson et al. 2013). In both taxa, the anterior teeth have conical crowns, whereas the more posterior teeth have blunt crowns. *Chaohusaurus* and *Utatsusaurus* each have only about 40 presacral vertebrae. The trunk of *Chaohusaurus* is slender, and Motani (2005) characterized its overall body shape as that of a "lizard with flippers." The fore fins of *Chaohusaurus* and *Utatsusaurus* retain distinct limb bones with constricted shafts and five digits, which do not include additional phalanges. In *Utatsusaurus*, the limb bones are already somewhat flattened. *Chaohusaurus* and *Utatsusaurus* probably still employed undulating motions of the trunk and tail for swimming (Motani 2005), but the presence of a small dorsal tail fin in at least *Chaohusaurus* indicates increased use of the tail. *Chaohusaurus* is most closely related to *Grippia*, from the Early Triassic (Olenekian) of Svalbard. Motani (1999) recognized a clade Grippidia, which is distinguished by rounded posterior tooth crowns and multiple tooth rows in the maxilla. The phylogenetic analysis by C. Ji et al. (2015) found a basal clade Grippioidea that includes *Utatsusaurus* and *Grippia* and placed *Chaohusaurus* as the sister-taxon of Ichthyosauria.

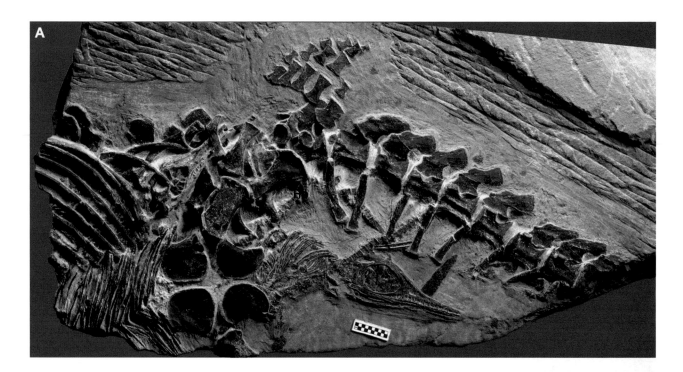

Figure 6.26. Mode of birth in two ichthyosauriforms. **A,** *Chaohusaurus geishanensis*, with fetus "born" head first; **B,** *Stenopterygius quadriscissus*, with fetus "born" tail first. Length of adult 2.2 m. In both cases, the fetus was probably expelled from the maternal body cavity by decomposition gasses. A, courtesy of Ryosuke Motani; B, courtesy of Rainer Schoch.

ICHTHYOPTERYGIA: EOICHTHYOSAURIA: ICHTHYOSAURIA

Ichthyosauria (from Greek *ichthys*, fish, and *sauros*, lizard) comprises all eoichthyosaurians more closely related to *Ichthyosaurus communis* than to *Grippia longirostris* (Motani 1999). It is characterized by a suite of synapomorphies including the laterally facing external naris; dorsal margin of the orbit formed by the prefrontal and postfrontal; postorbital excluded from the margin of the upper tempo-ral fenestra in lateral view; and ulna without a postaxial portion of the shaft (McGowan and Motani 2003).

The Middle Triassic *Cymbospondylus*, from the Anisian of Nevada and the Anisian-Ladinian of Switzerland, reached a total length of at least 9 m. Its proportionately small skull is robust and has small orbits (Merriam 1908). N. B. Fröbisch et al. (2006) reported the presence of a postparietal or neomorph bone at the posterior end of the skull roof. *Cymbospondylus* has at least 55 presacral vertebrae (Motani 1999).

Figure 6.27. Skeleton of the basal ichthyopterygian *Chaohusaurus geishanensis* in ventral view. Courtesy of Ryosuke Motani.

Mixosauria is characterized by several shared derived features: premaxilla posteriorly pointed, barely reaching the external naris; sagittal crest extending anteriorly to the external nares; upper temporal fenestrae with large anterior depressions ("terraces") extending to the nasals; and pubis more than twice the size of the ischium (Motani 1999). *Mixosaurus* has definitely been recorded from the Middle Triassic (Anisian-Ladinian) of China, Germany, Italy, Poland, and Switzerland (Fig. 6.28). The closely related *Phalarodon* is known from the Middle Triassic (Anisian-Ladinian) of Nevada, China, and Svalbard. Both taxa attained total lengths of rarely more than 1 m. *Mixosaurus* has a long, slender snout and large orbits. The posterior tooth crowns are rounded rather than conical (C. Ji et al. 2015). The vertebrae bear tall neural spines. The limb bones of *Mixosaurus* are shorter and broader than those in earlier ichthyopterygians. The fins each have five digits, but there are up to 10 phalanges per digit in the fore fin and up to eight per digit in the hind fin.

ICHTHYOPTERYGIA: EOICHTHYOSAURIA: ICHTHYOSAURIA: MERRIAMOSAURIA

Merriamosauria (named for the American paleontologist John Merriam and Greek *sauros*, lizard) comprises the most recent common ancestor of *Shastasaurus pacificus* and *Ichthyosaurus communis* and all descendants of that ancestor (Motani 1999). Shared derived features for this clade include the more extensive contact between the scapula and coracoid and the absence of an ossified metacarpal I.

Shastasauria is a clade of medium-sized to gigantic ichthyosaurs that ranged from the Middle to the Late Triassic and attained a worldwide distribution. Diagnostic features include the almost quadrangular humerus and the presence of more than 55 presacral vertebrae (Motani 1999). Shastasaurians lack a pronounced tail bend. *Shonisaurus*, from the Late Triassic (Carnian) of Nevada, attained a skull length of more than 2 m and a total length of more than 15 m (Camp 1980). Its long snout bears large teeth that have swollen bases. A tall parietal crest separates the upper temporal openings. The orbits of *Shonisaurus* are rather large. The long fore fins and hind fins have pronounced hyperphalangy. The phalanges along the posterior margin of the fore fin are notched. The Late Triassic *Shastasaurus*, from the Carnian-Norian of western North America and the Carnian of China, included the largest known ichthyosaur, *S. sikkanniensis* from the Norian of British Columbia,

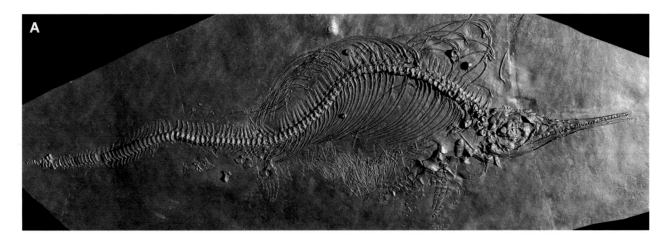

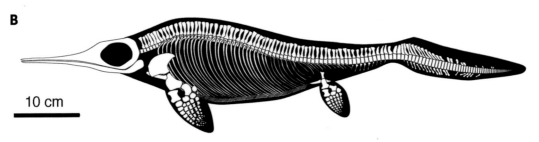

Figure 6.28. A, skeleton of the mixosaurid *Mixosaurus cornalianus* in lateral view. Length c. 75 cm. **B**, reconstructed skeleton of *Mixosaurus cornalianus*. A, courtesy of Rainer Schoch; B, modified from McGowan and Motani (2003).

which reached a total length of more than 20 m (Nicholls and Manabe 2004; Sander et al. 2011). *Shastasaurus* lacks teeth. Sander et al. (2011) interpreted it as a suction feeder, but Motani et al. (2013) argued that *Shastasaurus* lacks the requisite features of the skull and hyoid for this particular mode of feeding. *Thalattoarchon*, from the Middle Triassic (Anisian) of Nevada, was a shastasaurid with an estimated total length of nearly 9 m (N. B. Fröbisch et al. 2013). Although it resembles *Cymbospondylus* in overall appearance, its robust skull is nearly twice as large relative to total length. The teeth of *Thalattoarchon* have large, labiolingually flattened crowns with two distinct cutting edges, indicating predatory habits.

ICHTHYOPTERYGIA: EOICHTHYOSAURIA: ICHTHYOSAURIA: MERRIAMOSAURIA: EUICHTHYOSAURIA

Euichthyosauria (from Greek *eu*, true, *ichthys*, fish, and *sauros*, lizard) comprises all merriamosaurians more closely related to *Ichthyosaurus communis* than to *Shastasaurus pacificus* (Motani 1999; C. Ji et al. 2015). The teeth

are not attached to the jaws by bone, and the humerus has nearly equally long articular facets for the radius and ulna.

ICHTHYOPTERYGIA: EOICHTHYOSAURIA: ICHTHYOSAURIA: EUICHTHYOSAURIA: PARVIPELVIA

Parvipelvia (from Latin *parvus*, small, and *pelvis*, basin) comprises all post-Triassic euichthyosaurians along with a few Late Triassic taxa (Motani 1999). Motani (1999) listed several shared derived features for this clade: scapular blade straight; radius broader than long; and pubis more slender than ischium. C. Ji et al. (2015) used different synapomorphies including the absence of a dorsal lamina on the maxilla and the absence of a contact between the maxilla and external naris. The proportionately small pelvic girdle of parvipelvians indicates that much of the tail musculature was connected primarily to the trunk rather than attached to the pelvic girdle (and small hind fin) (Motani 2005). This possibly reflects a shift from undulating body motion during swimming to an oscillatory mode using a tail fluke, which would also account for the increase in the length of the tail region.

Few ichthyosaurian remains are known from the latest Triassic (Rhaetian), but there is an exceptionally rich fossil record with hundreds of skeletons from the Early Jurassic of Europe, particularly southern England and southern Germany. Some of these specimens even preserve approximate body outlines in the form of microbial mats that later became lithified (Martill 1987; Motani 2005).

Parvipelvians have unusually large orbits, which are reflected by a modified arrangement of the surrounding bones of the skull (Fig. 6.29). The dentition is homodont with teeth that have conical crowns and are uniformly set in grooves in the jaws (Motani 1997). In the vertebral column, the neural spines are typically low. Successive neural spines contact each other along a single, medially placed joint surface, which likely formed through fusion of the zygapophyseal facets. The presacral vertebrae have round double facets for contact with the proximal heads of the ribs; toward the posterior end of the vertebral series, these facets fuse. The shoulder girdle has a long scapular blade, a coracoid with a notch or embayment in its anterior margin, and a small, short, and T-shaped interclavicle. The humerus has a constricted shaft and a flat, expanded distal end. The short bones of the forearm are not separated by a space as in most reptiles. The elements of the distal region of the fore fin all are flattened and lack perichondral ossification (Caldwell 1997a), making it often difficult to distinguish individual sets of bones such as the carpus. All digits have many additional phalanges. The pelvic girdle is greatly reduced in size. Its bones are elongated and "waisted" rather than plate-like. The hind fins are modified in a manner similar to that of the fore fins.

Temnodontosauroidea comprises the last common ancestor of *Temnodontosaurus platyodon* and *Leptonectes tenuirostris* and all descendants of that ancestor (McGowan and Motani 2003).

The Early Jurassic (Hettangian-Toarcian) *Temnodontosaurus*, known from England and Germany, attained a total length of at least 9 m (Figs. 6.29B, 6.30). Its long snout is constricted anterior to the orbits and holds massive teeth with conical crowns that have two or three cutting edges. The temporal region is proportionately long. The vertebral count anterior to the tail bend approaches 90 or more. The fore fin of *Temnodontosaurus* is long and has only three primary digits. Both fore and hind fins have pronounced hyperphalangy.

Leptonectidae comprises the most recent common ancestor of *Eurhinosaurus longirostris* and *Leptonectes tenuirostris* and all descendants of that ancestor (Motani 1999). The snout is slender and holds proportionately small teeth. *Leptonectes*, from the latest Triassic (Rhaetian) to Early Jurassic (Pliensbachian) of England and France, has large round orbits and a considerably elongated and slender snout (Fig. 6.31). The snout of the closely related *Eurhinosaurus*, which attained a total length of about

Figure 6.29. Skulls of the parvipelvians **A**, *Eurhinosaurus longirostris,* and **B**, *Temnodontosaurus trigonodon* in lateral view. Note the difference in length between the rostral portion of the cranium and the mandible in *Eurhinosaurus*. No scale provided. Courtesy of Rainer Schoch.

7 m, from the Early Jurassic (Toarcian) of England and Germany, has a pronounced overbite, with the upper jaw up to more than twice as long as the lower (Fig. 6.29A). *Eurhinosaurus* possibly used its snout like the present-day swordfish (*Xiphias*), slashing through schools of prey fish. Its medium-sized teeth have slender crowns. The fore fin of *Eurhinosaurus* has a radius that is much larger than the ulna and four primary digits. Both pairs of fins are long and slender.

Suevoleviathan, from the Early Jurassic (Toarcian) of Germany (Maisch 1998), is a large-sized basal parvipelvian with distally splayed digits and an accessory digit in its fore fin. It reached a length of over 4 m.

EUICHTHYOSAURIA: PARVIPELVIA: THUNNOSAURIA

Thunnosauria (from Latin *thunnus*, tuna, and Greek *sauros*, lizard) comprises the most recent common ancestor of *Stenopterygius quadriscissus* and *Ichthyosaurus communis* and all descendants of that ancestor (Motani 1999; C. Ji et al. 2015). In these particularly "fish-like" ichthyosaurs, the fore fin is at least twice as long as the hind fin. *Ichthyosaurus*, from the latest Triassic (Rhaetian) to Early Jurassic (Hettangian-Pliensbachian) of Bel-

gium, Switzerland, and the United Kingdom, has a coracoid with prominent anterior and posterior notches and a fore fin with at least five primary digits, sometimes with an accessory digit inserted between the first and second (Lomax and Massare 2018). The best-known early thunnosaurian is *Stenopterygius*, from the Early Jurassic (Toarcian) of England, France, Germany, Luxembourg, and Switzerland (Fig. 6.32). It attained a total length of at least 3.5 m. Its slender snout bears small conical teeth or lacks teeth altogether. The neural spines are rather tall. The fore fin of *Stenopterygius* has four primary digits and frequently an accessory digit (C. Ji et al. 2015). It shows pronounced hyperphalangy, and at least some bones along the anterior margin of the fin are notched. In the pelvic girdle, the ischium and pubis are fused together. The late-surviving basal thunnosaurian *Malawania*, from the Early Cretaceous (Hauterivian-Barremian) of Iraq (Fischer et al. 2013), lacks the derived features of other Late Jurassic and Cretaceous ichthyosaurs and is most closely related to *Ichthyosaurus*.

Few ichthyosaurian fossils have been reported from the Middle Jurassic to date. Most of the Late Jurassic and Cretaceous ichthyosaurs are referred to Ophthalmosauridae, which is characterized by the structure of the ba-

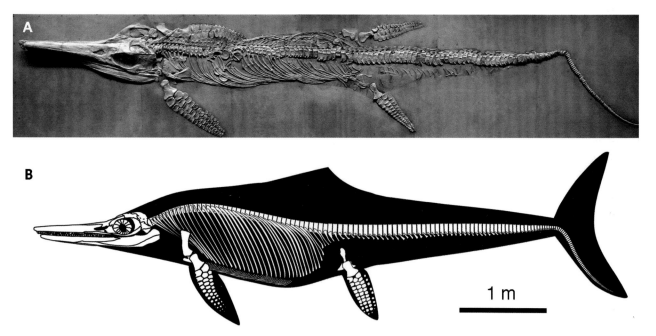

Figure 6.30. A, skeleton of *Temnodontosaurus trigonodon*. No scale provided. **B**, reconstructed skeleton of *Temnodontosaurus platyodon*. Neural arches were omitted. A, courtesy of Rainer Schoch; B, modified from McGowan and Motani (2003).

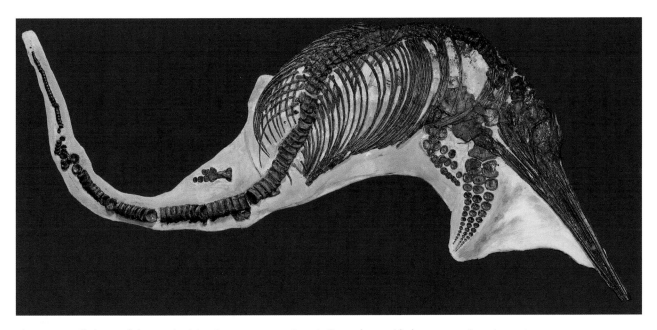

Figure 6.31. Skeleton of the parvipelvian *Leptonectes tenuirostris*. No scale provided. Courtesy of Royal Ontario Museum.

sioccipital, the presence of a facet on the humerus for an anterior accessory bone in the forearm, and the absence of notches on the bones of the fore fin (Fischer et al. 2012). *Ophthalmosaurus*, from the Middle to Late Jurassic (Callovian-Kimmeridgian) of Argentina, England, France, Mexico, Russia, and the western United States, has enormous orbits and small, loosely attached teeth. The distal end of the humerus bears three articular facets. The broad fore fin has accessory digits anteriorly and posteriorly. The ischium and pubis are fused to each other. *Ophthalmosaurus* attained a total length of up to 6 m. *Brachypterygius*, from the Late Jurassic (Kimmeridgian) of England and the Late Jurassic (Tithonian) of Russia, is distinguished by a distal facet on the humerus for contact with the intermedium. It further differs from *Ophthalmosaurus* in having small orbits, a long snout, and numerous robust teeth (McGowan 1997).

Historically, all Cretaceous ichthyosaurs were identified as *Platypterygius*. This taxon, reported from Argentina, Australia, Colombia, England, France, Germany, India, Russia, and western North America, is especially characterized by pronounced polydactyly, with two or three accessory digits anterior to the primary axis of the limb and at least two accessory digits posterior to this axis (Motani 1999). The skull has a long snout and a relatively small orbit. The stapes is large and has a round head. Recent studies on Early Cretaceous and early Late Cretaceous ichthyosaurs indicate a greater taxonomic diversity among these marine reptiles than previously assumed (Fischer et al. 2012, 2014). *Platypterygius* and its close relatives share apomorphies such as tooth roots that are quadrangular in transverse section, the presence of a large deltopectoral crest on the humerus, and the presence of postaxial accessory digits in the hind fin (Fischer et al. 2012).

Additional Clades of Triassic Marine Reptiles

In addition to ichthyosauromorphs and sauropterygians, several other groups of reptiles adopted a marine mode of life during the Triassic Period. None of these lineages diversified to the extent that ichthyosauromorphs and sauropterygians did, and they all appear to be restricted to the Triassic Period. They are of interest because they represent additional body plans for life in the sea and may prove important for sorting out the phylogenetic relationships of sauropterygians and ichthyosauromorphs.

Helveticosaurus and *Eusaurosphargis*

Helveticosaurus, from the Middle Triassic (Anisian-Ladinian) of Switzerland, was originally interpreted as a placodont based on the structure of its dorsal vertebrae (B. Peyer 1955). However, it differs considerably from undisputed

Figure 6.32. Skeleton of the parvipelvian *Stenopterygius* sp. in lateral view, with body outline preserved by lithified bacteria. No scale provided. Courtesy of Rainer Schoch.

representatives of the Placodontia in most other skeletal features (Rieppel 1989c). Attaining a snout-vent length of about 2 m, *Helveticosaurus* has a dorsoventrally deep snout with long, slender, and slightly recurved teeth. The anterior portion of its maxilla holds a large caniniform tooth. The vertebral column of *Helveticosaurus* has at least 40 presacral vertebrae including 13 or 14 cervical vertebrae. The dorsal vertebrae have well-developed transverse processes. Unlike in placodonts and other sauropterygians, the clavicle contacts the anterior rather than the medial margin of the scapula. The limb bones of *Helveticosaurus* are poorly ossified, and both manus and pes exhibit hyperphalangy. The phylogenetic analysis by Müller (2004) recovered *Helveticosaurus* as the sister taxon to Sauropterygia, but X.-h. Chen et al. (2014b) found its relationships less clearly resolved.

Eusaurosphargis, from the Middle Triassic (Anisian-Ladinian) of Italy and Switzerland, also has a deep snout and robust mandible, but its teeth differ from those of *Helveticosaurus* in being homodont and having leaf-shaped crowns with distinct lingual "heels" (Nosotti and Rieppel 2003). The upper temporal fenestrae are relatively small (Scheyer et al. 2017). The dorsal vertebrae have low neural spines and long transverse processes. *Eusaurosphargis* bears rows of small, conical osteoderms that top the neural spines and ribs. It also has small ventrolateral osteoderms of various shapes that taper laterally and are associated with gastralia. The anterior dorsal ribs of *Eusaurosphargis* bear fan-shaped uncinate processes. Nosotti and Rieppel (2003) and C. Li et al. (2011) hypothesized *Eusaurosphargis* as the sister taxon of *Helveticosaurus*. Based on an exquisitely preserved, largely articulated skeleton of a juvenile, Scheyer et al. (2017) found *Eusaurosphargis* crownward of *Helveticosaurus* as a sister taxon to Sauropterygia. These authors interpreted *Eusaurosphargis* as a primarily terrestrial animal.

Saurosphargidae

Saurosphargidae comprises *Largocephalosaurus* and *Sinosaurosphargis*, from the Middle Triassic (Anisian) of Guizhou (China), and *Saurosphargis*, from the Middle Triassic (Anisian) of Poland (Nosotti and Rieppel 2003; C. Li et al. 2011, 2014). Members of this clade are distinguished by the formation of a closed "basket" by broad dorsal ribs that abut each other and by dorsal armor composed of numerous small osteoderms. The Chinese taxa have a distinctive combination of skeletal features. The external nares are positioned closer to the orbits than to the tip of the snout. The posterior margin of the cranium is deeply embayed. Unlike in basal eosauropterygians, the palate has an interpterygoid vacuity. The teeth have leaf-shaped crowns with convex labial and concave lingual surfaces. The short neck comprises nine vertebrae. The dorsal vertebrae have long transverse processes and low neural spines with expanded tips that supported the overlying osteoderms. In *Sinosaurosphargis*, the dorsal osteoderms form a complete carapace over the neck, trunk, and proximal portions of the limbs (Chun et al. 2014). *Largocephalosaurus* reached a total length of about 2 m.

Chun et al. (2014) and X.-h. Chen et al. (2014b) placed Saurosphargidae close to Sauropterygia, based especially on the curved humerus and the contact between the clavicle and the anteromedial surface of the scapula. Scheyer et al. (2017) found *Sinosaurosphargis* as the sister taxon of Eosauropterygia.

Thalattosauriformes

Thalattosauriformes (from Greek *thalassa*, sea, and *sauros*, lizard, and Latin *forma*, form) ranged in time from the late Early Triassic (Olenekian) to the Late Triassic (Norian or Rhaetian) (Merriam 1905; Nicholls 1999; Müller 2005). Representatives of this clade are known from China, Europe, and western North America. They can be divided into Askeptosauroidea and Thalattosauria, but the vernacular "thalattosaurs" is used for the entire group. Thalattosauriformes is characterized by long premaxillae forming the rostrum of the cranium, the posterior embayment of the cranium so that the occiput is anterior to the jaw joints, and the presence of tall, posteriorly inclined neural spines on the caudal vertebrae (Nicholls 1999). The upper temporal fenestrae often form slit-like openings or are closed altogether, whereas the large lower temporal fenestrae are ventrally open (Nicholls 1999; Müller 2005). Thalattosaurs probably employed the long tail as the primary means for swimming.

The phylogenetic relationships of Thalattosauriformes remain unresolved. Rieppel (1987a) interpreted thalattosaurs as diapsid reptiles but could not confidently relate them to either Lepidosauromorpha or Archosauromorpha. Rieppel (1998) placed them as the sister taxon

of Sauropterygia. Depending on the phylogenetic evaluation of shared features that presumably reflect an aquatic mode of life, X.-h. Chen et al. (2014b) found Thalattosauriformes as closely related to both ichthyosauromorphs and sauropterygians, or only to ichthyosauromorphs, or not related to any other groups of Mesozoic marine reptiles. Scheyer et al. (2017) placed the clade crownward of Ichthyosauromorpha and closer to Sauropterygia.

THALATTOSAURIFORMES: ASKEPTOSAUROIDEA

Askeptosaurus, from the Middle Triassic (Anisian-Ladinian) of Italy and Switzerland (Müller 2005), *Anshunsaurus*, from the Late Triassic (Carnian) of Guizhou (China) (J. Liu and Rieppel 2005; Fig. 6.33), and *Endennasaurus*, from the Late Triassic (Norian) of Italy (Müller et al. 2005) differ from most other known thalattosaurs in a number of derived features, including the absence of fusion between the postfrontal and postorbital, presence of a posteriorly elongated posterolateral process of the frontal, absence of teeth on the pterygoid and vomer, and the high number of cervical vertebrae (Müller 2005). The long, parallel-sided snout has a blunt tip and bears recurved teeth. *Askeptosaurus* attained a total length of about 2.5 m and has 13 cervical vertebrae. Its trunk is long and rather slender. The pelvis has a thyroid fenestra. The limbs are proportionately short but well developed, and both carpus and tarsus are well ossified. These features, together with the presence of distinct ungual phalanges, suggest that *Askeptosaurus* and its close relatives were still capable of locomotion on land (Müller 2005). *Endennasaurus* is distinguished by its long, tapering, and edentulous snout (Müller et al. 2005).

THALATTOSAURIFORMES: THALATTOSAURIA

Thalattosauria differs from Askeptosauroidea in the presence of proportionately shorter limbs with broader limb bones, which indicate that this group was more extensively adapted for an aquatic mode of life than *Askeptosaurus* (Nicholls 1999; Müller 2005). Müller (2005) lists a number of diagnostic apomorphies for Thalattosauria including ventral deflection of the snout; maxilla short anteroposteriorly and tall dorsoventrally; absence of a contact between the nasal and prefrontal; posterior cervical and dorsal vertebrae with tall neural spines; and scapula slender and elongated. Thalattosauria also differs from Askeptosauroidea in the presence of blunt, sometimes button-

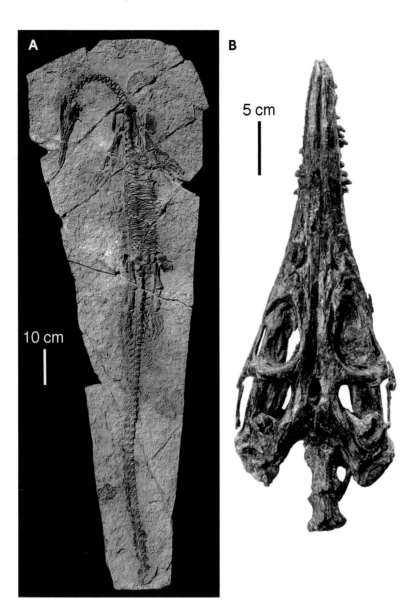

Figure 6.33. A, skeleton of the askeptosauroid *Anshunsaurus huangguoshuensis* in dorsal view; **B**, skull of *Anshunsaurus huangguoshuensis* in dorsal view. Courtesy of Jun Liu.

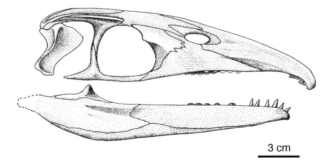

Figure 6.34. Skull of the thalattosauroid *Thalattosaurus alexandrae* in lateral view. Combined from illustrations in Nicholls (1999).

shaped tooth crowns (Fig. 6.34). The long tail of thalattosaurians is laterally flattened with tall neural spines and long hemal arches, which indicate that it was important for aquatic propulsion. *Concavispina*, from the Late Triassic (Carnian) of Guizhou (China), has an extraordinarily long tail with more than 114 caudal vertebrae (J. Liu et al. 2013). *Clarazia*, from the Middle Triassic (Anisian-Ladinian) of Switzerland, has a proportionately shorter snout with sharply ventrally deflected premaxillae (B. Peyer 1936; Rieppel 1987a). *Thalattosaurus*, from the Late Triassic (Carnian-Norian) of California, has much less deflected premaxillae (Merriam 1905; Nicholls 1999; Fig. 6.34). It has slit-like upper temporal fenestrae, but *Clarazia* lacks

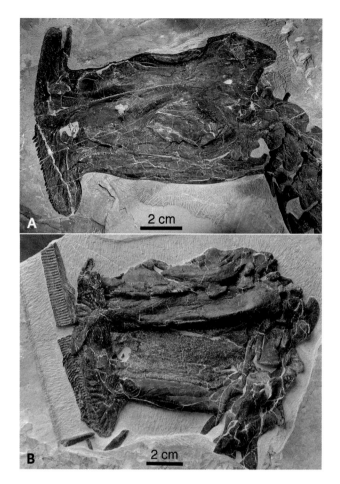

Figure 6.35. Skulls of the enigmatic Triassic marine reptile *Atopodentatus unicus* in **A**, dorsal, and **B**, palatal views. Note the greatly expanded anterior end of the snout. Courtesy of Nick Fraser.

any traces of these openings. *Thalattosaurus* probably attained a total length of about 2 m, whereas *Clarazia* probably reached only half that length. Thalattosaurs apparently preferred nearshore settings (Müller 2005).

Atopodentatus

Atopodentatus, from the Middle Triassic (Anisian) of Yunnan (China) (L. Cheng et al. 2014), is possibly related to Saurosphargidae and Eosauropterygia. It attained a total length of 2.7 m. Its skull is small relative to the heavily ossified postcranial skeleton. It was originally described as having a ventrally deflected snout, but more recent finds demonstrate that *Atopodentatus* has laterally greatly extended processes of the premaxillae, maxillae, and dentaries, resulting in a broadly T-shaped outline of the skull in plan view (C. Li et al. 2016; Fig. 6.35). The premaxillary teeth are peg-like and more robust that the densely spaced needle-like teeth in the maxillae and dentaries. The premaxillae overhang the dentaries and the opposing tooth rows form a comb-like structure. C. Li et al. (2016) considered *Atopodentatus* the oldest marine herbivorous reptile. They argued that it used its premaxillary dentition for scraping algae off surfaces. In their reconstruction, the animal then sucked a watery mix with the scraped-off algae into its mouth and the water could be filtered and subsequently expelled through the needle-like maxillary and dentary teeth.

Lepidosauromorpha

Rhynchocephalians, Squamates, and Their Relatives

Gauthier et al. (1988a) proposed the name Lepidosauromorpha (from Greek *lepis*, scale, *sauros*, lizard, and *morphe*, form) for all saurian reptiles that share a more recent common ancestor with the present-day tuatara (*Sphenodon*) and squamates rather than with crocodylians and birds. In the original definition, Lepidosauromorpha also included "Younginiformes," but Laurin (1991) and others have argued that "Younginiformes" is not part of the crown group Sauria.

Laurin (1991) provided a list of synapomorphies for Lepidosauromorpha: the presence of a prominent lateral conch on the quadrate (which supports the tympanic membrane); prominent retroarticular process formed entirely by the prearticular; interclavicle gracile, with slender lateral processes; humerus with a fully enclosed ectepicondylar foramen; and presence of a process on the ventromedial edge of the fourth distal tarsal fitting under the astragalus medial to the contact between the calcaneum and the fourth distal tarsal.

The early evolutionary history of lepidosauromorph reptiles is still poorly documented. Ezcurra et al. (2014) and Ezcurra (2016) most recently reviewed the known fossil record of late Permian (Lopingian) and Early Triassic saurians. The fossil record of early lepidosauromorph reptiles remains scant because representatives of this clade are typically small-bodied and their delicate bones are unlikely to survive the vagaries of fossilization.

The geologically oldest known undisputed record of lepidosauromorphs is *Paliguana*, based on a partial skull from the Early Triassic of South Africa (Ezcurra et al. 2014; Fig. 7.1). It already has a lateral conch on its (in lateral view bowed) quadrate (Carroll 1975; Ezcurra et al. 2014). Carroll (1975, 1977, 1988) placed *Saurosternoni*, based on a skeleton without the skull from the late Permian (Lopingian) or Early Triassic of South Africa, and *Palaeagama*, known from an almost complete but poorly preserved skeleton from the late Permian (Lopingian) or Early Triassic of South Africa, with *Paliguana* in a single family. However, more recent phylogenetic analyses (Müller 2004; Bickelmann et al. 2009; Jones et al. 2013; Ezcurra et al. 2014) found *Saurosternon* and *Palaeagama* outside Sauria and not closely related to *Paliguana*.

Various early Mesozoic diapsid reptiles have been interpreted as closely related to Lepidosauria. However, published phylogenetic analyses differ regarding the phylogenetic positions of these taxa (S. E. Evans and Borsuk-Białynicka 2009; Renesto and Bernardi 2014).

Kuehneosauridae, with *Icarosaurus* from the Late Triassic (Norian) of New Jersey and *Kuehneosaurus* and *Kuehneosuchus* from the Late Triassic of England, is characterized by

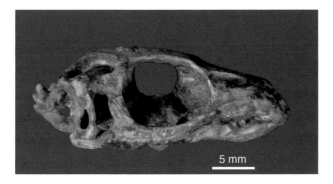

Figure 7.1. Partial skull of the basal lepidosauromorph *Paliguana whitei* in lateral view. Courtesy of Martín Ezcurra.

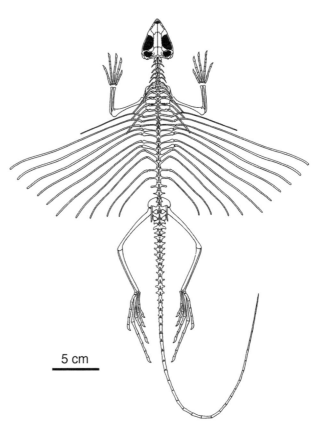

Figure 7.2. Reconstructed skeleton of the gliding ?lepidosauromorph *Kuehneosaurus latus* in dorsal view. Note the greatly elongated dorsal ribs. Drawing courtesy of the late Pamela Robinson.

greatly elongated, almost straight dorsal ribs, which could be folded back against the trunk and presumably supported a gliding membrane in life (P. L. Robinson 1962; Colbert 1970; Fig. 7.2). The dorsal vertebrae have prominent, blade-like transverse processes for the attachment of the ribs. *Icarosaurus* has a snout-vent length of about 10 cm. McGuire and Dudley (2011) estimated its gliding abilities as comparable to those of present-day flying lizards (*Draco*). Stein et al. (2008) reconstructed *Kuehneosuchus* (with its particularly elongated trunk ribs) as a glider (albeit not as good as *Icarosaurus*; McGuire and Dudley 2011) and *Kuehneosaurus* (which has proportionately shorter ribs) as capable of parachuting. There are still no detailed anatomical studies on *Kuehneosaurus* and *Kuehneosuchus*, and their taxonomic status remains uncertain. Stein et al. (2008) even raised the possibility that they represent a single, sexually dimorphic taxon. The cranium of *Kuehneosaurus* is characterized by the presence of medially positioned, confluent external nares, a rather large lacrimal, subthecodont tooth implantation, a ventrally open lower temporal fenestra, the absence of a postparietal and tabular, and the absence of teeth on the transverse flange of the pterygoid (S. E. Evans 2009). P. L. Robinson (1962) originally suggested that the quadrate was mobile, as in squamates, but S. E. Evans (2009) noted that the presence of a firm contact between the quadrate and pterygoid rules out mobility. The pelvic girdle of *Icarosaurus* shares the presence of a large thyroid fenestra with lepidosaurs, but this feature is also present in various archosauromorph reptiles (Chapter 8). The forelimbs appear to be distinctly shorter than the hind

limbs. S. E. Evans (2009) considered *Pamelina*, from the Early Triassic (Olenekian) of Poland, closely related to Kuehneosauridae based on various cranial apomorphies, but the referred dorsal vertebrae have short transverse processes, and there is no evidence for long trunk ribs. Gauthier et al. (1988a) and S. E. Evans (2009) considered kuehneosaurids, originally interpreted as basal lizards (P. L. Robinson 1962; Colbert 1970), stem-lepidosaurs, but the phylogenetic position of this group remains unresolved. A number of skeletal features, including the presence of a ventral process of the opisthotic, suggest archosauromorph rather than lepidosauromorph affinities (Benton 1985; Pritchard and Nesbitt 2017).

S. E. Evans and Borsuk-Białynicka (2009) placed *Sophineta*, from the Early Triassic (Olenekian) of Poland, as the sister taxon to Lepidosauria (see also Simões et al. 2018). The quadrate is immobile and lacks a lateral conch. The skull of *Sophineta* has a rather short snout.

The maxilla has a deep facial process. Tooth implantation is pleurodont. The lower temporal fenestra is open ventrally.

Megachirella, from the Middle Triassic (Anisian) of Italy, has a proportionately large skull (Renesto and Bernardi 2014). Its lower temporal fenestra is open ventrally. The quadrate forms a distinct lateral conch. *Megachirella* has stout forelimbs. The humerus has an entepicondylar foramen. The manual digits bear large, recurved unguals. Renesto and Bernardi (2014) interpreted *Megachirella* as a possible stem-lepidosaur. Based on detailed restudy of the material, Simões et al. (2018) found this taxon as a stem-squamate based on derived features such as a triradiate squamosal, the presence of a distinct alar process on the prootic, the well-developed radial condyle of the humerus, and the apparent fusion of the first distal carpal with metacarpal I.

Marmoretta, from the Middle Jurassic (Bathonian) of England and Scotland and Late Jurassic (Kimmeridgian) of Portugal, has a transversely broad and dorsoventrally slightly flattened cranium, which attained a length of up to 4 cm (S. E. Evans 1991; Waldman and S. E. Evans 1994). Its maxilla has a long anterior (subnarial) process. Tooth implantation is weakly pleurodont. The fused parietals bear a prominent median crest. The quadrate lacks a lateral conch. The slender lower jaw has a distinct coronoid process but lacks a retroarticular process. S. E. Evans (1991) surmised that *Marmoretta* was possibly semiaquatic. A closely related taxon, *Fraxinisaura*, is known from the Middle Triassic (Ladinian) of Germany (Schoch and Sues 2018b). The phylogenetic analysis by Simões et al. (2018) recovered *Marmoretta* as a stem-squamate.

Tamaulipasaurus, from the Middle Jurassic (Bathonian-Callovian) of Mexico, is a possible lepidosauromorph (J. M. Clark and Hernandez 1994). The structure of its well-ossified skull, especially the tiny orbits and extensive fusion of the braincase elements, is consistent with a fossorial mode of life. Tooth implantation is weakly pleurodont. Derived features shared by *Tamaulipasaurus* with squamates include the absence of a descending process on the squamosal, the presence of a well-developed conch on the quadrate, and procoelous cervical vertebrae that have the neural arches fused to the centra (J. M. Clark and Hernandez 1994).

Lepidosauria

Gauthier et al. (1988a) defined Lepidosauria (from Greek *lepis*, scale, and *sauros*, lizard) as comprising the most recent common ancestor of *Sphenodon* and Squamata and all descendants of that ancestor.

Present-day lepidosaurs have a transverse (rather than longitudinal) cloacal slit. The tip of the fleshy tongue is notched. The eye has several derived features including the reduction or absence of the ciliary processes, the attachment of the tendon of the nictitating membrane (which protects the eye) to the orbital wall, and the presence of a cartilaginous disk (tarsus) in the lower eyelid (Gauthier et al. 1988a). Extant lepidosaurs periodically shed their skin in its entirety (ecdysis). Many taxa have a row of modified scales along the middle of the back.

Cranial synapomorphies for Lepidosauria include the long anterior extent of the squamosal; lacrimal small or absent; the absence of a foramen in the stapes; the absence of teeth on the parasphenoid; well-developed dorsum sellae on the parabasisphenoid with paired canals for passage of the abducens nerve; mandibular facet for the craniomandibular joint formed solely by the articular; and superficial attachment of the teeth to the jaw elements (Gauthier et al. 1988a). For detailed surveys of the cranial structure in extant lepidosaurs, readers are referred to the comprehensive surveys by Cundall and Irish (2008), Evans (2008), and Gans and Montero (2008).

Shared derived features in the vertebral column of lepidosaurs include the presence of zygosphene-zygantrum accessory articulations between successive neural arches and the presence of fracture planes in or between the caudal vertebrae for tail shedding (autotomy). Hoffstetter and Gasc (1969) reviewed the postcranial axial skeleton of present-day lepidosaurs in detail.

The various portions forming the sternum in lepidosaurs fuse early during ontogeny. The xiphisternum is small or absent altogether. The humerus has an ectepicondylar foramen. The manus has symmetrical metacarpals. The bones of the pelvic girdle are fused in adults. The ilium has a posterodorsally inclined, narrow iliac blade, and a distinct pubic flange. The anterodorsal surface of the pubis is anteroventrally inclined. The tibial condyle of the femur is larger than the fibular condyle. The latter has a vertical recess for the proximal end of the fibula. In the tarsus, the lateral centrale is fused to the astragalus and the astragalus and calcaneum fuse into a single bone prior to the attainment of the animal's adult size (Gauthier et al. 1988a). The tarsus lacks the first and fifth distal tarsals. Metatarsal V is "hooked," with an angled proximal articular surface. It bears lateral and medial tubercles on its plantar (ventral) surface, on which the gastrocnemius muscle inserts, facilitating flexion of the pes at the ankle (P. L. Robinson 1975; D. Brinkman 1980, 1981). Secondary ossifications or calcified cartilages (epiphyses) form the articular ends and some muscular processes on the limb bones of lepidosaurs. The epiphyses are initially separated from the shafts of the bones by cartilaginous gaps but subsequently fuse to the shafts, terminating growth (Haines 1969). A. P. Russell and Bauer (2008) provided a detailed review of the appendicular skeleton in extant lepidosaurs.

Some structures (e.g., hyoid apparatus) that are commonly developed in cartilage in other diapsid reptiles become calcified in lepidosaurs well before these animals attain maximum body size (Gauthier et al. 1988a). Finally, most lepidosaurian bones are made up of dense, lamellar, and avascular bone of periosteal origin (Gauthier et al. 1988a).

Lepidosauria: Rhynchocephalia

The tuatara (*Sphenodon punctatus*) of New Zealand is the sole present-day representative of a lepidosaurian clade that was diverse and widespread during the Mesozoic Era. Günther (1867) first noted that *Sphenodon* differs from lizards in many features. He placed it in its own group, Rhynchocephalia (from Greek *rhynchos*, beak, snout, and *kephale*, head), named based on the chisel-like teeth in the premaxillae, which become fused to these bones and form a "beak"-like structure. The cranium of *Sphenodon* differs from those of squamates in the presence of a complete lower temporal bar and a quadratojugal. The quadrate is immobile and lacks a lateral conch. In its postcranial skeleton, *Sphenodon* differs from squamates in the presence of amphicoelous vertebrae, gastralia, and uncinate processes on the trunk ribs. Based on these and other features, the tuatara became a textbook example of a "living fossil." In recent decades, however, numerous discoveries of early Mesozoic rhynchocephalians have shown that various putatively plesiomorphic features of

Sphenodon, particularly the complete lower temporal bar and the absence of a tympanic membrane, represent evolutionary reversals (Whiteside 1986; Jones 2008). Moazen et al. (2009) argued that a complete lower temporal bar possibly provides support for the quadrate when it is subjected to forces generated during forceful biting.

Rhynchocephalia comprises the most recent common ancestor of *Sphenodon punctatus* and *Gephyrosaurus bridensis* and all descendants of that ancestor (Gauthier et al. 1988c). Gauthier et al. (2012) listed a suite of diagnostic synapomorphies for this clade: palatine with a lateral row of enlarged teeth, which typically extends more or less parallel to the maxillary tooth row and is separated from the latter by a deep groove; postfrontal overlapping the parietal table dorsally; dentary with a prominent posterior extension that terminates well behind the coronoid eminence on the lateral surface of the lower jaw; and the absence of a splenial.

Gephyrosaurus, from the Late Triassic of England (Whiteside and Duffin 2017) and the Early Jurassic (Hettangian-Sinemurian) of Wales (S. E. Evans 1980, 1981), is the sister taxon to all other rhynchocephalians, which are united as Sphenodontia (Gauthier et al. 1988a). All teeth have pleurodont implantation. *Gephyrosaurus* retains a small lacrimal. The lower temporal bar is incomplete. The quadrate has a distinct lateral conch. Whiteside and Duffin (2017) and Whiteside et al. (2017) described several related taxa from the Late Triassic (Rhaetian) of England and Switzerland, respectively.

LEPIDOSAURIA: RHYNCHOCEPHALIA: SPHENODONTIA

Sphenodontia (from Greek *sphen*, wedge, and *odous* (*odon*), tooth) is characterized by several synapomorphies (Gauthier et al. 2012) such as the absence of a lacrimal; postfrontal broadly overlapping the parietal dorsally; greatly enlarged teeth at the anterior end of the palatine tooth row; lower jaw with a distinct coronoid process and a short retroarticular process; and the absence of a tubercle on the ilium. Sphenodontians lack tooth replacement and instead add new teeth at the back of the tooth rows.

Diphydontosaurus, from the Late Triassic of England, resembles *Gephyrosaurus* in many features but its jaws have pleurodont teeth anteriorly and acrodont teeth more posteriorly (Whiteside 1986; Whiteside and Duffin 2017). The oldest known rhynchocephalian fossils, from the Middle Triassic (Ladinian) of Germany, resemble *Diphydontosaurus* (Jones et al. 2013). *Planocephalosaurus*, also from the Late Triassic of England, has a fully acrodont dentition (Fraser 1982; Fig. 7.3). It attained a total length of about 15 cm. Unlike in *Sphenodon*, the premaxillae of these basal rhynchocephalians bear several small conical teeth. The dentary holds up to 40 teeth. Jaw motion was exclusively vertical (orthal). The lower temporal bar is incomplete in *Diphydontosaurus* and *Planocephalosaurus*.

Clevosauridae, with *Brachyrhinodon*, from the Late Triassic (?Carnian) of Scotland, and *Clevosaurus*, from the Late Triassic of Brazil, Nova Scotia, and England and the Early Jurassic of Nova Scotia (Canada), South Africa, Wales, and Yunnan (China), is characterized by a small number of fully acrodont teeth, which have conical cusps with distinct flanges (extending posteriorly on the maxillary teeth and anteriorly on the dentary teeth) (P. L. Robinson 1973; Fraser 1988; Fig. 7.4). Adults have only a single chisel-like tooth in each premaxilla. The anterior (premaxillary) process of the maxilla is short or absent. The dorsal process of the jugal meets the squamosal approximately at the midpoint of the upper temporal bar. The lower temporal bar is complete but slender. *Clevosaurus* reached a total length of up to 25 cm.

Pleurosauridae was the first group of lepidosaurs that adapted to an aquatic mode of life in nearshore marine settings. It includes *Palaeopleurosaurus*, from the Early Jurassic (Toarcian) of Germany, and *Pleurosaurus*, from the

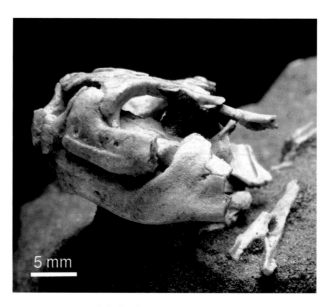

Figure 7.4. Partial skull of the clevosaurid *Clevosaurus bairdi* in oblique anterolateral view.

Late Jurassic (Kimmeridgian-Tithonian) of France and Germany (Cocude-Michel 1963; Carroll 1985; Carroll and Wild 1994; Fig. 7.5). The tapered snout of pleurosaurs is proportionally much longer than in other sphenodontians. Each premaxilla in adult specimens has a single chisel-like tooth. The maxillary teeth bear anterior flanges. The external nares are positioned well back on the snout. The lower temporal bar is incomplete (*Palaeopleurosaurus*) or absent (*Pleurosaurus*). The long and slender trunk region of *Pleurosaurus* comprises up to 57 vertebrae. The greatly elongated tail (Fig. 7.5A) has up to 120 caudal vertebrae. These caudals lack the fracture planes that would facilitate shedding of the tail (autotomy) and are common in terrestrial lepidosaurs. The limbs are proportionately short. Pleurosaurs presumably employed lateral undulation of the long body and tail for swimming. *Pleurosaurus* attained a total length of up to 1.5 m. *Palaeopleurosaurus* is structurally intermediate between land-dwelling sphenodontians and *Pleurosaurus*, with a less elongated trunk with 37 vertebrae and with proportionately short but still well-developed limbs (Carroll 1985). Bever and Norell (2017) interpreted *Vadasaurus*, from the Late Jurassic (Tithonian) of Germany, as closely related to Pleurosauridae. It already has a somewhat elongated tail (with more than 40 vertebrae) and reduced ossification in the limbs.

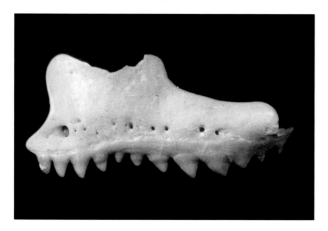

Figure 7.3. Left maxilla of the sphenodontian *Planocephalosaurus robinsonae* in lateral view. Length 8.2 mm. Courtesy of Nick Fraser.

Sapheosaurus, from the Late Jurassic (Kimmeridgian) of France (Fig. 7.6), and *Kallimodon*, from the Late Jurassic (Kimmeridgian-Tithonian) of France and Germany (Fig. 7.7), are semi- or fully aquatic (Cocude-Michel 1963;

Figure 7.5. A, skeleton of the pleurosaurid *Pleurosaurus goldfussi* in dorsal view. Note the greatly elongated dorsal and caudal regions and proportionately small limbs. Length 73 cm. **B**, skull of *Pleurosaurus goldfussi* in lateral view. Length c. 8.6 cm. B, modified from Carroll and Wild (1994).

Fabre 1981) and closely related to Pleurosauridae. Their bodies are less obviously modified for this mode of life than those of pleurosaurs. The cranium has a long postorbital region with long, narrow upper temporal fenestrae and a broad upper temporal bar (Reynoso 2000). Another aquatic sphenodontian, *Ankylosphenodon*, from the Early Cretaceous (Albian) of Mexico (Reynoso 2000), has distinctly pachyostotic vertebrae and ribs.

Homoeosaurus, from the Late Jurassic (Tithonian) of Germany and the Late Jurassic to Early Cretaceous (Kimmeridgian-Berriasian) of France (Cocude-Michel 1963; Fabre 1981), closely resembles *Sapheosaurus* and its relatives in the presence of posterior flanges on the maxillary teeth and the presence of long tubera on the ischia (Gauthier et al. 1988a). It attained a total length of up to 20 cm.

Sphenodontinae includes the extant *Sphenodon* (Figs. 1.1B, 7.8, 14.3). The lower temporal bar is complete and robust, whereas the upper temporal bar is slender (Fig. 7.8A), unlike the condition in more basal sphenodontians (Jones 2008). The quadrate is not emarginated posteriorly and lacks a lateral conch. The oldest sphenodontines are from the Early Jurassic (Pliensbachian) of Mexico (Reynoso 1996, 2005; Reynoso and Clark 1998). *Oenosaurus*, from the Late Jurassic (Tithonian) of Germany, is noteworthy for its highly modified dentition (Rauhut et al. 2012a). Its jaws support large tooth plates that formed through fusion of many closely packed tiny dentine tubes (Fig. 7.9) and were probably used for crushing food items.

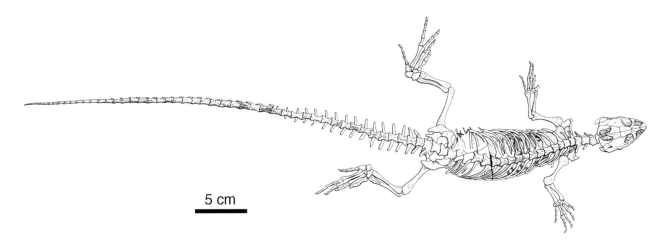

5 cm

Figure 7.6. Skeleton of the sphenodontian *Sapheosaurus thiollierei* in dorsal view. Redrawn by Fabre (1981) from Lortet (1892).

Figure 7.7. Skeleton of the sphenodontian *Kallimodon pulchellus* in ventral view. Courtesy of Oliver Rauhut.

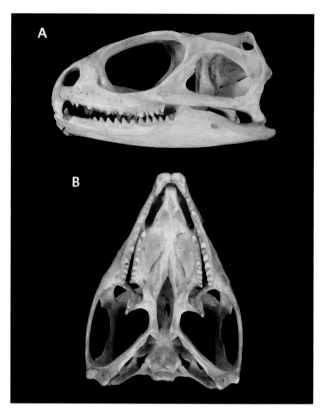

Figure 7.8. Skull of the tuatara (*Sphenodon punctatus*) in **A**, lateral, and **B**, palatal views. Courtesy of Chris Bell.

Figure 7.9. Partial cranium of the sphenodontid *Oenosaurus muehlheimensis* in palatal view. Note the massive maxillary tooth plates. Courtesy of Oliver Rauhut.

The maxillary teeth of most sphenodontines have conical crowns with small posterior flanges. The dentary teeth are pyramid-shaped with crests extending from the apex anterolaterally and anteromedially. As in *Kallimodon* and *Sapheosaurus*, *Sphenodon* and its close relatives employ proal jaw motion, in which the dentary teeth bite between those on the maxillae and palatines and then the mandible moves forward to facilitate crushing and shearing of food (Gorniak et al. 1982). Today *Sphenodon* survives only on several small islands off the coast of New Zealand, but attempts have been made to reintroduce it to other parts of its original range (Cree 2014). It feeds on a wide range of prey, ranging from insects to juvenile seabirds (which are consumed only by adults). The tuatara's slow metabolism as well as its long lifespan and reproductive cycle may be adaptations related to the cool climate in which it lives (Gans 1983). *Sphenodon* is represented only by a single species, *S. punctatus*. Some authors have recognized a second species, *S. guentheri*, but Hay et al. (2010) argued that the genetic differences do not justify this specific separation. Male tuatara can reach a total length of 65 cm, whereas females are smaller. Isolated jaw elements from the Neogene (Miocene) of New Zealand appear indistinguishable from those of extant *Sphenodon* (Jones et al. 2009). Apesteguía and Jones (2012) reported a fragment of a maxilla with teeth closely resembling those of *Sphenodon* from the Late Cretaceous (Campanian-Maastrichtian) of Argentina.

Opisthodontia (Apesteguía and Novas 2003) differs from other sphenodontines in the presence of robust jaws and thickly enameled teeth. The dentary teeth have transversely broad, bulbous crowns, which have anteromedially extending flanges and frequently show heavy wear. The upper temporal bar is deep, including a much expanded jugal, and there is no lower temporal bar. Based on the structure of their teeth, at least derived opisthodontians were probably herbivorous. *Opisthias*, from the Late Jurassic (Kimmeridgian-Tithonian) of the western United States, has pyramidal but only slightly broadened lower teeth (Fig. 7.10). Opisthodontians with transversely broad teeth include *Toxolophosaurus*, from the Early Cretaceous (Aptian) of Montana (Throckmorton et al. 1981); *Kaikaifilusaurus* (including *Priosphenodon*), from the Late Cretaceous (Cenomanian-Turonian) of Argentina (Apesteguía and Novas 2003); and *Sphenotitan*, from the Late Triassic (Norian) of Argentina (Martínez et al. 2013). *Kaikaifilusaurus* attained a skull length of 15 cm and an estimated total length of about 1 m.

Lepidosauria: Squamata

Surpassed in the number of species only by birds among present-day tetrapods, Squamata (from Latin *squama*, scale) is the second most diverse group, with 10,418 formally named species (as of July 2018; http://www.reptile-database.org). Today squamates are found on all continents except Antarctica and in most habitats except the polar regions and high-altitude mountain settings.

Squamata were traditionally divided into "lizards" ("Lacertilia" or "Sauria") and snakes (Serpentes or Ophidia) (e.g., Romer 1956). Some authors recognized worm lizards (Amphisbaenia) as a third major group (Gans 1978). However, phylogenetic analyses have consistently shown that Serpentes and Amphisbaenia are merely highly derived clades of "lizards" (Gauthier et al. 1988a, 2012; Townsend et al. 2004; Vidal and Hedges 2005; Conrad 2008).

The interrelationships of squamates, even at higher levels, remain contentious (Losos et al. 2012). In recent years, two major phylogenetic analyses based on morphological and paleontological data (Conrad 2008; Gauthier et al. 2012) supported an earlier study by Estes et al. (1988) that placed Iguania as the basal group of squamates and united the remaining clades (including Serpentes) as Scleroglossa (from Greek *scleros*, hard, rough, and Latin *glossa*, tongue) (Fig. 7.11A). Using molecular data, Lee et al. (2004) also found these two major clades. Among Scleroglossa, Gauthier et al. (2012) placed Gekkota as the sister taxon to a clade Autarchoglossa (from Greek *autarkes*, free, and Latin *glossa*, tongue), which comprises the most recent common ancestor of Anguimorpha and Scincomorpha and all descendants of that ancestor. There are major differences in tongue structure and

5 mm

mode of prey capture between Iguania and Scleroglossa (Schwenk 2000; Losos et al. 2012). Iguanians employ their tongues to capture prey (as does *Sphenodon*), whereas all other squamates typically use their jaws for this purpose.

By contrast, phylogenetic analyses based on DNA sequences and a range of genes (Vidal and Hedges 2005; Wiens et al. 2012; Pyron et al. 2013) have consistently recovered a different pattern of squamate interrelationships. Gekkota is the basal clade of Squamata, followed (from base to crown) by Scincoidea, Lacertoidea, Amphisbaenia, Serpentes, Anguimorpha, and Iguania (Fig. 7.11B). The difference in the placement of Iguania between this hypothesis and the one based on morphology and paleontology is particularly striking. Vidal and Hedges (2005) proposed a new clade Toxicofera (from Latin *toxicum*, poison, and *fero*, bear, carry) for the reception of Anguimorpha, Iguania, and Serpentes. Key support is provided by a set of genes that are involved in toxin production. Under this hypothesis, venom use

evolved once in the common ancestor of Toxicofera rather than, as traditionally assumed, independently in anguimorph lizards and derived snakes. Reeder et al. (2015) listed several morphological features as synapomorphies for Toxicofera, but none of these is unique to that clade. Gauthier et al. (2012) noted that a remarkably high degree of homoplasy in morphological features would be required to accommodate the Toxicofera hypothesis. Furthermore, Hargreaves et al. (2014) argued that the majority of genes used to support Toxicofera are expressed in various body tissues rather than just in the venom systems and probably represent general maintenance genes. These authors also hypothesized multiple origins of venom use among squamates.

Squamata comprises the most recent common ancestor of Iguania and Scleroglossa and all descendants of that ancestor (Estes et al. 1988; Figs. 7.12, 7.13). Reynoso (1998), Gauthier et al. (2012), and Simões et al. (2018) found *Huehuecuetzpalli*, from the Early Cretaceous (Albian) of Mexico, as a stem-squamate. It shares numerous derived features with Squamata including the absence of a posterior process of the jugal; proximal head of the quadrate pivoting on the tapering tip of the squamosal; quadratojugal absent; overlap between the quadrate and pterygoid narrow or absent; epipterygoid rod-like; angular not extending posteriorly to the jaw joint; coronoid eminence of the lower jaw formed exclusively by the coronoid bone; scapulocoracoid emarginated between the scapula and coracoid; coracoid with an anterior emargination; symphyseal portion of the pubis tapered, not expanded distally; and joint between the fibula and astragalus involving most of the distal end of the fibula (Gauthier et al. 2012). *Huehuecuetzpalli* probably attained a total length of up to 30 cm. Some of its features, such as the posterior position of the external nares and the elongate premaxilla, are autapomorphies for this taxon.

Huehuecuetzpalli and Squamata share the absence of a lower temporal bar, along with the absence of the quadratojugal and of the ventral process of the squamosal, and the presence of a mobile quadrate. The quadrate is connected to the paroccipital process of the braincase by a cartilaginous element (intercalar) and to the squamosal and/or supratemporal in squamates. Its proximal head articulates with the squamosal (P. L. Robinson 1967). The structure of these contacts facilitates fore-and-aft motion and rotation of the quadrate (streptostyly).

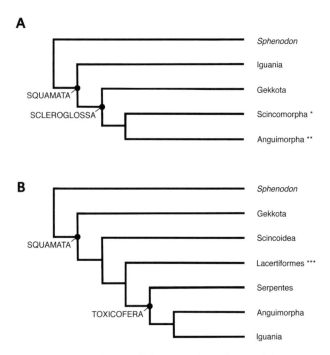

Figure 7.11. Conflicting phylogenetic hypotheses of the interrelationships of the major clades of Squamata. **A**, hypothesis based on morphology, and **B**, hypothesis based on molecules. Asterisk in **A** denotes inclusion of Amphisbaenia in Scincomorpha, and double asterisk denotes inclusion of Serpentes in Anguimorpha; triple asterisk in **B** denotes inclusion of Amphisbaenia in Lacertiformes. Dots denote node-based clades.

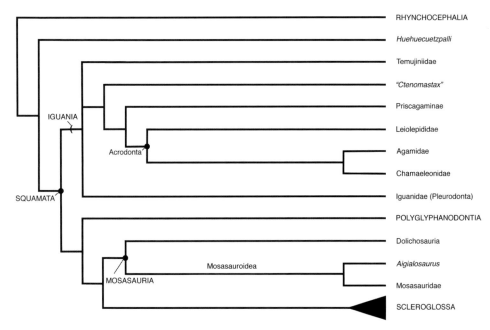

RHYNCHOCEPHALIA

Huehuecuetzpalli

Temujiniidae

"Ctenomastax"

Priscagaminae

Leiolepididae

Agamidae

Chamaeleonidae

Iguanidae (Pleurodonta)

POLYGLYPHANODONTIA

Dolichosauria

Aigialosaurus

Mosasauridae

SCLEROGLOSSA

Figure 7.12. Phylogenetic hypothesis of the interrelationships of Squamata based on Gauthier et al. (2012). Rhynchocephalia and *Huehuecuetzpalli* served as outgroup taxa. Dots denote node-based clades and parentheses denote stem-based clades.

The biomechanical significance of streptostyly has been discussed at length (K. K. Smith 1980, 1982; Metzger 2002; S. E. Evans 2008). Various functions have been proposed, ranging from increasing the mechanical advantage of certain adductor jaw muscles (especially the pterygoideus; K. K. Smith 1980) to increasing gape and facilitating the transport into and/or processing of food in the mouth. P. L. Robinson (1973) suggested that streptostyly facilitates accurate tooth occlusion in agamid lizards. As Metzger (2002) noted, no single explanation seems to account adequately for the presence of streptostyly in all groups of squamates. Some functional inferences, such as increased mechanical advantage for the pterygoideus muscle, now appear unlikely in light of more recent experimental research. Streptostyly may well prove to be a feature that has different functions in different clades of squamates.

In addition to streptostyly, squamate crania also have other potentially mobile intracranial joints (Versluys 1912; Frazzetta 1962, 1983; S. E. Evans 2008). Many squamates have a transverse hinge joint between the frontal and parietal bones, along which the snout can be raised and lowered (mesokinesis). This motion is linked to dorsoventral flexion between the more posterior bones of the palate. Mobility can also be present between the braincase and adjoining portions of the skull roof and palate (metakinesis).

Mobility at intracranial joints has been experimentally documented in some taxa of lizards (e.g., *Varanus*, K. K.

Smith 1982; *Gekko*, Herrel et al. 1999) but is absent in others (e.g., *Iguana*), even though the structure of the joints in the latter indicate potential for such mobility. There even exists variation among closely related taxa; for example, among agamid lizards, *Uromastyx* shows evidence of intracranial mobility, whereas *Agama* does not (Metzger 2002). Furthermore, the potential for cranial mobility can vary with age, with adults losing kinesis in some squamates (e.g., *Tupinambis*; Versluys 1912).

Mesokinesis appears to be important in the capture and manipulation of prey. For example, it can alter the angle between the upper and lower jaws and thus facilitate proper alignment of the tooth crowns during biting as well as subsequent disengagement of the teeth from the prey (Metzger 2002). The role of metakinesis is less obvious and difficult to examine experimentally. It possibly relates to such factors as providing room for growth between the basicranium and neighboring cranial bones. S. E. Evans (2008) surmised that metakinesis is linked to other forms of intracranial kinesis.

Frazzetta (1962) proposed a simple four-joint (quadric-crank) linkage model to describe intracranial mobility in lizards. He divided the cranium into maxillary and occipital segments, each of which can be further partitioned into smaller units. Anterodorsal excursion of the quadrate moves the bones of the palate forward, raising the snout at the mesokinetic and hypokinetic joints. Furthermore, combined with mesokinesis, rotation about the

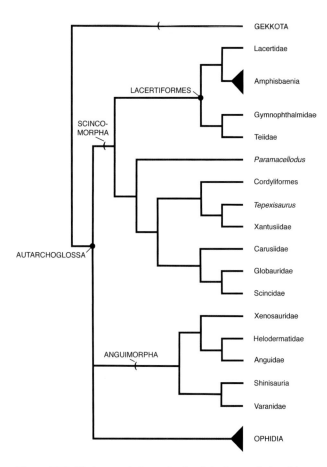

Figure 7.13. Phylogenetic hypothesis of the interrelationships of Autarchoglossa based mainly on Gauthier et al. (2012) but with topology of Anguimorpha modified. Gekkota serves as the outgroup taxon. Dots denote node-based clades and parentheses denote stem-based clades.

metakinetic axis depresses the parietal region relative to the remainder of the cranium. Posteroventral rotation of the quadrate pushes the snout down through ventro-flexion at the hypokinetic and mesokinetic joints when the base of the cranium shifts posteriorly. This motion may be associated with lateral excursion of the quadrate, because the basipterygoid processes of the parabasisphe-noid would force the pterygoid bones laterally. Metzger (2002) cautioned that Frazzetta's model is applicable only in those squamates that have mesokinesis, metakinesis, and streptostyly. Much more additional experimental re-search is needed to assess actual mobility at the various intracranial joints and its functional significance in present-day squamates.

Gauthier et al. (2012) listed 13 unambiguous synapo-morphies for Squamata but noted that many additional derived features (including soft-tissue ones) distinguish

crown-group squamates from the most recent common ancestor of crown-group lepidosaurs (e.g., Estes et al. 1988). Uniquely derived character states for Squamata are the long supratemporal process of the parietal that extends almost to the proximal head of the quadrate posterolaterally (Fig. 7.15) and the absence of a second distal tarsal. Other squamate synapomorphies absent in *Huehuecuetzpalli* include the fusion of the premaxillae; procoelous vertebrae; ulna with an enlarged distal epiph-ysis that fits into a depression on the ulnare; presence of a styloid process on the distal end of the radius; and fusion of the astragalus and calcaneum into a single element (Gauthier et al. 2012).

SQUAMATA: IGUANIA

Gauthier et al. (2012) listed a suite of diagnostic synapo-morphies for crown-group Iguania (from Taino *iwana*, iguana). They considered the presence of a loose sutural contact between the postorbital and squamosal (Fig. 7.16) unique to this clade. Another diagnostic feature is the an-teromedial process of the coronoid bone fitting into a groove below the posterior end of the dentary tooth row and wrapping around the ventral margin of the tooth-bearing border of the dentary. Additional synapomor-phies include the presence of a prefrontal boss (which is often not evident in small specimens) and the position of the septa for caudal autotomy behind the transverse processes of the caudal vertebrae. K. T. Smith (2009a) presented an analysis of iguanian interrelationships that differs in detail from that by Gauthier et al. (2012). He proposed a different set of synapomorphies for Iguania, including the presence of six or seven premaxillary teeth, the reduced contribution of the postfrontal to the orbital margin, and the pineal foramen becoming partially bor-dered by the fused frontals.

Present-day iguanians comprise two clades, the largely New World Iguanidae (Pleurodonta) and the Old World Acrodonta (Schulte et al. 2003). The early diversification of Iguania is still the subject of much discussion (see below), and this overview mainly follows Gauthier et al. (2012).

SQUAMATA: IGUANIA: IGUANIDAE

Saichangurvel (Conrad and Norell 2007; Fig. 7.14) and *Temujinia* (Gao and Norell 2000), both from the Late Cretaceous (Campanian) of Mongolia, are probably stem-iguanids (Gauthier et al. 2012). The latter authors recognized a clade Temujiniidae, which is characterized especially by the extension of the squamosal to the level of the epipterygoid and the absence of a dorsal process of the squamosal. DeMar et al. (2017) referred *Magnuviator*, from the Late Cretaceous (Campanian) of Montana, to Temujiniidae. *Magnuviator* attained an estimated snout–vent length of more than 20 cm.

Another clade closely related to Iguanidae is Isodontosauridae, known only from the Late Cretaceous (Campanian) of China and Mongolia (Alifanov 2000). DeMar et al. (2017) found it as the sister group to Iguanidae (Pleurodonta). Diagnostic synapomorphies include the retention of a frontoparietal fontanelle in adults, the position of the infraorbital canal entirely within the palatine, and the extent of the adductor fossa formed by the surangular on the lateral surface of the lower jaw (Gauthier et al. 2012).

There is still little agreement concerning the definitions and diagnoses of the constituent clades of Iguanidae/Iguania (Frost and Etheridge 1989; K. T. Smith 2009a; Daza et al. 2012; Gauthier et al. 2012; DeMar et al. 2017). Frost and Etheridge (1989; see also Frost et al. 2001) divided Iguania into a series of family-level groupings, but this classification has not been generally adopted.

The present-day geographic distribution of Iguaninae ranges from the American Southwest to the Antilles and to southern Brazil and Paraguay. In addition, *Brachylophus* (Fijian iguanas) occurs on various Pacific islands including Fiji and Tonga, and *Amblyrhynchus* (marine iguana) on the Galápagos Islands. *Armandisaurus*, from the Neogene (Miocene) of New Mexico (Norell and de Queiroz 1991), is closely related to the extant *Dipsosaurus* (desert iguana). K. T. Smith (2011b) also considered *Queironius*, from the Paleogene (Eocene) of North Dakota, a close relative of *Dipsosaurus*.

Present-day Crotaphytinae comprises *Crotaphytus* (collared lizards; Fig. 7.17A) and *Gambelia* (leopard lizards). It occurs from Oregon to the Mississippi and to northern Mexico. Crotaphytines feed on arthropods and small vertebrates and reach snout–vent lengths from 10 to 14 cm. *Gambelia* is first known from the Neogene-Quaternary (Pliocene-Pleistocene) of California (Norell 1989).

Phrynosomatinae is a diverse, mainly North American clade comprising *Phrynosoma* (horned lizards), *Uma* (fringe-toed lizards), *Sceloporus* (spiny lizards), and various related taxa. The cranium of *Phrynosoma* (Fig. 7.17B) is distinguished by the presence of often prominent "horns" on the parietal and squamosal. *Phrynosoma* subsists on insects and lives in arid to semiarid settings. The earliest record of *Phrynosoma* is from the Neogene (Pliocene) of California (Norell 1989). *Uma* feeds on arthropods as well as on plant material. It has lamellae on its digits for moving across and through sand. *Sceloporus*

Figure 7.14. Skeleton of the stem-iguanid *Saichangurvel davidsoni* in dorsal view. Length of skull 2.5 cm. Courtesy of Mick Ellison.

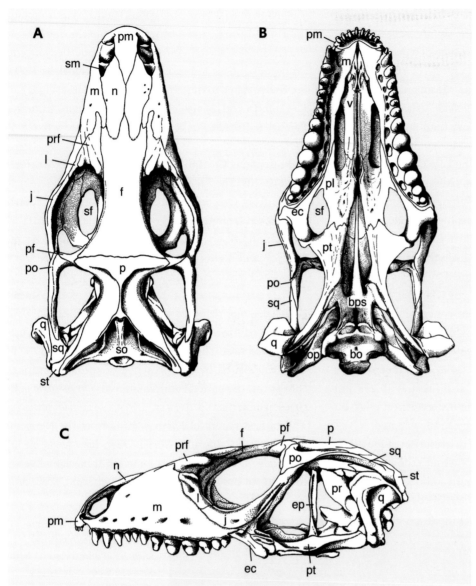

Figure 7.15. Cranium of a tegu (*Tupinambis teguixin*) in **A**, dorsal; **B**, ventral; and **C**, lateral views. Abbreviations: bo, basioccipital; bps, basiparasphenoid; d, dentary; ec, ectopterygoid; ep, epipterygoid; f, frontal; j, jugal; l, lacrimal; m, maxilla; n, nasal; op, opisthotic; p, parietal; pf, postfrontal; pl, palatine; pm, premaxilla; po, postorbital; pr, prootic; prf, prefrontal; pt, pterygoid; q, quadrate; sf, suborbital fenestra; sm, septomaxilla; so, supraoccipital; st, supratemporal; v, vomer. Modified from Barberena et al. (1970).

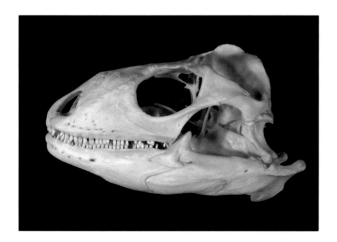

Figure 7.16. Skull of a green iguana (*Iguana iguana*) in lateral view.

occupies a wide range of habitats from the southern regions of the United States to Central America.

Oplurinae (Madagascar swifts) comprises *Chalarodon* and *Oplurus* from Madagascar and, in the case of the latter, also from the Comoros Islands. They have prominently keeled scales and, in some forms, spiny tails.

Liolaeminae have a present-day distribution ranging from Peru to Argentina, Brazil, and Uruguay. The earliest record of *Liolaemus* is from the Neogene (Miocene) of Argentina (Daza et al. 2012).

Extant Leiocephalinae is restricted to the Bahamas, Cuba, and Hispaniola. Various fossils from the Neogene of Florida, Nebraska, and Wyoming have been assigned

Figure 7.17. Iguanidae. **A**, collared lizard (*Crotaphytus collaris*; Crotaphytinae); **B**, roundtail horned lizard (*Phrynosoma modestum*; Phrynosomatinae). A, courtesy of Division of Amphibians and Reptiles, National Museum of Natural History; B, courtesy of Laurie Vitt.

Figure 7.18. Iguanidae. **A**, green anole (*Anolis carolinensis*; Polychrotinae); **B**, broad-headed woodlizard (*Enyaloides laticeps*; Hoplocercinae). A, courtesy of Division of Amphibians and Reptiles, National Museum of Natural History; B, courtesy of Laurie Vitt.

to *Leiocephalus* (Estes 1983), but Norell (1989) accepted only a Miocene record from Florida as correctly referred to this taxon.

Tropidurinae is a morphologically and ecologically diverse clade of iguanian lizards distributed throughout much of South America into Argentina and occurring also on the Galápagos Islands. Most species subsist on insects but some also frequently feed on plants. Daza et al. (2012) found *Uquiasaurus*, from the Neogene (Pliocene) of Argentina, as the sister taxon to a clade comprising Leiocephalinae, Liolaeminae, and Tropidurinae (Frost and Etheridge 1989; K. T. Smith 2009b).

Traditionally, Polychrotinae included *Anolis*, which is represented by more than 250 described extant species ranging from the southern United States to Argentina and more than 150 species from the Caribbean (Fig. 7.18A). However, recent molecular studies have separated *Polychrus* from *Anolis* and placed the latter in a separate clade, Dactyloidae (e.g., Pyron et al. 2013). The two groupings also differ in various cranial features (e.g., K. T. Smith

2009b). Because of its abundance and spectacular diversity, *Anolis* has become the subject of a vast body of studies on lizard behavior, ecology, and evolutionary biology. Losos (2011) provided a comprehensive overview of this work. Neogene (Miocene) specimens of *Anolis* are preserved in amber from the Dominican Republic (Sherratt et al. 2015) and Mexico (Lazzell 1965). The former records document the presence of already at least four different ecomorphs. Present-day anoles are typically sexually dimorphic and range in snout-vent length from 3 cm to nearly 20 cm. They feed on insects and other arthropods. *Afairiguana* (Conrad et al. 2007) and *Anolbanolis* (K. T. Smith 2009b), both from the Paleogene (Eocene) of Wyoming, and a still unnamed form from the Paleogene (Eocene) of Germany (K. T. Smith et al. 2018) are among the oldest known relatives of Polychrotinae.

Leiosaurinae is a small clade of iguanid lizards from Argentina, Brazil, and Chile. *Enyalius* comprises mostly forest-dwelling forms, whereas other taxa such as *Pristidactylus* are mainly terrestrial. *Pristidactylus* is first known from the Neogene (Miocene) of Argentina (Albino and Brizuela 2014).

Hoplocercinae (Fig. 7.18B) is represented by three extant genera and has a geographic range from Panama to Brazil. Hoplocercines subsist on insects and reach snout-vent lengths from 9 to 15 cm. *Cypressaurus*, from the Paleogene (Eocene) of North Dakota and Saskatchewan, is a possible stem-hoplocercine (K. T. Smith 2011b).

Corytophaninae (casquehead lizards) today ranges from Mexico to Ecuador and Venezuela. It is especially distinguished by the presence of a prominent, blade-like median crest on the parietal region of the skull roof (only present in males of *Basiliscus*). Corytophanines are largely arboreal. Extant forms range in snout-vent length from 9 to 20 cm. *Basiliscus* (basilisks) has long hind limbs and toes that allow it to run bipedally and across water. Conrad (2015) found *Geiseltaliellus*, from the Paleogene (Eocene) of France and Germany (K. T. Smith 2009a; Fig. 7.19A), and *Babibasiliscus*, from the Paleogene (Eocene) of Wyoming, as closely related to the extant *Basiliscus* (Fig. 7.19B). Attaining a snout-vent length of up to 9 cm, *Geiseltaliellus* was arboreal. Although it has a distinct median crest, *Geiseltaliellus* lacks the prominent posteriorly projecting extension. Caudal autotomy occurred between rather than in the caudal vertebrae (K. T. Smith and Wuttke 2012; K. T. Smith et al. 2018).

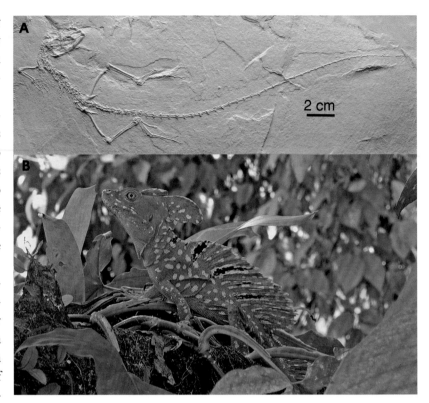

Figure 7.19. A, skeleton of the stem-basiliscine *Geiseltaliellus maarius* in dorsal view (surface coated for photography); **B**, plumed basilisk (*Basiliscus plumifrons*). A, courtesy of Krister Smith; B, courtesy of Laurie Vitt.

SQUAMATA: IGUANIA: ACRODONTA

Gauthier et al. (2012) listed among the diagnostic features for crown-group Acrodonta (from Greek *akron*, peak, top, and *odous* (*odon*), tooth) the presence of a "wing-shaped" ventral extension on the quadrate flange of the pterygoid and the acrodont marginal dentition. These authors cautioned against simply attributing isolated jaw elements with acrodont teeth to Acrodonta and noted that the presence of enlarged anterior teeth on the maxilla may prove a more reliable criterion for such referrals. Isolated maxillae with both features are first known from the Paleogene (Eocene) of Europe and North America (K. T. Smith 2009a).

K. T. Smith (2009a), Gauthier et al. (2012), DeMar et al. (2017), and Simões et al. (2018) hypothesized Priscagaminae, from the Late Cretaceous of Mongolia, as the sister taxon to crown-group Acrodonta. DeMar et al. (2017) added "*Ctenomastax*" (preoccupied name), from the Late Cretaceous (Campanian) of Mongolia (Gao and Norell 2000), as a sister taxon to this grouping. "*Ctenomastax*" and Priscagaminae share several unambiguous apomorphies, including the presence of four to six premaxillary teeth, caniniform anterior maxillary teeth that are distinctly larger than the neighboring teeth, and V-shaped wear facets produced by the maxillary teeth on the lateral surface of the dentary between the dentary teeth.

Priscagaminae is diagnosed by various derived features such as the position of the acute ventral margin of the jugal in a longitudinal groove on the dorsal surface of the supradental shelf of the maxilla and the rugose sculpturing on the bony bar formed by the jugal and postorbital (Gauthier et al. 2012). It includes *Mimeosaurus* and *Priscagama*, both from the Late Cretaceous (Campanian) of Mongolia (Borsuk-Białynicka and Moody 1984).

Among present-day acrodontans, K. T. Smith and Gauthier (2013) recognized a clade Leiolepididae that includes *Leiolepis* (butterfly lizards) and *Uromastyx* (spiny-tailed lizards). Unambiguous synapomorphies for Leiolepididae include the presence of a lateral flange on the septomaxilla, the jugal ramus of the postorbital not extending to the level of the proximal head of the quadrate, and the jugal broadly overla[ping the suborbital portion of the posterior process of the maxilla (Gauthier et al. 2012). *Leiolepis* ranges throughout Southeast Asia. *Uromastyx* occurs from North Africa and the Middle East to Iran and can attain a total length of up to about 90 cm. Leiolepididae has been placed at the base of crown-group Acrodonta (Gauthier et al. 2012) or as the sister taxon to Agamidae (Daza et al. 2012; Simões et al. 2015). Head et al. (2013) considered a large-sized acrodontan, *Barbaturex*, from the Paleogene (Eocene) of Myanmar, closely related to *Uromastyx*. *Barbaturex* attained an estimated snout–vent length of about 1 m. It has acrodont, triangular posterior teeth with continuous wear facets and prominent, anteromedially extending ridges on the dentary. Simões et al. (2015) considered *Gueragama*, from the Late Cretaceous of Brazil, closely related to leiolepidids based on the structure of its coronoid process. K. T. Smith (2011a) suggested that *Tinosaurus*, first described from the Paleogene (Eocene) of Wyoming, is possibly related to *Leiolepis*. *Tinosaurus* has also been widely reported from the Paleogene (Paleocene-Eocene) of East and South Asia, but the phylogenetic relationships of many of these records remain uncertain (K. T. Smith et al. 2011).

Gauthier et al. (2012) and K. T. Smith and Gauthier (2013) united Agamidae (Fig. 7.20A) and Chamaeleonidae (Fig. 7.20B) based on a suite of synapomorphies. Three of these are unique: presence of a contact between the postorbital and ectopterygoid; lacrimal foramen enlarged; and the posterior extension of the posterior process of the ectopterygoid beyond the coronoid apex.

Chamaeleonidae (Figs. 7.20B, 7.21) is readily distinguished by many shared derived features (Rieppel 1987b;

Figure 7.20. Acrodonta. **A**, agama (*Agama agama*; Agamidae); **B**, Cape dwarf chameleon (*Bradypodion pumilum*; Chamaeleonidae). Courtesy of Wolfgang Wüster.

Gauthier et al. 2012). In the cranium, the septomaxilla is absent, a descending lamina of the parietal extends posterior to the ascending portion of the supraoccipital, and the proximal head of the quadrate bluntly abuts the squamosal and supratemporal. The parietal forms an often prominent casque (Fig. 21B). The dermal bones of the cranium bear distinct tubercular sculpturing. The pectoral girdle lacks an interclavicle. The carpus has an intercarpal joint, in which the central carpal or the lateral centrale form the ball and the radiale, ulnare, and pisiform make up the socket. Similarly, the tarsus has an intertarsal joint, in which distal tarsal 4 forms the ball and the astragalocalcaneum bears the socket. In the manus, digits I-III oppose digits IV and V, whereas, in the pes, digits I and II oppose digits III-V. This configuration of the digits (zygodactyly; Figs. 7.20B, 7.21A) facilitates grasping tree branches and other narrow supports. The blade of the ilium is oriented vertically and has a laterally flattened dorsal end. Capable of rapidly changing

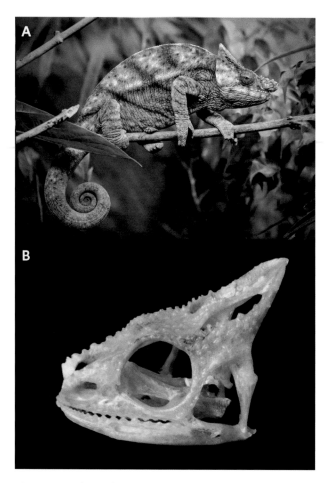

Figure 7.21. Chamaeleonid *Calumma parsonii.* **A**, animal moving along a branch; **B**, skull in lateral view. A, from Wikipedia (photo by Steve Wilson)—CC BY 2.0; B, courtesy of Pavel Zuber.

body coloration, extant chameleons are sit-and-wait predators that can independently move their eyes on "turrets" formed by fusion of the upper and lower eyelids and use their long projectile tongue to capture prey (Schwenk 2000). The oldest known records of *Chamaeleo* are from the Neogene (Miocene) of the Czech Republic, Germany, and Kenya (Estes 1983; Čerňanský 2010). Daza et al. (2016) reported a possible stem-chamaeleonid from mid-Cretaceous (Albian-Cenomanian) amber from Myanmar. Present-day chameleons occur in Spain, Africa (except for the Sahara), parts of the Arabian Peninsula, India, Madagascar, the Seychelles, Sri Lanka, and neighboring islands. They have snout-vent lengths ranging from 16 mm to more than 60 cm. Chameleons are usually but not always arboreal.

Agamidae is characterized especially by the presence of an epiotic foramen in the braincase, the lateral overlap of the postorbital by a vertical lappet of the parietal (forming the anteromedial margin of the upper temporal fenestra), and the posterodorsal slope of the retro-articular process to its distal tip (Gauthier et al. 2012). It comprises mostly terrestrial omnivores, but some are arboreal or have semiaquatic habits. *Draco* (flying lizards), found today in Southeast Asia and southwestern India, has greatly elongated dorsal ribs that support a skin membrane for gliding (McGuire and Dudley 2011). Its snout-vent length ranges from 6 to 15 cm. Elongated trunk ribs are also present in the possible acrodontan *Xianglong*, from the Early Cretaceous (Barremian) of Liaoning (China) (P. Li et al. 2007). Extant agamids can attain total lengths of up to more than 1 m. The oldest known representatives of this group have been reported from the Paleogene (Eocene) of Europe (e.g., Augé and R. Smith 1997) and India (e.g., Rana et al. 2013). Present-day agamids occur on the Arabian Peninsula; in most of Africa, Central, South, and Southeast Asia; and in Indonesia, Australia, and on the Philippines and neighboring islands. They mostly subsist on insects and other arthropods.

SQUAMATA: POLYGLYPHANODONTIA

Polyglyphanodontia (from Greek *polys*, many, *glyphein*, carve, and *odous* (*odon*), tooth) is known from the Late Cretaceous of Mongolia (Alifanov 2000) and western North America (Nydam et al. 2007). Its phylogenetic position remains uncertain. Polyglyphanodontians resemble extant teiid lizards and have often been considered closely related to or even placed in Teiioidea (Nydam et al. 2007 [Borioteiioidea]; Conrad 2008; Simões et al. 2018). Gauthier et al. (2012) demonstrated that they lacked various synapomorphies for crown-group Scleroglossa and found them outside that clade. Reeder et al. (2015) recovered Polyglyphanodontia as the sister group of Iguania and noted that polyglyphanodontians appear intermediate in their skeletal structure between iguanians and scleroglossans as traditionally conceived.

Gauthier et al. (2012) listed several diagnostic derived features for Polyglyphanodontia, such as the vertically expanded contact between the premaxillary processes of the maxillae, the great depth of the jugal below the orbit, and the presence of a deep, extensive adductor fossa on the surangular. Longrich et al. (2012a) also interpreted the V-shaped dentary symphysis, the long splenial slotting into the subdental shelf, and subapical tooth implantation as diagnostic for this clade. *Gilmoreteius* (formerly

Figure 7.22. Skeleton of the polyglyphanodontian *Polyglyphan-odon sternbergi* in dorsal view. Courtesy of Department of Paleobiology, National Museum of Natural History.

"*Macrocephalosaurus*"), from the Late Cretaceous (Campanian) of Mongolia (Sulimski 1975), and *Polyglyphanodon*, from the Late Cretaceous (Maastrichtian) of Utah (Gilmore 1942; Figs. 7.22, 7.23), share the absence of a supratemporal, the narrow participation of the pterygoid in the margin of the suborbital fenestra, and the splenial not extending posteriorly beyond the apex of the coronoid process (Gauthier et al. 2012). *Polyglyphanodon* attained a total length of about 1 m. Its more posterior teeth have transversely broadened crowns with finely serrated central blades, and the upper and lower teeth interdigitated during occlusion (Nydam and Cifelli 2005). *Chamops*, from the Late Cretaceous (Campanian-Maastrichtian) of western North America (Gao and Fox 1996), and closely related taxa with tricuspid, unexpanded tooth crowns represent another grouping among Polyglyphanodontia (Longrich et al. 2012a). The polyglyphanodontian *Tianyusaurus*, from the Late Cretaceous of Henan and Jiangxi (China), has a complete lower temporal bar and a fixed quadrate (Mo et al. 2010).

Sineoamphisbaena, from the Late Cretaceous (Campanian) of Inner Mongolia (China), was originally identified as the oldest and most basal amphisbaenian (Wu et al. 1993). Subsequent phylogenetic studies found no support for this hypothesis, and Kearney (2003) reinterpreted this taxon as a possibly burrowing polyglyphanodontian. The robust cranium of *Sineoamphisbaena* (length about 2.5 cm)

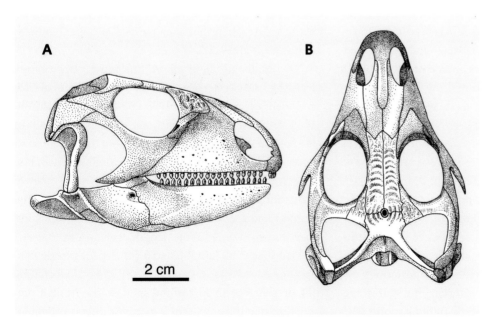

Figure 7.23. Reconstructed skull of *Polyglyphanodon sternbergi* in **A**, lateral, and **B**, dorsal views. Modified from Gilmore (1942) and Sulimski (1975).

is broader than long and retains upper temporal arches and fenestrae.

SQUAMATA: MOSASAURIA

Mosasauria (from Latin *Mosa*, the Roman name for the Maas (or Meuse) River in the present-day Netherlands, and Greek *sauros*, lizard) represents a diverse clade of squamates that adapted to life in the sea during the Late Cretaceous. Mosasaurs were apex predators in Late Cretaceous marine ecosystems, with some attaining total lengths of at least 13 m. They became extinct at the end of the Maastrichtian. Mosasaurian remains are known from all continents including Antarctica and from various islands including New Zealand.

Traditionally, Mosasauria were placed in Anguimorpha, and mosasaurians were interpreted as marine varanoid lizards (McDowell and Bogert 1954; deBraga and Carroll 1993; Lee 2009). As Cuvier (1824) already observed, the skulls of mosasaurs resemble those of varanoids. However, Caldwell (2012) argued that most if not all alleged synapomorphies shared by these two groups are misinterpreted or questionable. Cope (1869) hypothesized a close relationship between snakes and mosasaurs and established Pythonomorpha for the reception of these two groups. Lee (1997b) and Caldwell (1999) resurrected his hypothesis and presented new anatomical evidence in support. However, the phylogenetic analyses by Conrad (2008) and Gauthier et al. (2012) found no support for the monophyly of Pythonomorpha. The former recovered mosasaurs among Varanoidea, whereas the latter authors found them as a sister group of Scleroglossa close to the root of Squamata. Reeder et al. (2015) found a clade comprising Mosasauria and Serpentes outside Anguimorpha. It is possible that the modifications of many skeletal features in mosasaurs for a fully aquatic mode of life obscure the phylogenetic relationships of this group (see Chapter 6).

Gauthier et al. (2012) recognized a clade Mosasauria, comprising Dolichosauridae and Mosasauroidea, and listed a suite of diagnostic derived features including the posterior extension of the nasal process of the premaxilla more than halfway to the frontal; fusion of the postfrontal and postorbital; no fusion between the constituent bones of the braincase; quadrate massive, bowed anteriorly (in lateral view) and laterally (in posterior view) and forming a prominent, ventrally curving suprastapedial process; presence of an intramandibular joint with a ball on the angular fitting into a cup on the splenial; marginal tooth crowns separated from each other by wide gaps; the reduced scapulocoracoid; and separate astragalus and calcaneum ossifications.

Dolichosauridae comprises a group of early Late Cretaceous aquatic squamates, none of which apparently exceeded 1 m in total length. *Dolichosaurus*, from the Late Cretaceous (Cenomanian) of England (Caldwell 2000), and *Aphanizocnemus*, from the Late Cretaceous (Cenomanian) of Lebanon (Dal Sasso and Pinna 1997), have long necks, trunks, and tails and proportionately small limbs. Gauthier et al. (2012) listed as shared derived features for Dolichosauridae the presence of at least eight cervical and a total of 33 to 39 presacral vertebrae. Caldwell (2000) inferred that dolichosaurs used lateral undulation of the elongated body and tail for swimming. The forelimbs of *Aphanizocnemus* are longer than its hind limbs. There are several other taxa of Late Cretaceous marine squamates that are probably related to Dolichosauridae. *Pontosaurus*, from the Cenomanian-Turonian of Croatia and the Cenomanian of Lebanon (Caldwell 2006), attained a total length of about 1 m and has a greatly elongated tail. Caldwell (2006) hypothesized a clade comprising Dolichosauridae, *Pontosaurus*, and related taxa as well as snakes, but Gauthier et al. (2012) found no support for such a grouping.

Mosasauroidea (mosasaurs) comprises *Aigialosaurus* and Mosasauridae. Diagnostic shared derived features for this clade include the extension of the ventral process of the postorbital lateral to the apex of the dorsal process of the jugal and the penultimate phalanges of the manus being shorter than the phalanges just proximal to them (Gauthier et al. 2012).

The limbs of *Aigialosaurus* (including *Opetiosaurus*), from the Late Cretaceous (Cenomanian-Turonian) of Croatia (deBraga and Carroll 1993; Dutchak and Caldwell 2006, 2009), lack the modifications for an aquatic mode of life present in derived Mosasauridae, but they are proportionately shorter than in most other limbed squamates. However, the skull of *Aigialosaurus* shares with those of Mosasauridae derived features such as the fusion of the frontals and the structure of the intramandibular joint. *Aigialosaurus* attained a total length of about 1 m.

Gauthier et al. (2012) listed a suite of diagnostic synapomorphies for Mosasauridae (Figs. 7.24, 7.25) including the presence of a supraorbital shelf on the frontal; prefrontal extending posteriorly to midorbit; postorbital in-

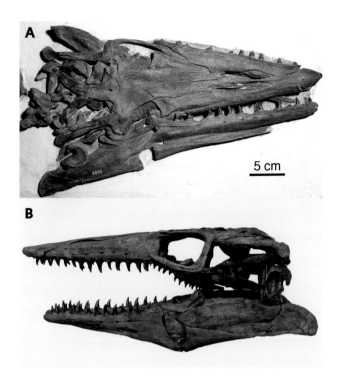

Figure 7.24. A, skull of the mosasaurine *Clidastes liodontus* in dorsal view; **B**, reconstructed skull of the tylosaurine *Tylosaurus proriger* in lateral view. Length of skull c. 1.6 m. A, courtesy of Department of Paleobiology, National Museum of Natural History; B, courtesy of Mike Triebold.

serting in a long, V-shaped trough on the dorsal surface of the squamosal, turning posterolaterally more posteriorly; limb bones lacking epiphyses; pelvic bones separate and lacking sutural contacts with each other; and hyperphalangy in more than one digit of the pes. Mosasaurid teeth are anchored to the jaws by a distinctive combination of tissues: a layer of acellular cementum between the tooth root and cellular cementum; a robust cone formed of cellular cementum; mineralized periodontal ligaments; and ridges of bone between the individual teeth (Luan et al. 2009). The tooth implantation superficially appears thecodont but differs in structure from the truly thecodont implantation present, for example, in crocodylians.

Bell and Polcyn (2005) and Caldwell and Palci (2007) distinguished two major lineages among Mosasauridae. Each independently evolved features of the pelvis and hind limbs suitable for a fully aquatic mode of life. In each lineage, the basal taxa have limbs suitable for facultative terrestrial locomotion (plesiopedal) and an unmodified pelvic structure (plesiopelvic), whereas the more derived members modified their limbs for aquatic locomotion

(hydropedal) and have pelves that lack sacra and have modified ilia (hydropelvic). Some mosasaurs (e.g., *Platecarpus*) have paddle-like limbs with distally spreading digits, whereas others (e.g., *Plotosaurus*) have distally tapering, flipper-like limbs with more tightly spaced digits (Russell 1967; Caldwell 1996). Most mosasaurs lack any bony connection between the vertebral column and pelvic girdle. Thus, Caldwell and Palci (2007) argued that they were no longer able to support their bodies on land. This is consistent with recent discoveries indicating that mosasaurs were live-bearing (e.g., Field et al. 2015).

Makádi et al. (2012) also recognized two major groupings among Mosasauridae. One comprises *Dallasaurus* (Turonian of Texas; Polcyn and Bell 2005) and Mosasaurinae. The second includes Tethysaurinae, Halisaurinae, Tylosaurinae + Plioplatecarpinae, *Carsosaurus* (Cenomanian-Turonian of Croatia; Caldwell et al. 1995), and *Komensaurus* (Cenomanian of Slovenia; Caldwell and Palci 2007). These groupings have not yet been formally defined and diagnosed.

Mosasauridae has a rich and diverse fossil record, dating back to the Coniacian (Russell 1967). It attained a worldwide distribution. Some taxa (e.g., *Tylosaurus*) could attain total lengths of at least 13 m, whereas *Dallasaurus* is about 1 m long. Mosasaurs fed on a wide range of animals (Massare 1987; Everhart 2004). *Globidens*, from the Late Cretaceous (Campanian) of Alabama, Kansas, and South Dakota and the Late Cretaceous (Maastrichtian) of various regions of Africa and Syria (Bardet et al. 2005), has robust, bulbous tooth crowns suitable for feeding on hard-shelled invertebrates. *Pannoniasaurus*, from the Late Cretaceous (Santonian) of Hungary, is noteworthy for its occurrence in a freshwater environment (Makádi et al. 2012).

Undulation of the body and the long tail of mosasauroids was presumably the primary means for swimming, with the limbs serving for steering. Lindgren et al. (2007, 2010) reconstructed the distal portion of the tail in *Plotosaurus* (Camp 1942) and in *Platecarpus* (Fig. 7.25) as deflected and supporting a distinct tail fluke. Lindgren et al. (2013) found evidence for a smaller terminal caudal fin in *Prognathodon*.

Using oxygen isotope ratios in tooth enamel, Harrell et al. (2016) argued that mosasaurs could maintain elevated body temperatures and possibly were even endothermic. Lindgren et al. (2014) inferred that an unidentified

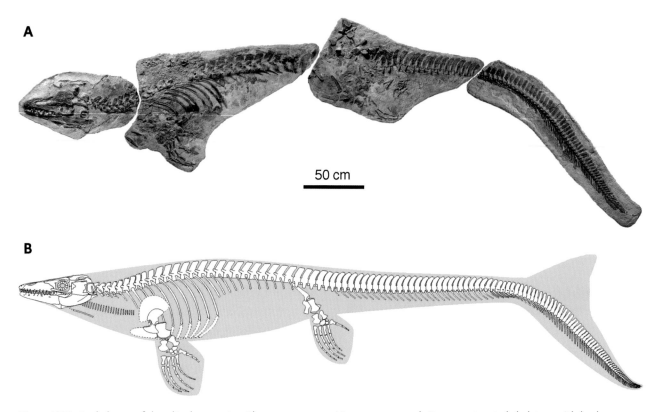

Figure 7.25. A, skeleton of the plioplatecarpine *Platecarpus tympaniticus* as preserved; **B**, reconstructed skeleton with body silhouette. Broken lines denote cartilaginous structures. From Lindgren et al. (2010)—CC BY 2.5.

mosasaur had countershading, with a dark back and a light-colored underside.

SQUAMATA: DIBAMIDAE

The phylogenetic position of this clade of "snake-like" squamates is uncertain. Pyron et al. (2013) placed Dibamidae at the root of Squamata, whereas other authors considered it related to Gekkota (e.g., Wu et al. 1996; Wiens et al. 2012) or Scincomorpha (Camp 1923). Dibamids have only small, flap-like hind limbs and lack any trace of forelimbs. The tubular cranium lacks postorbital and upper temporal arches (S. E. Evans 2008). The eyes are much reduced. Rieppel et al. (2008) and Gauthier et al. (2012) listed various synapomorphies for Dibamidae including the nasal abutting the maxilla; postparietal projection of the parietal extending medially; entire lateral margin of the vomer in sutural contact with the medial edge of the palatal shelf of the maxilla; scroll-like palatines forming choanal tubes; optic foramen absent; cultriform process absent; four cervical vertebrae; and tongue without a distal notch. Dibamids have a disjunct present-day

geographic distribution, with *Dibamus* occurring in Indonesia and the Philippines and *Anelytropsis* in Mexico. They have no known fossil record, and little is known about their biology. Dibamids have snout-vent lengths ranging from 5 to 20 cm.

SQUAMATA: GEKKOTA

Gekkota (from Malayan *gekoq*, tokay lizard) were long considered the most basal lineage of squamates (e.g., Camp 1923) because of the presence of various features considered plesiomorphies, such as the presence of amphicoelous vertebrae in many gekkotans and the presence of a stapedial fenestra. Later morphological studies reinterpreted the evolutionary polarities of these features (e.g., Rieppel 1988). Phylogenetic analyses based on DNA sequences and gene data, however, have consistently recovered Gekkota as the basal clade of Squamata (Vidal and Hedges 2009; Wiens et al. 2012; Pyron et al. 2013; see also Simões et al. 2018).

Gauthier et al. (2012) listed a suite of unambiguous synapomorphies for Gekkota, including fused frontals;

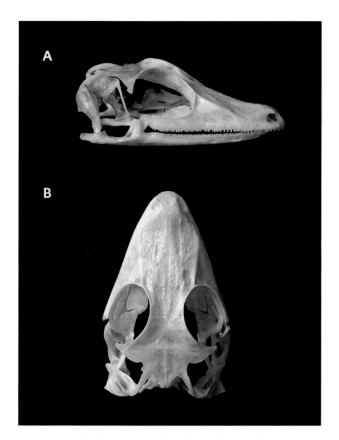

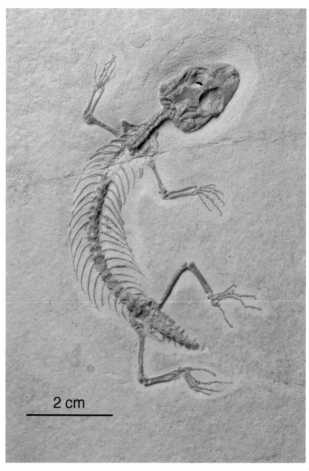

Figure 7.26. Skull of a giant leaf-tailed gecko (*Uroplatus fimbriatus*; Gekkonidae) in **A**, lateral, and **B**, dorsal views. Courtesy of Pavel Zuber.

Figure 7.27. Skeleton of the stem-gekkotan *Eichstaettisaurus schroederi* in dorsal view. Most of the tail had been shed and was in the process of regeneration. Courtesy of Helmut Tischlinger.

suborbital processes of the frontals extending beneath the brain and fused to each other along the midline ventrally; parietal foramen absent; lacrimal duct enclosed in the prefrontal except ventrally; jugal lacking the dorsal process or jugal absent altogether; Meckelian canal on the dentary closed anterior to the splenial; and angular absent. The cranium is low dorsoventrally and has large orbits (Fig. 7.26). It lacks postorbital and upper temporal bars. The cranial bones are typically thin and lack sculpturing. Many but not all gekkotans have amphicoelous rather than procoelous centra. Additional plesiomorphic features include the presence of a continuous second branchial arch in some taxa and the presence of a stapedial foramen (Rieppel 1988).

Eichstaettisaurus, from the Late Jurassic (Tithonian) of Germany (Fig. 7.27) and the Early Cretaceous (Albian) of Italy, is generally considered an early stem-gekkotan (Gauthier et al. 2012; Simões et al. 2017, 2018). It shares several unambiguous apomorphies with crown-group gekkotans including well-developed subolfactory pro-

cesses of the frontals (but not in contact and fusing ventromedially as in the latter); temporal ramus of the squamosal in contact with the supratemporal process of the parietal; and autotomy septa on the caudal vertebrae positioned behind the caudal ribs (Gauthier et al. 2012). *Eichstaettisaurus* retains postorbital and supratemporal bars. *Ardeosaurus*, also from the Late Jurassic (Tithonian) of Germany, is closely related to *Eichstaettisaurus* (Simões et al. 2017). These stem-gekkotans each attained a snout-vent length of 8 cm and have limbs with features that suggest climbing (scansorial) habits.

Gobekko, from the Late Cretaceous (Campanian) of Mongolia, is a basal crown-group gekkotan (Daza et al. 2013, 2014). It has plesiomorphic features such as a parietal foramen and paired frontals with subolfactory processes that do not contact each other along the midline. The separation between the latter processes is still more pronounced in an unnamed stem-gekkotan from early Late Cretaceous (Cenomanian) amber in Myanmar (Daza et al. 2016). K. T. Smith et al. (2018) reported on a gekkotan from the Paleogene (Eocene) of Germany that shows a combination of apomorphic features such as the absence of a parietal foramen with plesiomorphic traits such as a well-developed jugal and supratemporal bar.

Extant geckos vary considerably in size, ranging from 1.6 cm to 37 cm in snout-vent length. They subsist mainly on insects and other invertebrates. Most are nocturnal, and many are scansorial. Except for Eublepharidae (eyelid geckos), all present-day gekkotans lack mobile eyelids and instead have a fixed "spectacle" covering the eye. They clean this scale by licking it with the tongue and replace it during each molt (Gauthier et al. 2012).

Many geckos have lamellar toe pads composed of sets of hundreds of closely packed, hair-like keratinous structures (setae). Together, these sets of setae establish close contact with the substrate through intermolecular attractive forces (Autumn et al. 2000) and allow these lizards to cling to and move on vertical surfaces. By contrast, ground-dwelling geckos have simpler, pointed digits. The Southeast Asian *Ptychozoon* (gliding geckos) has skin folds, especially along the sides of the trunk, that allow it to parachute and glide between trees (A. P. Russell et al. 2001).

Molecular data strongly support recognition of two clades among extant Gekkota, Pygopodoidea and Gekkonoidea (Han et al. 2004; Gamble et al. 2008; Pyron et al. 2013; Fig. 7.28). Daza et al. (2014) noted that the various subclades of these two groupings have for the most part not yet been diagnosed based on skeletal features, making the placement of extinct gekkotans difficult.

SQUAMATA: GEKKOTA: PYGOPODOIDEA

Pygopodoidea (from Greek *pyx*, rump, and *pous*, foot) comprises Carphodactylidae, Diplodactylidae, and the "snake-like" Pygopodidae (flap-footed lizards). Pygopodidae is noteworthy for the absence of forelimbs and the reduction of the hind limbs to flap-like structures (Fig. 7.28B). The cranium is typically more elongate than those in other gekkotans and lacks temporal arches. Pygopodids occur in Australia and Papua New Guinea. Some species are burrowers, whereas others are terrestrial or arboreal. Most feed on arthropods, but *Lialis* preys on other liz-

Figure 7.28. Gekkota. **A**, thick-toed gecko (*Pachydactylus oshaughnessyi*; Gekkonidae); **B**, marble-faced delma (*Delma australis*; Pygopodidae). A, from Wikipedia (photo by Ryan Van Huyssteen)—CC BY-SA 4.0; B, from Wikipedia (photo by Matt Clancy)—CC BY 2.0.

ards. The earliest record of *Pygopus* is from the Neogene (Miocene) of Australia (M. N. Hutchinson 1997). The sister taxon of Pygopodidae, Carphodactylidae, is known only from Australia and has no known fossil record. Carphodactylidae and Pygopodidae share the presence of an L-shaped squamosal (Daza et al. 2014). The sister taxon to this clade is Diplodactylidae, the oldest known occurrence of which is from the Neogene (Miocene) of New Zealand (Lee et al. 2009). Extant diplodactylids occur in Australia, New Caledonia, and New Zealand.

SQUAMATA: GEKKOTA: GEKKONOIDEA

The molecular-based phylogenetic analysis by Han et al. (2004) divided Gekkonoidea into Eublepharidae, Gekkonidae, Phyllodactylidae, and Sphaerodactylidae (Figs. 7.28A, 7.29). Probably the oldest gekkonoid is *Yantarogekko*, based on a head and anterior portion of the body preserved in Paleogene (Eocene) amber from Russia (Bauer et al. 2005). It differs from Eublepharidae in the presence of a spectacle and the presence of digital pads with well-developed lamellae, but it cannot be definitively placed among Gekkonoidea (Daza et al. 2014).

Figure 7.29. Gekkota. **A**, Texas banded gecko (*Coleonyx brevis*; Eublepharidae); **B**, mourning gecko (*Lepidodactylus lugubris*; Gekkonidae). A, courtesy of Laurie Vitt; B, courtesy of Division of Amphibians and Reptiles, National Museum of Natural History.

Among Sphaerodactylidae, the earliest occurrences of *Euleptes* (European leaf-toed gecko) are from the Neogene (Miocene) of the Czech Republic, France, Germany, and Slovakia. *Sphaerodactylus* (ball-finger geckos) is well-documented from partial and complete individuals preserved in Miocene amber from the Dominican Republic (Daza and Bauer 2012; Daza et al. 2014). Daza and Bauer (2012) listed various diagnostic skeletal features for Sphaerodactylidae, including a strongly hooked metatarsal V. Eublepharidae now occurs in the American Southwest (Fig. 7.29A), northern Central America, western Sub-Saharan Africa, South Asia, and on the Ryukyu Islands and Borneo. Extant Gekkonidae (Figs. 7.28A, 7.29B) has a wide geographic distribution in southern Europe, the Middle East, Africa, Madagascar, much of Asia, Indonesia, Papua New Guinea, the Philippines, Japan, the Comoros, Australia, and from the southern United States to

Brazil and Chile. Phyllodactylidae now occurs in southern Europe, the Middle East and the Arabian Peninsula, Africa north of the equator, and in the Americas from Mexico to Argentina. Sphaerodactylidae ranges today from Mexico to Peru and Brazil and occurs in the Caribbean, southwest and Central Asia, southern Europe, Morocco, and Algeria, on much of the Arabian Peninsula and in neighboring regions of East Africa.

SQUAMATA: AUTARCHOGLOSSA

Autarchoglossa (from Greek *autos*, self, and *arkein*, sufficient, and Latin *glossa*, tongue) comprises the most recent common ancestor of Anguimorpha and Scincomorpha and all descendants of that ancestor (Estes et al. 1988). Gauthier et al. (2012) listed a suite of diagnostic synapomorphies for this clade: dermal sculpturing on the frontal and parietal; presence of a posteroventral process on the jugal; presence of a posterodorsally extending ridge on the medial surface of the facial process of the maxilla that delimits the nasolacrimal fossa anteriorly and dorsally; enclosure of the ethmoidal nerve in the anterior half of the septomaxilla; vomeronasal organ encapsulated by the vomer medially and posteriorly; ectopterygoid with a slot that clasps the maxilla laterally; and rectus abdominis muscles inserting into hinges between the transverse rows formed by the ventral scales. Molecular-based phylogenetic analyses (Vidal and Hedges 2005, 2009; Pyron et al. 2013) have found no support for the monophyly of Autarchoglossa (see also Reeder et al. 2015).

Autarchoglossans often rely on the vomeronasal organ rather than vision for prey detection. They frequently have elongated bodies, which, along with the attachment of the abdominal musculature to the ventral skin, repeatedly led to the evolution of snake-like body plans in this clade (Wiens et al. 2006; Gauthier et al. 2012).

SQUAMATA: AUTARCHOGLOSSA: SCINCOMORPHA

Camp (1923) recognized a grouping Scincomorpha (from Latin *scincus*, a kind of lizard, and Greek *morphe*, form) including Cordylidae, Gerrhosauridae, Gymnophthalmidae, Lacertidae, Scincidae, and Teiidae. The phylogenetic analysis by Gauthier et al. (2012) found a crown group Scincomorpha and listed a suite of diagnostic synapomorphies, including the nasal abutting the maxilla; frontal in contact with the maxilla, separating the nasal from the prefrontal; choanal fossa fully developed on the palatine to the end of the bone; clavicle fenestrated; clavicle greatly expanded proximally; and symphyseal process of the pubis directed anteromedially. The authors noted that statistical support for this hypothesis is not robust because of a lack of data for various specialized fossorial taxa and extinct stem-taxa. Molecular-based studies (e.g., Vidal and Hedges 2005; Wiens et al. 2010) recovered Scincomorpha as a paraphyletic grouping, with Lacertiformes and Scincoidea forming successive sister taxa to Toxicofera.

AUTARCHOGLOSSA: SCINCOMORPHA: LACERTIFORMES

Lacertiformes (from Latin *lacerta*, a kind of lizard, and *forma*, form) comprises Lacertidae, Amphisbaenia, and Teiioidea. Müller et al. (2011) and numerous other authors (e.g., Pyron et al. 2013) found Amphisbaenia most closely related to Lacertidae, which makes the former part of Lacertiformes. Camp (1923) considered

Figure 7.30. Lacertidae. **A**, Italian wall lizard (*Podarcis sicula*); **B**, grass lizard (*Takydromus* sp.). A, courtesy of Laurie Vitt; B, courtesy of Division of Amphibians and Reptiles, National Museum of Natural History.

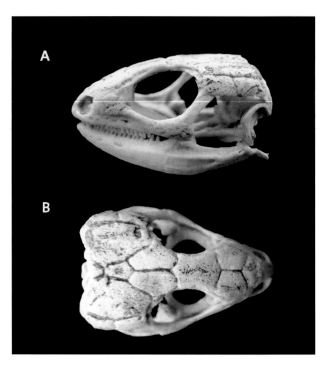

Figure 7.31. Lacertidae. Skull of a sand lizard (*Lacerta agilis*) in **A**, lateral, and **B**, dorsal views. Note the cranial osteoderms. Courtesy of Pavel Zuber.

amphisbaenians related to teiids. Gauthier et al. (2012) did not group amphisbaenians with lacertiforms, but K. T. Smith and Gauthier (2013) recognized Lacertiformes as comprising Lacertidae, Amphisbaenia, Teiidae, and Gymnophthalmidae.

Lacertidae (wall or true lizards) ranges throughout Africa and Eurasia into Southeast Asia (Arnold et al. 2007; Figs. 7.30, 7.31). Most are ground- or rock-dwelling, diurnal forms that subsist on insects. Gauthier et al. (2012) listed a suite of diagnostic synapomorphies for Lacertidae, including the postfrontal nearly covering the entire upper temporal fenestra; postorbital nearly excluding the squamosal from the upper temporal fenestra; ectopterygoid with a prominent posterior process; single osteoderms inside the supraorbital scales; and single osteoderms in the cheek scales. The stem-lacertid *Cryptolacerta*, from the Paleogene (Eocene) of Germany, attained a snout–vent length of about 7 cm. It has a robust skull and proportionately very short limbs. It possibly burrowed in leaf litter or soil (Müller et al. 2011). *Succinilacerta*, from Paleogene (Eocene) amber from Russia, is possibly a crown-group lacertid based on its external features (Borsuk-Białynicka et al. 1999). *Lacerta* occurs today in Europe, from Turkey into Central Asia, and in the Middle East (Arnold et al. 2007). Its earliest record is from the Paleogene (Oligocene) of France (Augé 1988). *Gallotia* comprises subfossil and extant large-bodied lacertids endemic to the Canary Islands. Species of this taxon could reach a total length of about 1 m (Barahona et al. 2000), and they include a lot of plant material in their diet.

Čerňanský and K. T. Smith (2017) recognized a new clade Eolacertidae, which they found as the sister group of Lacertidae in their phylogenetic analysis. Eolacertidae comprises *Eolacerta* and *Stefanika*, both from the Paleogene (Eocene) of Germany. Diagnostic synapomorphies for this clade include an osteoderm attached to the angle of the jugal and the partial overlap of the parietal by the postfrontal (Čerňanský and K. T. Smith 2017). *Eolacerta* attained a snout–vent length of over 30 cm and has a very long tail (K. T. Smith et al. 2012). The skull roof is strongly ornamented. Daza et al. (2016) reported a possible stem-lacertiform preserved in early Late Cretaceous (Cenomanian) amber from Myanmar.

Teiioidea is generally considered the New World ecological equivalent of its Old World sister group Lacertidae (Estes et al. 1988). Gauthier et al. (2012) listed several diagnostic derived features for Teiioidea, including the prootic forming part of the medial opening of the recessus scalae tympani; distal part of the lingual process of the hyoid detached; prearticular with a distinct crest; coracoid with a posterior emargination; and humerus lacking an ectepicondylar foramen. Teiioidea can be divided into Teiidae and Gymnophthalmidae.

Present-day Teiidae ("macroteiids") ranges throughout the western and southern United States through Mexico and Central America into much of South America (except for the southernmost part of that continent and the High Andes) and is also present on many Caribbean islands (Fig. 7.32). Gauthier et al. (2012) enumerated various diagnostic synapomorphies, including postorbital overlapping the squamosal laterally; temporal muscles originating dorsally on the parietal table and the supratemporal process of the parietal; palatine with a deep, well-demarcated dorsal canal, ending in an enclosed fossa; ectopterygoid close to or in anterolateral contact with the palatine (Fig. 7.15); and pubic tubercle positioned closer

Figure 7.32. Teiidae. **A**, giant ameiva (*Ameiva ameiva*); **B**, forest whiptail (*Kentropyx pelviceps*). Courtesy of Laurie Vitt.

to the symphysis than to the acetabulum. Morphological and molecular data support recognition of two subclades, Tupinambinae and Teiinae (Presch 1983; Giugliano et al. 2007). *Tupinambis* (tegus) attains a total length of more than 1 m. The earliest occurrence of Tupinambinae is from the Neogene (Miocene) of Colombia (R. M. Sullivan and Estes 1997). Extant Teiinae includes *Ameiva* (junglerunners), which occurs in Panama, much of South America, and in the Caribbean (Fig. 7.32A), and its close relative *Cnemidophorus* (whiptail lizards), which ranges from Idaho to Argentina. The earliest record of *Cnemidophorus* is from the Neogene (Miocene) of Florida (Estes 1983).

Gymnophthalmidae ("microteiids") comprises small-sized teioids with typically long, slender bodies and rather short limbs, which are nearly absent in some taxa. Most have a transparent "window" in the lower eyelid through which they can see when the eye is closed. Diagnostic synapomorphies for this clade include a transverse flange of the vomer that rises vertically and contacts the septomaxilla to enclose the vomeronasal organ posteriorly and the strongly dorsally angled second ceratobranchial (Gauthier et al. 2012). Gymnophthalmidae ranges from southern Central America to much of South America east of the Andes. It has no known fossil record. Gymnophthalmids are mostly diurnal and subsist on insects.

Codrea et al. (2017) recently proposed a new teiioid grouping, Barbateiidae, from the Late Cretaceous (Maastrichtian) of Romania, but did not provide a phylogenetic analysis in support of this classification. The skull roof has pronounced dermal sculpturing.

AUTARCHOGLOSSA: SCINCOMORPHA: LACERTIFORMES: AMPHISBAENIA

Amphisbaenia (worm lizards; from Greek *amphisbaina*, a mythical snake that can move both forward and backward) include 196 extant species (as of July 2018; http://www.reptile-database.org). A sister-group relationship between Lacertidae and Amphisbaenia has long been strongly supported by molecular data (Townsend et al. 2004; Vidal and Hedges 2005, 2009; Pyron et al. 2013; see also Simões et al. 2018). Müller et al. (2011) interpreted the stem-lacertid *Cryptolacerta*, from the Paleogene (Eocene) of Germany, as further supporting this hypothesis, but Gauthier et al. (2012) and Longrich et al. (2015) did not support this hypothesis. Tałanda (2016) argued for a close relationship between Amphisbaenia and the probably burrowing scincomorph *Slavoia*, from the Late Cretaceous (Campanian) of Mongolia.

Amphisbaenians (worm lizards) have elongate, cylindrical bodies (Fig. 7.33) with the skin on the trunk divided in rings (annuli), with two rings per vertebra except for only one in *Blanus*. The trunk comprises 82 to about 175 vertebrae, but the tail has fewer than 40 vertebrae (Gans 1978). The vertebrae lack zygosphene-zygantrum accessory articulations. The skin of present-day amphisbaenians is covered by smooth, not deeply imbricated, scales and moves independently of the trunk. Most extant burrowing squamates lack the left lung, but amphisbaenians do not have a right lung. Most present-day amphisbaenians lack external traces of limbs. The sole exception is *Bipes*, which has well-developed forelimbs (Fig. 7.33A).

Gauthier et al. (2012) provided a list of synapomorphies for Amphisbaenia, including the presence of a greatly enlarged median tooth on the fused premaxillae;

Figure 7.33. Amphisbaenia. **A**, Mexican mole lizard (*Bipes biporus*; Bipedidae). Note the well-developed forelimbs and absence of hind limbs. **B**, black-and-white worm lizard (*Amphisbaena fuliginosa*; Amphisbaenidae). A, courtesy of Adam G. Clause; B, courtesy of Laurie Vitt.

nasal shorter than the frontal; frontoparietal suture deeply interdigitated; quadrate ramus of the pterygoid short, tightly wrapping around the posteromedial surface of the quadrate; ectopterygoid much enlarged medially, closing the suborbital fenestra; orbitosphenoid expanded to the floor of the braincase and enclosing the optic foramen; and basal tubera bearing large apophyseal ossifications. Pyron et al. (2013) found strong support for the monophyly of Amphisbaenia from molecular data.

Most amphisbaenians use their heavily ossified skulls (Figs. 7.34, 7.35) for digging. Well-developed neck and trunk muscles generate the requisite forward thrust for burrowing. Forms with different head shapes employ different modes of burrowing (Gans 1974, 1978). All amphisbaenians have small or vestigial eyes and lack external ears. The cranium lacks temporal arches and any potentially mobile sutural contacts. The endocranial cavity is fully enclosed in bone, with complex sutural contacts between its constituent elements.

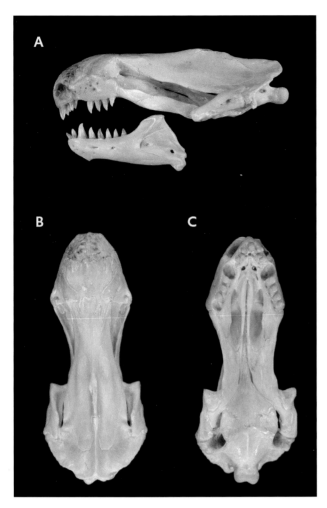

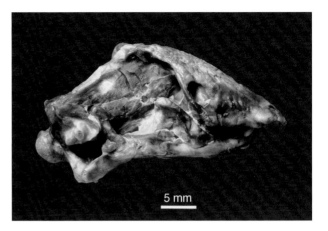

Figure 7.35. Skull of the rhineurid *Spathorhynchus fossorium* in lateral view. Courtesy of the late Don Baird.

Figure 7.34. A, skull of a red worm lizard (*Amphisbaena alba*) in lateral view; **B–C**, cranium of *Amphisbaena alba* in **B**, dorsal, and **C**, palatal views. Courtesy of Chris Bell.

The basal clade of Amphisbaenia, Rhineuridae, is first recorded from the Paleogene (Paleocene) and has an extensive fossil record in North America (Hembree 2007; Longrich et al. 2015). Diagnostic features include the (in lateral view) ventrally deflected retroarticular process and the anteromedial process of the coronoid fitting into a groove below the dentary tooth row. Some rhineurids such as *Spathorhynchus*, from the Paleogene (Eocene) of Wyoming (Fig. 7.35), retain a bony bar behind the orbit (Berman 1973; Müller et al. 2016). The sole extant representative of Rhineuridae is the Florida worm lizard (*Rhineura floridana*), which feeds on invertebrates and attains a snout-vent length of up to 38 cm.

The oldest stem-amphisbaenian, *Chthonophis*, from the Paleogene (Paleocene) of Montana, has a short, deep lower jaw with a robust mandibular symphysis and only seven or eight teeth (Longrich et al. 2015). *Oligodontosaurus*, from the Paleogene (Paleocene) of Montana, has an enlarged last dentary tooth and a short coronoid eminence (Estes 1983; Longrich et al. 2015). The phylogenetic analysis by Longrich et al. (2015) found both *Chthonophis* and *Oligodontosaurus* more closely related to crown-group amphisbaenians than to Rhineuridae.

Longrich et al. (2015) united all derived amphisbaenians as Amphisbaeniformes. Within Amphisbaeniformes, Blanidae and Bipedidae form the basal grouping based on molecular data (Longrich et al. 2015). The Mediterranean worm lizards (*Blanus*) occur today on the Iberian Peninsula and in Turkey, Syria, Lebanon, and Morocco. *Blanus* was already widely distributed through Europe by the Neogene (Miocene) (Bolet et al. 2014). It has reduced clavicles and rudiments of femora (Kearney 2003). Čerňanský et al. (2015) described a blanid, *Cuvieribaena*, from the Paleogene (Eocene) of France. Longrich et al. (2015) also referred Polyodontobaenidae, from the Paleogene (Paleocene) of Belgium and France (Folie et al. 2013), to Blanidae.

The only known present-day representative of Bipedidae is *Bipes* (mole lizards), which occurs in Baja California and southwestern Mexico. It is distinguished from other amphisbaenians by the presence of well-developed pectoral girdle and forelimbs (Gans 1978; Fig. 7.33A) and by complete fusion of the frontals and parietals (Kearney 2003; Gauthier et al. 2012). *Bipes* attains a snout-vent length of up to 24 cm and feeds on arthropods. *Anniealexandria*, from the Paleogene (Eocene) of Wyoming and possibly France, is a stem-bipedid (K. T. Smith 2009b; Augé 2012).

The remaining groups within Amphisbaeniformes form a clade Afrobaenia, which is supported by both molecular data (Pyron et al. 2013) and skeletal features. The latter include the posterior extension of the dentary close to the posterior surangular foramen on the lateral surface of the lower jaw, the formation of the coronoid eminence by the coronoid and the surangular, and the retroarticular process being short or absent (Gauthier et al. 2012). *Todrasaurus*, from the Paleogene (Paleocene) of Morocco, is a possible stem-afrobaenian (Augé and Rage 2006). Cadeidae is represented only by the extant *Cadea* (keel-headed worm lizards) from Cuba and is separated from other amphisbaeniforms based on molecular data (Vidal and Hedges 2009). *Cadea* resembles *Blanus* in various cranial features (Čerňanský et al. 2014). Amphisbaenidae (worm lizards) has by far the greatest species diversity among amphisbaenians, with more than 100 present-day species (Gans 1978; Figs. 7.33B, 7.34). Shared derived features include a descending lamina of the nasal that extends below the level of the suture between the nasal and maxilla, the absence of a suprastapedial process on the quadrate, and the absence of basipterygoid processes (Gauthier et al. 2012). The present-day geographic distribution of this clade encompasses South America and the Caribbean, the Mediterranean region, and Sub-Saharan Africa. Amphisbaenidae includes *Listromycter* and *Lophocranion*, both from the Neogene (Miocene) of Kenya (Charig and Gans 1990; Longrich et al. 2015). Extant representatives attain snout-vent lengths ranging from 9 cm to 72 cm. Trogonophidae (spade-headed worm lizards) occurs today in North Africa, Turkey, the Arabian Peninsula, and Iran. Snout-vent lengths range from 8 to 24 cm. Diagnostic features for this clade include the acrodont marginal dentition and the absence of the anteromedial and posteromedial processes of the coronoid bone (Gauthier et al. 2012). *Trogonophis* has been reported from the Neogene (Pliocene) of Morocco (Bailon 2000).

AUTARCHOGLOSSA: SCINCOMORPHA: SCINCOIDEA

Gauthier et al. (2012) interpreted *Paramacellodus*, from the Early Cretaceous (Berriasian) of England, the Early Cretaceous (Barremian) of Morocco and Spain, the Late Jurassic (Kimmeridgian) of Utah and Wyoming, and the Middle Jurassic (Bathonian) of Scotland (Hoffstetter 1967; S. E. Evans and Chure 1998), and *Parmeosaurus*, from the Late Cretaceous (Campanian) of Mongolia (Gao and Norell 2000), as basal stem-scincoids. This assessment was based primarily on the posterodorsal extension of the coronoid process of the dentary between the lateral and medial processes of the coronoid. *Paramacellodus* also shares two unambiguous apomorphies with crown-group scincoids: a postparietal projection near the midline of the parietal and the maxilla covering the jugal below the orbit in lateral view (Gauthier et al. 2012). It and various closely related taxa (Paramacellodidae; Estes 1983) attained a wide geographic distribution during the Jurassic and Early Cretaceous (Richter 1994; S. E. Evans and Chure 1998: Broschinski 1999). The bodies of some but not all paramacellodids (Hoffstetter 1967; Richter 1994) and *Parmeosaurus* are covered by rectangular osteoderms.

Xantusiidae (night lizards) occurs today in North and Central America and on Cuba (Fig. 7.36A). These small-bodied lizards lack movable eyelids. The lower eyelid has a clear scale through which the lizard can see. Xantusiids feed primarily on insects. Gauthier et al. (2012) listed a suite of synapomorphies for this clade, including the dentary terminating posteriorly well behind the coronoid apex; splenial fused to the dentary; Meckel's canal closed and fused anterior to the splenial; and the absence of an angular. *Palaeoxantusia*, from the Paleogene (Paleocene) of Montana and Wyoming and the Paleogene (Eocene) of California, North Dakota, Utah, and Wyoming, and *Palepidophyma*, from the Paleogene (Eocene) of Wyoming, are the oldest xantusiids (K. T. Smith and Gauthier 2013). The phylogenetic analysis by Gauthier et al. (2012) recovered *Tepexisaurus*, from the Early Cretaceous (Albian) of Mexico, as a stem-xantusiid rather than as a stem-scincoid (Reynoso and Callison 2000). Nydam and Fitzpatrick (2009) postulated a close phylogenetic relationship between Contogeniidae and Xantusiidae. Contogeniidae comprises several taxa of stem-xantusiids from the Late Cretaceous and Paleogene (Paleocene) of the western United States. One particularly characteristic feature of this group is the presence of truncated tooth crowns with apical grooves.

Figure 7.36. A, granite night lizard (*Xantusia henshawi*; Xantusiidae); **B**, black-lined plated lizard (*Gerrhosaurus nigrolineatus*; Gerrhosauridae). Courtesy of Laurie Vitt.

Xantusiidae is the sister group to Cordyliformes, which comprises Cordylidae (girdled lizards) and Gerrhosauridae (African plated lizards), both of which occur only in Sub-Saharan Africa today. Gauthier et al. (2012) listed various shared derived features for this clade, including the postorbital completely occluding the upper temporal fenestra; parietal covering nearly the entire occiput in dorsal view; temporal ramus of the squamosal in broad contact with the supratemporal process of the parietal; and ectopterygoid close to or in contact with the palatine anteriorly, excluding the maxilla from the suborbital fenestra.

Diagnostic synapomorphies for Cordyliformes include the postorbital nearly excluding the squamosal from the upper temporal fenestra; parietal with a bifid process that clasps the supraoccipital crest; single osteoderms inside the supraorbital scales; and compound osteoderms in the gular scales (Gauthier et al. 2012). The

placement of Cordyliformes in Scincoidea is also supported by molecular data (e.g., Pyron et al. 2013). Gauthier et al. (2012) listed a suite of shared derived features for Cordylidae, including the presence of a lateral flange on the septomaxilla; the dorsal overlap of the squamosal by the postorbital; palatine barely overlapping the pterygoid along a nearly transverse joint; and clavicle rod-like. Extant cordylids have body and tail scales arranged in distinct whorls and rings, with the dorsal and ventral scales typically separated by a ventrolateral fold. Present-day representatives of this clade are restricted mainly to southern Africa but extend as far north as Ethiopia. They are diurnal and predominantly insectivorous. In Europe, Cordylidae was present from the Paleogene (Oligocene) to the Neogene (Miocene) (Čerňanský 2012). Krause et al. (2003) tentatively referred *Konkasaurus*, from the Late Cretaceous (Maastrichtian) of Madagascar, to Cordylidae, but the position of this taxon among Scincoidea remains uncertain. Diagnostic synapomorphies for Gerrhosauridae include the presence of a palpebral osteoderm below the supraorbital scales, compound osteoderms in the ventral scales, and 28 or more presacral vertebrae (Gauthier et al. 2012). Gerrhosaurids have slender bodies and long tails (Fig. 7.36B). They range widely through Sub-Saharan Africa and Madagascar and mainly subsist on insects and other arthropods. The oldest record of *Gerrhosaurus* is from the Neogene (Miocene) of Kenya (Estes 1962).

The phylogenetic analyses by Conrad (2008) and Müller et al. (2011) placed *Ornatocephalus*, from the Paleogene (Eocene) of Germany (Weber 2004), and *Sakurasaurus*, from the Early Cretaceous (Valanginian-Hauterivian) of Japan (S. E. Evans and Manabe 1999), close to Cordyliformes. *Ornatocephalus*, which attained a total length of almost 1 m, has an extraordinarily long (70 cm), possibly prehensile tail and features of the limbs that suggest arboreal habits (Weber 2004; K. T. Smith et al. 2018). Its proportionately large skull bears extensive covering of osteoderms including small bony occipital spines and osteoderms in the supraorbital scales.

Gauthier et al. (2012) found a clade comprising Scincidae and two groupings of Cretaceous stem-scincids, Carusiidae and Globauridae. In support of this hypothesis, they listed several synapomorphies such as the presence of paired premaxillae and the presence of a supratemporal shelf on the postfrontal extending over the anterodorsal portion of the upper temporal fenestra.

Gauthier et al. (2012) recovered Carusiidae as the sister taxon to a clade comprising Globauridae and Scincidae. Carusiidae comprises *Carusia* and *Myrmecodaptria*, both from the Late Cretaceous (Campanian) of Mongolia (Gao and Norell 1998, 2000). Diagnostic synapomorphies for this clade include the absence of the lacrimal, the fusion of the frontals, and the fusion between the postorbital and postfrontal (Gauthier et al. 2012).

Gauthier et al. (2012) united three taxa from the Late Cretaceous (Campanian) of Mongolia (Gao and Norell 2000) as Globauridae, which they interpreted as the sister taxon to Scincidae (see also Alifanov 2000). Globauridae is diagnosed by several shared derived features such as a moderately interdigitated suture between the frontal and parietal, the posterior process of the maxilla being restricted to the anterior half of the orbit, and the position of the anterior inferior alveolar foramen dorsal or posterodorsal to that of the anterior mylohyoid foramen in the splenial (Gauthier et al. 2012).

Scincidae (skinks) is the taxonomically most diverse group of present-day lizards, with some 1,400 species occupying a wide range of habitats. Hedges (2014) divided this clade into a series of family-level taxa based on molecular and morphological data. Skinks are highly variable in body form and in body size, ranging from 2.7 to 35 cm in snout-vent length (Fig. 7.37). Most feed on arthropods and other invertebrates, but some forms prey on small vertebrates, and others subsist on plants. Skinks occur on all continents except Antarctica and on many oceanic islands, not ranging much above about 60° N latitude. One set of diagnostic synapomorphies for Scincidae concerns the compound nature of the osteoderms inside the prominent supraorbital, cheek, and gular scales. Gauthier et al. (2012) listed additional derived features, including postorbital narrowing anteriorly; the presence of a pronounced medial ridge on the jugal, with the base of this feature projecting laterally behind the ectopterygoid in dorsal view; palatines in contact with each other anteriorly; and the presence of a palpebral bone below the supraorbital scales. Many skinks have reduced limbs or lack them altogether. The Australian *Lerista* presents a particularly striking example of limb reduction, ranging from surface dwellers with the original full complement of five manual and

Figure 7.37. Scincidae. **A**, narrow-banded skink (*Eremiascincus fasciolatus*); **B**, Great Plains skink (*Plestiodon obsoletus*). A, courtesy of Division of Amphibians and Reptiles, National Museum of Natural History; B, courtesy of Laurie Vitt.

five pedal digits to burrowing forms without limbs (Greer 1989).

Scincinae comprises mainly forms with cylindrical bodies and proportionately small legs. Gauthier et al. (2012) united two extant taxa of highly derived limbless burrowing skinks, Acontinae, from southern Africa, and *Feylinia*, from Central and West Africa, in a clade (see also Conrad 2008). However, Pyron et al. (2013) considered them only distantly related to each other and referred *Feylinia* to Scincinae. Lygosominae occurs mainly in Australia, but some taxa are found elsewhere, including on many oceanic islands. *Sphenomorphus* is found in Mexico and Central America. The limbs are often reduced in size, but only a few lygosomines lack limbs altogether.

Scincinae has a poorly known fossil record to date. In Australia and New Zealand, the oldest records date from the Neogene (Miocene) (M. N. Hutchinson 1992; Lee et al. 2009). The scincine *Eumeces* is first known from the Neogene (Miocene) of Morocco, and numerous occurrences of *Plestiodon* (formerly included in *Eumeces*) have been reported from the Neogene (Miocene-Pliocene) of the United States (Estes 1983).

AUTARCHOGLOSSA: ANGUIMORPHA

Anguimorpha (from Latin *anguis*, a kind of snake, and Greek *morphe*, form) is the second major clade of Autarchoglossa. Traditionally, it comprised Anguidae, Varanoidea (Platynota; typically including Helodermatidae), and Xenosauridae (usually including *Xenosaurus* and *Shinisaurus*). Many researchers also placed snakes and mosasaurians (the latter as marine varanoids) in Anguimorpha (e.g., McDowell and Bogert 1954; Rieppel 1980a). The phylogenetic analysis by Gauthier et al. (2012) included all snake-like squamates except for pygopodoid gekkotans in this grouping (see also Estes et al. 1988). However, as these authors discussed in detail, this result is problematic. Reeder et al. (2015) found snakes and mosasaurians outside Anguimorpha and amphisbaenians not closely related to snakes. Conrad (2008) united skinks with most of the "fossorial" autarchoglossans including snakes in a clade Scinciformes. Because of these conflicting hypotheses of relationships, it is difficult to define and diagnose Anguimorpha. Widely used synapomorphies for Anguimorpha are the notched anterior portion of the tongue is separated by a transverse fold from and can be retracted into a sheath formed by the posterior portion (Rieppel 1980a) and the absence of the second ceratobranchial (Gauthier et al. 2012). Anguimorphs have head (cephalic) and dorsal body osteoderms, each in an epidermal scale. The frequently cited "anguimorph" pattern of tooth replacement (McDowell and Bogert 1954), in which replacement teeth develop posterolingual rather than lingual to the functional teeth, is problematic in terms of both morphology and phylogenetic distribution and thus is no longer considered diagnostically useful (Rieppel 1980a; Gauthier et al. 2012). Molecular-based phylogenetic analyses (Vidal and Hedges 2005, 2009; Wiens et al. 2012; Pyron et al. 2013) support the monophyly of Anguimorpha but exclude snakes from this clade.

Anguimorpha has a rich and varied fossil record (Estes 1983). Various Jurassic and Early Cretaceous autarchoglossans have been referred to this clade, but most are poorly known. A possible anguimorph, *Parviraptor*, from the Middle Jurassic (Bathonian) to Early Cretaceous (Berriasian) of England and the Late Jurassic (Kimmeridgian) of Colorado (S. E. Evans 1994, 1998; S. E. Evans and Chure 1998), has been assigned to various squamate clades and even considered a stem-snake (Caldwell et al. 2015). *Dorsetisaurus*, from the Early Cretaceous (Berriasian) of England (Hoffstetter 1967) and possibly the Late Jurassic (Kimmeridgian) of Colorado, Utah, and Wyoming, as well as Portugal (Estes 1983; S. E. Evans and Chure 1998), is generally considered a basal anguimorph (e.g., Conrad 2008). It has lanceolate tooth crowns, which (unlike those of many other anguimorphs) lack any trace of infolding or striations at the base of the crown. Fernandez et al. (2015) reported small hard-shelled eggs containing exquisitely preserved skeletal remains of squamate embryos from the Early Cretaceous of Thailand. They interpreted these embronic remains as representing a previously unrecognized lineage of Anguimorpha.

AUTARCHOGLOSSA: ANGUIMORPHA: NEOANGUIMORPHA

The molecular-based analysis by Vidal and Hedges (2009) found a clade, Neoanguimorpha (from Greek *neos*, new, Latin *anguis*, a kind of snake, and Greek *morphe*, form), comprising Anguidae, Helodermatidae, and Xenosauridae, and related taxa. Wiens et al. (2012) and Reeder et al. (2015) recovered the same topology.

Anguidae has a wide present-day geographic distribution in Asia, Europe, the Americas, and in the Caribbean. Diagnostic synapomorphies include the strongly medially and ventrally twisted retroarticular process, the position of the anterior inferior alveolar foramen between the dentary and splenial, single osteoderms in the ventral scales, and osteoderms investing the caudal scales (Gauthier et al. 2012).

Extant Anguinae occurs in Eurasia, North Africa, and North America (Fig. 7.38A). Their lower-level classification remains contentious (Klembara 2012). One present-day species of *Pseudopus*, ranging from southeastern Europe to Kazakhstan and Kyrgyzstan, attains a total length of up to 1.2 m (most of which comprises the long tail). The oldest records of *Pseudopus* are from the Neogene (Miocene) of the Czech Republic, France, Germany, and Poland (Klembara 2012). Conrad (2008) found *Ophisauriscus*, from the Paleogene (Eocene) of France and Ger-

Figure 7.38. Anguidae. **A**, eastern glass lizard (*Ophisaurus ventralis*; Anguininae); **B**, California legless lizard (*Anniella pulchra*; Anniellinae). A, courtesy of Division of Amphibians and Reptiles, National Museum of Natural History; B, courtesy of Laurie Vitt.

lizards) and Diploglossinae (galliwasps) and listed several diagnostic synapomorphies in support including the presence of a deep choanal groove on the ventral surface of the palatine and bicuspid posterior tooth crowns. Gerrhonotines are characterized by the possession of large plate-like scales and superficially resemble small crocodylians. The oldest stem-gerrhonotine is from the Paleogene (Eocene) of Wyoming (K. T. Smith and Gauthier 2013). The earliest record of the crown group is *Elgaria* from the Neogene (Miocene) of Wyoming (Scarpetta 2018). Diploglossines superficially resemble skinks in having long bodies and tails, which are covered by smooth scales, and short limbs. They occur today in Mexico, Central and South America, and in the Caribbean. *Eodiploglossus*, from the Paleogene (Eocene) of Wyoming, is the oldest stem-diploglossine (Gauthier 1982).

Glyptosaurinae is readily characterized by extensive dermal armor with tuberculate, polygonal osteoderms (R. M. Sullivan 1979). The body osteoderms imbricate from front to back but interdigitate laterally. An additional diagnostic synapomorphy is the fusion of the postorbital and postfrontal (Gauthier et al. 2012). Glyptosaurinae was widely distributed in North America during the Paleogene (Paleocene-Oligocene) and is also known from Europe and Asia. The Eocene-Oligocene *Helodermoides* attained a skull length of up to 10 cm, and its cranium is covered by loosely attached, bulbous and hexagonal osteoderms (R. M. Sullivan 1979; Fig. 7.39). Glyptosaurines were presumably omnivorous or herbivorous (Gauthier 1982).

McDowell and Bogert (1954) argued that the present-day *Xenosaurus* (knob-scaled lizards) and *Shinisaurus* (Chinese crocodile lizard) are closely related to each other even though they differ considerably in overall appearance (Fig. 7.40A). Subsequently, Estes et al. (1988) defined and diagnosed a clade Xenosauridae that included both taxa, and Gauthier et al. (2012) also recovered this grouping. However, most recent phylogenetic analyses (e.g., Conrad 2008; Bhullar 2011; Pyron et al. 2013; Conrad et al. 2014) placed *Xenosaurus* (Fig. 7.40A) and *Shinisaurus* (Fig. 7.40B) in different subclades of Anguimorpha.

Xenosauridae is diagnosed by the presence of incipient cusps on the "shoulders" of the crowns in the more posterior maxillary and dentary teeth and the presence of a prominent ventral projection of the maxillary pro-

many (R. M. Sullivan et al. 1999; Augé 2005), as the sister taxon of *Pseudopus*. *Ophisauriscus* attained a total length of over 50 cm, with up to 65 vertebrae anterior to the vent and tiny limbs (K. T. Smith et al. 2018). The body is completely covered by overlapping osteoderms.

Most recent studies have included *Anniella* (California legless lizard; Fig. 7.38B) in Anguinae (Gao and Norell 1998; Conrad 2008; but see Gauthier et al. 2012). *Anniella* occurs today on the coast of central California and western Baja California. It is a limbless burrower and can reach a total length of more than 25 cm. The earliest record of *Anniella* is from the Neogene (Miocene) of California (Gauthier 1980). *Apodosauriscus*, from the Paleogene (Eocene) of Wyoming, is the oldest taxon closely related to *Anniella* (Gauthier 1982; K. T. Smith and Gauthier 2013).

Gauthier et al. (2012) recognized a strictly New World anguid subclade that comprises Gerrhonotinae (alligator

ming, is the oldest close relative of *Xenosaurus* (Gauthier 1982; Bhullar 2011).

Helodermatidae is noteworthy as the only group of extant lepidosaurs other than colubroid snakes that employs venom. Diagnostic shared derived features for this clade include the steeply rising narial margin of the maxilla; suborbital processes of the frontals meeting in a ventromedian suture; absence of a postorbital; absence of a palpebral ossification; low number of dentary teeth; and dentary tooth crowns with anterolingual grooves for venom delivery (Pregill et al. 1986; Bhullar and K. T. Smith 2008; Gauthier et al. 2012). The head and dorsal surface of the body are covered by thick, typically hexagonal osteoderms. Some of the cranial osteoderms are fused to the underlying bones whereas others are embedded in the skin. The body is proportionately long, and the tail is short. *Heloderma horridum* (beaded lizard) and *Heloderma suspectum* (Gila monster; Fig. 7.41) occur in the American Southwest, Mexico, and Guatemala. Both feed on a variety of small vertebrates and on eggs. *Heloderma horridum* can attain a total length of up to 1 m. The earliest record of *Heloderma* is from the Neogene (Miocene) of Texas (Stevens 1977). *Eurheloderma*, from the Paleogene (Eocene or Oligocene) of France and the Paleogene (Eocene) of Germany (Hoffstetter 1957; Augé 2005; K. T. Smith et al. 2018), is the oldest undisputed helodermatid; a parietal resembling that of *Eurheloderma* has been reported from the Paleogene (Paleocene) of Wyoming (Pregill et al. 1986). *Eurheloderma* closely resembles *Heloderma* in many features but apparently lacks the distinctive beady scalation of the latter (K. T. Smith et al.

Figure 7.40. A, flathead knob-scaled lizard (*Xenosaurus platyceps*; Xenosauridae); **B**, Chinese crocodile lizard (*Shinisaurus crocodilurus*; Shinisauridae). A, courtesy of Laurie Vitt; B, from Wikipedia (photo by spacebirdy)—CC BY-SA 3.0

cess of the ectopterygoid (Conrad 2008; Bhullar 2011). *Xenosaurus* has a broad postorbital and squamosal (Rieppel 1980a). Occurring in Mexico and Guatemala, it is crepuscular and lives in narrow crevices in rock and tree stumps. *Restes*, from the Paleogene (Paleocene) of Wyo-

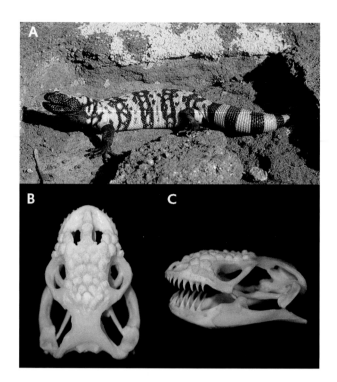

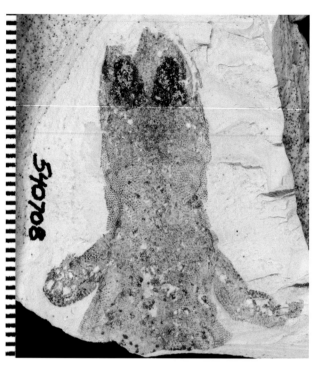

Figure 7.41. A, Gila monster (*Heloderma suspectum*; Helodermatidae); **B-C**, skull of *Heloderma suspectum* in **B**, dorsal, and **C**, lateral views. Note the osteoderms fused to the cranium; other osteoderms covering the head and neck were unfused and removed during preparation. A, courtesy of Laurie Vitt; B-C, courtesy of Alexandra Laube.

Fig. 7.42. Fossilized skin of the head, neck, forelimbs, and anterior trunk region of a Paleogene (Eocene) shinisaurian in dorsal view. Scale in millimeters. Courtesy of Department of Paleobiology, National Museum of Natural History.

2018). The phylogenetic relationships of various Cretaceous-age taxa from North America that were previously considered closely related to Helodermatidae (e.g., *Paraderma*; Gao and Fox 1996) remain uncertain.

AUTARCHOGLOSSA: ANGUIMORPHA: PALEOANGUIMORPHA

Both molecular- and morphology-based phylogenetic analyses (Townsend et al. 2004; Wiens et al. 2012; Pyron et al. 2013; Conrad et al. 2014) indicate that *Shinisaurus* (Fig. 7.40B) is more closely related to *Lanthanotus* and Varanidae (Paleoanguimorpha) and that Helodermatidae is more closely related to Anguidae and related taxa including *Xenosaurus* (Neoanguimorpha).

Several anguimorph taxa from the Late Cretaceous (Campanian) of Mongolia—*Gobiderma* (Borsuk-Białynicka 1984; Conrad et al. 2011), *Estesia* (Norell et al. 1992; Norell and Gao 1997), *Aiolosaurus* (Gao and Norell 2000), and *Ovoo* (Norell et al. 2007)—are stem-varanoids (Gauthier

et al. 2012), but their precise phylogenetic positions remain uncertain.

AUTARCHOGLOSSA: ANGUIMORPHA: PALEOANGUIMORPHA: SHINISAURIA

Conrad (2008) defined a clade Shinisauria comprising all taxa more closely related to *Shinisaurus crocodilurus* than to *Anguis fragilis*, *Heloderma horridum*, and *Varanus varius*. Conrad et al. (2014) listed various diagnostic apomorphies for Shinisauria including the presence of a dorsolateral tuberosity on the prefrontal, the participation of the prefrontal in the margin of the external narial fenestra, and the presence of a reduced prootic crest. *Shinisaurus*, from southern China and Vietnam (Fig. 7.40B), inhabits small streams and feeds on small aquatic animals. The oldest shinisaurids are from the Paleogene (Eocene) of Wyoming (Conrad 2006; Conrad et al. 2014; Fig. 7.42). *Merkurosaurus*, from the Neogene (Miocene) of the Czech Republic and Germany, is the sister taxon of Shinisauridae (Klembara 2008). *Dalinghosaurus*, from the Early Cretaceous (Barremian-Aptian) of Liaoning (China)

(S. E. Evans and Wang 2005), is the oldest shinisaurian (Conrad et al. 2014). *Necrosaurus*, from the Paleogene (Eocene-Oligocene) of France and Germany (Augé 2005; Conrad 2008; K. T. Smith et al. 2018), is possibly more closely related to Shinisauria than to Varanidae (K. T. Smith 2017). K. T. Smith et al. (2018) noted that *Necrosaurus feisti* lacks the posterior shift of the external nares present in Varanidae.

AUTARCHOGLOSSA: ANGUIMORPHA: PALEOANGUIMORPHA: VARANIDAE

As traditionally defined, Varanoidea comprised Helodermatidae (Gila monster and beaded lizard), *Lanthanotus* (earless monitor), and Varanidae (monitor lizards). Conrad (2008) defined Varanidae as comprising the last common ancestor of *Varanus varius* (lace monitor) and *Lanthanotus borneensis* (earless monitor) and all its descendants. Diagnostic synapomorphies for this clade include the absence of a contact between the frontal and the maxilla, the anterior position of the basal tubera so that the crista tuberalis is inclined posterodorsally, and the convex ventral margin of the dentary (Norell et al. 2007).

Lanthanotus is found today only in the Malaysian state of Sarawak on Borneo. Attaining a total length of up to 45 cm, it has a long neck and trunk and short legs. The cranium has a complete postorbital bony bar but lacks an upper temporal arch (Rieppel 1980a). The prefrontal and postfrontal meet on the dorsal margin of the orbit. Little is known about the life history of *Lanthanotus* other than that it burrows and is semiaquatic. *Lanthanotus* has no known fossil record. Conrad (2008) considered *Cherminotus*, from the Late Cretaceous (Campanian) of Mongolia (Borsuk-Białynicka 1984), a possible stem-lanthanotine.

Varaninae has a wide present-day geographic distribution encompassing Africa, Asia, Australia, and Papua New Guinea (Mertens 1942; Pianka and King 2004; Fig. 7.43). The fossil record indicates that it attained a much wider geographic distribution (including Europe and North America) earlier during the Cenozoic. Extant varanines include terrestrial, semiaquatic, and semi- to fully arboreal carnivores. They range in total length from 20 cm to 3 m. One still poorly known species of *Varanus*, long placed in a separate genus, *Megalania*, from the Quaternary (Pleistocene) of Australia and some neighboring

Figure 7.43. Varanidae. **A**, perentie (*Varanus giganteus*); **B**, Komodo dragon (*Varanus komodoensis*). A, from Wikipedia (photo by Greg Goebel)—CC BY-SA 2.0; B, courtesy of Division of Amphibians and Reptiles, National Museum of Natural History.

islands, possibly attained a total length of at least 5 m. The cranium of *Varanus* (Fig. 7.44) has an incomplete postorbital bar. The prefrontal and postfrontal are separated from each other on the orbital margin. The antorbital portion of the cranium is elongate. Larger-bodied species of *Varanus* have recurved, labiolingually flattened tooth crowns with serrated cutting edges, whereas some other forms have distinctly heterodont dentitions. Fry et al. (2009) argued that *Varanus komodoensis* (Komodo dragon; Fig. 7.43B) uses venom during predation and inferred the same for "*Megalania*," but Hargreaves et al. (2014) questioned the evidence for these claims. The oldest records of *Varanus* are from the Neogene (Miocene) (Clos 1995; Ivanov et al. 2018). *Saniwa*, from the Paleogene (Eocene) of Wyoming and Belgium (Figs. 7.45, 7.46), is the sister taxon of *Varanus* (Rieppel and Grande 2007; Gauthier et al. 2012). It attained a length of 1.3 m.

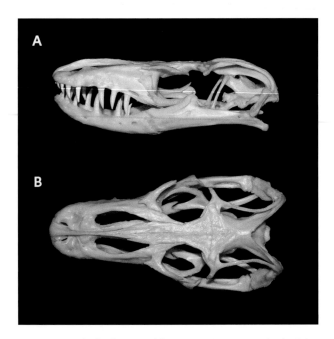

Figure 7.44. Skull of a crocodile monitor (*Varanus salvadorii*) in **A**, lateral, and **B**, dorsal views. Courtesy of Pavel Zuber.

SQUAMATA: OPHIDIA

Snakes are readily distinguished by their highly mobile skulls, long, flexible bodies with large numbers of vertebrae, and the absence of limbs. This discussion follows Lee and Caldwell (2000) in using the name Serpentes (from Latin *serpens*, snake) for the crown group containing all extant snakes, and the name Ophidia (from Greek *ophis*, snake) for the total group (Pan-Serpentes; Head 2015) encompassing Serpentes and various extinct taxa closely related to this clade (Fig. 7.47).

Snakes have heads that are small relative to total body length, yet they swallow often large prey items whole. Greatly increased mobility between the cranial bones makes this mode of feeding possible (Rieppel 1980b; Cundall and Irish 2008; Fig. 7.48). The jaws are suspended from the braincase by ligaments. The frontals and parietals grow downward to form a secondary sidewall of the braincase in derived snakes. The snout is connected to the braincase just anterior to the orbit along a joint between the nasal and frontal bones (prokinesis). This

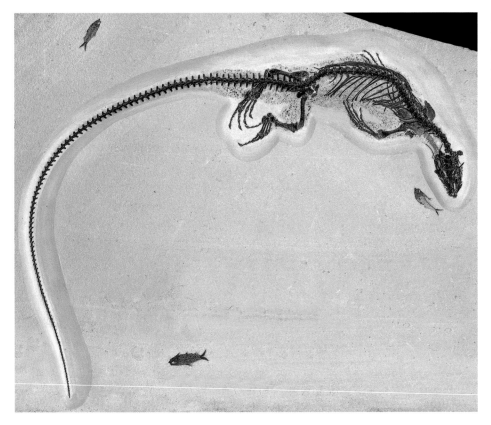

Figure 7.45. Skeleton of the varanid *Saniwa ensidens* in dorsal view, with three associated fish. Length 1.31 m. Courtesy of Lance Grande.

Figure 7.46. Skull of *Saniwa ensidens* in dorsal view. Courtesy of Lance Grande.

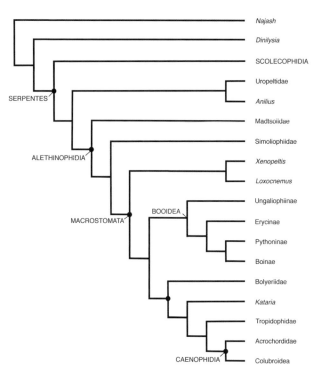

Figure 7.47. Phylogenetic hypothesis of the interrelationships of Ophidia. Based on Scanferla et al. (2013) except for the sister-taxon relationship of *Xenopeltis* and *Loxocemus* hypothesized by Gauthier et al. (2012). Dots denote node-based clades.

contact facilitates raising of the snout. Snakes lack the mesokinetic joint between the frontal and parietal present in other squamates. The quadrates connect the mandibular rami to the braincase. In more derived snakes, the quadrate extends posteroventrally and is suspended from the posteriorly projecting elongate supratemporal, which greatly increases the gape. The left and right upper jaw and palatal bones can move independently of each other in more derived snakes. In the latter, cranial mobility is further increased by reduced contacts between the palatal elements and between the nasals and frontals (Rieppel 1980b). The maxillae are used to secure the prey during the initial strike. The tooth-bearing palatines and pterygoids move independently to pull the prey into the mouth (Gans 1974). Unlike in more basal squamates, snakes lack a synovial joint between the pterygoid and parabasisphenoid. Each mandibular ramus has a joint between the dentary and postdentary bones, which allows it to bend outward during the intake of prey. The lower jaws are connected by an intermandibular ligament at their anterior (symphyseal) ends. In derived snakes, they can be spread far apart anteriorly as large prey is pulled into the mouth.

Extant snakes lack mobile eyelids. Instead, each eye is covered by a transparent scale (brille). Snakes have no tympanic membrane and perceive only low-frequency vibrations that are conducted either through the quadrate or through the lower jaw. They use the long, forked, and retractable tongue for chemoreception. This flicking tongue captures airborne molecules from the environment and transmits them to the vomeronasal organ in the roof of the mouth. Differences in the intensity of the molecular signal between the two prongs of the forked tip of the tongue allow snakes to identify the location of the source of the molecules (Schwenk 1994).

Snakes have from 160 to more than 400 vertebrae (Hoffstetter and Gasc 1969). Most of the precaudal vertebrae bear ribs. The cloacal region of the vertebral column has ribs that terminate distally in forked structures (lymphapophyses) and enclose a large lymph heart on either flank. In addition to the pre- and postzygapophyses, the neural arches each have paired anterior pedicles with ventrolaterally facing articular facets that fit into posterior recesses on the preceding neural arch (zygosphene-zygantrum accessory articulation; Fig. 7.49). The structure of the vertebrae, combined with a well-developed and highly differentiated axial musculature, increases lateral and dorsoventral flexibility as well as control of body movements (Gasc 1974). The tail is proportionately short in snakes, generally comprising no more than about 25 percent of the total length of the animal. The caudal

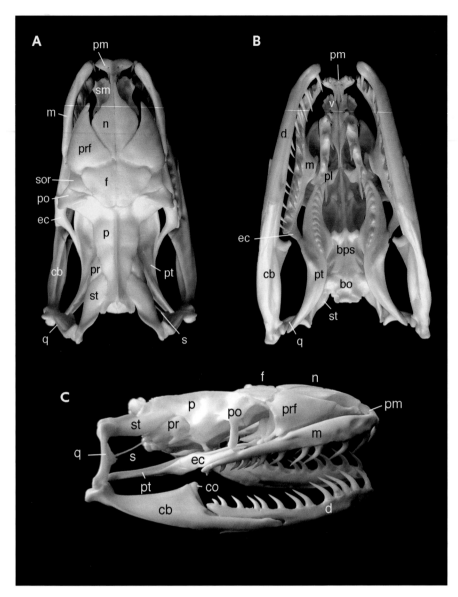

Figure 7.48. Skull of a blood python (*Python curtus*) in **A**, dorsal; **B**, ventral; and **C**, lateral views. Abbreviations: bo, basioccipital; bps, basiparasphenoid; cb, compound bone of the lower jaw; co, coronoid; d, dentary; ec, ectopterygoid; f, frontal; j, jugal; l, lacrimal; m, maxilla; n, nasal; p, parietal; pl, palatine; pm, premaxilla; po, postorbital; pr, prootic; prf, prefrontal; pt, pterygoid; q, quadrate; s, stapes; sm, septomaxilla; sor, supraorbital; st, supratemporal; v, vomer. Photographs courtesy of Pavel Zuber.

vertebrae often bear paired ventral bony processes (hemapophyses) that surround the blood vessels supplying the tail.

It is well established that *Hox* genes regulate the development of the different regions of the axial skeleton in vertebrates (Wellik 2007). Earlier studies posited that the snake body plan, with its structurally uniform precloacal vertebral column, reflected progressive expansion of the domains of specific *Hox* gene expression along the body axis, suppressing regional differentiation as well as the development of forelimbs (Cohn and Tickle 1999). Woltering (2012) suggested that retention of the standard *Hox* domains in vertebrates with alterations in gene ex-

pression prevented the differentiation of distinct axial regions. Head and Polly (2015) argued that the precursors of snakes never had a highly regionalized axial skeleton. They considered the latter condition independently derived in archosauriform reptiles and mammaliaform synapsids, respectively. Head and Polly (2015) hypothesized that the loss of the appendicular skeleton and increases in the number of vertebrae independent of *Hox* domain boundaries in the axial skeleton can account for the development of the ophidian body plan.

All snakes lack any traces of the shoulder girdle and forelimbs, and most also lack even rudiments of the pelvis and hind limbs. A few Cretaceous taxa have well-

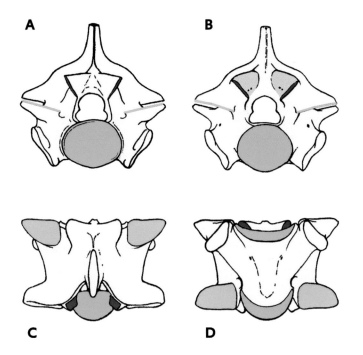

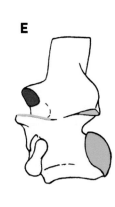

Figure 7.49. Articular features of a snake vertebra (exemplified by *Python*). **A**, anterior; **B**, posterior; **C**, dorsal; **D**, ventral; and **E**, lateral views. Pre- and postzygapophyseal facets, blue; articular surfaces of vertebral centrum, green; zygosphene-zygantrum accessory articular facets, red. Outlines of vertebra from Hoffstetter and Gasc (1969).

developed hind limbs (Rage and Escuillié 2000; see below). Among extant snakes, pythons have externally exposed, spur-like vestiges of the hind limbs, which are larger in males and are used during courtship.

The phylogenetic relationships of snakes and the origin of their body plan remain far from resolved (Rieppel 1988; Gauthier et al. 2012). Based on anatomical and paleontological data, various hypotheses of relationships have been proposed. Over the years, several clades of squamates have been interpreted as the closest relatives of snakes—varanoids (e.g., McDowell and Bogert 1954), burrowing scincomorphs (Senn and Northcutt 1973), and amphisbaenians (Rage 1982; Conrad 2008). Currently, two phylogenetic hypotheses dominate discussions. Gauthier et al. (2012) recognized a clade of fossorial squamates, which they informally named "Krypteia" (from Greek *kryptos*, hidden) and comprises snakes, amphisbaenians, and various other burrowing forms such as dibamids. The competing hypothesis considers snakes closely related to mosasaurians (Cope 1869; Caldwell and Lee 1997; Caldwell 1999; Lee 2005, 2009; Simões et al. 2018). The former supports the traditional view that snakes evolved from land-dwelling, fossorial precursors, whereas the hypothesis linking snakes and mosasaurians implies a possibly aquatic origin for this group. The loss of limbs and elongation of

the body in snakes can be accounted for under either hypothesis. Extant burrowing lizards (e.g., Scincidae) show varying degrees of limb loss and body elongation, but some mosasaurians (e.g., Dolichosauridae) also have distinctly elongated bodies and proportionately short limbs.

As noted earlier, phylogenetic analyses based on molecular data have consistently placed snakes with anguimorphs and iguanians in a clade Toxicofera (Vidal and Hedges 2005, 2009). However, this hypothesis has little morphological support, and Hargreaves et al. (2014) questioned the genetic evidence for it. Reeder et al. (2015) found snakes as the sister taxon of mosasaurs and both groups outside anguimorphs. They noted that the basal scolecophidian snakes are burrowers and thus rejected the hypothesis of a marine origin for snakes. Based on an analysis of ancestral character states in snakes, Hsiang et al. (2015) reconstructed a ground-dwelling anguimorph ancestor for this group that subsisted on small, soft-bodied prey and may have been semi-fossorial. Miralles et al. (2018) supported the hypothesis of a burrowing origin of snakes and suggested that worm snakes (Scolecophidia) might represent extant survivors of the early radiation of fossorial snakes.

The phylogenetic positions of many extinct ophidian taxa remain uncertain because they are usually based only on isolated vertebrae, which were traditionally assigned to particular groups based on their overall similarity to those of various present-day snakes. However, many vertebral features are subject to the relative position of individual vertebrae along the column, independently evolved in multiple lineages, and their phylogenetic significance has yet to be tested rigorously (Head 2015). Consequently, our understanding of early snake diversity remains tantalizingly incomplete.

STEM-SNAKES

Several taxa of Mesozoic squamates have been interpreted as stem-snakes, but the incomplete nature of these records renders such interpretations controversial. Caldwell et al. (2015) identified mostly disassociated cranial and postcranial bones from the Middle Jurassic (Bathonian) to Early Cretaceous (Berriasian) of England, the Late Jurassic (Kimmeridgian) of Portugal, and the Late Jurassic (Kimmeridgian) of Colorado as representing stem-snakes. These fossils have various snake-like features but, because of their fragmentary nature, their phylogenetic interpretation will likely remain contentious until more complete material becomes available.

Tetrapodophis, from the Early Cretaceous (Aptian) of Brazil (Martill et al. 2015), has attracted much interest as a potential stem-snake. Attaining a total length of almost 20 cm, this taxon has a long body (with about 160 presacral vertebrae) and long tail (with about 112 caudals). It has four small limbs and weakly developed limb girdles. Martill et al. (2015) interpreted *Tetrapodophis* as a burrowing animal, but the morphometric analysis by Lee et al. (2016) challenged this interpretation. Furthermore, certain features of the limbs are more consistent with aquatic habits. The phylogenetic position of *Tetrapodophis* remains uncertain, but there are no undisputed features to suggest a close relationship to snakes.

Coniophis, known only from the Late Cretaceous (Maastrichtian) of Montana and the Late Cretaceous (Maastrichtian) and Paleogene (Paleocene-Eocene) of Wyoming, is documented by isolated vertebrae and questionably attributed jaw remains (Longrich et al. 2012a). This small-sized stem-snake has often been associated with basal crown-group snakes such as *Anilius*. Two vertebrae referred to *Coniophis*, from the mid-Cretaceous (Albian-Cenomanian) of Utah (J. D. Gardner and Cifelli 1999), currently represent the oldest securely dated record of Ophidia (Head 2015).

Two Late Cretaceous taxa of terrestrial snakes, *Najash* and *Dinilysia*, were placed as successively more closely related to Serpentes in the phylogenetic analyses by Gauthier et al. (2012) and Zaher and Scanferla (2012). Earlier studies (e.g., Conrad 2008) placed *Dinilysia* closer to the most derived clade of snakes, Alethinophidia. *Najash*, from the Late Cretaceous (Cenomanian-Turonian) of Argentina, is known from postcranial remains including an articulated postcranial skeleton and referred cranial bones (Apesteguía and Zaher 2006; Zaher et al. 2009; Fig. 7.50). Distinctive features of this taxon include the presence of two sacral vertebrae (although Palci et al. [2013] recognized only one), unfused pelvic bones, and a robust femur with a prominent, blade-like trochanter.

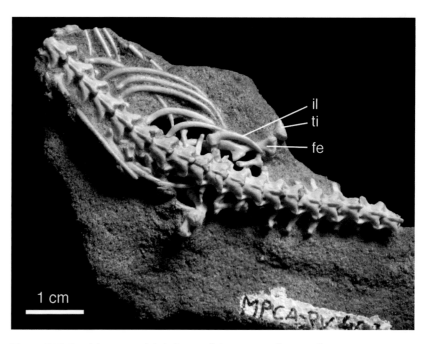

Figure 7.50. Partial postcranial skeleton of the stem-snake *Najash rionegrina* showing pelvic and hind-limb elements. Abbreviations: fe, femur; il, ilium; ti, tibia. Courtesy of Sebastian Apesteguía.

Palci et al. (2013) questioned the referral of some cranial material to *Najash*, but the latter does show features to be expected in stem-snakes.

Dinilysia, from the Late Cretaceous (Santonian-Campanian) of Argentina, is represented by several skulls and some postcranial remains (Estes et al. 1970; Caldwell and Albino 2002; Zaher and Scanferla 2012). It attained an estimated total length of about 1.5 m and has a proportionately large (about 10 cm long) skull (Fig. 7.51). Gauthier et al. (2012) recognized a clade comprising *Dinilysia* and Serpentes supported by various synapomorphies: nasal overlapping only a narrow horizontal dorsal shelf on the frontal; supraoccipital overlapping the otoccipital on the midline; foramen for the optic nerve (cranial nerve II) partially or fully enclosed by the frontal; and the median contact between the otoccipitals on the dorsal margin of the foramen magnum.

OPHIDIA: SERPENTES

Gauthier et al. (2012) listed a suite of diagnostic derived features for crown-group snakes including the anteroposteriorly narrow frontal, with a blunt process for the prefrontal off the lateral base of the subolfactory process extending into a socket on the prefrontal; crista tuberalis and crista prootica surrounding the footplate of the stapes and the lateral opening of the recessus scalae tympani; cranial trabeculae remaining paired throughout ontogeny (unlike the condition in other amniotes, in which the trabeculae fuse medially); and anterior ends of the dentaries smoothly rounded and lacking a distinct symphyseal area.

OPHIDIA: SERPENTES: SCOLECOPHIDIA

Scolecophidia (blind snakes; from Greek *skolex*, worm, and *ophis*, serpent) comprises three groups of small-sized, superficially worm-like snakes, most of which are fossorial. The skull is highly modified (Fig. 7.52). Gauthier et al. (2012) listed a suite of diagnostic synapomorphies for Scolecophidia, including parietals lacking a sagittal crest; choanal process of the palatine curved and finger-like; and supraoccipital without a nuchal crest. The frontals overlap the nasals (rather than the reverse, as in other snakes). The short mandibular rami form a symphysis anteriorly, unlike in more derived snakes. The jaw joint is placed far forward. The eyes of most scolecophidians are reduced in size and even are lost in some taxa. Parts of the pelvic girdle are retained in most scolecophidians, but there are no external traces of hind limbs. Extant scolecophidians feed exclusively on soft-bodied invertebrates, particularly termites and ants. The earliest record of this clade is from the Paleogene (Paleocene) (Mead 2013).

Scolecophidia has traditionally been divided into three family-level groupings (but see Vidal et al. 2009). Gauthier et al. (2012) found a clade comprising Leptotyphlopidae (Fig. 7.53A) and Typhlopidae (Fig. 7.53B) and listed several synapomorphies in its support, including the absence of an ectopterygoid; supraoccipital paired; vomerine process of the palatine extending lateral

Figure 7.51. Partial cranium (including right maxilla) of the stem-snake *Dinilysia patagonica* in dorsal view. Courtesy of Alessandro Palci.

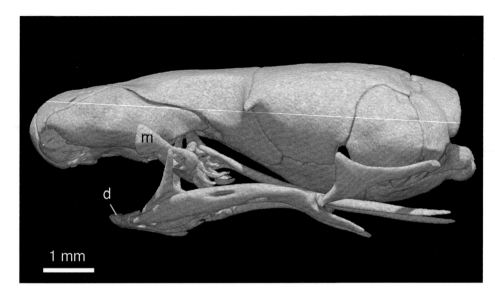

Figure 7.52. Skull of the scolecophidian *Typhlops jamaicensis* in lateral view (three-dimensional reconstruction from X-ray computed tomography). Abbreviations: d, dentary; m, maxilla. Note the toothed maxilla and edentulous dentary. Courtesy of Jessie Maisano and Digimorph.

Figure 7.53. Scolecophidia. **A**, seven-striped blind snake (*Siagonodon septemstriatus*; Leptotyphlopidae); **B**, Trinidad blind snake (*Typhlops squamosus*; Typhlopidae). Courtesy of Laurie Vitt.

to the posteromedial process of the vomer; and dorsal origin of the adductor jaw muscles restricted to the parietal. Extant Leptotyphlopidae (thread or slender blind snakes) is present in the subtropical and tropical regions of Africa and the Americas and in Southwest Asia. The largest forms attain a total length of up to 46 cm. Typhlopidae (blind snakes) occurs today in southeastern Europe and the Near East and in the tropical regions of Africa, Asia, and the Americas; and in Australia, Indonesia, and on Papua New Guinea and in the Philippines. Typhlopids differ from leptotyphlopids in the presence of a greater number of scale rows around the body and the absence of dentary teeth. The African typhlopid *Afrotyphlops schlegelii* (Schlegel's giant blind snake) is the largest extant scolecophidian, attaining a total length of 95 cm.

Anomalepididae (dawn blind snakes) is poorly known and occurs today in Panama and parts of South America. The maxilla is free of the palatine and suspended from the rod-like prefrontal, which forms loose contacts with the frontal and maxilla, respectively (Gauthier et al. 2012). Anomalepidids also differ from other scolecophidians in the possession of teeth in both the dentary and maxilla, as well as the presence of the ectopterygoid and supratemporal in most anomalepidids (Rieppel et al. 2009). Total lengths range from 15 to 40 cm. Based on molecular data, Miralles et al. (2018) hypothesized that Scolecophidia is paraphyletic, with Anomalepididae being the sister group of Alethinophidia.

OPHIDIA: SERPENTES: ALETHINOPHIDIA

Nopcsa (1923) first proposed Alethinophidia (from Greek *alethinos*, genuine, and *ophis*, serpent) for the reception of all extant snakes except Scolecophidia. Conrad (2008) formally defined this clade. Diagnostic shared derived features for Alethinophidia include the presence of a distinct medial pillar of the frontal, the presence of a tooth-bearing anterior process of the palatine, and the presence of a neomorph bone ("laterosphenoid" or "ophidiosphenoid") in the anterolateral wall of the braincase (Underwood 1967; Gauthier et al. 2012).

Unlike scolecophidians, alethinophidians are morphologically highly varied and have a wide range of modes of life (Greene 1997). Traditionally, phylogenetic analyses based on morphological data or combining morphological and molecular data have found two clades among Alethinophidia, Anilioidea and Macrostomata (Rieppel 1988; Lee and Scanlon 2002). However, some molecular studies hypothesized that the macrostomatan condition of a large gape evolved or was lost independently multiple times (e.g., Vidal et al. 2009). This review follows Gauthier et al. (2012) in recognizing Macrostomata as a morphologically well-supported clade.

BASAL ALETHINOPHIDIA

Aniliidae is represented by *Anilius* (red pipe snake; Fig. 7.54), which occurs throughout tropical northern South America today. *Anilius* is fossorial and typically attains a total length of up to 90 cm. It feeds on other reptiles and on fish. The oldest stem-aniliid is *Australophis*, from the Late Cretaceous (Campanian-Maastrichtian) of Argentina (Gómez et al. 2008; Head 2015).

Uropeltidae (including Cylindrophiidae and Anomochilidae; Gower et al. 2005) is a clade of basal alethinophidians known from southern India and Southeast Asia. These snakes are fossorial and subsist on a variety of invertebrates and small vertebrates. Cundall et al. (1993) and Scanlon and Lee (2011) listed diagnostic shared derived characters for Uropeltidae such as the broad, edentulous anterior process of the palatine and the presence of a distinct flange on the retroarticular process. *Anomochilus* and *Cylindrophis* (pipe and dwarf pipe snakes) occur today on Sri Lanka and in parts of Southeast Asia and China. Unlike other uropeltids, both taxa have cloa-

Figure 7.54. American pipesnake (*Anilius scytale*; Aniliidae). Courtesy of Laurie Vitt.

cal spurs and vestigial pelvic bones. *Cylindrophis* attains a total length of up to 35 cm. *Uropeltis* (shield-tail snakes) is restricted to southern India and Sri Lanka today and reaches a total length of up to 42 cm. Uropeltidae has no known fossil record.

OPHIDIA: SERPENTES: ALETHINOPHIDIA: MACROSTOMATA

Gauthier et al. (2012) defined Macrostomata (from Greek *makros*, large, and *stoma*, mouth) as the clade comprising the last common ancestor of *Loxocnemus bicolor* (Mexican burrowing snake), *Xenopeltis unicolor* (Asian sunbeam snake), and *Coluber constrictor* (black racer) and all descendants of that ancestor. They identified 24 diagnostic synapomorphies, including the absence of the facial process of the maxilla; supratemporal lying dorsally on the parietal; quadrate height 70–74 percent of the height of the braincase; vertical lamina of the vomer nearly completely separating the olfactory chambers along with the septomaxilla and nasal; palatine teeth enlarged, similar in size to the marginal teeth; pterygoid contacting the palatine in a tongue-in-groove joint; pterygoid teeth enlarged, similar in size to the marginal teeth; anterior end of the ectopterygoid located dorsal to the maxilla, invading the dorsal surface of the latter to varying degrees; nuchal crest on the supraoccipital extending laterally onto the otoccipital; dorsum sellae enclosed in a distinct fossa walled by the basisphenoid

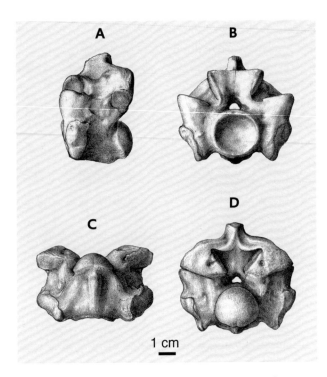

Figure 7.55. Trunk vertebra of the madtsoiid *Gigantophis garstini* in **A**, lateral; **B**, anterior; **C**, ventral; and **D**, posterior views. From Andrews (1906).

laterally and ventrally and the parasphenoid rostrum anteriorly; and coronoid eminence formed mainly by the surangular (Fig. 7.55).

Scanferla et al. (2013) found *Kataria*, from the Paleogene (Paleocene) of Bolivia, among "derived" macrostomatans. Its elongate maxilla has a heterodont tooth row and a posterior tooth set off by a short diastema.

ALETHINOPHIDIA: MACROSTOMATA: *XENOPELTIS* + *LOXOCNEMUS* AND BOLYERIIDAE

Many morphology-based phylogenetic analyses have found the extant *Xenopeltis* (sunbeam snakes) and *Loxocnemus* (Mexican burrowing python) as successive sister taxa to other macrostomatans (e.g., Cundall et al. 1993; Zaher and Scanferla 2012). Molecular-based analyses (Slowinski and Lawson 2002; Pyron et al. 2013) found *Xenopeltis* and *Loxocnemus* as successive sister taxa of Pythoninae. Gauthier et al. (2012) united *Xenopeltis* and *Loxocnemus* in a clade for which they cited several synapomorphies: development of the stylohyal process of the quadrate as an oval disk on the medial surface of

the quadrate; long overlapping contact of the ectopterygoid with the pterygoid; and the presence of one to three teeth in the premaxilla.

Xenopeltis is known from Southeast Asia and Indonesia. It has distinctly iridescent scales and typically attains a total length of about 1 m. *Xenopeltis* subsists on small vertebrates and burrows in mud. *Loxocnemus* occurs from southeastern Mexico to Costa Rica. It is nocturnal, apparently fossorial, and preys on small vertebrates. *Loxocnemus* can attain a total length of more than 1 m. Skeletal remains from the Paleogene (Oligocene) of North Dakota assigned to *Ogmophis* (which was originally based only on isolated vertebrae) share with *Loxocnemus* various apomorphies in the structure of the vertebrae and some cranial bones (K. T. Smith 2013).

Bolyeriidae (Mascarene boas) is known from two extant taxa, *Bolyeria* and *Cesarea*, from Mauritius and adjoining islets. *Bolyeria* apparently became extinct sometime after 1975. All vestiges of the pelvic girdle are lacking. *Xenophidion* (spinejaw snakes), from peninsular Malaysia and Sabah (Borneo), is closely related to Bolyeriidae (Lawson et al. 2004; Gauthier et al. 2012). A unique apomorphy shared by Bolyeriidae and *Xenophidion* is the presence of a divided maxilla with a hinge joint between the anterior and posterior portions of this jaw element. Additional synapomorphies include the presence of a ventral crest on the basioccipital and the fusion of the medial frontal pillar with the subolfactory process (Gauthier et al. 2012). *Xenophidion* has a large anterior dentary tooth but lacks upper teeth. Bolyeriidae has no known fossil record and is probably close to the base of Macrostomata (Gauthier et al. 2012).

ALETHINOPHIDIA: MACROSTOMATA: MADTSOIIDAE AND SIMOLIOPHIIDAE

The phylogenetic placement of these two distinctive clades of extinct snakes has long been controversial.

Madtsoiidae is known from the Late Cretaceous (Maastrichtian) of Argentina, India, Madagascar, Romania, and Spain, the Late Cretaceous of Niger, the Paleogene (Paleocene) of Brazil and Pakistan, the Paleogene (Eocene) of Argentina and Egypt (Fig. 7.55), and the Paleogene (Eocene) to Quaternary (Pleistocene) of Australia (Scanlon 2003, 2005, 2006; Rage 1998; LaDuke et al. 2010; J. A. Wilson et al. 2010; Mohabey et al. 2011; Vasile

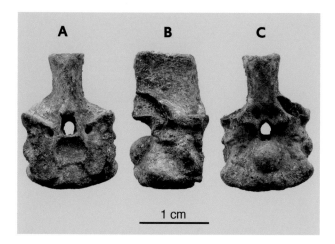

Figure 7.56. Trunk vertebra of the simoliophiid *Simoliophis rochebrunei* in **A**, anterior; **B**, lateral; and **C**, posterior views. Courtesy of Romain Houssineau (Ouest-Paléo).

et al. 2013; Rage et al. 2014). These snakes show considerable variation in body size, with total lengths ranging from about 1 m to possibly 9 m. Scanlon (2006) listed a suite of vertebral features for Madtsoiidae, but the phylogenetic significance of these traits is not clear (Mohabey et al. 2011). One possibly apomorphy shared by at least some large-bodied madtsoiid taxa is the presence of large parazygantral foramina within distinct fossae on the neural arches. Madtsoiids were apparently mainly land-dwelling predators. A skeleton of *Sanajeh*, from the Late Cretaceous (Maastrichtian) of India, was found in association with a clutch of eggs attributable to sauropod dinosaurs, coiled around one egg and adjacent to bones of a hatchling sauropod (J. A. Wilson et al. 2010). Although most recent studies placed Madtsoiidae among Alethinophidia (e.g., Rieppel et al. 2002; Zaher and Scanferla 2012; Vasile et al. 2013), some phylogenetic analyses (e.g., Lee and Scanlon 2002; Scanlon 2006) hypothesized them as stem-snakes that lack various macrostomatan cranial features.

Simoliophiidae is known only from the Late Cretaceous (Cenomanian) of Egypt, France, Israel, Lebanon, Libya, and Morocco (Tchernov et al. 2000; Rage and Escuillié 2000; Rieppel and Head 2004; Houssaye et al. 2011; Rage et al. 2016; Figs. 7.56–7.58). Representatives of this clade lived in shallow-water marine settings in the western region of Tethys and attained total lengths ranging from 50 cm to 1.5 m. Cranial synapomorphies for this clade include a well-developed supratemporal process of the parietal, the reduced height of the facial process of the maxilla, and the coronoid eminence being formed only by the coronoid bone (Gauthier et al. 2012). The elongate body (with 144–156 presacral vertebrae; Gauthier et al. 2012) and the short tail are laterally flattened. Houssaye (2013b) observed the presence of both pachyostosis and osteosclerosis in most dorsal vertebrae and ribs; these conditions are more prominently developed in some simoliophiid taxa (Fig. 7.56) than in others. At least *Eupodophis* (Fig. 7.57) and *Haasiophis* have short but well-developed hind limbs that include a set of tarsal bones but lack metatarsals or phalanges (Rage and Escuillié 2000; Head and Rieppel 2004; Houssaye et al. 2011).

Several studies found Simoliophiidae as stem-snakes (e.g., Caldwell and Lee 1997; Lee and Scanlon 2002), whereas other phylogenetic analyses recovered this clade among or close to Macrostomata (e.g., Tchernov et al. 2000; Wilson et al. 2010; Zaher and Scanferla 2012). Gauthier et al. (2012) hypothesized a sister-group relationship between Simoliophiidae and Henophidia. In addition to vestigial hind limbs, simoliophiids also have other plesiomorphic features that are absent in other alethinophidians (Gauthier et al. 2012).

ALETHINOPHIDIA: MACROSTOMATA: HENOPHIDIA

Gauthier et al. (2012) used the name Henophidia (from Greek *henos*, old, and *ophis*, snake) for a clade comprising all crown-group macrostomatans except *Xenopeltis*, *Loxocnemus*, and Bolyeriidae. They listed four diagnostic synapomorphies for this clade: dorsal lamina of the nasal in point contact with the frontal; stylohyal process of the quadrate forming an oval disk on the medial surface of the quadrate; parasphenoid rostrum shaped like an I-beam in transverse section, strongly compressed laterally, and abruptly narrowing at the trabeculae; and posterior position of the single mental foramen on the dentary (Fig. 7.59).

ALETHINOPHIDIA: MACROSTOMATA: HENOPHIDIA: BOOIDEA

Extant Booidea include Boinae (boas), Erycinae (sand boas), Ungaliophiinae (Central American dwarf boas), and Pythoninae (pythons) (Fig. 7.60). Booid interrelationships remain contentious.

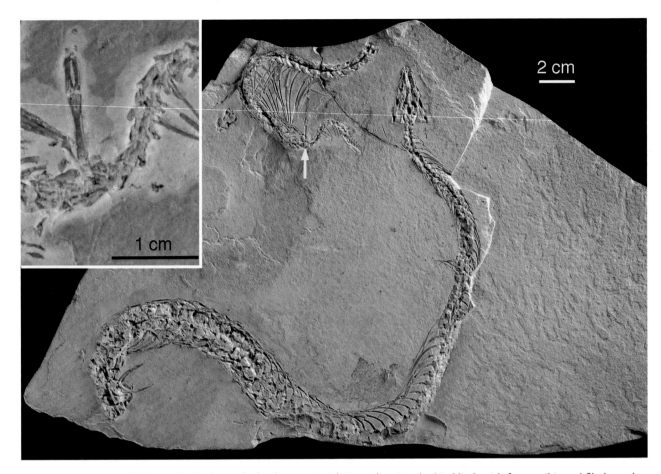

Figure 7.57. Skeleton of the simoliophiid *Eupodophis descouensi* with inset showing the hind limb with femur, tibia and fibula, and tarsal bones. Courtesy of the late Jean-Claude Rage and François Escuillié.

Ungaliophiinae is represented today by *Ungaliophis* (Neotropical dwarf boas) and *Exiliboa* (Oaxacan dwarf boa). Gauthier et al. (2012) listed a suite of shared derived features for the latter two taxa, including a posteriorly undivided palatal shelf of the premaxilla; height of the maxillary tooth crowns constant throughout the tooth row; subolfactory process of the frontal not in contact with the parasphenoid rostrum; and the absence of a coronoid bone. K. T. Smith (2013) added the absence of hemapophyses and their replacement by hemal keels throughout the caudal vertebral column as diagnostic for Ungaliophiinae. He also described a stem-ungaliophiine from the Paleogene (Oligocene) of North Dakota, which he referred to *Calamagras* (which was originally based only on isolated vertebrae).

The monophyly of Erycinae is questionable (K. T. Smith 2013; Head 2015). One synapomorphy generally considered diagnostic for at least a subset of erycines is the presence of complex accessory processes, along with the absence of zygosphene-zygantrum articulations, on the more posterior caudal vertebrae (Szyndlar 1994). Present-day erycines occur in western North America and range from Central Africa eastward to China. The oldest stem-erycines are from the Paleogene (Eocene) of France and Germany (Rage 1977; K. T. Smith et al. 2018). Numerous fossils from the Paleogene and Neogene of Europe were referred to Erycinae (Hoffstetter and Rage 1972; Szyndlar 1994; Szyndlar and Schleich 1994), but Rage and Szyndlar (2005) questioned the validity of many of these assignments. The oldest undisputed records of erycines from North America are from the Neogene (Miocene) (K. T. Smith 2013). Gauthier et al. (2012) found a clade comprising Ungaliophiinae and Erycinae.

Among Boidae, *Messelophis* and *Rieppelophis*, both from the Paleogene (Eocene) of Germany (Baszio 2004; Scanferla et al. 2016), are represented by complete, articulated

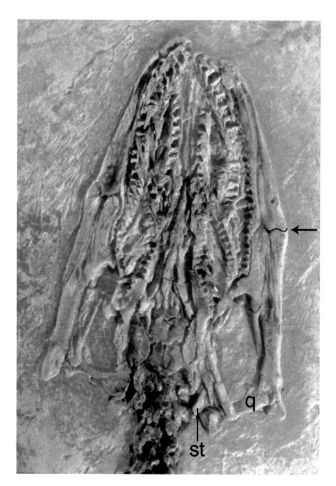

Figure 7.58. Skull of the simoliophiid *Haasiophis terrasanctus* in ventral view. Abbreviations: q, quadrate; st, supratemporal temporal. Arrow points to intramandibular articulation. Length of right mandibular ramus 2.7 cm. Courtesy of Olivier Rieppel.

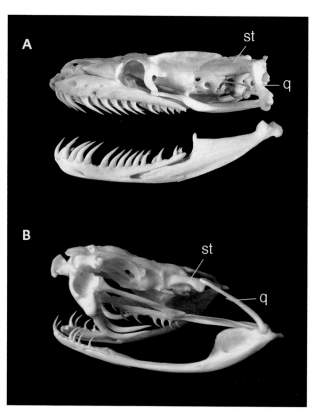

Figure 7.59. Skulls of henophidian snakes in lateral view. **A**, blood python (*Python curtus*); **B**, timber rattlesnake (*Crotalus horridus*). Abbreviations: q, quadrate; st, supratemporal. Courtesy of Pavel Zuber.

skeletons. Each attained a total length of only about 40 cm. Placement of these two small-sized taxa among Boidae is supported by shared derived features such as an expanded lateral flange of the prefrontal, the presence of a distinct, finger-like medial foot process of the prefrontal, and the presence of a well-developed surangular crest (Scanferla et al. 2016).

Calabaria (burrowing "python"), which ranges from Liberia to the Democratic Republic of the Congo today, is considered the most basal extant member of Boidae and lays eggs, unlike more derived boids, which are live-bearing (Eckstut et al. 2009).

Pythoninae (pythons) comprises mostly large-sized snakes, which can attain a total length of more than 6 m, although adults of most species reach lengths of less 4 m.

Extant pythonines occur in Sub-Saharan Africa, South and Southeast Asia, and Australia (Fig. 7.60A). The oldest undisputed records of Pythoninae date from the Neogene (Miocene) (Scanlon 2001; Rage 2013). *Python* is first known from the Miocene of Pakistan (Head 2005). Gauthier et al. (2012) enumerated several diagnostic synapomorphies for Pythoninae, including the long internasal processes of the premaxillae clasped between the descending laminae of the nasals; lacrimal duct completely enclosed in the prefrontal; trigeminal foramen enclosed fully by the prootic; and the presence of a supraciliary bone. The supraoccipital crest is low or absent. Vestiges of the pelvic elements and hind limbs are present, the latter forming cloacal spurs. Many pythons have interlabial pits with infrared receptors on the snout. These snakes mostly prey on birds and mammals, which they immobilize and kill by constriction. Although Pythoninae and Boinae have traditionally been

Figure 7.60. Boidae. **A**, blood python (*Python curtus*; Pythoninae); **B**, boa constrictor (*Boa constrictor*; Boinae). A, courtesy of Division of Amphibians and Reptiles, National Museum of Natural History; B, courtesy of Laurie Vitt.

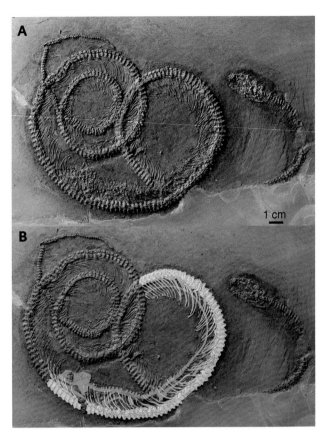

Figure 7.61. A, skeleton of the boine *Palaeopython fischeri*, preserved with gut content; **B**, photo of this specimen (with vertebrae rendered in white) showing the skeleton of the ingested stem-basilicine *Geiseltaliellus maarius* (in orange) and a cuticle fragment of an insect ingested by the lizard (in blue). Total length of snake skeleton 1.03 m. Courtesy of Krister Smith.

united based on morphological data, molecular-based hypotheses have posited their separation (e.g., Slowinski and Lawson 2002).

Most representatives of Boinae (Figs. 7.60B, 7.61) prey on reptiles, birds, and mammals, which, like Pythoninae, they immobilize and kill by constriction. The supraoccipital bears a pronounced crest. Gauthier et al. (2012) listed several diagnostic apomorphies for Boinae, including the medial flange of the frontal between the olfactory tracts slanting forward; palatine with a short tooth-bearing process (with up to five teeth); and right opening for the Vidian canal larger than the left. Most boines are arboreal, but *Eunectes* (anacondas) is semi-aquatic. *Eunectes murinus* reaches a verified total length of at least 5 m and is the heaviest present-day snake. Extant boines occur in the tropical Americas including the West Indies, on Madagascar, and on islands of the Southwest Pacific. The oldest stem-boine, *Titanoboa*, from the

Paleogene (Paleocene) of Colombia (Head et al. 2009; Head 2015; Fig. 7.62), is the largest known snake, with an estimated total length of up to about 13 m. The oldest record of *Corallus* (tree boas) is from the Paleogene (Paleocene) of Brazil (Rage 2001) and that of *Eunectes* is from the Neogene (Miocene) of Colombia (Hoffstetter and Rage 1977). Molecular data support the monophyly of extant New World boas (Burbrink 2005).

"DERIVED MACROSTOMATA"

Tropidophiidae (wood and eyelash boas) includes two extant genera of small to medium-sized terrestrial or arboreal snakes (total length of up to 1 m, but usually smaller) known from the Caribbean, especially Cuba, and several regions in Central and South America. Gauthier et al. (2012) found this clade as the sister group of Caenophidia. A key synapomorphy in support of this hypothesis is the

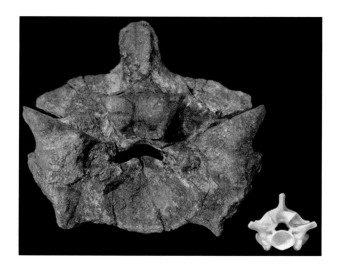

Figure 7.62. Trunk vertebra of the stem-boine *Titanoboa cerrejonensis* compared to that of a 3.4 m long extant *Boa constrictor*. Both vertebrae are shown in anterior view. Courtesy of Smithsonian Institution.

presence of an ectopterygoid flange on the maxilla. Additional derived features include the absence of a supratemporal process on the parietal, the expansion of the suborbital process of the maxilla below the contact with the ectopterygoid, the absence of a coronoid bone, the presence of a retroarticular process, and the presence of vertebral pedicles throughout the vertebral column (Gauthier et al. 2012). By contrast, phylogenetic analyses based on mitochondrial and nuclear genes (Slownski and Lawson 2002; Vidal et al. 2009) support grouping Tropidophiidae with Aniliidae (Amerophidia; Vidal et al. 2009).

Gauthier et al. (2012) listed several diagnostic synapomorphies for Tropidophiidae, including the presence of a supraorbital shelf of the frontal, which is medially demarcated by a long furrow on the dorsal aspect of the bone, and the lateral splaying of the vomerine process of the palatine to buttress the vomer posteriorly. Present-day representatives predominantly feed on lizards and other small vertebrates. The oldest record of Tropidophiidae is *Szyndlaria*, from the Paleogene (Eocene) of France (Rage and Augé 2010).

ALETHINOPHIDIA: MACROSTOMATA: HENOPHIDIA: CAENOPHIDIA

Caenophidia (from Greek *kainos*, new, and *ophis*, snake) represents the most diverse group of snakes. Diagnostic synapomorphies for this clade include the absence of a

suture between the nasal and prefrontal; fusion of the medial pillar of the frontal with the subolfactory process; maxillary process of the ectopterygoid expanding anteriorly to three times the width of the shaft of the ectopterygoid; absence of a coronoid eminence; absence of all pelvic elements; and absence of a femur (Gauthier et al. 2012).

CAENOPHIDIA: ACROCHORDIDAE

Present-day file snakes (*Acrochordus*) attain total lengths between 1 and 2.7 m and occur in brackish and marine settings ranging from Malaysia and Thailand through Indonesia to Papua New Guinea and northern Australia. Extant file snakes are characterized by loose-fitting skin with small nonoverlapping, granular scales and unusually low growth, metabolic, and reproductive rates (Sanders et al. 2011). Its considerable gape enables *Acrochordus* to prey even on large fish. The earliest records of *Acrochordus* are from the Neogene (Miocene) of South and Southeast Asia (Head 2005; Head et al. 2007).

Both molecular- and morphology-based phylogenetic analyses (e.g., Sanders et al. 2010; Gauthier et al. 2012) support a clade comprising Acrochordidae and Colubroidea. Gauthier et al. (2012) listed several shared derived features for this grouping including nasal shorter than the frontal; prefrontal short anteroposteriorly; and vomeronasal organ completely encapsulated by the vomer posteriorly.

CAENOPHIDIA: PALAEOPHIIDAE + NIGEROPHIIDAE

Rage (1984) and other authors have associated Palaeophiidae, a poorly known group of extinct marine snakes, with Acrochordidae. However, Sanders et al. (2011) found no support for such a relationship. In the absence of known cranial elements, the phylogenetic position of this group remains unresolved. Palaeophiids have been recorded from the Late Cretaceous (Maastrichtian) to Paleogene (Paleocene) of Morocco and the Paleogene (Eocene) of Belgium, Ecuador, Egypt, England, India, Kazakhstan, Libya, Mali, Nigeria, Ukraine, Uzbekistan, and the southeastern and southern United States (Rage et al. 2003). They are known only from their characteristic vertebrae and ribs, which are assigned to two form taxa, *Palaeophis* and *Pterosphenus*. Palaeophiids ranged in estimated total length from about 1.3 to more

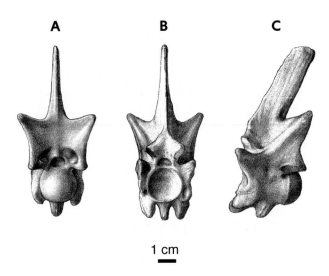

A   **B**   **C**

1 cm

Figure 7.63. Trunk vertebra of the palaeophiid *Pterosphenus schweinfurthi* in **A**, posterior; **B**, anterior; and **C**, lateral views. From Andrews (1906).

than 9 m. They lived in nearshore marine settings but could move into brackish and freshwater settings (Houssaye et al. 2013). The vertebrae are characterized by the presence of variably developed, vertically extending processes (pterapophyses) on the dorsal margins of the postzygapophyses (particularly prominent in *Pterosphenus*) and the condyle being offset by a gradual anterior constriction (Head et al. 2005; Fig. 7.63). Especially the anterior and midtrunk regions of the vertebral column in paleophiids show osteosclerosis, which presumably aided in controlling buoyancy (Houssaye et al. 2013). The ribs are not strongly curved, indicating a laterally flattened trunk. *Archaeophis*, from the Paleogene (Eocene) of Italy, and a possibly related taxon from the Paleogene (Eocene) of Turkmenistan have been interpreted as closely related to Palaeophiidae (Rage 1984), but this hypothesis has not yet been rigorously tested.

Rage (1984) also associated another poorly known group of snakes, Nigerophiidae, with Acrochordidae, but again there is no clear support for this placement. Nigerophiids have vertebral features suggesting aquatic habits. They are known from the Late Cretaceous (Maastrichtian) of Madagascar, Sudan, and possibly India, the Paleogene (Paleocene) of Niger, and possibly the Paleogene (Eocene) of Belgium and Kazakhstan (LaDuke et al. 2010; Pritchard et al. 2014). *Kelyophis*, from the Late Cretaceous (Maastrichtian) of Madagascar, probably attained a total length of less than 1 m (LaDuke et al. 2010).

CAENOPHIDIA: COLUBROIDEA

Colubroidea (from Latin *coluber*, a kind of snake) is by far the most diverse clade of Alethinophidia, encompassing more than 80 percent of all present-day snake species. It includes all extant venomous snakes. Snake venoms are complex mixtures of mostly proteins and peptides. Their primary function is to immobilize or kill prey, but some compounds may also aid in digestion. The use of venom for defense is probably secondary (Greene 1997). The venom glands of snakes are developed in the upper jaw, unlike in the condition in helodermatid anguimorphs, wherein the venom gland is confined to the lower jaw (Fry et al. 2006). Viperidae and Elapidae have venom glands with specialized compressor muscles. The development of a venom-delivery system possibly allowed colubroids to reorganize their musculoskeletal system for faster locomotion, because they no longer required axial musculature for prey constriction (Savitzky 1980; Kuch et al. 2006). Colubroids have longer axial muscle-tendon units that span many vertebrae.

Present-day colubroids occur on every continent except Antarctica and occupy a wide range of habitats. Gauthier et al. (2012) found a unique if ambiguous synapomorphy for extant representatives of Colubroidea, the presence of spike-like mineralizations on the hemipenes. They also listed several synapomorphies in the structure of the skull, including the absence of a suboptic shelf on the frontal, the undivided posteromedial head of the postorbital, and the angular being taller anteriorly, with a finger-like process extending over Meckel's canal. The oldest undisputed records of Colubroidea are from the Paleogene (Eocene) of India (Rage et al. 2008) and Myanmar (Head et al. 2005). Head et al. (2005) noted a distinctive combination of derived vertebral features for colubroid snakes: circular condyle and cotylus; cotylus with well-developed ventrolateral processes; presence of a distinct hemal keel; and prezygapophyses with well-developed accessory processes.

Traditionally, Colubroidea have been divided into four family-level groupings: Atractaspididae, Viperidae, Elapidae, and Colubridae. However, the interrelationships of the various colubroid lineages are far from

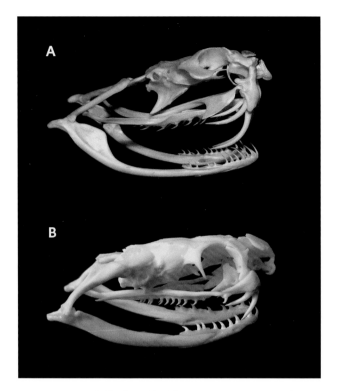

Figure 7.64. **A**, skull of a puff adder (*Bitis arietans*; Viperidae) in lateral view; **B**, skull of a king cobra (*Ophiophagus hannah*; Elapidae) in lateral view. Note multiple replacement fangs. Courtesy of Pavel Zuber.

Figure 7.65. Viperidae. **A**, European adder (*Vipera berus*; Viperinae); **B**, Brazilian lancehead (*Bothrops moojeni*; Crotalinae). A, courtesy of Wolfgang Wüster; B, courtesy of Laurie Vitt.

resolved. Based on molecular-based phylogenetic studies, Burbank and Crother (2011) proposed a different classification with additional family-level taxa: Xenodermatidae (correctly Xenodermidae), Homalopsidae, Pareatidae, Colubridae (comprising Calamariinae, Colubrinae, Natricinae, Pseudoxenodontinae, and Dipsadinae), Elapidae (with Elapinae and Hydrophiinae), Lamprophiidae (divided into Atractaspidinae, Lamprophiinae, Psammophiinae, and Pseudoxyrhophiinae), and Viperidae (comprising Azemiopinae, Crotalinae, and Viperinae). Several of these groupings have not yet been diagnosed in terms of skeletal features and thus are not further discussed here. Burbrink and Crother (2011) provided a detailed survey of the major groups of present-day colubroids.

Three important colubroid groupings with known fossil records—Viperidae, Elapidae, and Colubridae—are discussed here. The phylogenetic analysis by Gauthier et al. (2012) found them as a clade diagnosed by several shared derived features including the presence of a

thickened, anterolaterally projecting flange on the subolfactory process of the frontal; the frontal clasping the prefrontal in a notch; Meckel's canal closed and fused anterior to the splenial; and the presence of 168-180 presacral vertebrae.

The most diagnostic synapomorphy for Viperidae (vipers and pit vipers) is the presence of a long, tubular fang for venom delivery on the mobile block-like and otherwise toothless maxilla (Fig. 7.64A). This fang projects vertically during biting and can be folded back against the palate when the mouth is closed (solenoglyphous condition). Viperids occur on most continents except Australia and Antarctica and are not found on oceanic islands.

Viperidae includes Viperinae and Crotalinae (Figs. 7.65, 7.66). Viperinae occurs today in Africa, Europe, and Asia. It includes adders (*Vipera*; Fig. 7.65A), horned vipers (*Cerastes*), and puff adders (*Bitis*; Fig. 7.64A). Viperines range from about 70 cm to more than 2 m in total length.

Figure 7.66. Timber rattlesnake (*Crotalus horridus*; Crotalinae). Courtesy of Division of Amphibians and Reptiles, National Museum of Natural History.

Figure 7.67. Elapidae. **A**, king cobra (*Ophiophagus hannah*; Elapinae); **B**, sea krait (*Laticauda colubrina*; Hydrophiinae). Note the laterally flattened tail. **C**, krait (*Bungarus fasciatus*; Elapinae). A, courtesy of Wolfgang Wüster; B-C, courtesy of Division of Amphibians and Reptiles, National Museum of Natural History.

The oldest record of fangs referable to Viperinae is from the Neogene (Miocene) of Germany (Kuch et al. 2006). *Vipera* is first reported from the Neogene (Miocene) of Germany, Hungary, Morocco, and Ukraine (Rage 1984, 2013). Crotalinae occurs in the Americas and in Southwest and South Asia today. It includes moccasins (*Agkistrodon*), pit vipers (e.g., *Bothrops*; Fig. 7.65B), and rattlesnakes (*Crotalus* and *Sistrurus*; Fig. 7.66). The sides of the snout bear prominent sensory (loreal) pits between the nostrils and eyes that serve in infrared detection of prey. Crotalines vary considerably in body size, with the largest extant species attaining a total length of more than 3.5 m. Their fossil record in North America extends back to the Neogene (Miocene) (Parmley and Holman 1995).

Elapidae (Fig. 7.67) is characterized by the presence of a grooved or tubular fixed fang on the anterior portion of the maxilla (proteroglyphous condition) (Fig. 7.64B). Gauthier et al. (2012) found one unique derived feature, the presence of a V-shaped notch between the dorsal and ventral processes of the vomerine septum, that is shared at least by *Micrurus fulvius* (common coral snake) and *Laticauda colubrina* (banded sea krait). The interrelationships of elapid snakes remain contentious, but the group has traditionally been divided into Elapinae and Hydrophiinae. Extant elapines are mostly terrestrial, with the exceptions of mambas (*Dendroaspis*) and tree cobras (*Pseudohaje*), both of which are arboreal, and the aquatic water cobras (*Boulengerina*). They include cobras (*Naja*; *Ophiophagus*, Fig. 7.67A) and kraits (*Bungarus*, Fig. 7.67C) and generally prey on vertebrates. The largest present-day elapid, *Ophiophagus hannah* (king cobra), can attain a total length of 5.8 m whereas one species of *Drysdalia* only has a total length of 18 cm. Extant representatives occur from southern North America to southern South America and through Africa (except for the Sahara) to southern Asia and Australia. The oldest record of Elapinae in Africa is from the Paleogene (Oligocene) of Tanzania (McCartney et al. 2014). In Australia and Europe, this lineage extends back to at least the Neogene (Miocene) (Rage 1984, 2013; Scanlon et al. 2003; Kuch et al. 2006). The earliest record of *Bungarus* is from the Neogene (Miocene) of Pakistan (Head 2005).

Hydrophiinae includes land-dwelling elapids such as death adders (*Acanthophis*) and taipans (*Oxyuranus*) as well as sea snakes (Scanlon et al. 2003). The terrestrial hydrophiines prey on vertebrates such as lizards. The aquatic forms are found in the Indian and Pacific Ocean and subsist on fish. They have laterally flattened bodies and paddle-like tails and lack large ventral scales. Hydrophiines comprise two lineages that independently adapted to life in the ocean quite recently in geological time (Sanders et al. 2008). Sea kraits (*Laticauda*; Fig. 7.67B) frequently move onto land and lay their eggs there. By contrast, true sea snakes such as *Hydrophis* are no longer capable of moving onto land and bear live young in the water. Scanlon et al. (2003) reported a possible occurrence of *Laticauda* in the Neogene (Miocene) of Australia.

Colubridae (Fig. 7.68) is the most diverse group of snakes in terms of body form and modes of life and has

Figure 7.68. Colubridae. **A**, western ratsnake (*Pantherophis obsoletus*; Colubrinae); **B**, grass snake *Natrix natrix* (Natricinae). A, courtesy of Division of Amphibians and Reptiles, National Museum of Natural History; B, from Wikipedia (photo by Piet Spaans)—CC BY-SA 2.5.

a nearly worldwide distribution today. Many colubrids lack fangs and associated venom glands, but some taxa, such as *Dispholidus* (boomslang), have fixed fangs for venom delivery at the posterior end of the maxilla (opisthoglyphous condition). The latter condition apparently evolved independently in several colubrid lineages (Greene 1997). Gauthier et al. (2012) listed several diagnostic synapomorphies for Colubridae, including the descending lamina of the subolfactory process of the frontal tightly clasping the parasphenoid dorsolaterally; presence of a supraorbital shelf on the frontal, which is medially demarcated by a long furrow on the dorsal surface of the bone; trigeminal foramen enclosed by the prootic; and pterygoid with teeth extending onto its quadrate ramus. Colubridae is divided into Colubrinae (Fig. 7.68A) and Natricinae (Fig. 7.68B). In Europe, the earliest colubrid is known from the Paleogene (Oligocene) of France, and the clade was already diverse by Miocene times (Rage 2013). In North America, the oldest colubrids are from the Neogene (Miocene) (Parmley and Holman 1985). The interrelationships of Colubridae remain uncertain.

Natricinae is present widely throughout Eurasia and ranges from North America to northern Central America. Some representatives are semiaquatic, but most have terrestrial habits. Natricines range in total length from about 16 cm to 2 m. Gauthier et al. (2012) listed several diagnostic synapomorphies for Natricinae: nasal small, often cruciform; maxilla with enlarged teeth posteriorly; and lateral ramus of the maxillary process of the ectopterygoid large, rectangular, and extending directly laterally.

Gauthier et al. (2012) found a clade Colubrinae in their phylogenetic analysis. Diagnostic shared derived features are the posterolateral protrusion of the posterodorsolateral corner of the frontal in dorsal view and the posteriorly extending vertebral pedicles in the anterior half of the vertebral column. Colubrines are varied in body form and range in total length from 16 cm to 3.7 m. Extant species occupy a wide variety of habitats and are known from all continents except Antarctica.

8   Archosauromorpha

The Ruling Reptiles and Their Relatives

Archosauria (from Greek *archon*, ruler, and *sauros*, lizard) includes many of the most spectacular reptiles alive during the Mesozoic Era. Only two archosaurian clades, crocodylians and birds, have survived to the present day. In addition, various mostly Triassic groups of diapsid reptiles are not descendants of the most recent common ancestor of birds and crocodylians but are more closely related to archosaurs than to lepidosaurs. Benton (1985) and Gauthier et al. (1988a) united these taxa with Archosauria and its closest relatives (Archosauriformes) as Archosauromorpha (from Archosauria and Greek *morphe*, form).

Dilkes (1998) defined Archosauromorpha as comprising *Protorosaurus* and all saurians more closely related to *Protorosaurus* than to Lepidosauria. Laurin (1991) presented a list of synapomorphies for this clade: premaxilla extending posterodorsal to the external naris, excluding the maxilla from the narial margin; external nares positioned close to the midline of the cranium; at least seven cervical vertebrae; and centra of at least some of the cervical and anterior dorsal vertebrae parallelogram-shaped in lateral view, with the anterior articular surface of the centrum positioned higher than the posterior one. Ezcurra (2016) also listed the presence of various bony laminae on the cervical and anterior dorsal vertebrae as diagnostic features for Archosauromorpha. The humerus lacks an entepicondylar foramen. The carpus lacks a medial centrale. In the tarsus, the astragalus and calcaneum meet along a complex concave-convex surface, which encloses a vascular foramen. D. Brinkman (1980) surmised that this articular surface facilitates contact between the calcaneum and the distal tarsals when the hind limb assumes a more upright position. The calcaneum bears a lateral tuber, which presumably served as a pulley over which the tendon of the gastrocnemius muscle passed to its insertion on the lateral surface of metatarsal V, increasing the propulsive moment arm of this muscle (C. Sullivan 2010). The tarsus has no fifth distal tarsal. The proximal end of metatarsal V projects medially and contacts the fourth distal tarsal. Lepidosaurs also have a "hooked" metatarsal V, but this bone bears distinct plantar tubercles and probably evolved independently from the corresponding element in archosauromorphs (P. L. Robinson 1975; D. Brinkman 1981).

The interrelationships of archosauromorph reptiles other than Archosauriformes and their precise relationships to the latter group remain contentious (Fig. 8.1). Whereas the individual clades can be readily diagnosed, there are relatively few features that unambiguously relate them to each other (Dilkes 1998; Ezcurra et al. 2014; Pritchard et al. 2015; Ezcurra 2016).

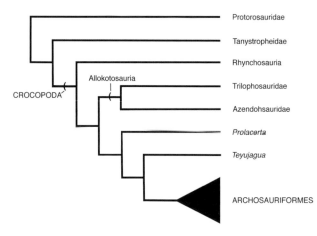

Figure 8.1. Phylogenetic hypothesis of the interrelationships of Archosauromorpha. Parentheses denote stem-based clades. Based on Nesbitt et al. (2015) with addition from Ezcurra (2016).

Archosauromorpha: Protorosauridae

Camp (1945) and other authors united a number of vaguely lizard-like late Permian (Lopingian) and Triassic reptiles in a group variously known as Protorosauria (from Greek *proteros*, earlier, and *sauros*, lizard; based on the late Permian (Lopingian) *Protorosaurus*), Prolacertilia (from Latin *pro*, before, and *lacerta*, lizard; based on the Early Triassic *Prolacerta*), or Prolacertiformes. Kuhn-Schnyder (1962) and Wild (1973) interpreted the Triassic taxa *Macrocnemus* and *Tanystropheus* as early lepidosaurs, based primarily on the assumed presence of a streptostylic quadrate. However, Rieppel and Gronowski (1981) demonstrated that the quadrate in *Macrocnemus* is held in place by a ventral process of the squamosal and is immobile. Gow (1975) first hypothesized a closer relationship between *Prolacerta* and Archosauria. Pritchard et al. (2015) found *Protorosaurus* as the sister taxon to all other archosauromorphs and referred most "protorosaurs" to Tanystropheidae.

Protorosauridae comprises all taxa more closely related to *Protorosaurus speneri* than to *Tanystropheus longobardicus*, *Prolacerta broomi*, *Sharovipteryx mirabilis*, and *Varanus komodoensis* (modified from Ezcurra et al. 2014). *Protorosaurus*, from the late Permian (Lopingian: Wuchiapingian) of England and Germany, reached a total length of up to 1.5 m (Gottmann-Quesada and Sander 2009). Its cranium (Fig. 8.2) has an incomplete lower temporal bar. The maxilla apparently enters into the margin of the external naris. *Protorosaurus* has straight, conical tooth

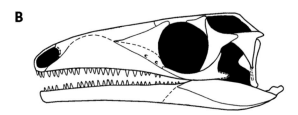

Figure 8.2. A, skull of the protorosaurid *Protorosaurus speneri* as preserved; **B**, reconstructed skull of *Protorosaurus speneri*. A, courtesy of Rainer Schoch; B, combined from Gottmann-Quesada and Sander (2009) and personal observations.

crowns. Its more posterior cervical vertebrae have elongate centra (Fig. 8.3). It has the distinction of being the first fossil reptile ever reported: Spener (1710) reported on a skeleton found in Thuringia (Germany) in 1706 and interpreted it as that of a fossil crocodile. At least two specimens of *Protorosaurus* preserve gut contents, which indicate that this reptile fed on ovules of ferns and conifers even though its dentition would suggest carnivorous habits (Munk and Sues 1993). Ezcurra et al. (2014) described postcranial bones of another protorosaurid, *Aenigmastropheus* from the late Permian (Lopingian) of Tanzania. The phylogenetic analysis by Ezcurra (2016) recovered *Protorosaurus* and *Aenigmastropheus* as the most basal archosauromorphs. Both share the close medial placement of the zygapophyses on the more anterior dorsal vertebrae and a strongly developed olecranon process of the ulna.

Czatkowiella, from the Early Triassic (Olenekian) of Poland, resembles *Protorosaurus* in the possession of elongate cervical vertebrae and long, slender cervical ribs, as well as in the absence of a posterolateral process on the premaxilla (Borsuk-Białynicka and S. E. Evans 2009).

Figure 8.3. Skeleton (cast) of *Protorosaurus speneri*. Associated fish fossil outlined in yellow. Courtesy of Rainer Schoch.

Archosauromorpha: Tanystropheidae

Tanystropheidae comprises the most recent common ancestor of *Macrocnemus*, *Tanystropheus*, and *Langobardisaurus* and all descendants of that ancestor (Dilkes 1998). It attained a wide geographic distribution during the Triassic. Synapomorphies for this clade include the flattened, expanded apices of the neural spines, the distinctly posterodorsally curved blade of the scapula, and the prominent posterior process of the ischium (Pritchard et al. 2015). The elongate neck has long cervical ribs that typically extend across intervertebral contacts (Rieppel et al. 2008). The Middle Triassic *Macrocnemus*, from the Anisian-Ladinian of Germany and Switzerland and the Ladinian of Yunnan (China) (B. Peyer 1937; Rieppel 1989; Jiang et al. 2011), attained a total length of up to 1 m (Fig. 8.4). Its longest cervical vertebrae are slightly more than twice as long as the dorsals. The hind limb is considerably longer than the forelimb, and Rieppel (1989b) considered facultative bipedalism possible during rapid locomotion. Like *Protorosaurus*, *Macrocnemus* has a "hooked" metatarsal V. *Tanystropheus*, from the Middle and Late Triassic (Anisian-Norian) of Italy, the Middle Triassic (Anisian-Ladinian) of Germany, Romania, and Switzerland, the Middle Triassic (Anisian-Ladinian) of

Figure 8.4. Skeleton with extensive skin impressions of a juvenile specimen of the tanystropheid *Macrocnemus bassanii*. Skull length 3.8 cm and snout-vent length c. 17 cm. Courtesy of Heinz Furrer.

China, and the Middle Triassic (Anisian) of Israel, is distinguished by the unusually long neck, which, in adult individuals, can reach more than twice the length of the trunk (B. Peyer 1931a; Wild 1973; Nosotti 2007). Nine of the 12 or 13 cervical vertebrae are greatly elongated (Fig. 8.5), with the ninth being the longest in *T. longobar-*

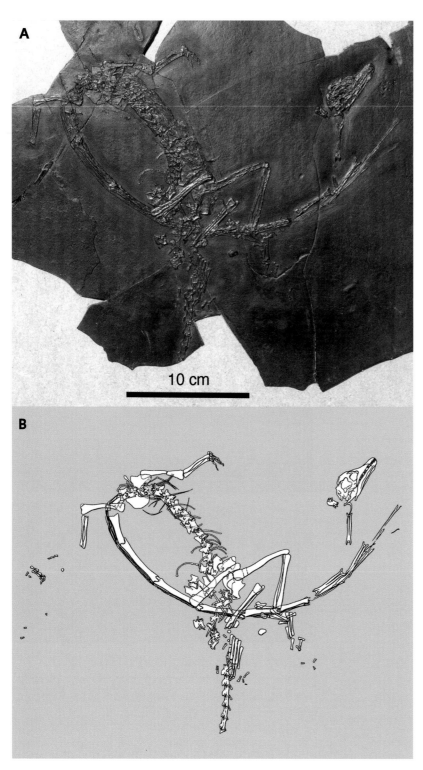

Figure 8.5. Skeleton of a small individual of the tanystropheid *Tanystropheus longobardicus*. **A**, specimen as preserved; **B**, outline drawing of the same specimen. A, courtesy of Heinz Furrer; B, modified from B. Peyer (1931a).

dicus. Their neural spines are low. The long cervical ribs, especially in the midcervical region, have slender shafts extending parallel to the ventral surfaces of the centra. The function of this enormously elongated but rather inflexible neck is problematic (Nosotti 2007). Gut contents show that adults of *Tanystropheus* fed on fish and cephalopods (Wild 1973). Juveniles of *Tanystropheus* have tricuspid teeth at the back of the jaws, whereas adult individuals have only simple conical tooth crowns, indicating an ontogenetic change in diet. As in *Macrocnemus*, the lower temporal opening in *Tanystropheus* is open ventrally. Metatarsal V is "hooked" but short in *Tanystropheus*. *Tanystropheus* attained a total length of at least 5 m. Adult individuals probably hunted in nearshore marine settings, using undulation of the tail and trunk assisted by the hind limbs for swimming (Nosotti 2007).

The cervical vertebrae of *Augustaburiania*, from the Early Triassic (Olenekian) of Russia, and *Protanystropheus*, from the Middle Triassic (Anisian) of Germany, the Netherlands, and Poland, are long but not nearly as elongated as in *Tanystropheus* (Sennikov 2011). *Tanytrachelos*, from the Late Triassic (Norian) of the United States, attained a total length of about 30 cm and lacks elongated cervicals (Olsen 1979; Pritchard et al. 2015). A peculiar feature shared by at least *Tanystropheus* and *Tanytrachelos* is the presence of calcified cartilaginous rods in the pelvic region of some specimens of these taxa. Wild (1973) interpreted the rods as possibly related to lizard-like hemipenes. However, Böhme (1988) disagreed with this interpretation, and the function of these structures remains uncertain. *Langobardisaurus*, from the Late Triassic (Norian) of Italy, is closely related to *Tanytrachelos* (Pritchard et al. 2015) and is noteworthy for its heterodont dentition with tricuspid "cheek" teeth and a large crushing tooth with minute cusps at the posterior end of each dentary and maxillary tooth rows (Renesto and Dalla Vecchia 2000).

Dinocephalosaurus, from the Middle Triassic (Anisian) of Guizhou (China), resembles *Tanystropheus* in the extraordinary elongation of its neck. Unlike in the latter, however, this lengthening is due to a substantial increase in the number of cervical vertebrae (27) rather than the elongation of individual cervicals (Rieppel et al. 2008). Rieppel et al. (2008) argued that *Dinocephalosaurus* was more adapted to an aquatic mode of life than was *Tanystropheus*. Liu et al. (2017) demonstrated that it was viviparous.

Ozimek, from the Late Triassic (Norian) of Poland (Dzik and Sulej 2016), is possibly related to Tanystropheidae based on its elongated cervical vertebrae and the short, posterodorsally curved scapula. However, its limbs are very long and slender, which led Dzik and Sulej (2016) to compare it to the enigmatic gliding reptile *Sharovipteryx*, from the Middle-Late Triassic (Ladinian-Carnian) of Kyrgyzstan, and even infer the presence of gliding membranes, as in the latter. The plate-like coracoids of *Ozimek* are greatly enlarged and meet ventromedially.

Archosauromorpha: Crocopoda

Ezcurra (2016) proposed a clade Crocopoda (from Greek *krokodeilos*, a kind of lizard, and *pous*, foot) for all taxa more closely related to *Azendohsaurus madagaskarensis*, *Trilophosaurus buettneri*, *Rhynchosaurus articeps*, and *Proterosuchus fergusi* than to *Protorosaurus speneri* and *Tanystropheus longobardicus*. Diagnostic synapomorphies include the presence of a prominent tuber on the calcaneum, ankylothecodont tooth implantation, and the presence of a supratemporal fossa immediately anterior or medial to the upper temporal fenestra.

Archosauromorpha: Crocopoda: Rhynchosauria

Rhynchosauria (from Greek *rhynchos*, snout, and *sauros*, lizard) comprises all taxa more closely related to *Rhyncho-

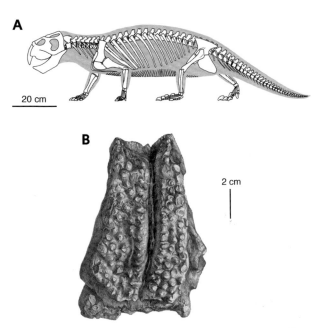

Figure 8.6. A, reconstructed skeleton of the hyperodapedontine rhynchosaurid *Hyperodapedon gordoni* in lateral view; **B**, right maxillary tooth plate of a very large specimen of *Hyperodapedon huxleyi* in occlusal view. Anterior is toward the top of the illustration. A, courtesy of Michael Benton; B, from Lydekker (1885).

saurus articeps than to *Trilophosaurus buettneri*, *Prolacerta broomi*, or *Crocodylus niloticus* (Ezcurra 2016). In all rhynchosaurs for which the skull is known, the premaxillae have long, ventrally projecting bony processes that form a prominent "beak" overhanging the mandibular symphysis (Fig. 8.7). Initially, these processes were homologized with the incisor-like teeth that are fused to the premaxillae in adult tuataras (*Sphenodon*), and thus rhynchosaurs were grouped with *Sphenodon* and its relatives (Rhynchocephalia). However, this similarity is superficial, and no compelling evidence links rhynchosaurs to rhynchocephalians (Carroll 1977).

Dilkes (1998) listed several diagnostic features for Rhynchosauria including the presence of a single, median external narial opening, a contact between the premaxilla and prefrontal, and the presence of usually two or more rows of maxillary teeth (Figs. 8.6B, 8.7). Rhynchosaurs ranged in time from the Early Triassic (Induan-Olenekian) to the Late Triassic (Norian). They are best known from the Southern Hemisphere, where they were the most

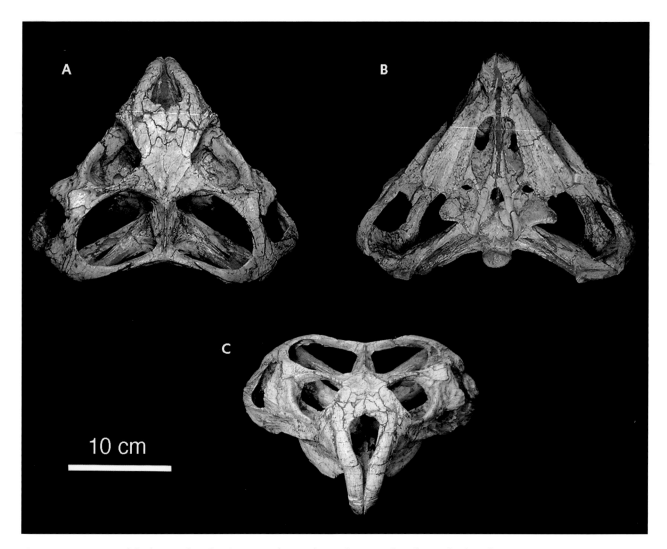

Figure 8.7. Cranium of the hyperodapedontine *Teyumbaita sulcognathus* in **A**, dorsal; **B**, palatal; and C, anterior views. Courtesy of Cesar Schultz.

common herbivores in many Middle and early Late Triassic tetrapod communities (e.g., Martínez et al. 2013b).

The basal rhynchosaur *Mesosuchus*, from the Middle Triassic (Anisian) of South Africa, has a premaxilla bearing two teeth, and the maxilla and dentary each have only a single row of marginal teeth (Dilkes 1998). The teeth sit in shallow sockets. The palatine and pterygoid both lack teeth. *Howesia* and *Eohyosaurus*, also from the early Middle Triassic (Anisian) of South Africa, both have several rows of small conical teeth on the mediolaterally broad maxilla and dentary (Dilkes 1995; Butler et al. 2015). The lower temporal bar is incomplete in all three taxa. These basal rhynchosaurs each attained a total length of less than 1 m.

Rhynchosauridae comprises the most recent common ancestor of *Rhynchosaurus*, "*Scaphonyx*," *Stenaulorhynchus*, and *Hyperodapedon* and all descendants of that ancestor (Dilkes 1998). It ranges in time from the Middle Triassic (Anisian) to the Late Triassic (Norian). Synapomorphies of Rhynchosauridae include the blade-and-groove occlusion with the dentary blade fitting into a maxillary groove; the raised bony rim around the orbit; posterior process of the jugal extending ventrolaterally; long medial process of the squamosal forming much of the posterior margin of the upper temporal fenestra; and the absence of teeth on the palatine, anterior ramus of the pterygoid, and vomer (Butler et al. 2015; Figs. 8.6B, 8.7). This blade-and-groove occlusion facilitated cutting and shredding of fibrous plant fodder (Benton 1983; Reisz and Sues 2000). Old teeth were not replaced but became worn down with use, and new teeth were added at the back of the jaws. Individual teeth are set in sockets and firmly attached to the jawbones by secondary bone (ankylothecodont implantation; Chatterjee 1974). Unlike in basal rhynchosaurs, the lower temporal bar is complete.

One subclade of Rhynchosauridae, Hyperodapedontinae is well documented from the Late Triassic (Carnian-Norian) of Argentina, Brazil, India, and Scotland (Chatterjee 1974; Benton 1983, 1990; Montefeltro et al. 2010) and is also known from North America. *Hyperodapedon* (Fig. 8.6A) attained a total length of at least 2 m. Presumably related to the development of powerful adductor jaw muscles, the transverse width of the posterior portion of the dorsoventrally deep cranium exceeds its anteroposterior length (Fig. 8.7). Benton (1983) argued that *Hyperodapedon* could have employed its hind limbs, especially the proportionately large feet with their blade-like unguals, for digging.

Archosauromorpha: Crocopoda: Allokotosauria

Allokotosauria (from Greek *allokotos*, foreign, strange, and *sauros*, lizard) is a group, only recently recognized, of presumably mostly herbivorous Triassic archosauromorphs. Nesbitt et al. (2015) defined it as the least inclusive clade containing *Azendohsaurus madagaskarensis* and *Trilophosaurus buettneri* but not *Tanystropheus longobardicus*,

Proterosuchus fergusi, *Protorosaurus speneri*, or *Rhynchosaurus articeps*. Allokotosauria is characterized by a number of synapomorphies including the expanded and hooked posterior side of the proximal head of the quadrate and the presence of a prominent tubercle above the glenoid facet of the scapula (Nesbitt et al. 2015). The oldest known representative of this clade is *Coelodontognathus*, from the Early Triassic (Olenekian) of Russia (Ezcurra 2016).

The dorsoventrally deep cranium of *Trilophosaurus*, from the Late Triassic (Carnian-Norian) of Arizona, New Mexico, and Texas (Gregory 1945; Spielmann et al. 2008; Fig. 8.8A), has only upper temporal fenestrae, which are separated by a tall, narrow parietal crest medially. The temporal bar is deep dorsoventrally and lacks any trace of an opening or a ventral emargination. Both features indicate substantial development of the adductor jaw muscles. The anterior ends of the snout and mandible of *Trilophosaurus* lack teeth and were probably covered by a keratinous beak in life. The teeth of the maxilla and the more posterior portion of the dentary (Fig. 8.8B) have transversely broad crowns that bear three cusps linked by ridges and interlock like cogs when the jaws come into contact. The tooth implantation is ankylothecodont. The femur of *Trilophosaurus* bears a prominent internal trochanter, which extends along the proximal third of this bone. The ungual phalanges of the proportionately large manus and pes are mediolaterally flattened and strongly curved. *Trilophosaurus* attained a total length of up to 2.5 m (Gregory 1945).

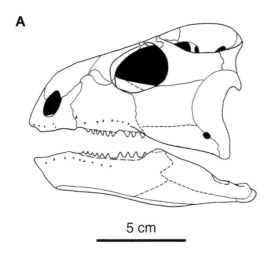

A

5 cm

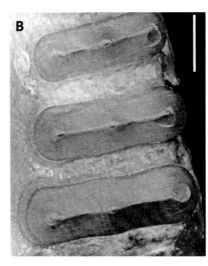

B

Figure 8.8. A, reconstruction of the skull of the trilophosaurid *Trilophosaurus buettneri* in lateral view. **B**, occlusal view of three molariform dentary teeth of *Trilophosaurus buettneri*. Scale bar = 1 mm. A, modified from Sues and Fraser (2010); B, digitally modified from an SEM photograph in Heckert (2004).

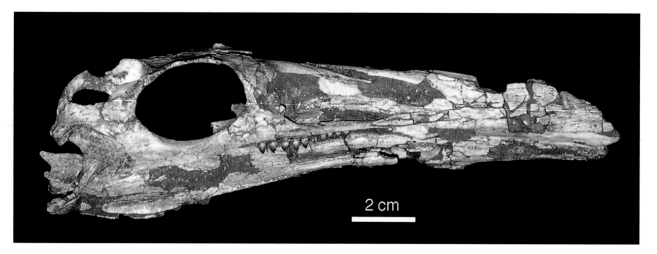

Figure 8.9. Skull of the trilophosaurid *Teraterpeton hrynewichorum* in lateral view.

The closely related *Spinosuchus*, from the Late Triassic (Norian) of Texas, has cervicodorsal vertebrae with long, rod-like neural spines that have thin, sheet-like lateral expansions and dorsal vertebrae with greatly elongated, mediolaterally flattened neural spines that terminate in expanded triangular apices (Spielmann et al. 2009).

Teraterpeton, from the Late Triassic (Carnian) of Nova Scotia, is also closely related to *Trilophosaurus* and shares with it the presence of only upper temporal fenestrae and deep, imperforate temporal bars (Sues 2003; Fig. 8.9). Furthermore, the anterior portions of the upper and lower jaws lack teeth and probably supported a beak in life. Unlike in *Trilophosaurus*, however, the snout of *Teraterpeton* is long and narrow and the external naris is long and low. The maxilla and the more posterior portion of the dentary have inset rows of transversely broad teeth, each with a single tall cusp and a basined "heel." The palatine of *Teraterpeton* bears a row of similar teeth extending parallel to the maxillary row. The ungual phalanges are strongly flattened mediolaterally. The tail is short.

Azendohsaurus, from the Late Triassic (Carnian) of Morocco, was originally identified as an ornithischian dinosaur (Duituit 1972) based on the similarity of its teeth to those of basal ornithischians. However, more recent discoveries of well-preserved remains of the skull and postcranial skeleton from the Late Triassic (Carnian) of Madagascar established that *Azendohsaurus* is not a dinosaur but an allokotosaurian archosauromorph that independently developed tooth crowns with denticulated cutting edges (Flynn et al. 2010; Nesbitt et al. 2015). The tooth crowns of *Azendohsaurus* are expanded mesiodistally and are leaf-shaped. Those of the lower teeth are taller and bear nearly twice as many denticles as those of the upper teeth. The teeth are fused to their alveoli. The external nares were apparently confluent. The lower temporal bar of *Azendohsaurus* is incomplete. The palatine bears a single row of large teeth and the pterygoid four rows of smaller teeth. The palatal teeth closely resemble the marginal ones. The structure of the skull and dentition indicates that *Azendohsaurus* was an herbivore (Flynn et al. 2010). *Azendohsaurus* reached a total length of at least 2 m. The long cervical vertebrae bear small epipophyses on the postzygapophyses (Nesbitt et al. 2015). The posterior cervicals, anterior dorsals, and sacrals have hyposphene-hypantrum accessory articulations, which are otherwise present only in archosaurian reptiles. *Shringasaurus*, from the Middle Triassic (Anisian) of India (Sengupta et al. 2017), is closely related to *Azendohsaurus* and attains a total length of 3 to 4 m. It is noteworthy for the sexually dimorphic presence of a pair of prominent, anterodorsally extending supraorbital horns formed by the frontals.

Archosauromorpha: Crocopoda: *Prolacerta + Teyujagua*

Prolacerta, from the Early Triassic (Induan) of South Africa and Antarctica (Fig. 8.10), resembles *Protorosaurus* and related taxa in many features of its skeleton (Camp 1945; Gow 1975). However, phylogenetic analyses (Benton 1985; Dilkes 1998; Modesto and Sues 2004; Ezcurra 2016) have consistently placed it closer to archosauriforms than to other "protorosaurians." Ezcurra (2016) listed the pres-

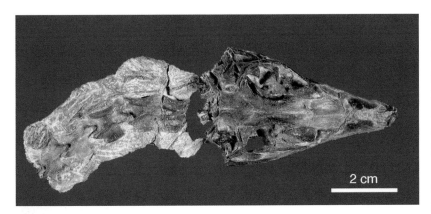

Figure 8.10. Skull (in dorsal view) and anterior cervical vertebrae of the prolacertid *Prolacerta broomi*. Courtesy of the late Don Baird.

ence of two rows of large foramina on the dentary, starting at its anteroventral end, and the presence of postaxial cervical intercentra as synapomorphies for *Prolacerta* and Archosauriformes. The tooth crowns of *Prolacerta* resemble those of Archosauriformes in being recurved and labiolingually flattened, but they lack serrated cutting edges. *Prolacerta* attained a total length of 60 to 70 cm. *Kadimakara*, from the Early Triassic (Induan) of Australia, is closely related to *Prolacerta* (Ezcurra 2016).

Teyujagua, from the Early Triassic of Brazil, differs from *Prolacerta* in the presence of serrated distal cutting edges on its teeth, the dorsomedially positioned, confluent external nares, the presence of a lateral shelf on the surangular, and the presence of an external mandibular fenestra (Pinheiro et al. 2016).

Archosauromorpha: Choristodera

Choristodera (from Greek *choristos*, separated, and *dere*, neck) is a clade of superficially often crocodile-like reptiles that inhabited freshwater environments in Europe, Asia, and North America. Dilkes (1998) defined it as comprising the most recent common ancestor of *Lazarussuchus*, *Cteniogenys*, and *Champsosaurus* and all descendants of that ancestor. Choristoderans ranged in time from the Middle Jurassic (Bathonian) to the Neogene. Storrs et al. (1996) referred the poorly known *Pachystropheus*, from the Late Triassic (Rhaetian) of England and Germany, to this group, but the current lack of clearly attributable cranial remains leaves this placement uncertain.

The phylogenetic position of Choristodera remains uncertain. Dilkes (1998) placed it as the sister group of Sauria. Müller (2004) considered choristoderans the most

basal lepidosauromorphs, whereas Gauthier et al. (1988b) interpreted them as the most basal archosauromorphs. The phylogenetic analysis by Ezcurra (2016) recovered Choristodera as part of Sauria in an unresolved trichotomy with Archosauromorpha and Lepidosauromorpha.

Choristodera is diagnosed by a suite of synapomorphies (S. E. Evans 1990; Gao and Fox 2005; Ezcurra 2016): skull typically strongly dorsoventrally flattened; external nares confluent medially; nasals fused along the midline of the cranium; prefrontals contacting each other along the midline of the cranium; new (neomorph) bone present between the skull roof, pterygoid, and quadrate; sacral and caudal ribs not fused to their corresponding vertebrae; and postcranial bones (especially the ribs) pachyostotic in more derived choristoderans.

Lazarussuchus, from the Paleogene (Paleocene) of France and Germany and the Neogene (Miocene) of the Czech Republic, was interpreted as a late-surviving basal choristoderan (Hecht 1992). Matsumoto et al. (2013) listed additional features in support of this hypothesis. *Lazarussuchus* differs from other known choristoderans in features such as the position of the external nares high on the snout and the short nasals separating the prefrontals anteriorly (Matsumoto et al. 2013). It reached a total length of about 30 cm.

Cteniogenys, from the Middle Jurassic (Bathonian) of England and Scotland, the Late Jurassic (Kimmeridgian) of Portugal, and the Late Jurassic (Kimmeridgian-Tithonian) of Utah and Wyoming (S. E. Evans 1990), attained a total length of about 30 cm. It is distinguished from other choristoderans by features such as low tooth crowns without enamel infolding at their bases and separate postorbital and postfrontal bones that enter into the margin of the orbit. Furthermore, the proportionately large orbits of *Cteniogenys* face laterally rather than dorsally, the upper temporal fenestrae are relatively small, and the vertebral centra are proportionately longer than in more derived choristoderans.

Monjurosuchidae, with *Monjurosuchus* (Fig. 8.11) and *Philydrosaurus*, both from the Early Cretaceous (Barremian-Aptian) of Liaoning (China), is characterized

Figure 8.11. Skeleton of an immature individual of the monjurosuchid *Monjurosuchus splendens* in dorsal view with skin impressions. Courtesy of Mick Ellison.

by the absence of a lower temporal fenestra, the upper temporal fenestra being smaller than the orbit, and the presence of a prominent posterior process on the ischium (Gao et al. 2000; Gao and Fox 2005). *Monjurosuchus* reached a snout-vent length of up to 30 cm. Hyphalosauridae, with *Hyphalosaurus*, from the Early Cretaceous (Aptian) of Liaoning (China), and *Shokawa*, from the Early Cretaceous (Valanginian) of Japan, share proportionately small skulls, the closure of the lower temporal fenestra in adults, greatly elongated necks with 16–24 cervical vertebrae, and triangular, spike-like neural spines on the caudal vertebrae (Gao and Ksepka 2008).

Neochoristodera is a clade of Cretaceous and Paleocene choristoderans that is diagnosed by numerous shared derived features including: long, fused nasals; orbit small, dorsally facing; posteriorly broad upper temporal fenestra larger than the orbit; infolding of the enamel at the bases of the tooth crowns; and vertebral centra anteroposteriorly short and spool-like (Gao and Fox 2005). *Champsosaurus*, from the Late Cretaceous (Campanian) to Paleocene of western North America and the Paleocene of France, has a long snout with numerous slender, conical teeth and a transversely broad, somewhat bulbous temporal region with large upper temporal fenestrae (Erickson 1972; Gao and Fox 1998; Fig. 8.12). The dentaries form a long mandibular symphysis, in which the splenials also participate. Metatarsal V is not "hooked" (Gao and Fox 1998). *Champsosaurus* attained a total length of up to 2 m. By contrast, the larger *Simoedosaurus*, from the

Paleocene of France and western North America, has a shorter and broader snout, a more triangular outline of the skull in plan view, and a short mandibular symphysis (Russell-Sigogneau and Russell 1975; Erickson 1987). Furthermore, the postorbital and postfrontal are fused to each other, and the alveoli are broad transversely. *Simoedosaurus* shares with *Tchoiria*, from the Early Cretaceous (Aptian-Albian) of Mongolia, and *Ikechosaurus*, from the Early Cretaceous of Inner Mongolia (China), the presence of paired rows of palatal teeth and the short contact between the parietal and the fused postorbital and postfrontal (Ksepka et al. 2005). S. E. Evans and Hecht (1993) interpreted the more aquatic, gharial-like *Champsosaurus* (Fig. 8.12) as possibly feeding on fish and the more crocodile-like *Simoedosaurus* (which is found together with *Champsosaurus* at several locations) as subsisting on larger prey.

Possible Archosauromorphs

Three taxa of distinctive small reptiles from the Triassic have been tentatively assigned to Archosauromorpha. However, the known specimens have few features that support these placements, and many details of the structure of the poorly preserved skeletons remain controversial.

Fraser et al. (2007) reported two partial skeletons of a small-sized reptile, *Mecistotrachelos*, from the Late Triassic (Norian) of Virginia. *Mecistotrachelos* has eight pairs of greatly elongated trunk ribs (2–8) that articulate with

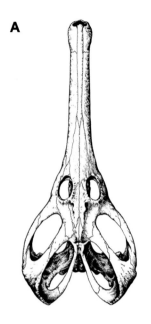

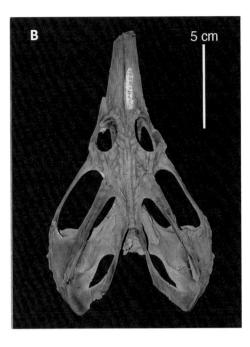

Figure 8.12. **A**, reconstructed cranium of the champsosaurid *Champsosaurus gigas* in dorsal view; **B**, cranium of *Champsosaurus lindoei* (lacking the tip of the snout) in dorsal view. A, modified from Erickson (1972); B, courtesy of Don Brinkman.

Figure 8.13. Partial skeleton of the possible archosauromorph *Longisquama insignis* with impressions of greatly elongated dorsal scales. Courtesy of Robert Reisz and Diane Scott.

prominent transverse processes on the dorsal vertebrae. These ribs apparently supported a gliding membrane in life, similar to the condition in the present-day gliding lizards of the genus *Draco*. The skull of *Mecistotrachelos* (length about 2 cm) has a tapered snout bearing numerous small teeth. The cervical vertebrae have elongate centra. The forelimb is somewhat shorter than the hind limb. The slender hind limb has a long, straight femur, a distal limb segment that is longer than the femur, and a short pes.

Sharov (1970, 1971) first reported discoveries of two taxa of distinctive small reptiles from the Middle to Late Triassic (Ladinian-Carnian) of Kyrgyzstan. One form, *Longisquama*, is known from a skull and articulated anterior part of the postcranial skeleton and is readily distinguished by the presence of a fan-like arrangement of long scales along its back (Sharov 1970; Voigt et al. 2009; Fig. 8.13). Each scale is shaped like a hockey stick, with an expanded and flattened distal end. It had a probably pliable, ridged outer layer that enclosed a central

Figure 8.14. Skeleton of the possible archosauromorph *Sharovipteryx mirabilis* with impressions of skin membranes between the hind limbs and along the sides of the trunk. Total length c. 24 cm. Courtesy of Nick Fraser.

space. The apparently diapsid skull of *Longisquama* (length about 2.5 cm) has small, conical marginal teeth. The neck is short and comprises seven vertebrae. Manual digit IV is as long as the humerus. The elongated penultimate manual phalanges and the pointed, curved unguals suggest arboreal habits for *Longisquama* (Unwin et al. 2000).

The other taxon, *Sharovipteryx*, is known from a single articulated skeleton with extensive skin impressions (Sharov 1971; Unwin et al. 2000; Fig. 8.14). This specimen has a total length of about 24 cm. Its long and narrow skull has large orbits. The neck and especially the tail are long. The slender limb bones are for the most part hollow. The forelimb was apparently short, whereas the hind limbs are greatly elongated. The femur is as long as the trunk, and the tibia is slightly longer than the femur. An extensive membrane extended between the hind legs and the base of the tail. A second, smaller membrane was developed along the trunk and back to

the knee. Gans et al. (1987) and Dyke et al. (2006) reconstructed *Sharovipteryx* as a glider, which may have used its tail for counterbalance and its head and neck for steering. Unwin et al. (2000) suggested a possible relationship between *Sharovipteryx* and "protorosaurian" archosauromorphs.

Archosauromorpha: Crocopoda: Archosauriformes

For many years, dinosaurs, pterosaurs, and crocodylians, as well as an assortment of variously related, mostly Triassic reptiles were grouped together as Archosauria (Romer 1956, 1966). Those forms that could not be readily assigned to the former three clades were traditionally united as Thecodontia (from Greek *theke*, container, and *odous* (*odon*), tooth) in reference to their tooth implantation. In time, Thecodontia came to be seen as the ancestral stock from which all other archosaurs arose (Romer 1956, 1966; Charig 1976). Gauthier and Padian (1985) and Gauthier (1986) first argued that Archosauria can be divided into two major lineages, one leading to crocodylians and the other to dinosaurs and their descendants, birds. Many Triassic-age taxa can confidently be assigned to either lineage, but others, although clearly related to Archosauria, are more basal than the most recent common ancestor of crocodylians and birds and thus are not part of the crown group (Gauthier et al. 1988a). They are grouped together with archosaurs as Archosauriformes.

Nesbitt (2011) defined Archosauriformes (Fig. 8.15) as the least inclusive clade containing *Crocodylus niloticus* and *Proterosuchus fergusi*. This clade is readily distin-

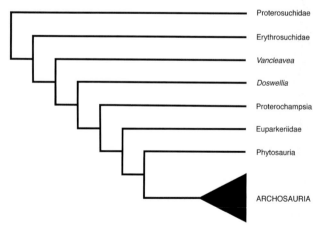

Figure 8.15. Phylogenetic hypothesis of the interrelationships of Archosauriformes. Based primarily on Nesbitt (2011).

guished by a number of synapomorphies: the presence of antorbital and external mandibular fenestrae; orbits dorsoventrally tall and anteroposteriorly narrow; marginal teeth deeply implanted in the jaw elements; tooth crowns with typically serrated cutting edges; alveolar walls usually forming interdental plates on the lingual surfaces of the jawbones; contact between the jugal and quadratojugal; and the presence of an ossified laterosphenoid (Nesbitt 2011). Gauthier (1994) argued that most of these features are functionally related to feeding on larger prey. He hypothesized that the development of new cranial and mandibular openings as well as the increased ossification of the sidewall of the braincase are related to the substantial development of the adductor jaw muscles for fast and powerful biting. The eyeball was probably confined to the dorsal portion of the orbit, leaving space below for the anterior expansion of the pterygoideus muscle, which would have facilitated rapid jaw closure.

Archosauriformes: Proterosuchidae

Proterosuchidae comprises all taxa more closely related to *Proterosuchus fergusi* than to *Erythrosuchus africanus*, *Crocodylus niloticus*, or *Passer domesticus* (Ezcurra et al. 2013). It is considered the basal clade of Archosauriformes

and includes the oldest known representative of this group, *Archosaurus*, from the late Permian (Lopingian: Changhsingian) of Russia (Ezcurra et al. 2013, 2014). *Proterosuchus*, from the Early Triassic (Induan) of South Africa, Antarctica, China, and India (Fig. 8.16), attained a skull length of up to 50 cm and a total length of up to 3.5 m (Ezcurra and Butler 2015). Ezcurra (2016) listed a number of shared derived features for Proterosuchidae including the presence of a diastema between the premaxillary and maxollary teeth; maxilla extending posteriorly to the level of or behing the posterior margin of the orbit in mature individuals; presence of a dorsomedial pineal fossa on the parietal; and a transverse groove on the proximal surface of the femur. *Proterosuchus* retains palatal teeth and a supratemporal bone. Proterosuchids are often portrayed as crocodile-like semiaquatic predators, but there exists no compelling evidence in support of this view (Cruickshank 1972).

Archosauriformes: Erythrosuchidae

Erythrosuchidae comprises all taxa more closely related to *Erythrosuchus africanus* than to *Proterosuchus fergusi*, *Crocodylus niloticus*, or *Passer domesticus* (Ezcurra et al. 2010). It includes several taxa of large-bodied archosauriforms with proportionately large heads and short

Figure 8.16. Skull of the proterosuchid *Proterosuchus alexanderi* in lateral view. Courtesy of Martín Ezcurra.

Figure 8.17. Reconstructed skeleton of the erythrosuchid *Garjainia prima* in oblique lateral view. Length c. 2.8 m. Courtesy of Igor Novikov.

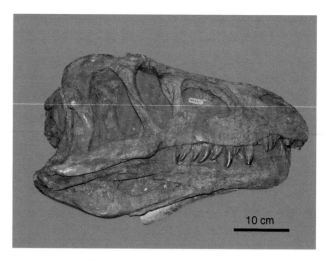

Figure 8.18. Skull of the erythrosuchid *Erythrosuchus africanus* in lateral view. Courtesy of Martín Ezcurra.

necks (Fig. 8.17) from the Early Triassic (Olenekian) and Middle Triassic (Anisian) of China, Russia, and southern Africa (Ezcurra et al. 2013). Diagnostic features for Erythrosuchidae are the presence of a flange of the nasal extending between the premaxilla and maxilla, the presence of three distinct posterior processes on the dentary, and the proportionately long deltopectoral crest on the humerus (Gower et al. 2014). The skull is dorsoventrally deep and transversely narrow (Fig. 8.18). The triradiate rather than plate-like pelvic girdle and the presence of an incipient fourth trochanter and small internal trochanter on the femur indicate that the femur could be adducted to a somewhat greater extent than in *Proterosuchus* (C. Sullivan 2015). *Erythrosuchus*, from the Middle Triassic (Anisian) of South Africa and Namibia, attained a total length of up to 5 m and a skull length of up to 1 m (Gower 2003). Erythrosuchids represent the oldest group of large-bodied predatory reptiles on land.

Archosauriformes: Proterochampsidae— Doswelliidae— *Vancleavea*

Several lineages of non-archosaurian archosauriform reptiles of Middle to Late Triassic age apparently were at least semiaquatic. Various phylogenetic studies have placed them closer to Archosauria than to Erythrosuchidae (Nesbitt 2011; Ezcurra 2016; Sookias 2016). Ezcurra (2016) united these groups as Proterochampsia but neither Nesbitt (2011) nor Sookias (2016) found support for this clade.

Ezcurra (2016) proposed a clade Eucrocopoda for all taxa more closely related to *Euparkeria capensis*, *Proterochampsa barrionuevoi*, *Doswellia kaltenbachi*, *Parasuchus*

hislopi, *Passer domesticus*, or *Crocodylus niloticus* than to *Proterosuchus fergusi* or *Erythrosuchus africanus*. He noted shared features such as a well-ossified proximal head and a fourth trochanter on the femur that are absent in more basal crocopods. As the interrelationships of basal archosauriforms are not well-resolved, and there is important but as yet undescribed new fossil material, this taxonomic category is not adopted here.

Proterochampsidae is the least inclusive clade comprising *Chanaresuchus bonapartei* and *Proterochampsa barrionuevoi* but not *Euparkeria capensis*, *Passer domesticus*, or *Crocodylus niloticus* (Trotteyn et al. 2013). It comprises superficially crocodile-like forms with proportionately large skulls (Fig. 8.19) from the Late Triassic (Carnian-Norian) of Argentina and Brazil. The dorsal margin of the orbit forms a raised ridge or shelf. The premaxilla is often somewhat deflected. Some taxa have a median dorsal row of small osteoderms. The dorsoventrally low, heavily ornamented skull (with a length of up to 42 cm) and dorsally facing external nares of *Proterochampsa* led Sill (1967) to interpret this taxon as an early crocodylian, but there are no synapomorphies to support this interpretation.

Doswelliidae is the least inclusive clade comprising *Doswellia kaltenbachi* but not *Proterochampsa barrionuevoi*, *Erythrosuchus africanus*, *Crocodylus niloticus*, or *Passer domesticus*. It is known from the Middle Triassic (Ladinian) of Germany (Schoch and Sues 2014) and the Late Triassic (Carnian-Norian) of the United States. The clade is

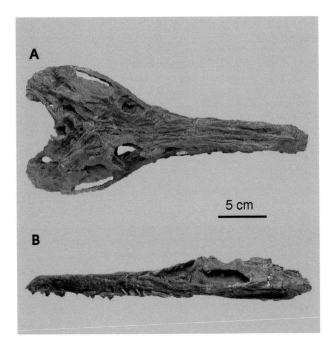

Figure 8.19. Cranium of the proterochampsid *Pseudochampsa ischigualastensis* in **A**, dorsal, and **B**, lateral views. From Trotteyn and Ezcurra (2014)—CC BY 4.0.

characterized by extensive dorsal dermal armor extending from the back of the head to at least the base of the tail. The osteoderms have coarsely reticulate, deeply incised ornamentation composed of pits of subequal size and shape and a mostly smooth anterior articular lamina, which was overlapped by the preceding osteoderm. *Doswellia*, from the Late Triassic (Carnian-Norian) of New Mexico, Texas, Utah, and Virginia, has a superficially crocodile-like skull with dorsally facing orbits and a transversely narrow snout bearing numerous slender teeth (Dilkes and Sues 2009; Heckert et al. 2012). Its transversely broad trunk bore at least 10 longitudinal rows of osteoderms in the posterior dorsal and sacral region. *Doswellia* attained a total length of about 2 m.

Vancleavea, from the Late Triassic (Norian-Rhaetian) of Arizona and New Mexico, has an anteroposteriorly short but dorsoventrally deep skull that lacks antorbital, upper temporal, and external mandibular fenestrae (Nesbitt et al. 2009a; Fig. 8.20). The maxilla and dentary each hold a large, caniniform tooth. There is no lacrimal. The elongate body of *Vancleavea* is completely covered in dermal armor composed of four types of overlapping osteoderms. A fifth type of elongate osteoderms forms a crest-like dorsal fringe on the tail. *Vancleavea* reached a

total length of about 1.2 m. Its limbs are proportionately small. Nesbitt et al. (2009) reconstructed *Vancleavea* as a semiaquatic carnivore that employed its long and dorsoventrally deep tail for sculling.

C. Li et al. (2016) reported a closely related taxon, *Litorosuchus*, from the Middle Triassic (Ladinian) of Yunnan (China). Aside from its occurrence in nearshore marine strata, it has a number of features that are consistent with an aquatic mode of life (e.g., long and deep tail). *Litorosuchus* attained a total length of about 2 m. It has various plesiomorphic features that are absent in *Vancleavea*, including antorbital, upper temporal, and external mandibular fenestrae (C. Li et al. 2016).

Archosauriformes: *Triopticus*

Triopticus, from the Late Triassic (Carnian-Norian) of Texas (Stocker et al. 2016), is known only from a partial cranium, and its phylogenetic position remains unresolved. It differs from other known archosauriform reptiles in the ossification of the interorbital septum, the fusion of the bones framing the orbit, and the presence of a deep median pit on the posterodorsal surface of the cranium. Five rugose bony bosses surround the pit. The posterior two bosses form a shelf overhanging the occiput. The skull roof of *Triopticus* superficially resembles that of dome-headed dinosaurs (Chapter 12) and represents an intriguing example of convergent evolution.

Archosauriformes: Euparkeriidae

Euparkeriidae is the most inclusive clade containing *Euparkeria capensis* but not *Crocodylus niloticus* or *Passer*

Figure 8.20. Skull of the vancleaveid *Vancleavea campi* in lateral view. Note the heterodont dentition and the absence of antorbital, external mandibular, and upper temporal fenestrae.

domesticus (Sookias 2016). This clade is known from the Middle Triassic (Anisian) of South Africa, the Early Triassic (Olenekian) of Poland, and the Middle Triassic of China (Sookias et al. 2014; Sookias 2016). Sookias (2016) listed a suite of diagnostic features for Euparkeriidae: the presence of four premaxillary teeth; exoccipital with two foramina for the passage of the hypoglossal nerve (cranial nerve XII); the vertical orientation of the parabasisphenoid; the presence of distinct postparietals; two paramedian rows of dorsal osteoderms; and dorsal osteo-derms, each with a longitudinal keel and a tapered anterior process.

Euparkeria, from the Middle Triassic (Anisian) of South Africa (Fig. 8.21), was long considered close to the ancestry of all later archosaurs (e.g., Romer 1966) and shares a number of apomorphies with them (Nesbitt 2011). Ezcurra (2016) found *Euparkeria* as the sister taxon to a clade comprising Proterochampsia and Archosauria. However, Nesbitt (2011) and Sookias (2016) recovered Euparkeriidae as most closely related to Archosauria.

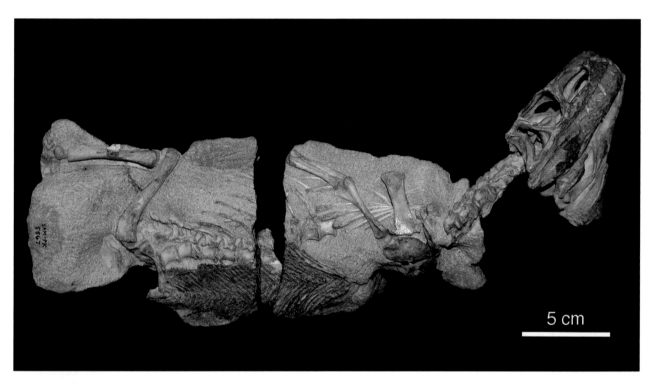

Figure 8.21. Skeleton of the euparkeriid *Euparkeria capensis*. Courtesy of Martín Ezcurra.

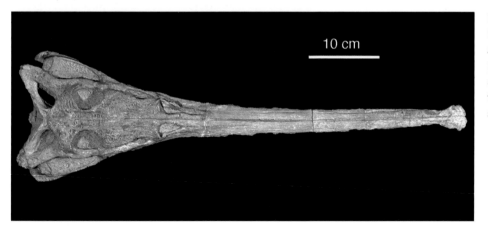

Figure 8.22. Cranium of the basal phytosaur *Ebrachosuchus neukami* in dorsal view. Note position of external nares anterior to the antorbital fenestrae. From Lautenschlager and Butler (2016)—CC BY 4.0.

Attaining a total length of about 1 m, *Euparkeria* retained a sprawling limb posture (C. Sullivan 2015).

Archosauriformes: Phytosauria

Phytosauria is the most inclusive clade containing *Rutiodon carolinensis* but not *Aetosaurus ferratus*, *Rauisuchus tiradentes*, *Prestosuchus chiniquensis*, *Ornithosuchus woodwardi*, and *Crocodylus niloticus* (Sereno 2005). It is known from the Middle Triassic of Yunnan (China) and the Late Triassic of Canada, Europe, Greenland, India, Madagascar, the United States, and possibly Brazil (Stocker and Butler 2013; Kammerer et al. 2015; Stocker et al. 2017). Phytosaurs were long considered part of Archosauria based particularly on the structure of the ankle joint, which they share with pseudosuchians (Sereno 1991a; Ezcurra 2016). However, Nesbitt (2011) and Stocker et al. (2017) found phytosaurs outside Archosauria.

Phytosauria (from Greek *phyton*, plant, and *sauros*, lizard) is a misnomer based on an early misinterpretation of sandstone infillings in the pulp cavities of teeth on fragmentary fossils from Germany as blunt tooth crowns. In fact, phytosaurs were amphibious predators similar in overall appearance, and presumably comparable in habits, to present-day crocodylians. Gut contents of two skeletons of the basal phytosaur *Parasuchus* from the Late Triassic of India include skeletal remains of an archosauromorph reptile, and one of them also preserves cranial bones of a rhynchosaur (Chatterjee 1978). The positions of the orbits and external nares on the dorsal surface of the cranium suggest that phytosaurs laid in wait just below the surface of the water.

Cranial synapomorphies of phytosaurs include a long snout in which the premaxilla is longer than the maxilla and holds at least six teeth; the presence of a distinct bone ("septomaxilla") anterior to the nasals and surrounded by the premaxillae; external nares positioned well back from the tip of the snout on the dorsal surface of the cranium; quadratojugal more or less triangular; long mandibular symphysis formed by the dentaries and splenials; and a markedly heterodont dentition (Stocker and Butler 2013). In the postcranial skeleton, the anterior margin of the coracoid is deeply notched and the humerus bears an ectepicondylar flange. Phytosaurs have two paramedian rows of roughly triangular osteoderms over the trunk and the long tail, and rounded armor plates covering the limbs. Furthermore, a shield of osteoderms covered the throat region ventrally.

The classification of phytosaurs was long in a chaotic state, especially due to considerable individual variation among the many known skulls. Recent studies (Stocker 2010; Stocker and Butler 2013; Kammerer et al. 2015) have clarified this situation.

The stem-phytosaur *Diandongosuchus*, from the Middle Triassic (Ladinian) of Yunnan (China), differs from the

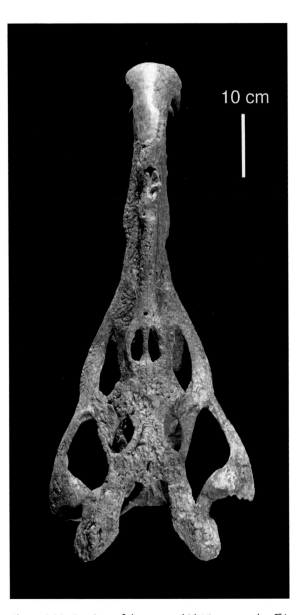

Figure 8.23. Cranium of the parasuchid *Nicrosaurus kapffi* in dorsal view. Note position of external nares between the antorbital fenestrae. Courtesy of Rainer Schoch.

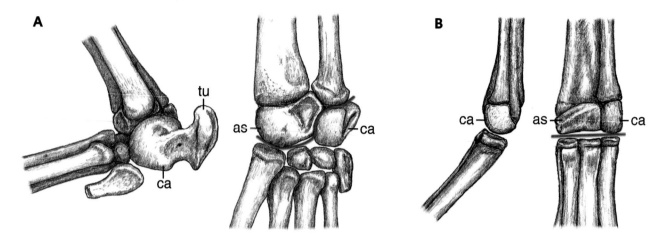

Figure 8.24. Comparison of **A**, "crocodile-normal," and **B**, mesotarsal ankle joint. Both are shown in lateral and dorsal view. Red line indicates zone of flexion. Abbreviations: as astragalus; ca, calcaneum; tu, calcaneal tuber. Courtesy of Jeff Martz.

more derived phytosaurs in the posterior extension of the anterodorsal process of the premaxilla well behind the external naris, the presence of nine teeth in the premaxilla, the presence of a longitudinal ridge on the lateral surface of the jugal, and the presence of more than one pair of dorsal osteoderms per vertebra (Stocker et al. 2017). It was found in nearshore marine deposits.

Parasuchidae comprises the most recent common ancestor of *Wannia scurriensis*, *Parasuchus hislopi*, and *Mystriosuchus planirostris*, and all descendants of that ancestor (Stocker et al. 2017). The best-known basal representative is *Parasuchus,* from the Late Triassic (Carnian-Norian) of Germany, India, Morocco, and Poland as well as Pennsylvania, Texas, and Wyoming (Kammerer et al. 2015). Another basal taxon, *Ebrachosuchus*, from the Late Triassic (Carnian) of Germany (Fig. 8.22), differs from *Parasuchus* especially in the much greater elongation of the snout and higher tooth count (Kammerer et al. 2015). Derived parasuchids share the more posterior position of the external nares relative to the antorbital fenestrae (Stocker and Butler 2013; Fig. 8.23). *Smilosuchus,* from the Late Triassic (Norian) of Arizona, attained skull lengths of well over 1 m and a total length of more than 7 m (Camp 1930; Long and Murry 1995). Although most parasuchids lived near or in freshwater, at least *Mystriosuchus* ventured into lagoonal and coastal marine settings (Gozzi and Renesto 2003).

For many years the structure of the ankle was considered a key diagnostic feature uniting crocodylians, most other pseudosuchians, and phytosaurs. The crurotarsal or "crocodile-normal" ankle (Fig. 8.24) is characterized by a distinct process on the astragalus that fits into a deep socket and contacts an adjacent bony flange on the calcaneum, facilitating considerable motion between these two tarsal bones (Krebs 1963, 1974; Thulborn 1980; Chatterjee 1982; Sereno and Arcucci 1989; Sereno 1991a). If the pes is flexed dorsally, the astragalus functions as part of the distal portion of the leg (crus). The calcaneum, together with the distal tarsals and the pes, can rotate on the astragalus through an arc of about 45 degrees (Fig. 8.24A). It bears a prominent tuber with an expanded distal end that bears a vertical groove for the tendons of the gastrocnemius, the major flexor muscle of the foot, and increases the leverage of that muscle (C. Sullivan 2015). In most other reptiles, the ankle joint extends between the proximal and distal tarsals (mesotarsal ankle: Fig. 8.24B). However, recent discoveries have established that even basal members of the avemetatarsalian lineage have a "crocodile-normal" ankle structure (see Chapter 10).

9 Pseudosuchia

Crocodile-Line Archosaurs

Archosauria (from Greek *archon*, ruler, and *sauros*, lizard) is the least inclusive clade containing *Crocodylus niloticus* (Nile crocodile) and *Passer domesticus* (domestic sparrow) (Sereno 2005). Nesbitt (2011) listed a suite of shared derived features for this crown group including a contact between the palatal processes of the maxillae along the midline of the cranium; lagenar (cochlear) recess elongate and tubular; antorbital fossa on the lacrimal and portions of the maxilla; posteroventral portion of the coracoid "swollen"; and a divided tibial facet on the astragalus. The oldest undisputed archosaur known to date is *Xilousuchus*, from the Early Triassic (Olenekian) of Shanxi (China) (Nesbitt et al. 2011).

Pseudosuchia

Gauthier and Padian (1985) distinguished two principal lineages among Archosauria, one leading to crocodylians (Fig. 9.1) and the other leading to birds. The former is named Pseudosuchia (from Greek *pseudes*, false, and *souchos*, crocodile) and is defined as the most inclusive clade containing *Crocodylus niloticus* but not *Passer domesticus* (Sereno 2005).

Pseudosuchia: Suchia

Suchia (from Greek *souchos*, crocodile) is the least inclusive clade containing *Ornithosuchus woodwardi*, *Tarjadia ruthae*, *Aetosaurus ferratus*, *Rauisuchus tiradentes*, *Prestosuchus chiniquensis*, and *Crocodylus niloticus* (modified from Nesbitt [2011]). Krebs (1974) originally proposed this name for Crocodylia and other pseudosuchians with a "crocodile-normal" ankle.

PSEUDOSUCHIA: SUCHIA: ORNITHOSUCHIDAE

Ornithosuchidae comprises the most recent common ancestor of *Ornithosuchus*, *Riojasuchus*, and *Venaticosuchus* and all descendants of that ancestor (Sereno 1991a). It is known from the Late Triassic (Carnian-Norian) of Argentina and the Late Triassic of Scotland. This clade is characterized by various derived cranial features including a bulbous and slightly ventrally deflected anterior end of the snout, the presence of a distinct gap (diastema) between the premaxillary and maxillary teeth, and the tapered ventral end of the orbit bounded by the dorsal processes of the jugal (Sereno 1991a; Nesbitt 2011; Baczko and Ezcurra 2013; Ezcurra 2016; Fig. 9.2). The pelvis has a perforated acetabulum. The tarsus of ornithosuchids is unique among pseudosuchians in the presence of a peg on the calcaneum that fits into a concavity on the astragalus ("crocodile-reverse" ankle; Chatterjee 1982). *Ornithosuchus*, from the Late Triassic of Scotland, attained a total length of 2.2 m (Baczko and Ezcurra 2016). The dermal armor comprises paired

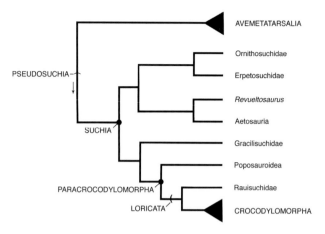

Figure 9.1. Phylogenetic hypothesis of the interrelationships of selected taxa of Pseudosuchia. Dots denote node-based clades and parentheses denote stem-based clades. Based on Nesbitt (2011) and Ezcurra et al. (2017).

Figure 9.2. Skull of the ornithosuchid *Riojasuchus tenuisceps* in lateral view. From Baczko and Desojo (2016)—CC BY 4.0.

rows of cervical and dorsal osteoderms and a single row over the tail. The forelimb is shorter than the hind limb, and the manus appears to be more suited for grasping than for locomotion. Thus, ornithosuchids were likely at least facultatively bipedal (Walker 1964; Baczko and Ezcurra 2013).

PSEUDOSUCHIA: SUCHIA: ERPETOSUCHIDAE

Erpetosuchidae is the most inclusive clade containing *Erpetosuchus granti* but not *Passer domesticus, Postosuchus kirkpatricki, Crocodylus niloticus, Ornithosuchus woodwardi,* or *Aetosaurus ferratus* (Nesbitt and Butler 2013). This clade is characterized by the snout being broader transversely than deep dorsoventrally, the restriction of the tooth row to the anterior half of the maxilla, the distinct anterior elongation of the antorbital fossa, and the ventrolat-

erally facing posterior process of the jugal (Ezcurra et al. 2017). The dorsal armor consists of four longitudinal rows of sculptured osteoderms. Erpetosuchidae was originally known only from the small-bodied *Erpetosuchus,* from the Late Triassic of Scotland (Benton and Walker 2002) and the Late Triassic (Norian) of Connecticut (Olsen et al. 2001), but recent studies have shown that this clade attained a much wider spatiotemporal distribution as well as greater range in body size (Nesbitt and Butler 2013; Ezcurra et al. 2017). *Tarjadia,* from the Late Triassic (Carnian) of Argentina, reached a total length of about 3 m and has a strongly ornamented, thick skull roof. The phylogenetic analysis by Ezcurra et al. (2017) found Erpetosuchidae forming a clade with Ornithosuchidae and Aetosauria.

PSEUDOSUCHIA: SUCHIA: *REVUELTOSAURUS* + AETOSAURIA

Most early archosauriforms have labiolingually flattened tooth crowns that have serrated cutting edges (carinae) and indicate carnivorous habits. However, one as yet unnamed clade of pseudosuchians has leaf-shaped tooth crowns that often bear marginal denticles, which suggest omnivory or herbivory. It includes *Revueltosaurus,* from the Late Triassic (Norian) of Arizona and New Mexico, and Aetosauria. Isolated teeth of *Revueltosaurus* so closely resemble those of some ornithischian dinosaurs that they were originally referred to that group. However, discoveries of as yet mostly undescribed skeletal remains established that *Revueltosaurus* is a pseudosuchian most closely related to aetosaurs (Parker et al. 2005; Nesbitt 2011). *Revueltosaurus* and Aetosauria share the possession of extensive dermal armor composed of longitudinal rows of plate-like osteoderms over the dorsal and ventral surfaces of the body and tail (Nesbitt 2011). The dorsal surface of each dorsal osteoderm has a smooth, thick, and bar-like anterior portion and an extensively sculptured posterior region. Smaller armor elements cover the limbs. *Revueltosaurus* and aetosaurs also share several synapomorphies in the structure of the cranium, including the postorbital bar being formed largely by the postorbital and a lateral slot on the jugal for contact with the posterior process of the maxilla (Nesbitt 2011).

Aetosauria (from Greek *aetos,* eagle, and *sauros,* lizard) is the most inclusive clade containing *Aetosaurus ferratus* and *Desmatosuchus haplocerus,* but not *Rutiodon carolinensis, Postosuchus kirkpatricki, Prestosuchus chiniquensis,*

Poposaurus gracilis, *Gracilisuchus stipanicicorum*, *Crocodylus niloticus*, and *Revueltosaurus callenderi* (Parker 2007). It is readily distinguished by the absence of teeth on the anterior portions of the premaxilla and dentary, the sharply tapered anterior end of the dentary, the anteroventral angling of the quadrate, and the lateral exposure of the upper temporal fenestra (Walker 1961; Parker 2007, 2016; Nesbitt 2011; Desojo et al. 2013: Fig. 9.3A). The skull is proportionately small. The tip of the snout is blunt and, in some taxa, slightly upturned. The maxillary and most of the dentary teeth in most aetosaurs have simple, mesiodistally broad crowns. Aetosaurs presumably were omnivorous or herbivorous. The ilium is distinctly deflected ventrolaterally, overhanging the somewhat inturned head of the femur fitting into the acetabulum. This configuration facilitated adduction of the femur so that the knee joint faced nearly anteriorly. The slightly medially turned femoral head also allowed abduction of the femur, so that the hind limb could also adopt a more sprawling posture common in basal archosaurs (Parrish 1987; C. Sullivan 2015). The osteoderms bear prominent dorsal ornamentation that exhibits considerable variation among the various taxa (Long and Murry 1995; Desojo et al. 2013; Parker 2016; Fig. 9.3B). Aetosaurs are known from the Late Triassic (Carnian-

Norian) of Algeria, Argentina, Brazil, Canada, Chile, England, Germany, Greenland, India, Italy, Morocco, Scotland, and the United States (Desojo et al. 2013). *Aetosaurus*, from the Late Triassic (Norian) of Germany, Greenland, and Italy, attained a total length of up to 1.5 m (Schoch 2007). *Desmatosuchus*, from the Late Triassic (Carnian-Norian) of Arizona, New Mexico, and Texas, reached a length of more than 4 m and has lateral cervical osteoderms with large, recurved lateral spikes (Long and Murry 1995; Parker 2008; Fig. 9.4).

PSEUDOSUCHIA: SUCHIA: GRACILISUCHIDAE

Gracilisuchidae is the most inclusive clade containing *Gracilisuchus stipanicicorum* but not *Ornithosuchus woodwardi*, *Aetosaurus ferratus*, *Poposaurus gracilis*, *Postosuchus kirkpatricki*, *Rutiodon carolinensis*, *Erpetosuchus granti*, *Revueltosaurus callenderi*, *Crocodylus niloticus*, or *Passer domesticus* (Butler et al. 2014). It is known from the Late Triassic (Carnian) of Argentina and the Middle to Late Triassic (Anisian-Ladinian) of China. Butler et al. (2014) listed diagnostic apomorphies for this clade including the posterodorsal process of the premaxilla fitting into a slot on the lateral surface of the nasal, nasal participating in the dorsal margin of the antorbital fossa, and the presence of a posterodorsal process on the maxilla (Fig. 9.5). *Gracilisuchus*, from the Late Triassic (Carnian) of Argentina, attained a skull length of about 10 cm and a total length of about 60 cm (Romer 1972). In both *Gracilisuchus* and Ornithosuchidae, the acetabulum faces laterally and is shallow, and the femur lacks a distinct inturned head. Together, these features indicate a semi-upright posture (C. Sullivan 2015).

PSEUDOSUCHIA: SUCHIA: PARACROCODYLOMORPHA

Various taxa of Middle and Late Triassic pseudosuchians with craniodental features indicating predatory habits have been grouped together as Rauisuchia (e.g., Brusatte et al. 2010a). However, Nesbitt (2011) argued that these taxa do not a represent a clade but are successively more closely related to the clade comprising crocodylians and their closest relatives (Crocodylomorpha). He adopted Parrish's (1993) name, Paracrocodylomorpha (from Greek *para*, near, *krokodeilos*, a kind of lizard, and *morphe*, form), for the least inclusive clade containing *Crocodylus niloticus* and *Poposaurus gracilis*. Many

Figure 9.3. A, skull of the stagonolepidid *Paratypothorax andressorum* in dorsolateral view; **B**, skeleton of *Paratypothorax andressorum* as preserved. Courtesy of Rainer Schoch.

5 cm

20 cm

A

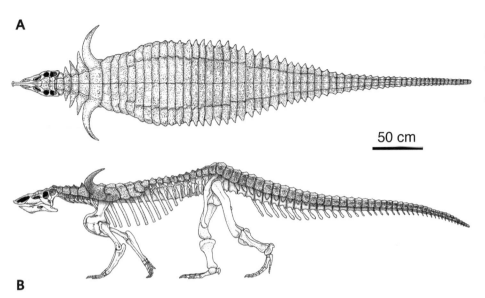

B

50 cm

Figure 9.4. Skeleton and dorsal armor of the stagonolepidid *Desmatosuchus spurensis* in **A**, dorsal, and **B**, lateral views. Drawing by Jeffrey Martz courtesy of William G. Parker.

paracrocodylomorph taxa are still poorly known and there exists no general consensus concerning their interrelationships (Nesbitt et al. 2013). A number of them have modifications of the pelvis and hind limb comparable to those in aetosaurs, resulting in a "pillar-like" orientation of the hind limb close to the body and an upright posture (Bonaparte 1984; Gauthier et al. 2011; C. Sullivan 2015). This anatomical interpretation is confirmed by narrow-track trackways (such as *Chirotherium*) that have been attributed to paracrocodylomorphs based on the close correspondence between the structure of individual footprints and skeletal features of the manus and pes.

PSEUDOSUCHIA: SUCHIA: PARACROCODYLOMORPHA: POPOSAUROIDEA

Poposauroidea (from the Popo Agie River in Wyoming, and Greek, *sauros*, lizard) is the most inclusive clade containing *Poposaurus gracilis* but not *Postosuchus kirkpatricki*, *Crocodylus niloticus*, *Ornithosuchus woodwardi*, or *Aetosaurus ferratus* (Nesbitt 2011). Diagnostic derived features for this clade include the maxilla entering into the margin of the external naris, the presence of the openings for the cerebral branches of the internal carotid artery on the ventral surface of the braincase, and the presence of three or more sacral vertebrae (Nesbitt et al. 2013). The basal poposauroid *Qianosuchus*, from the Middle Triassic (Anisian) of Guizhou (China), has osteoderms, but all other known poposauroids apparently lack dermal armor. *Qianosuchus* apparently lived in nearshore marine settings

and has features indicative of an at least semiaquatic mode of life, such as a laterally flattened tail (Li et al. 2006).

Among Poposauroidea, Ctenosauriscidae is the most inclusive clade containing *Ctenosauriscus koeneni* but not *Poposaurus gracilis*, *Effigia okeeffeae*, *Postosuchus kirkpatricki*, *Crocodylus niloticus*, *Ornithosuchus woodwardi*, or *Aetosaurus ferratus* (Butler et al. 2011). The key diagnostic apomorphy of this clade is the presence of unusually tall neural spines, which are distinctly curved in the dorsal region, on the presacral, sacral, and anterior caudal vertebrae (Fig. 9.6). The dorsal neural spines can be more than seven times taller than the corresponding centra. Ctenosauriscidae includes *Arizonasaurus*, from the Middle Triassic (Anisian) of Arizona and New Mexico (Nesbitt 2005), and *Ctenosauriscus*, from the Early to Middle Triassic (Olenekian-Anisian) of Germany (Butler et al. 2011; Fig. 9.6).

Poposaurus, from the Late Triassic (Norian) of Arizona, New Mexico, Texas, Utah, and Wyoming, has long, slender hind limbs, appressed metatarsals, a large tridactyl pes with hoof-like unguals, and forelimbs that are less than half as long as the hind limbs. This suite of features indicates that *Poposaurus* was an obligate biped (Gauthier et al. 2011; Nesbitt et al. 2013). The ilium has a prominent bony crest above the acetabulum, which forms a ventrally facing cup for contact with the nearly vertically aligned femur (C. Sullivan 2015).

Shuvosauridae is the least inclusive clade containing *Shuvosaurus inexpectatus* and *Sillosuchus longicervix* (Nesbitt 2011). Diagnostic apomorphies for this clade include the presence of deep pneumatic fossae on the anterior por-

A

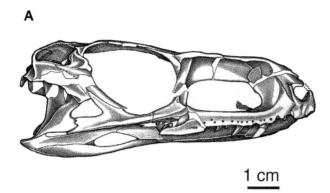

1 cm

B

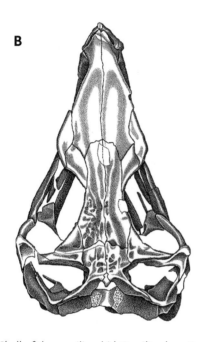

Figure 9.5. Skull of the gracilisuchid *Gracilisuchus stipanicicorum* in **A**, lateral, and **B**, dorsal views. Courtesy of Diane Scott.

10 cm

Figure 9.6. Partial vertebral column of the ctenosauriscid *Ctenosauriscus koeneni* in lateral view. Note the greatly elongated neural spines of the posterior cervical and dorsal vertebrae. From Butler et al. (2011)—CC BY 2.5.

tion of the cervical vertebrae, presence of a rimmed depression on the posterior portion of the cervical centra, and sacral ribs shared between two sacral vertebrae (Nesbitt 2011). *Shuvosaurus*, from the Late Triassic (Norian) of Texas, and the closely related *Effigia*, from the Late Triassic (Rhaetian) of New Mexico (Fig. 9.7), lack teeth, and a beak apparently covered the jaws in life. The shoulder girdle is robust but the forelimbs are proportionately small. The distal end of the pubis forms a prominent "boot." *Effigia* attained a total length of 2 m. Like *Poposaurus*, *Shuvosaurus* and *Effigia* were obligate bipeds with a more or less upright limb posture (Nesbitt 2007; Fig. 9.7).

PSEUDOSUCHIA: SUCHIA: PARACROCODYLOMORPHA: LORICATA

Loricata (from Latin *loricatus*, clad in armor) is the most inclusive clade containing *Crocodylus niloticus* but not *Poposaurus gracilis*, *Ornithosuchus woodwardi*, or *Aetosaurus ferratus* (Nesbitt 2011). In addition to Crocodylomorpha, it also includes Rauisuchidae and related taxa such as *Batrachotomus*, from the Middle Triassic (Ladinian) of Germany (Gower 1999; Fig. 9.8), and *Prestosuchus*, from the Late Triassic (Carnian) of Brazil (Fig. 9.9).

Rauisuchidae is the most inclusive clade containing *Rauisuchus tiradentes* but not *Aetosaurus ferratus*, *Prestosuchus chiniquensis*, *Poposaurus gracilis*, or *Crocodylus niloticus* (Nesbitt et al. 2013). It shares a suite of diagnostic apomorphies including the presence of a rugose ridge along the lateral edge of the nasal, the division of the lower temporal fenestra by contact between the squamosal and postorbital, and the presence of a rounded longitudinal ridge on the lateral surface of the jugal (Nesbitt 2011; Weinbaum 2011; Nesbitt et al. 2013). Rauisuchidae is known from the Late Triassic (Carnian) of Brazil and India and the Late Triassic (Norian) of Germany and Poland as well as Arizona, New Mexico, North Carolina, and Texas (Nesbitt et al. 2013). The best-known representative, *Postosuchus*, from the Late Triassic (Norian) of Arizona, North Carolina, and Texas, attained a skull length of about 60 cm (Weinbaum 2011) and an overall length of up to 6 m (Fig. 9.10). Although its manus is proportionately small the metacarpus suggests at least facultative quadrupedal locomotion (K. Peyer et al. 2008; Hutson and Hutson 2015). Gut contents from a specimen from North Carolina confirm that *Postosuchus* was an apex predator (K. Peyer et al. 2008).

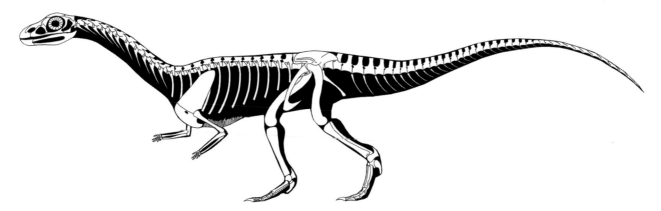

Figure 9.7. Reconstructed skeleton of the shuvosaurid *Effigia okeeffeae* in lateral view. Length c. 3 m. Courtesy of Sterling Nesbitt.

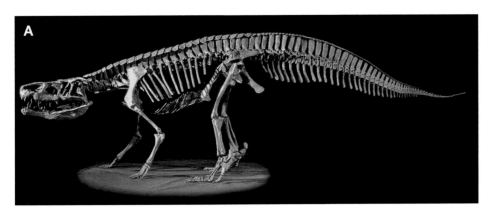

Figure 9.8. A, reconstructed skeleton of the basal loricatan *Batrachotomus kupferzellensis* in lateral view. Length up to 6 m. **B**, reconstructed skull of *Batrachotomus kupferzellensis*. A, courtesy of Rainer Schoch; B, from Gower (1999).

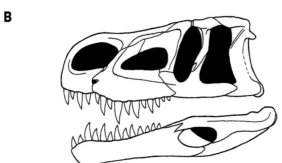

CROCODYLOMORPHA

Crocodylomorpha (from Greek *krokodeilos*, a kind of liz-ard, and *morphe*, form) is the sole clade of Pseudosuchia that has survived to the present day. Currently 25 extant species of crocodylians are recognized as valid. However, Crocodylia was far more diverse earlier during the Cenozoic Era.

Huxley (1875) united present-day crocodylians with a variety of Mesozoic crocodile-like reptiles. He used the position of the choanae on the secondary bony palate and the mode of articulation between successive verte-brae as the principal criteria to distinguish three major groups among crocodylians: Parasuchia, Mesosuchia, and Eusuchia. Of these, Parasuchia comprised only what are now known to be a phytosaur and an aetosaur, re-spectively, and the name was subsequently adopted for phytosaurs. Later classifications retained Huxley's Meso-suchia and Eusuchia and added Protosuchia, Sebeco-suchia, and Thalattosuchia (e.g., Romer 1956).

Following Benton and J. M. Clark (1988), most recent authors have restricted the name Crocodylia to crown-

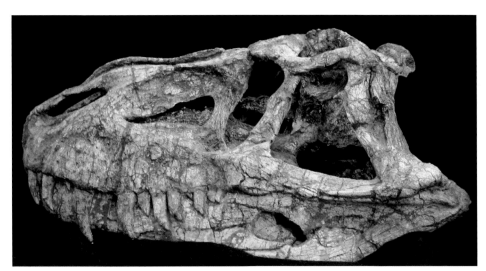

Figure 9.9. Skull of the basal loricatan *Prestosuchus chiniquensis* in lateral view. Length of skull c. 90 cm. Courtesy of Cesar Schultz.

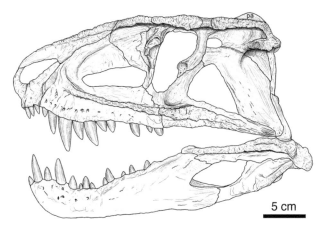

Figure 9.10. Reconstructed skull of the rauisuchid *Postosuchus kirkpatricki* in lateral view. Courtesy of Jonathan Weinbaum.

group crocodylians (see below). Many publications have used the spelling "Crocodilia," but the correct version is "Crocodylia," because the name of the nominal genus is *Crocodylus*. Crocodylia and various mostly Mesozoic relatives are united as Crocodyliformes. In turn, Crocodyliformes is grouped with a set of taxa of early Mesozoic pseudosuchians as Crocodylomorpha (Fig. 9.11). Nesbitt (2011) defined Crocodylomorpha as the most inclusive clade containing *Crocodylus niloticus* but not *Poposaurus gracilis*, *Gracilisuchus stipanicicorum*, *Prestosuchus chiniquensis*, or *Aetosaurus ferratus*. The oldest known crocodylomorph is *Trialestes*, from the Late Triassic (Carnian) of Argentina (Irmis et al. 2013).

Present-day crocodylians are semiaquatic ambush predators capable of both sprawling and semi-upright limb posture. By contrast, extinct crocodylomorphs are strikingly varied in their body plans and inferred modes of life. Some were terrestrial forms with fully upright limb posture. Others adapted to life in the open sea, modifying their forelimbs as paddles and developing a tail fin. Finally, there were crocodyliforms with features of the jaws and teeth that suggest omnivory and even herbivory. Cuvier (1825) first recognized that certain Mesozoic crocodyliforms differ from present-day crocodylians in their skeletal structure. However, most researchers were more impressed by similarities such as flattened snouts and inferred amphibious habits shared by the extant and the few fossil forms known at the time and assumed that the group had undergone little evolutionary change since the Mesozoic. This erroneous view of crocodylians as "living fossils" unfortunately persists to the present day.

Nesbitt (2011) considered *Postosuchus* and its close relatives most closely related to Crocodylomorpha, listing shared derived features such as the foreshortening of the lower temporal fenestra, certain features in the structure of the braincase, the posteriorly hooked head of the humerus, and the presence of a deep ventral groove on the coracoid.

A suite of synapomorphies is diagnostic for Crocodylomorpha (J. M. Clark et al. 2004; Nesbitt 2011): recess separating the premaxilla and maxilla on the side of the snout; absence of a postfrontal; quadrate and quadratojugal extending anterodorsally; proximal head of the quadrate contacting the squamosal and the prootic (or, in more derived taxa, the laterosphenoid); bones of the

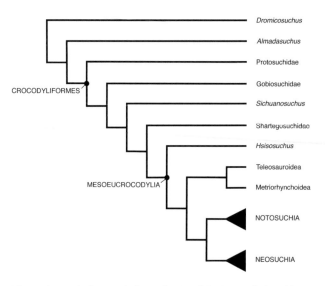

CROCODYLIFORMES

MESOEUCROCODYLIA

Dromicosuchus

Almadasuchus

Protosuchidae

Gobiosuchidae

Sichuanosuchus

Shartegusuchidae

Hsisosuchus

Teleosauroidea

Metriorhynchoidea

NOTOSUCHIA

NEOSUCHIA

Figure 9.11. Phylogenetic hypothesis of the interrelationships of selected taxa of Crocodylomorpha. Dots denote node-based clades. Combined from Clark (1994) and A. H. Turner (2015).

braincase and adjacent portions of the skull containing an elaborate system of air-filled (pneumatic) spaces, which originated from the ear region; and Eustachian passages form a uniquely complex system. The latter two features are likely related to hearing (Dufeau and Witmer 2015). The shoulder girdle lacks clavicles and has postero-ventrally elongated coracoids. The carpus has two elongated proximal elements, the radiale and ulnare. The dorsal dermal armor of early crocodylomorphs is composed of a pair of paramedian rows of osteoderms over the trunk and tail (paravertebral shield). Frey (1988) argued that longitudinal muscle groups attached to this armor and braced the vertebral column in the manner of a box-girder bridge. In extant crocodylians, the trunk region bears four rows of dorsal osteoderms medially and at least one lateral row of accessory osteoderms.

In basal crocodylomorphs, the ilium has a pronounced lateral crest above the acetabulum. This feature, along with the distinctly inturned head of the femur, indicates parasagittal locomotion and an upright limb posture, unlike in extant crocodylians but more similar to the condition in avemetatarsalians (Parrish 1987; C. Sullivan 2015). The upright posture and elongated distal limb bones indicate that these lightly built animals were fleet-footed pursuit predators. Bonaparte (1972) proposed a group "Sphenosuchia" for the reception of basal crocodylomorphs. However, phylogenetic analyses found that these crocodylomorphs do not represent a clade but

rather a series of taxa successively more closely related to Crocodyliformes (J. M. Clark and Sues 2002; J. M. Clark et al. 2004; Nesbitt 2011; Pol et al. 2013).

Basal crocodylomorphs typically attained a total length ranging from 1 to 1.5 m and have long and slender limbs. They ranged in time from the Late Triassic (Carnian) to the Late Jurassic (Kimmeridgian-Tithonian). The forelimb is shorter than the hind limb and the manus is small. *Dromicosuchus*, from the Late Triassic (Norian) of North Carolina (Sues et al. 2003; Fig. 9.12), and *Sphenosuchus*, from the Early Jurassic (Hettangian-Sinemurian) of South Africa (Walker 1990), are representative taxa. Some small-bodied forms, including *Terrestrisuchus*, from the Late Triassic (Norian-Rhaetian) of England, reached a total length of perhaps 50 cm (Crush 1984). By contrast, other taxa such as *Carnufex*, from the Late Triassic (Carnian) of North Carolina, attained total lengths of at least 3 m (Zanno et al. 2015).

Various "sphenosuchian" taxa, such as *Junggarsuchus*, from the Middle Jurassic (Bathonian-Callovian) of Xinjiang (China) (J. M. Clark et al. 2004; Fig. 9.13), are structurally intermediate between basal crocodylomorphs and crocodyliforms. Of these, *Almadasuchus*, from the Late Jurassic (Oxfordian) of Argentina, is most closely related to Crocodyliformes (Pol et al. 2013). Synapomorphies include the absence of basipterygoid processes, the subtriangular outline of the basisphenoid, and the absence of a dorsal ridge of the frontal (Pol et al. 2013). The cranium of *Almadasuchus* was strongly sutured and lacked any kinesis.

CROCODYLOMORPHA: CROCODYLIFORMES

Crocodyliformes (from Greek *krokodeilos*, a kind of lizard, and Latin *forma*, form) is the least inclusive clade containing *Crocodylus niloticus* and *Protosuchus richardsoni* (Nesbitt 2011). It is characterized by the presence of a vertical posterodorsal process of the premaxilla that is firmly sutured to the maxilla and the presence of osteoderms covering the limbs. As additional diagnostic features J. M. Clark et al. (2004) cited the distinctive ornamentation composed of irregular pits on the cranial bones, the broadening and flattening of the postorbital portion of the cranium (skull deck), and the anteroposterior expansion of the distal end of the coracoid. The antorbital fenestra and fossa (where present) are proportionately smaller than in basal crocodylomorphs. The quadrate is more horizontally aligned than in earlier forms

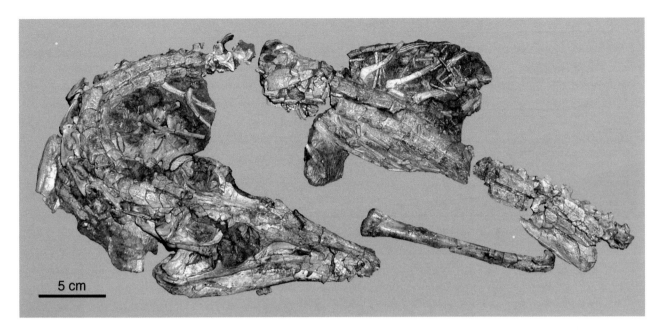

Figure 9.12. Partial skeleton of the crocodylomorph *Dromicosuchus grallator.*

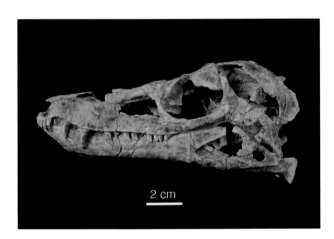

Figure 9.13. Skull of the crocodylomorph *Junggarsuchus sloani* in lateral view. Courtesy of Jim Clark.

and, like the pterygoid, is firmly sutured to the braincase. Presumably these changes help resist forces generated by the powerful adductor jaw musculature, particularly the pterygoideus muscle (Pol et al. 2013). Compared to the condition in other pseudosuchians, the jaw joints are located posterolaterally, often behind the occipital surface of the cranium. This position increases the gape as well as the moment arms of the adductor jaw muscles relative to the jaw joint (Schumacher 1973). A distinctive feature of the crocodyliform pelvic girdle is the absence of a bony contact between the pubis and ilium and the narrowing of the median symphysis between the pubic bones (Claes-

sens and Vickaryous 2013). The separation of the pubic bones from the remainder of the pelvis allows them to rotate ventrally to assist during respiration.

The oldest known crocodyliforms include *Protosuchus*, from the Late Triassic (Rhaetian) and Early Jurassic (Hettangian-Sinemurian) of North America (Colbert and Mook 1951; J. M. Clark 1986; Fig. 9.14) and the Early Jurassic (Hettangian-Sinemurian) of southern Africa (Busbey and Gow 1984; J. M. Clark 1986), and *Hemiprotosuchus*, from the Late Triassic (Norian) of Argentina (Bonaparte 1972). *Edentosuchus*, from the Early Cretaceous of Xinjiang (China), has molar-like posterior teeth bearing multiple cusps along the edges of their occlusal surfaces (Pol et al. 2004; Ösi 2014). *Protosuchus* attained a total length of up to 1 m. Protosuchidae is characterized by a few derived cranial features such as a narrow ventral contact between the quadrate and the fused exoccipital and opisthotic (otoccipital) and a deep groove along the ventral edge of the pterygoid ramus of the quadrate (Pol et al. 2004). Unlike in more derived crocodyliforms, there is no well-developed secondary bony palate. The dorsal dermal armor of *Protosuchus* and other early crocodyliforms comprises two paramedian rows of rectangular osteoderms in one-to-one correspondence to the underlying vertebrae (Frey 1988; Fig. 9.14). In addition, the ventral surfaces of the trunk and tail are covered by shields of small rectangular osteoderms.

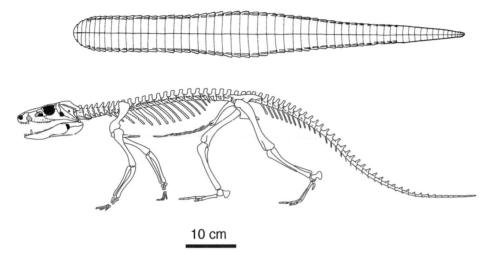

10 cm

Figure 9.14. Reconstructed skeleton of the protosuchid *Protosuchus richardsoni*. Dorsal dermal armor shown above the skeleton. Courtesy of Sterling Nesbitt with dorsal armor modified from Frey (1988).

The interrelationships of basal crocodyliforms and their relationships to more derived clades remain contentious. Some authors (Osmólska et al. 1997; Wu et al. 1997; Wilberg 2015) recognized a monophyletic grouping Protosuchia that also included certain geologically younger small-bodied taxa such as *Sichuanosuchus*, from the Late Jurassic and Early Cretaceous of Sichuan (China) (Fig. 9.15), and *Gobiosuchus*, from the Late Cretaceous (Campanian) of Mongolia. By contrast, Pol et al. (2004), Pol and Norell (2004), and Buscalioni (2017) considered the latter forms and others more closely related to more derived crocodyliforms (Mesoeucrocodylia) than to Protosuchidae. Shartegosuchidae is characterized by the absence of an external mandibular fenestra, pronounced sculpturing on the ventral surface of the pterygoid (also present in *Protosuchus*), and the crenulated, horizontal apices of the posterior dentary and maxillary tooth crowns (J. M. Clark 2011). It includes *Shartegosuchus* and *Nominosuchus*, both from the Late Jurassic of Mongolia (Storrs and Efimov 2000), and *Fruitachampsa*, from the Late Jurassic (Kimmeridgian-Tithonian) of Colorado (J. M. Clark 2011; Fig. 9.16). Buscalioni (2017) recognized another clade, Gobiosuchidae, to which she also referred *Cassissuchus*, from the Early Cretaceous (Barremian) of Spain, and *Zaraasuchus*, from the Late Cretaceous (Campanian) of Mongolia.

CROCODYLOMORPHA: CROCODYLIFORMES: MESOEUCROCODYLIA

The majority of crocodyliform taxa are united as Mesoeucrocodylia (from Greek *mesos*, middle, *eu*, true, and

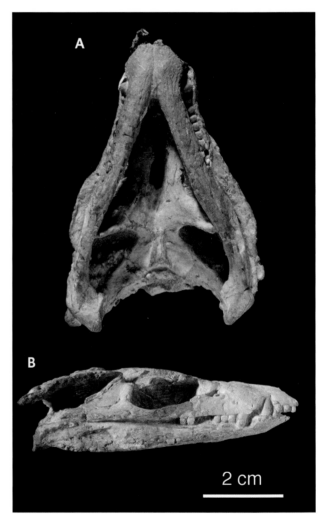

Figure 9.15. Skull of the crocodyliform *Sichuanosuchus shuhanensis* in **A**, palatal, and **B**, lateral views.

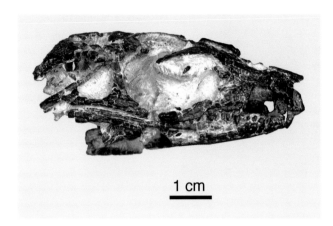

Figure 9.16. Skull of the shartegosuchid *Fruitachampsa callisoni* in lateral view. Courtesy of Jim Clark.

Crocodylia), which comprises Huxley's "Mesosuchia" (now considered a paraphyletic grouping) and Eusuchia (Benton and J. M. Clark 1988). Sereno et al. (2001) defined Mesoeucrocodylia as all crocodyliforms more closely related to *Crocodylus niloticus* than to *Protosuchus richardsoni*. This clade is primarily distinguished by the presence of a secondary bony palate formed by shelf-like processes of the maxillae and palatines. The pterygoids are typically fused behind the choanae.

Wu et al. (1994) interpreted *Hsisosuchus*, from the Late Jurassic of Sichuan (China), as a basal member of Mesoeucrocodylia, but Pol et al. (2014) placed it as the sister taxon of the latter. *Hsisosuchus* is distinguished by a number of apomorphies such as the absence of an external mandibular fenestra, the absence of a suborbital fenestra, and the presence of a prominent transverse ridge across the pterygoids.

CROCODYLOMORPHA: CROCODYLIFORMES: MESOEUCROCODYLIA: THALATTOSUCHIA

Thalattosuchia (from Greek *thalassa*, sea, and *souchos*, crocodile) adapted to life in the open sea early during the evolutionary history of Mesoeucrocodylia. It is defined as the most inclusive clade consisting of *Teleosaurus cadomensis* and *Metriorhynchus geoffroyii* but not *Pholidosaurus schaumburgensis*, *Goniopholis crassidens*, or *Dyrosaurus phosphaticus* (M. T. Young and Andrade 2009). Thalattosuchia ranged in time from the Early Jurassic (Toarcian) to the Early Cretaceous (Aptian) and is known from Europe, Africa, the Americas, and China (M. T. Young and Andrade 2009). It is characterized by a suite of cranial synapomor-

phies including the nearly tubular snout, absence of a contact between the nasal and premaxilla, the extension of the jugal anterior to the prefrontal, and the great elongation of the mandibular symphysis with extensive involvement of the splenials (M. T. Young and Andrade 2009). The hind limbs are considerably longer than the forelimbs. The femur lacks a fourth trochanter.

Thalattosuchia comprises two subclades, Teleosauroidea and Metriorhynchoidea. Teleosauroidea is the most inclusive clade consisting of *Teleosaurus cadomensis* but not *Metriorhynchus geoffroyii* (M. T. Young and Andrade 2009).

The teleosauroid *Steneosaurus* is known from many excellently preserved skulls and skeletons from the Early Jurassic (Toarcian-Pliensbachian) of Europe (Westphal 1962). Early Jurassic representatives of *Steneosaurus* reached total lengths of up to 3 m long, but one Middle Jurassic (Callovian) species attained an estimated length of 7 m (M. M. Johnson et al. 2015). Most teleosauroids lived in estuarine and nearshore marine settings and may still have been capable of locomotion on land. Shared derived features include the long snout, the transversely expanded anterior ends of the dentaries and premaxillae, the shallow mandibular symphysis, the oval external narial opening, the procumbent teeth in the anterior and middle regions of the maxilla, and the absence of cutting edges (carinae) on the tooth crowns (M. T. Young and Andrade 2009; Fig. 9.17). The length of the forelimb is only about half that of the hind limb (Fig. 9.18). Well-preserved specimens established that at least the pes had webbing between the digits. Nearly vertical zygapophyseal facets on the dorsal and anterior caudal vertebrae indicate a rigid trunk and base of the tail (Krebs 1962). Teleosauroids probably employed lateral movements of the more distal portion of the long tail for swimming. They have broad osteoderms in the trunk region, but the armor over the neck and much of the tail is less developed than in more basal crocodyliforms (Frey 1988). *Steneosaurus* and related forms have long, narrow jaws with slender teeth, indicating a diet of fish. Hua and de Buffrénil (1996) interpreted them as ambush hunters similar to the present-day gharial, using rapid sideways movements of the head to catch their prey. *Machimosaurus*, from the Late Jurassic (Oxfordian-Tithonian) of Europe and Ethiopia and the Early Cretaceous (Hauterivian) of Tunisia, attained a length of at least 7 m and has robust jaws with stout,

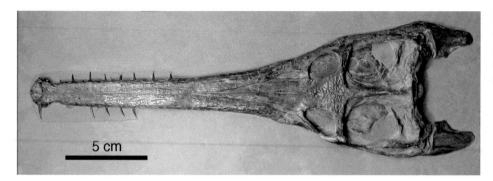

Figure 9.17. Skull of the teleosauroid *Steneosaurus bollensis* in dorsal view. Courtesy of Rodolfo Salas-Gismondi.

5 cm

Figure 9.18. Skeleton (missing most of its tail) of the teleosauroid *Platysuchus multiscrobiculatus* in dorsal view. Length as preserved c. 1.4 m. Courtesy of Heinrich Mallison.

conical, and blunt-tipped teeth. M. T. Young et al. (2014) speculated that turtles may have formed an important part of its diet.

Metriorhynchoidea is the most inclusive clade consisting of *Metriorhynchus geoffroyii* but not *Teleosaurus cadomensis* (M. T. Young and Andrade 2009). Representatives of this clade have skeletal features consistent with a fully pelagic mode of life. Metriorhynchoidea ranged in time from the Middle Jurassic (Bajocian) to the Early Cretaceous (Valanginian) and are known from Europe and the Americas. The orbits face laterally rather than dorsally as in teleosauroids such as *Steneosaurus*. A broad lateral lappet of the prefrontal overhangs the anterodorsal portion of the orbit (Figs. 9.19, 9.20). The premaxilla has only three teeth. Just anterior to the orbits the snout housed a pair of large glands that may have served for salt excretion (Fernández and Gasparini 2008). (Extant crocodylians except for alligators and caimans have salt-excreting glands on their tongues.) Derived metriorhynchoids lack dermal armor and have more streamlined bodies than other crocodyliforms (M. T. Young et al. 2010). The distal portion of the caudal vertebral column has a pronounced downward bend. Behind this bend the vertebrae supported the lower (ventral) lobe of a tail fin. Based on exceptionally well-preserved fossils, this fin also had a smaller, fleshy upper (dorsal) lobe. The forelimb is paddle-like, with short, flattened bones in the most derived metriorhynchids. The hind limb is longer and much less modified than the forelimb. Preserved stomach contents established that *Metriorhynchus*, from the Middle to Late

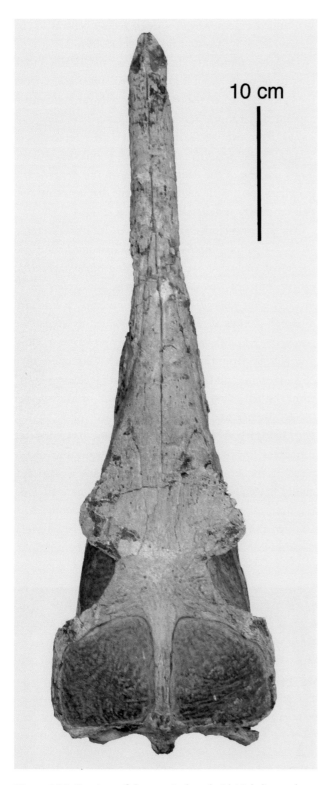

Figure 9.19. Cranium of the metriorhynchoid *Maledictosuchus riclaensis* in dorsal view. Courtesy of Rodolfo Salas-Gismondi.

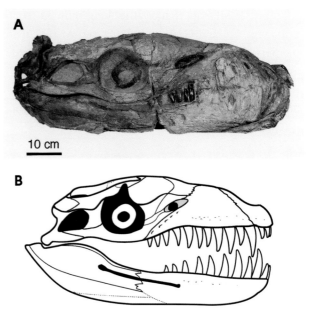

Figure 9.20. A, Skull of the metriorhynchoid *Dakosaurus andinensis* in lateral view. The anterior portion of snout is weathered. **B**, reconstruction of the skull of *Dakosaurus andinensis* in lateral view. A, courtesy of Zulma Gasparini; B, modified from Pol and Gasparini (2009).

Jurassic (Callovian-Kimmeridgian) of England and France, subsisted primarily on cephalopods and fish (Martill 1986). *Dakosaurus*, from the Late Jurassic to Early Cretaceous (Kimmeridgian-Berriasian) of Europe and the Late Jurassic (Tithonian) of Argentina, has an anteroposteriorly short and dorsoventrally deep snout with robust, labiolingually flattened teeth that have serrated cutting edges and tight tooth-to-tooth contact (Pol and Gasparini 2009; M. T. Young et al. 2012; Fig. 9.20). It attained a length of at least 4.5 m. Pierce et al. (2009a) reconstructed short-snouted metriorhynchids like *Dakosaurus* as capable of powerful biting as well as employing rolling of the entire body to kill and dismember large prey.

Because of their highly modified skeletal structure, the phylogenetic position of Thalattosuchia remains contentious. Several phylogenetic analyses placed thalattosuchians together with most other long-snouted mesoeucrocodylians except gharials (J. M. Clark 1994; Pol and Gasparini 2009; Andrade et al. 2011). Other studies found Thalattosuchia as the sister taxon to all other mesoeucrocodylians (J. M. Clark 1986; Larsson and Sues 2007; Sereno and Larsson 2009). Pierce et al. (2009b) noted that the cranial structure of thalattosuchians differs in detail from that of other long-snouted crocodyliforms. Pol and Gasparini (2009) and A. H. Turner and Sertich (2010) found Thalattosuchia among the derived mesoeucrocodylians (Neosuchia), whereas A. H. Turner and Pritchard (2015) recovered it outside that clade. Wilberg (2015) even considered it the sister group to Crocodyliformes.

CROCODYLOMORPHA: CROCODYLIFORMES: MESOEUCROCODYLIA: METASUCHIA

Benton and J. M. Clark (1988) united all mesoeucrocodylians more derived than Thalattosuchia as Metasuchia (from Greek *meta*, near, and *souchos*, crocodile). Sereno et al. (2001) defined Metasuchia as comprising the most recent common ancestor of *Notosuchus terrestris* and *Crocodylus niloticus* and all descendants of that ancestor. This clade can be divided into Notosuchia and Neosuchia.

MESOEUCROCODYLIA: METASUCHIA: NOTOSUCHIA

A number of phylogenetic analyses (Sereno and Larsson 2009; A. H. Turner and Sertich 2010; Pol et al. 2014) found strong support for a clade Notosuchia (from Greek *notios*, southern, and *souchos*, crocodile), but the interrelationships of its diverse constituent taxa differ considerably between these (and other) studies. This at least in part reflects the morphological diversity among notosuchians. They range from forms with a total length of 75 cm, some with cranial and dental features suggesting omnivory or even herbivory, to large-bodied predators attaining total lengths of up to 6 m. Sereno et al. (2001) defined Notosuchia as comprising all crocodyliforms more closely related to *Notosuchus terrestris* than to *Crocodylus niloticus*. It is a predominantly Gondwanan clade, with only a few records from Europe and Asia.

Notosuchia is characterized by a suite of synapomorphies including the absence of a posteroventral process of the dentary below the external mandibular fenestra; otic recess extending the full length of the postorbital; neural spines on the cervical vertebrae rod-like; and blade of the scapula greatly expanded (Pol et al. 2014).

One major subclade includes Uruguaysuchidae, Mahajangasuchidae, and Peirosauridae and is distinguished by various derived features such as the cylindrical rather than flattened postorbital bar, the presence of a large foramen at the contact between the premaxilla and maxilla, and the presence of a deep recess ventral to the glenoid facet of the coracoid (Pol et al. 2014).

Uruguaysuchids are short-snouted and attained total lengths of up to about 1 m. Of these, *Anatosuchus*, from the Early Cretaceous (Aptian-Albian) of Niger, has a much expanded, flat snout (Sereno and Larsson 2009). *Uruguaysuchus*, from the late Early or early Late Cretaceous of Uruguay, has incisiform teeth in the premaxilla and anterior portion of the dentary and posterior dentary and maxillary teeth with spatulate crowns that bear tiny denticles mesially and distally (Soto et al. 2011).

Mahajangasuchidae and Peirosauridae include larger-sized (2-6 m long) notosuchians. Mahajangasuchidae is the most inclusive clade containing *Mahajangasuchus insignis* but not *Notosuchus terrestris*, *Simosuchus clarki*, *Araripesuchus gomesii*, *Baurusuchus pachecoi*, *Peirosaurus torminni*, *Goniopholis crassidens*, *Pholidosaurus schaumbergensis*, or *Crocodylus niloticus* (Sereno and Larsson 2009). *Kaprosuchus*, from the Late Cretaceous (Cenomanian) of Niger (Sereno and Larsson 2009), and *Mahajangasuchus*, from the Late Cretaceous (Campanian) of Madagascar (A. H. Turner and Buckley 2008), have robust snouts with fused nasals (Fig. 9.21). *Kaprosuchus* has horn-like processes formed by parietals and squamosals at the back of the cranium and greatly enlarged, labiolingually flattened teeth in the premaxillae as well as in the anterior portions of the maxillae and dentaries. It attained a skull length of more than 50 cm. The interrelationships among Peirosauridae are still contentious.

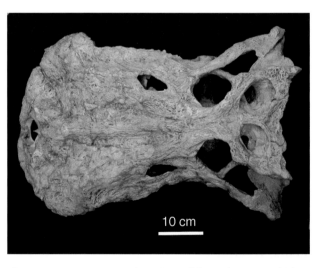

Figure 9.21. Reconstructed cranium of the mahajangasuchid *Mahajangasuchus insignis* in dorsal view. Courtesy of David Krause.

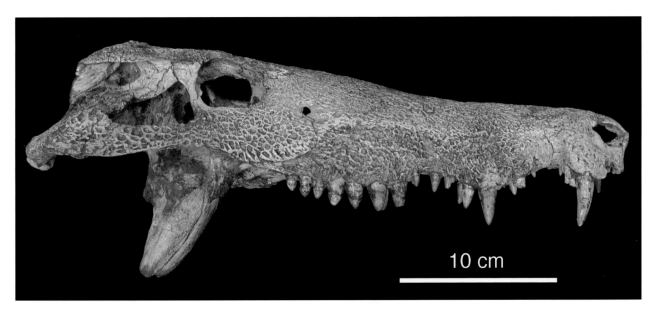

Figure 9.22. Cranium of the peirosaurid *Hamadasuchus* sp. in lateral view.

Representative taxa include *Hamadasuchus*, from the Late Cretaceous (Cenomanian) of Morocco (Larsson and Sues 2007; Fig. 9.22), and *Uberabasuchus*, from the Late Cretaceous (Campanian-Maastrichtian) of Brazil (Carvalho et al. 2004). They share transversely narrow but dorsoventrally deep snouts with a constriction between the premaxilla and maxilla on either side for the reception of a caniniform dentary tooth.

The phylogenetic analysis by Pol et al. (2014) supported a second subclade of Notosuchia, Ziphosuchia (Ortega et al. 2000). It is diagnosed by a number of synapomorphies such as posterior maxillary teeth that are much larger than the more anterior ones and a dorsoventrally deep, anteriorly tapering mandibular symphysis.

Simosuchus, from the Late Cretaceous (Maastrichtian) of Madagascar, attained a total length of 75 cm (Krause and Kley 2010). Its snout is short and broad, with a nearly flat anterior surface largely formed by the premaxillae (Fig. 9.23). Thus, the secondary bony palate is almost square in outline. The minute teeth of *Simosuchus* have broad and labiolingually flattened crowns, each with up to seven cusps on the maxillary teeth and up to nine on the dentary teeth. They suggest omnivorous or herbivorous habits. *Simosuchus* has extensively developed dermal armor, with distinct shields composed of multiple rows of osteoderms covering the trunk and tail dorsally and ventrally. Its robust limbs (except for the manus and pes) are also covered by osteoderms.

Notosuchus, from the Late Cretaceous (Santonian) of Argentina, has a short snout with anteriorly facing, confluent external nares. Its upper tooth row consists of two or three incisor-like teeth and an enlarged caniniform tooth in the premaxilla and six obliquely aligned molar-like teeth in the maxilla (Fiorelli and Calvo 2008). *Notosuchus* reached a total length of about 1.4 m.

Chimaerasuchus, from the Early Cretaceous (Aptian-Albian) of Hubei (China), has two procumbent teeth in each premaxilla and four molar-like teeth, each with three longitudinal rows of recurved cusps, in the maxilla (Wu and Sues 1996; Ösi 2014). Its molariform teeth closely resemble the postcanines in tritylodontid synapsids and are suitable for shredding high-fiber plant food.

Sphagesauridae shares typically obliquely aligned molariform teeth in the maxilla and dentary that have more or less triangular, thickly enameled crowns, each with a denticulated keel and apicobasal striations (Iori et al. 2013; Fig. 9.24). Wear facets on the molariform teeth indicate unilateral fore-and-aft jaw motion (Pol 2003). *Armadillosuchus*, from the Late Cretaceous (Turonian-Santonian) of Brazil, is noteworthy for its armadillo-like dorsal dermal armor, which has a rigid cervical shield composed of hexagonal osteoderms and a banded cervicodorsal portion (Marinho and Carvalho 2009).

J. E. Martin and de Lapparent de Broin (2016) proposed Candidodontidae for several small-bodied ziphosuchians with complex molar-like teeth. This clade includes

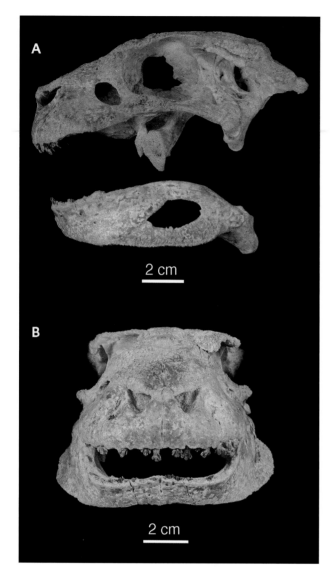

Figure 9.23. Skull of the notosuchian *Simosuchus clarki* in **A**, lateral, and **B**, anterior views. Courtesy of David Krause.

Candidodon, from the Early Cretaceous (Aptian-Albian) of Brazil (Carvalho 1994), *Lavocatchampsa*, from the Late Cretaceous (Cenomanian) of Morocco (J. E. Martin and de Lapparent de Broin 2016), and *Pakasuchus*, from the late Early or early Late Cretaceous of Tanzania (O'Connor et al. 2010).

Sebecosuchia (from Sebek, an ancient Egyptian crocodile-headed deity, and Greek *souchos*, crocodile) comprises the most recent common ancestor of Baurusuchidae and *Sebecus* and all descendants of that ancestor (Carvalho et al. 2004). It is characterized by numerous derived craniodental features such as the presence of a prominent lateral notch at the contact between the pre-

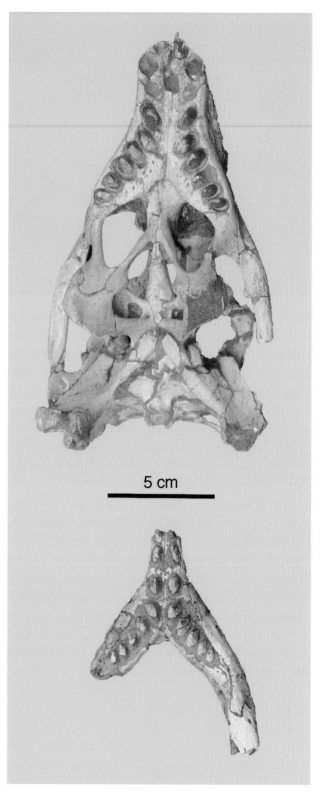

Figure 9.24. Cranium and mandible of the sphagesaurid *Caipirasuchus stenognathus* to show tooth rows in occlusal view. Courtesy of Rodolfo Salas-Gismondi.

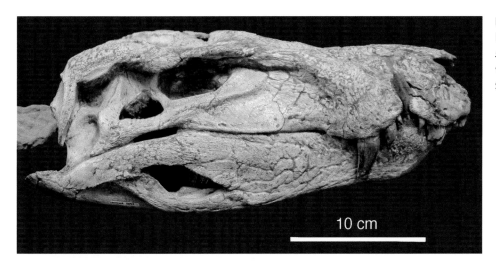

Figure 9.25. Skull of the baurusuchid *Baurusuchus salgadoensis* in lateral view. Courtesy of Rodolfo Salas-Gismondi.

maxilla and maxilla (for the reception of greatly enlarged dentary teeth), the expanded antorbital portion of the jugal, and anterior dentary teeth that are more than twice as tall as the other dentary teeth (Pol et al. 2012, 2014). The snout is dorsoventrally deep and transversely narrow with nearly vertical sides. The maxillary and dentary teeth closely resemble (and have sometimes been confused with) those of theropod dinosaurs in the strong labiolingual flattening of the crowns and presence of serrated mesial and distal cutting edges (ziphodont morphotype). Sebecosuchians were the only group of large-sized, apparently land-dwelling predators among crocodyliforms that existed both alongside theropod dinosaurs (Baurusuchidae) and in ecosystems with few or no other major predators (Sebecidae).

Baurusuchidae comprises the most recent common ancestor of *Baurusuchus* and *Stratiotosuchus* and all descendants of that ancestor (Carvalho et al. 2004). It is known from the Late Cretaceous of Argentina and Brazil (Carvalho et al. 2004, 2005, 2011). Among the diagnostic features for this clade are the reduced number of marginal teeth; the conspicuous notch between the premaxilla and maxilla for the reception of a dentary caniniform tooth; the enlarged third premaxillary tooth overhanging the dentary; and the dorsoventrally deep mandibular symphysis (Nascimento and Zaher 2010; Fig. 9.25). *Baurusuchus*, from the Late Cretaceous (Turonian-Santonian) of Brazil, attained a total length of up to 4 m and apparently had a fully upright limb posture (Carvalho et al. 2005; Nascimento and Zaher 2010).

Sebecidae, from the Paleogene (Paleocene) to Neogene (Miocene) of South America (Paolillo and Linares 2007),

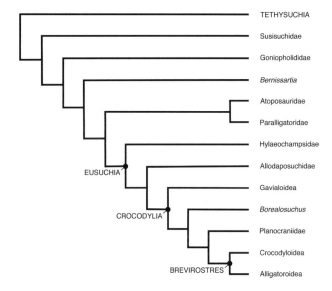

Figure 9.26. Phylogenetic hypothesis of the interrelationships of Neosuchia. Dots denote node-based clades. Based on Brochu (2003) and A. H. Turner (2015).

is the only known notosuchian group to survive the end-Cretaceous extinction event. *Bergisuchus*, from the Paleogene (Eocene) of Germany, and *Iberosuchus*, from the Paleogene (Eocene) of France, Portugal, and Spain, appear to be related to Sebecidae (Berg 1966; Pol et al. 2012). Pol et al. (2012) argued that *Sebecus*, from the Paleogene (Paleocene) of Argentina, had fully upright limb posture and terrestrial habits.

Dal Sasso et al. (2017) interpreted *Razanandrongobe*, a possible crocodyliform from the Middle Jurassic (Bathonian) of Madagascar, as the oldest known notosuchian to date. It has a dorsoventrally deep but transversely narrow, U-shaped snout with robust teeth and confluent, anteriorly facing external nares. If it is indeed a notosuchian, *Razanandrongobe* is noteworthy for its geological age and large body size, with a skull length of possibly up to 1 m.

MESOEUCROCODYLIA: METASUCHIA: NEOSUCHIA

Neosuchia (from Greek *neos*, new, and *souchos*, crocodile) comprises all crocodyliforms more closely related to *Crocodylus niloticus* than to *Notosuchus terrestris* (Sereno et al. 2001; Fig. 9.26). As previously noted, some authors also unite Thalattosuchia with other long-snouted crocodyliforms and place that grouping in Neosuchia. This would obviously affect how Neosuchia is defined and diagnosed. Here Neosuchia is diagnosed by the presence of a nearly vertically aligned maxilla, two sinusoidal waves of teeth in the maxilla and dentary, respectively, and the absence of an antorbital fenestra.

MESOEUCROCODYLIA: METASUCHIA: NEOSUCHIA: TETHYSUCHIA

Tethysuchia (from Greek *Tethys*, the goddess of the sea, and *souchos*, crocodile) comprises the most recent common ancestor of *Pholidosaurus purbeckensis* and *Dyrosaurus phosphaticus* and all descendants of that ancestor (Andrade et al. 2011). This clade is distinguished by cranial synapomorphies such as the absence of an antorbital fossa and premaxillae that project laterally and are broader transversely than the maxillae (in plan view). Various phylogenetic analyses (Sereno et al. 2001; Hastings et al. 2010, 2011; Andrade et al. 2011; A. H. Turner and Pritchard 2015) found some pholidosaurids more closely related to Dyrosauridae. However, J. E. Martin et al. (2014b) recovered a new grouping Coelognathosuchia with a subclade Pholidosauridae.

Dyrosauridae is characterized by the presence of elongate upper temporal fenestrae; posteriorly directed occipital tuberosities formed by the exoccipitals; the extensive participation of the exoccipitals in the formation of the occipital condyle; and the small size of the seventh dentary tooth and its position close to that of the eighth tooth (Jouve et al. 2006; Fig. 9.27). Dyrosaurids were widely distributed in Late Cretaceous to Paleogene estuarine and nearshore marine deposits across North and West Africa, in Bolivia, Brazil, Mexico, Myanmar,

Pakistan, Saudi Arabia, and the eastern United States (Jouve et al. 2006; Barbosa et al. 2008). More recently, Hastings et al. (2010, 2011) described two taxa, including the short-snouted *Cerrejonisuchus*, from Paleogene (Paleocene) freshwater strata in Colombia. Dyrosauridae diversified rapidly following the end-Cretaceous extinction event. They range in total length from more than 1 m to 8 m and show considerable variation in snout shapes and sizes as well as in tooth form, indicating a wide range of feeding preferences. *Dyrosaurus*, from the Paleogene (Eocene) of Algeria and Morocco, and *Hyposaurus*, from the Late Cretaceous (Maastrichtian) and

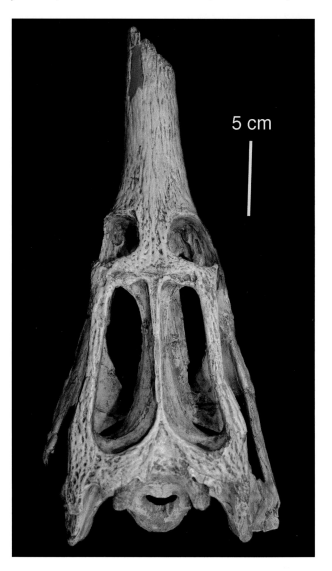

Figure 9.27. Partial cranium of the dyrosaurid *Rhabdognathus aslerensis* in dorsal view. Note elongated upper temporal fenestrae. Courtesy of Chris Brochu.

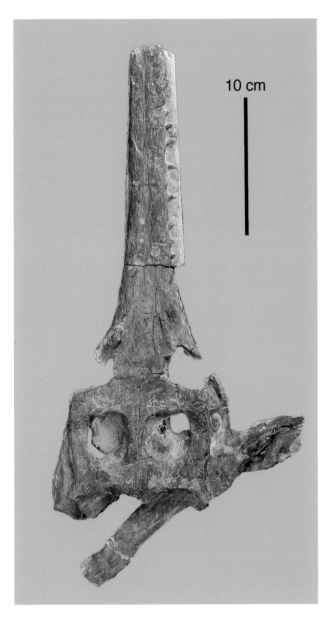

10 cm

Figure 9.28. Skull of the pholidosaurid *Pholidosaurus laevis*, with cranium exposed in dorsal view. Courtesy of Rodolfo Salas-Gismondi.

Paleogene (Paleocene) of the eastern United States, have long, slender snouts with pointed teeth, consistent with feeding on fish. By contrast, *Phosphatosaurus*, from the Paleogene (Eocene) of Mali, Niger, and Tunisia, has a robust snout with blunt-crowned teeth, which suggest a diet including prey such as turtles and mollusks (Buffetaut 1982). The dorsal and ventral dermal armor is reduced in dyrosaurids. The tail has tall neural spines and long chevrons. The well-developed forelimbs and hind

limbs are of equal length, and dyrosaurids were probably still capable of locomotion on land (Buffetaut 1982).

Pholidosauridae includes *Pholidosaurus*, from the Early Cretaceous (Berriasian) of England, France, and Germany (J. E. Martin et al. 2016a; Fig. 9.28); *Terminonaris*, from the Early to Late Cretaceous (Albian-Turonian) of western North America and possibly from the Late Cretaceous (Cenomanian) of Germany (Wu et al. 2001a); and *Sarcosuchus*, from the Early Cretaceous (Aptian-Albian) of Algeria, Brazil, and Niger (Taquet 1976; Sereno et al. 2001). Diagnostic features for this clade include the greatly elongated rostrum (about 70 percent of total skull length), the exclusion of the nasal from the posterior narial margin, and the presence of a posterolateral depression on the maxilla, which nearly extends to the jugal (J. E. Martin et al. 2016a). *Sarcosuchus* attained a skull length of 1.6 m and a total length of at least 11 m (Sereno et al. 2001). The snout is broad transversely in adults, unlike in other pholidosaurids. Its distinctly inflated anterior end surrounds the large external narial opening (Sereno et al. 2001). The mandible is shorter than the cranium, and the lower teeth would bite inside the upper teeth. The tooth crowns are massive and blunt. *Sarcosuchus* was probably a river-dweller capable of taking a wide range of prey (Sereno et al. 2001).

MESOEUCROCODYLIA: METASUCHIA: NEOSUCHIA: SUSISUCHIDAE

Salisbury et al. (2006) interpreted *Isisfordia*, from the Early or Late Cretaceous (Albian or Cenomanian) of Australia, as the oldest known eusuchian based on the structure of the choanae and the slightly procoelous cervical, dorsal, and anterior caudal vertebrae. Subsequently, A. H. Turner and Pritchard (2015) demonstrated that the choanae are bordered anteromedially by the palatines, not the pterygoids. They also hypothesized a more basal phylogenetic position for *Isisfordia* and the closely related *Susisuchus*, from the Early Cretaceous (Aptian) of Brazil (Salisbury et al. 2003). A. H. Turner and Pritchard (2015) grouped the two taxa together as Susisuchidae, which is distinguished especially by the pear-shaped outline of the external narial opening and the participation of the nasals in the margin of this opening. The snout is proportionately long and flattened. *Isisfordia* attained a total length of slightly more than 1 m.

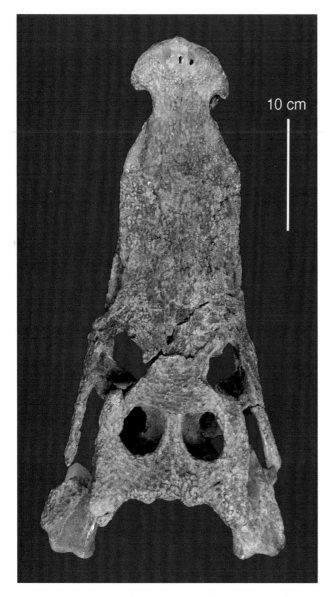

10 cm

Figure 9.29. Cranium of the goniopholidid *Amphicotylus lucasii* in dorsal view. Courtesy of Rodolfo Salas-Gismondi.

MESOEUCROCODYLIA: METASUCHIA: NEOSUCHIA: GONIOPHOLIDIDAE

Goniopholididae is distinguished especially by the presence of a prominent posterolateral depression on the maxilla near the alveolar margin (Pritchard et al. 2013; Fig. 9.29). A similar depression is present in Pholidosauridae but differs in detailed structure. Other possibly diagnostic features include the presence of a smooth ventral surface along the alveolar margin of the maxilla and the presence of a foramen at the lateral edge of the postorbital bar. Goniopholidids resemble present-day crocodylians in overall skull shape and inferred habits and attained total lengths between 2 and 4 m. The clade includes taxa with broad snouts such as *Goniopholis*, from the Early Cretaceous (Berriasian) of England and Germany (Andrade et al. 2011), and forms with more elongate and narrower snouts such as *Calsoyasuchus*, from the Early Jurassic (Pliensbachian) of Arizona (Tykoski et al. 2002), and *Eutretauranosuchus*, from the Late Jurassic (Kimmerigian-Tithonian) of Colorado and Wyoming (Pritchard et al. 2013).

MESOEUCROCODYLIA: METASUCHIA: NEOSUCHIA: EUSUCHIA

Eusuchia (from Greek *eu*, true, and *souchos*, crocodile) comprises the most recent common ancestor of *Hylaeochampsa vectiana*, *Crocodylus niloticus*, *Gavialis gangeticus*, and *Alligator mississippiensis*, and all descendants of that ancestor (Brochu 2003a). Huxley (1875) used two major diagnostic features for this clade: the extensive, complete secondary bony palate in which the choanae are completely surrounded by the pterygoids (Fig. 9.30), and the procoelous condition of most vertebrae with ball-and-socket joints between successive centra. Subsequent studies indicated that each of these traits evolved independently in other crocodyliforms (J. M. Clark et al. 2004; Rogers 2003). However, the combination of the two features, along with the presence of dorsal armor composed of four or more rows of osteoderms, is still considered diagnostic for Eusuchia (Salisbury et al. 2006).

Bernissartia, from the Early Cretaceous (Barremian) of Belgium and Spain, has long been considered either a basal eusuchian or a stem-eusuchian (Norell and J. M. Clark 1992; A. H. Turner and Pritchard 2013). However, its precise phylogenetic position remains uncertain. *Bernissartia* has procoelous caudals but amphicoelous presacral vertebrae. Its dorsal armor is composed of four rows of osteoderms, the median ones of which each bear two distinct keels on their dorsal surfaces (Frey 1988).

Recent studies of Mesozoic eusuchians, based on reassessments of previously reported taxa and discoveries of new ones, have generated competing new hypotheses regarding their interrelationships and their relationships to Crocodylia (Pol et al. 2009; Adams 2014; Narváez et al. 2015; A. H. Turner 2015).

Atoposauridae is a clade of small-bodied crocodyliforms with total lengths between 30 and 50 cm. Adams

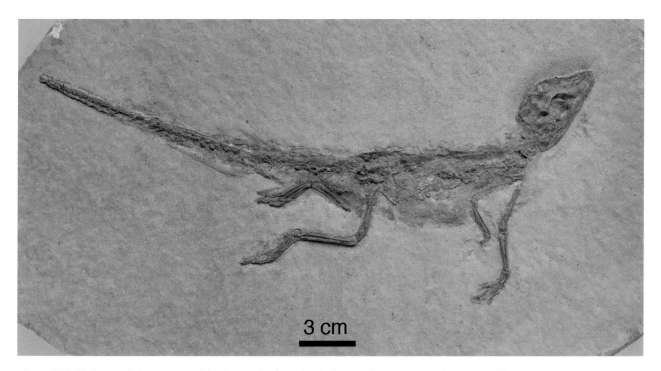

Figure 9.30. Skeleton of the atoposaurid *Alligatorellus bavaricus* in lateral view. Courtesy of Helmut Tischlinger.

(2014) found it as the basal group of Neosuchia, but A. H. Turner and Pritchard (2015) and Schwarz et al. (2017) placed it in Eusuchia and closer to Crocodylia. Atoposauridae includes *Alligatorellus* (Fig. 9.31), *Alligatorium,* and *Atoposaurus* from the Late Jurassic (Kimmeridgian-Tithonian) of France and Germany (Tennant and Mannion 2014); *Theriosuchus*, from the Middle Jurassic (Bajocian-Bathonian) of Scotland, the Early Cretaceous of England and Spain, and the Early Cretaceous of Thailand (Young et al. 2015; Tennant et al. 2016); and *Knoetschkesuchus*, from the Late Jurassic (Kimmeridgian) of Germany and Portugal (Schwarz et al. 2017). Tennant et al. (2016) accepted only the first three of these taxa as members of Atoposauridae, which they defined as a clade comprising all taxa more closely related to *Atoposaurus jourdani* than to *Crocodylus niloticus*. Tennant et al. (2016) listed various synapomorphies for Atoposauridae such as the complete division of the external nares by the anterior processes of the nasals, the projection of the premaxillary internarial bar anterior to the main bodies of the premaxillae, the otic aperture not being closed posteriorly by the quadrate and otoccipital, and the presence of a ventral depression on the dorsal margin of the postorbital. They also argued that *Theriosuchus* should be separated from Atoposauridae based on cranial features such

as the presence of a raised supraorbital ridge and the parallel anterior ends of the palatine bar between the suborbital fenestrae. However, the phylogenetic analysis by Schwarz et al. (2017) recovered *Theriosuchus* again as part of Atoposauridae. Tennant et al. (2016) also reinterpreted certain Late Cretaceous crocodyliforms originally referred to *Theriosuchus* (J. E. Martin et al. 2014a) as more derived neosuchians.

Paralligatoridae is characterized by a number of shared derived features such as a ventrally expanded medial condyle of the quadrate, which is separated from the lateral condyle by a deep groove, the anterior flaring of the bony bar between the suborbital fenestrae, and a broad and round olecranon process of the ulna (A. H. Turner 2015). The clade is known from the Early Cretaceous of Jilin (China) (Wu et al. 2001b) and the Late Cretaceous of China, Kazakhstan, Mongolia, Tadzhikistan, and Uzbekistan (Pol et al. 2009; A. H. Turner 2015; Kuzmin et al. 2019). Putative paralligatorids from the Late Jurassic (Oxfordian-Kimmeridgian) of Brazil (Montefeltro et al. 2013) and the Early Cretaceous (Aptian-Albian) of Montana and Texas (Adams 2014) do not belong to Paralligatoridae (Kuzmin et al. 2019). Adams (2014) found Paralligatoridae as the sister taxon to Eusuchia (including *Hylaeochampsa*) (see also Pol et al. [2009]). A. H. Turner (2015) hypothesized

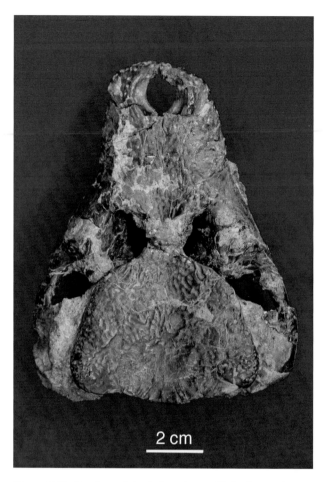

Figure 9.31. Cranium of the hylaeochampsid *Iharkutosuchus makadii* in dorsal view. Courtesy of Rodolfo Salas-Gismondi.

that Paralligatoridae and Hylaeochampsidae together form a clade that is the sister group to Crocodylia, but this grouping is only weakly supported.

Hylaeochampsidae is united by a number of synapomorphies such as the maxilla forming part of the lower temporal bar, a massive postorbital bar, a reduced or no quadratojugal spine, and the presence of a prominent knob on the ventral surface of the quadrate (Narváez et al. 2015). *Hylaeochampsa*, from the Early Cretaceous (Barremian) of England, and *Iharkutosuchus*, from the Late Cretaceous (Santonian) of Hungary (Fig. 9.32), have greatly enlarged, complex teeth suitable for crushing and grinding at the posterior end of the maxillary (and, in the latter taxon, also dentary) tooth row (Ösi 2014). The choanae are situated far back and completely surrounded by the pterygoids.

Allodaposuchidae comprises *Allodaposuchus precedens* and all crocodyliforms more closely related to it than to *Hylaeochampsa vectiana*, *Shamosuchus djadochtaensis*, *Borealosuchus sternbergii*, *Planocrania datangensis*, *Alligator mississippiensis*, *Crocodylus niloticus*, or *Gavialis gangeticus* (Narváez et al. 2015). Narváez et al. (2015) hypothesized it as the sister taxon of Hylaeochampsidae and, in turn, this grouping as most closely related to Crocodylia. Allodaposuchidae is diagnosed by the presence of a shallow fossa at the anteromedial corner of the upper temporal fenestra and the tenth dentary alveolus being the largest behind the fourth (Fig. 9.33). It is known from the Late Cretaceous (Maastrichtian) of France, Romania, and Spain (Puértolas-Pascual et al. 2014; Narváez et al. 2015). Allodaposuchidae and Hylaeochampsidae represent apparently endemic radiations of basal eusuchians in the Cretaceous of Europe, some of which resemble alligatoroid crocodylians in overall appearance and inferred habits.

Stomatosuchus, from the Late Cretaceous (Cenomanian) of Egypt, has an almost 2 m long skull with a greatly elongated, flat snout and unusually slender lower jaws that abruptly turn toward the symphysis anteriorly (Stromer 1925). Unfortunately, the only known specimen was destroyed during World War II. Sereno and Larsson (2009) described slender, U-shaped mandibular remains of a possibly closely related taxon, *Laganosuchus*, from the Late Cretaceous (Cenomanian) of Morocco and Niger. Too little is known about these "duck-faced" crocodyliforms to assess their phylogenetic relationships. Closely related (if not referable) to Stomatosuchidae are *Aegisuchus* and *Aegyptosuchus*, from the Late Cretaceous (Cenomanian) of Morocco and Egypt, respectively (Holliday and N. M. Gardner 2012). Both have flat crania with a distinctive ornamentation of the skull deck, especially a raised, roughened boss on the parietal, which is surrounded by grooves for blood vessels and likely had a specialized integumentary cover.

EUSUCHIA: CROCODYLIA

Brochu (1999, 2003a) defined Crocodylia (from Greek *krokodeilos*, a kind of lizard) as comprising the most recent common ancestor of *Alligator mississippiensis* (American alligator), *Crocodylus niloticus* (Nile crocodile), and *Gavialis gangeticus* (gharial), and all descendants of that ancestor. This definition has since been widely adopted (e.g., Grigg and Kirshner 2015).

Present-day crocodylians are largely restricted to the tropical realms of the world except for the two species

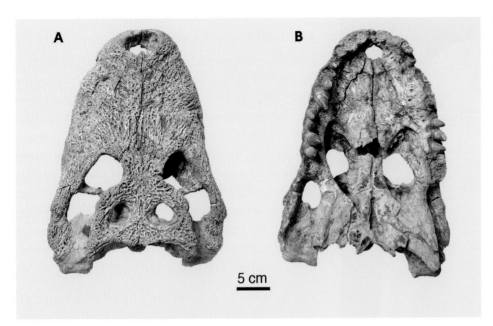

Figure 9.33. West African crocodile (*Crocodylus suchus*) employing "high walk" on its way to water. Courtesy of Wolfgang Wüster.

of *Alligator*, which live in temperate regions. They hunt in the water but spend much time on land in order to regulate their body temperature (Fig. 9.34). Hatchlings feed on arthropods and small fish, whereas more mature individuals subsist on larger fish and on land vertebrates. Extant species dismember large prey by rapidly rolling their entire body about its long axis ("death roll"). Another noteworthy aspect of the crocodylian body plan is the nature of the respiratory apparatus, which reduces reliance on costal aspiration. The liver divides the body cavity into an anterior thoracic and a posterior abdominal cavity. A diaphragmatic muscle originates from the posterior gastralia and pelvic region and inserts onto a thick sheet of connective tissue surrounding the liver and connecting the liver to the lungs. This muscle pulls the liver back, expanding the thoracic cavity and drawing air into the lungs (Gans and B. Clark 1976). In addition, movements of the ribs and rotation of the pubic bones contribute to lung ventilation (Farmer and Carrier 2000). Schachner et al. (2013a) demonstrated that airflow is unidirectional in crocodylians, much like in birds. However, as this condition is also found in varanid lizards, it either evolved independently in the latter or represents the ancestral state for Sauria (Schachner et al. 2013b).

Brochu (1999) listed a number of craniodental synapomorphies for Crocodylia: skull table with nearly horizontal lateral margins and distinct squamosal "prongs" posteriorly; exoccipital with a long process lateral to the cranioquadrate passage; upper temporal fenestra lacking a fossa at its anteromedial corner; lower jaw with an external mandibular fenestra and a posterodorsally extending retroarticular process; and anterior teeth of the dentary projecting anterodorsally rather than directly forward.

Crocodylia includes Alligatoroidea, Crocodyloidea, and Gavialoidea, all of which first appeared in the fossil record during the Late Cretaceous. There are currently 25 recognized extant species. However, Grigg and Kirshner (2015) noted that various data already suggest the existence of additional species, possibly bringing the total number close to 30. Morphology-based analyses find

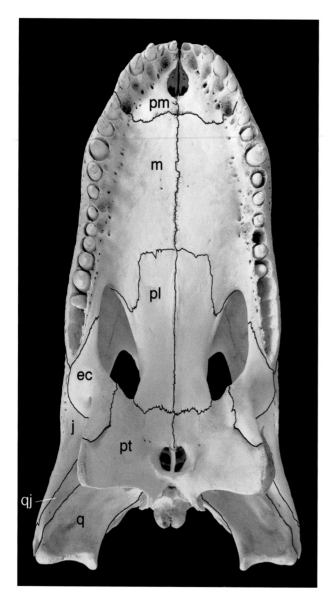

Figure 9.34. Cranium of an American alligator (*Alligator mississippiensis*) in ventral view with sutures highlighted. Note the position of the choanae entirely within the pterygoids. Abbreviations: ec, ectopterygoid; m, maxilla; pl, palatine; pm, premaxilla; pt, pterygoid; q, quadrate; qj, quadratojugal. Courtesy of Rodolfo Salas-Gismondi.

a clade comprising Alligatoroidea and Crocodyloidea (Brevirostres) as the sister group of Gavialoidea (Brochu 1997, 2003a). On the other hand, molecular data support a closer relationship between *Tomistoma* and *Gavialis* (Poe 1997; Harshman et al. 2003; McAliley et al. 2006; Oaks 2011). In addition to the three major subclades, Crocodylia includes a variety of extinct taxa that cannot be assigned to any of these groups.

EUSUCHIA: CROCODYLIA: *BOREALOSUCHUS*

Borealosuchus, from the Late Cretaceous (Maastrichtian) to Paleogene (Eocene) of North America, was long considered representative of the ancestral crocodylian body plan. Its constituent species were assigned to *Leidyosuchus*, which is now interpreted as a basal alligatoroid. Brochu (1997, 1999) showed that *Borealosuchus* is related to, but not a member of, Brevirostres and that its similarities to alligatoroids are either plesiomorphic or convergently acquired. *Borealosuchus* is distinguished especially by the broad lateral curvature of the upper tooth row behind the sixth or seventh maxillary tooth position. It attained a total length of up to 3 m.

EUSUCHIA: CROCODYLIA: PLANOCRANIIDAE

Planocraniidae comprises *Planocrania datangensis* and all crocodylians more closely related to it than to *Alligator mississippiensis*, *Crocodylus niloticus*, *Gavialis gangeticus*, *Borealosuchus sternbergii*, *Thoracosaurus macrorhynchus*, *Allodaposuchus precedens*, or *Hylaeochampsa vectiana* (Brochu 2013). It is known only from the Paleogene of China, France, Germany, Italy, Spain, and the United States (Berg 1966; Rossmann 2000; Brochu 2013). Planocraniidae is distinguished by the possession of a transversely narrow but dorsoventrally deep snout (Fig. 9.35) and labiolingually flattened maxillary and posterior dentary tooth crowns that typically have finely serrated cutting edges. Rossmann (2000) argued that *Boverisuchus* (formerly "*Pristichampsus*"), from the Paleogene (Eocene) of France, Germany, Italy, Spain, and the western United States, was predominantly terrestrial based on the structure of its limbs. It attained a total length of up to 3 m. Narváez et al. (2015) found *Borealosuchus* and Planocraniidae as successive sister taxa to Alligatoroidea and Crocodyloidea.

EUSUCHIA: CROCODYLIA: BREVIROSTRES

Brochu (1999, 2003a, 2013) used the name Brevirostres (from Latin *brevis*, short, and Latin *rostrum*, snout) for a clade comprising the most recent common ancestor of *Alligator mississippiensis* and *Crocodylus niloticus* and all descendants of that ancestor (Figs. 9.36, 9.37). Diagnostic derived features for this clade include the presence of a posterior notch in the suborbital fenestra; the invagination of the posterior margin of the otic aperture; the dis-

Figure 9.35. Cranium of the planocraniid *Boverisuchus vorax* in lateral view. Courtesy of Rodolfo Salas-Gismondi.

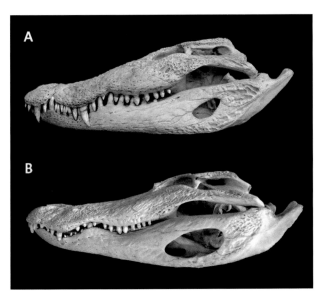

Figure 9.36. Skulls of **A**, a marsh crocodile (*Crocodylus palustris*), and **B**, an American alligator (*Alligator mississippiensis*) in lateral view. Note the differences in the occlusal relationships between the upper and lower teeth. Courtesy of Rodolfo Salas-Gismondi.

tinctly concave proximal edge of the deltopectoral process on the humerus; and the greatly reduced anterior process on the ilium (Brochu 1999).

EUSUCHIA: CROCODYLIA: BREVIROSTRES: ALLIGATOROIDEA

Alligatoroidea comprises all crocodylians more closely related to *Alligator mississippiensis* than to *Crocodylus ni-*

loticus or *Gavialis gangeticus* (Brochu 1999, 2003a). This clade is characterized by several synapomorphies: the subequal anterior processes of the surangular; the broad separation of the ectopterygoid by the maxilla from the posterior maxillary teeth; and the presence of a dorsally placed pneumatic foramen (foramen aërum) on the quadrate and a laterally positioned foramen aërum on the articular (Brochu 1997, 1999). Present-day alligators and crocodiles differ in the mode in which upper and lower teeth come in contact (Fig. 9.36). In alligators, the dentary teeth bite lingual to those of the premaxillae and maxillae, resulting in a characteristic overbite. By contrast, crocodiles have an enlarged tooth in each dentary that bites into a well-defined notch between the premaxilla and maxilla. However, the overbite in alligators and the notch between the premaxilla and maxilla together represent the plesiomorphic features for most crocodylians and their close relatives (Brochu 1999). Thus, Brochu (2003) hypothesized that alligatoroids lost the notch and crocodyloids the overbite during their respective evolutionary histories.

The basal alligatoroid *Leidyosuchus*, from the Late Cretaceous (Campanian) of Alberta, superficially resembles *Borealosuchus* and early crocodyloids. It has a transversely broad, flat snout with one or two dentary teeth biting into a deep notch between the premaxilla and maxilla (Brochu 1999; Wu 2001; Fig. 9.38).

Deinosuchus, from the Late Cretaceous (Campanian) of Montana, Texas, and the Atlantic Coastal Plain from New

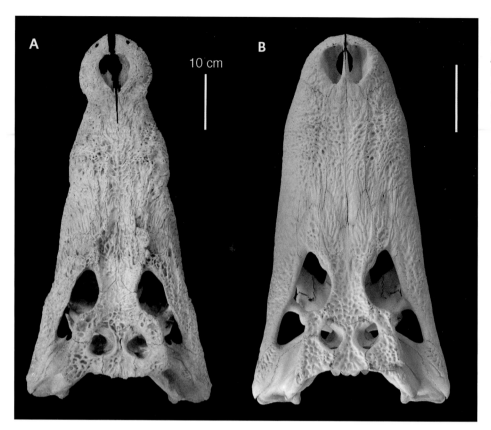

Figure 9.37. Crania of **A**, a saltwater crocodile (*Crocodylus porosus*), and **B**, an American alligator (*Alligator mississippiensis*) in dorsal view. Scale bars each equal 10 cm. Courtesy of Rodolfo Salas-Gismondi.

Jersey to Mississippi, attained a total length between 9 to 12 m (Erickson and Brochu 1999). The phylogenetic analysis by Brochu (1999) recovered it as a basal alligatoroid. The dorsal osteoderms of *Deinosuchus* are thick, unlike the flat plates in other crocodylians, and their external surfaces bear widely separated, deep pits. The up to 1.8 m long, massive skull has a broad snout with a much enlarged narial opening (Fig. 9.39).

All other alligatoroids share several apomorphies such as the considerable reduction or absence of a spine on the quadratojugal in adult individuals, the absence of a contact between the parietal and squamosal within the supratemporal fossa, and the fourth dentary tooth fitting into a closed pit between the premaxilla and maxilla (Brochu 1999). The basal member of this clade is *Diplocynodon*, which is known from the Paleogene (Eocene) to Neogene (Miocene or Pliocene) of Europe (Berg 1966; J. E. Martin 2010; J. E. Martin and Gross 2011). It has large fourth and fifth maxillary alveoli of identical size and enlarged, confluent third and fourth dentary alveoli.

More derived alligatoroids are grouped together as Globidonta, which Brochu (1999, 2003a) defined as comprising *Alligator mississippiensis* and all crocodylians more closely related to it than to *Diplocynodon ratelii*. This clade is diagnosed by several apomorphies such as the (in lateral view) plate-shaped axial intercentrum with prominent parapophyseal processes in mature individuals, the lateral surface of the prootic being largely concealed by the quadrate and laterosphenoid, and the division of the choanae by a prominent septum (Brochu 1999). Basal members of Globidonta do not exceed 1.5 m in total length and have short, broad, and blunt snouts with enlarged teeth at the back of the tooth rows. *Stangerochampsa*, from the Late Cretaceous (Maastrichtian) of Alberta, is one of the oldest known globidontans. *Allognathosuchus*, from the Paleogene (Eocene) of Wyoming and possibly Belgium, has posterior teeth with massive, bulbous crowns.

Alligatoridae comprises the most recent common ancestor of *Alligator mississippiensis* and *Caiman crocodilus* (spectacled caiman) and all descendants of that ancestor (Brochu 1999, 2003a). As diagnostic synapomorphies Brochu (1999) listed the position of the suture between the frontal and parietal entirely on the skull roof, the poste-

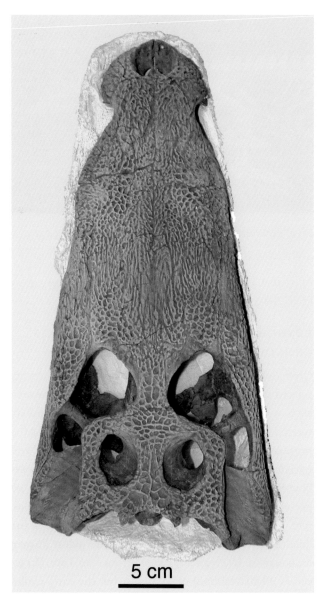

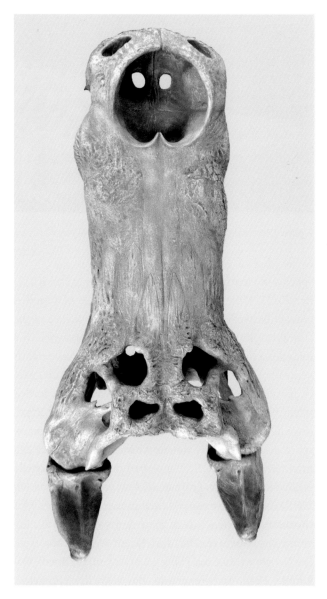

Figure 9.38. Cranium of the alligatoroid *Leidyosuchus canadensis* in dorsal view. Courtesy of Don Brinkman.

Figure 9.39. Reconstructed skull of the alligatoroid *Deinosuchus riograndensis* in dorsal view. Length c. 1.8 m. Courtesy of Rodolfo Salas-Gismondi.

rior maxillary process within the lacrimal, and the medial passage of the suture between the postorbital and squamosal ventral to the skull table. It is divided into Alligatorinae and Caimaninae. Unlike crocodylids, alligatorids are restricted to freshwater settings.

Alligatorinae comprises *Alligator mississippiensis* (Figs. 9.36B, 9.37B, 9.40) and all crocodylians more closely related to it to than to *Caiman crocodilus* (Brochu 1999, 2003a). It includes two extant species, *Alligator mississippiensis* (American alligator) in the southeastern United States

and *Alligator sinensis* (Chinese alligator) in eastern China (Grigg and Kirshner 2015). The oldest record of *Alligator* is from the Paleogene (Eocene) of Nebraska (Whiting and Hastings 2015). *Alligator mississippiensis* dates back to the Neogene (Miocene) (Whiting et al. 2016) and today ranges widely throughout the southeastern United States. It attains a total length of up to 4.6 m.

Caimaninae comprises *Caiman crocodilus* and all crocodylians more closely related to it than to *Alligator mississippiensis* (Brochu 1999, 2003a). It includes six extant

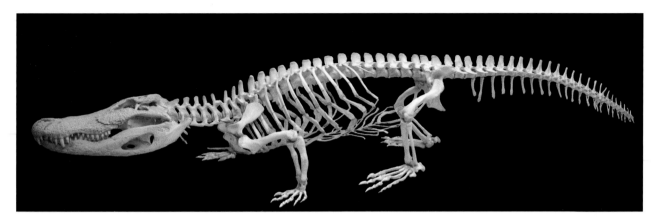

Figure 9.40. Skeleton of an American alligator (*Alligator mississippiensis*; Alligatorinae) in lateral view. Courtesy of Heinrich Mallison.

species in Central and South America (Grigg and Kirshner 2015). Brochu (2010) listed the absence of a medial contact between the splenials, the large exposure of the supraoccipital on the skull roof, and the presence of a long descending process of the exoccipital that extends ventrally to the basioccipital tubera among the diagnostic features for this clade. Caimanines are first recorded from the Paleogene (Paleocene) of Argentina and Brazil (Brochu 2011; Cidade et al. 2017). The subclade Jacarea, represented today by caimans (*Caiman*) and the black caiman (*Melanosuchus*), first appeared during the Neogene (Miocene). The dwarf caiman (*Paleosuchus*) has no known fossil record. Two extinct caimanines are noteworthy for the distinctive structure of their skulls and for attaining enormous body size. The "duck-faced" *Mourasuchus,* from the Neogene (Miocene-Pliocene) of Argentina, Brazil, Colombia, Peru, and Venezuela, has a greatly elongated, transversely broad, and flat snout and very slender mandibular rami with up to 40 small teeth in each jaw, superficially resembling that of *Stomatosuchus* (Price 1964; Langston 1965; Scheyer and Delfino 2016; Cidade et al. 2017; Fig. 9.41). It attained a skull length of up to 1.2 m and probably fed on small prey that it caught in large numbers (Cidade et al. 2017). The closely related *Purussaurus*, from the Neogene (Miocene) of Brazil, Colombia, Peru, and Venezuela, differs from *Mourasuchus* in having robust jaws, a transversely broad and dorsoventrallydeep snout with a very large median depression on the dorsal surface behind the external narial opening (Fig. 9.42), and large, stout teeth. It clearly was an apex predator, capable of extremely powerful biting (Aureliano et al. 2015). *Purussaurus* reached a skull length

of up to 1.4 m and a total length of up to 12 m (Aguilera et al. 2006; Scheyer and Delfino 2016).

EUSUCHIA: CROCODYLIA: BREVIROSTRES: CROCODYLOIDEA

Crocodyloidea comprises all crocodylians more closely related to *Crocodylus niloticus* than to *Alligator mississippiensis* or *Gavialis gangeticus* (Brochu 2003a). It is characterized by a suite of cranial synapomorphies such as the presence of a broad contact between the postorbital and parietal, the limited extension of the palatine process beyond the anterior end of the suborbital fenestra, the presence of an extended medial condyle on the quadrate, and the dorsal position of the foramen aërum in the quadrate (Brochu 1999; Fig. 9.37A). The oldest known crocodyloid is *Prodiplocynodon*, from the Late Cretaceous (Maastrichtian) of Montana, but the early evolutionary history of this clade remains poorly documented (Brochu 2000, 2003a). During the Paleogene, crocodyloids ranged widely across the Northern Hemisphere (Brochu 2000; Delfino and T. Smith 2009).

Crocodylidae comprises the most recent common ancestor of *Crocodylus niloticus, Osteolaemus tetraspis* (dwarf crocodile), and *Tomistoma schlegelii* (false or Malayan gharial) and all descendants of that ancestor (Brochu 2003a). Historically, many Cenozoic crocodylian and even Mesozoic crocodyliform fossils were identified as *Crocodylus* based on what turned out to be plesiomorphic features (Salisbury and Willis 1996; Brochu 2000). Molecular and paleontological data have since established that *Crocodylus* first appeared during the Neogene, with the oldest known fossil occurrences from the Miocene-Pliocene of Africa

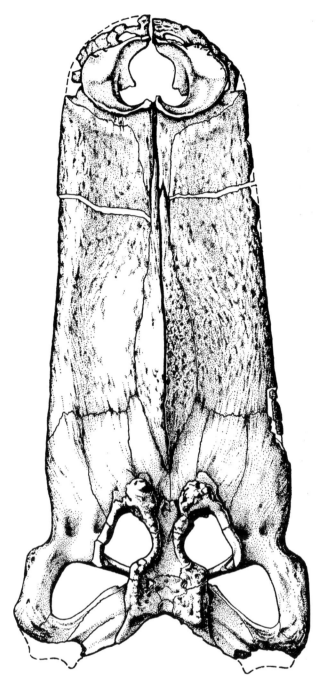

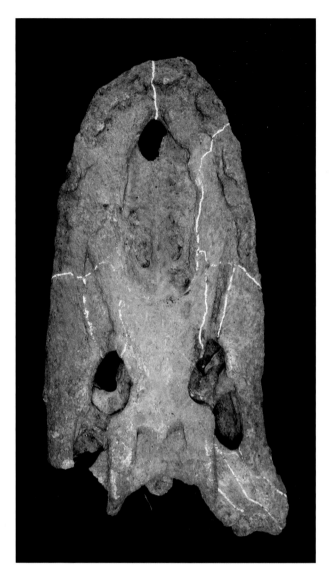

Figure 9.42. Cranium of the caimanine *Purussaurus mirandai* in dorsal view. Note the large dorsomedian depression behind the external narial opening. White lines trace identifiable sutures. Skull length 1.26 m. Courtesy of Torsten Scheyer.

Figure 9.41. Cranium of the caimanine *Mourasuchus amazonensis* in dorsal view. Midline length of cranium 1.14 m. From Price (1964)—CC BY 4.0.

and the Indian subcontinent (Oaks 2011; Brochu and Storrs 2012). The largest known species of *Crocodylus, C. thorbjarnarsoni*, from the Neogene (Pliocene) and Quaternary (Pleistocene) of Kenya, attained a total length of 7.5 m (Brochu and Storrs 2012; Fig. 9.43). By comparison, the largest present-day crocodylian taxon is *Crocodylus po-*

rosus (saltwater crocodile), from the Indo-Pacific region, with a reliably recorded maximum length of about 6.2 m. The extant species of *Crocodylus* living in the Western Hemisphere form a subclade of their own (Brochu 2000; Oaks 2011). Scheyer et al. (2013) reported a closely related extinct form from the Neogene (Pliocene) of Venezuela.

Phylogenetic studies by Brochu (1997, 2003a, 2007a) found a previously unrecognized African clade of Crocodylidae, Osteolaeminae, which he defined as comprising *Osteolaemus tetraspis* and all crocodylians more closely related to it than to *Crocodylus niloticus* (Brochu 2003a).

Attaining a total length of 1.5 to 2 m, *Osteolaemus* has a short, blunt snout. Dwarf crocodiles range from western Sub-Saharan Africa into western Central Africa. Extinct osteolaemines were more diverse in form and body size and include *Voay*, from the Quaternary (Holocene) of Madagascar, and *Rimasuchus*, from the Neogene (Miocene-Pliocene) and Quaternary (Pleistocene) of Chad, Egypt, Libya, Tunisia, and Uganda. *Voay* is distinguished by the presence of prominent "horns" formed by the upturned posterolateral corners of the squamosals and a deep snout (Brochu 2007a). It apparently vanished during the extinction of the native Malagasy megafauna following the arrival of humans on that island. *Rimasuchus* attained a skull length of 1 m and has a short, broad snout with robust teeth (Storrs 2003).

Finally, the unusual crocodylid *Euthecodon*, from the Neogene (Miocene) to Quaternary (Pleistocene) of North

Figure 9.43. Cranium of the crocodylid *Crocodylus thorbjarnarsoni* in dorsal view. Courtesy of Rodolfo Salas-Gismondi.

and East Africa, has a very long snout with laterally protruding alveoli and a proportionately small, squarish skull deck (Fig. 9.44). Its slender tooth crowns each bear mesial and distal keels (Storrs 2003). *Euthecodon* attained a skull length of up to 1.5 m and a total length of up to at least 7 m (Brochu and Storrs 2012). Brochu (2007a) referred it to Osteolaeminae, but Conrad et al. (2013) argued that *Euthecodon* is more closely related to the extant slender-snouted crocodile (*Mecistops cataphractus*), which is a member of Crocodylinae.

Tomistominae comprises *Tomistoma schlegelii* and crocodylians more closely related to it than to *Crocodylus niloticus* (Brochu 2003a). It dates back to the Eocene (Brochu 2007b). Brochu and Gingerich (2000) characterized Tomistominae by the presence of a deep but narrow splenial symphysis (unlike the broad symphysis in gavialoids) and a wedge-shaped, acute palatine process. Extant *Tomistoma* (Fig. 9.45A) occurs in freshwater lakes and rivers in Malaysia and parts of Indonesia and can reach a length of at least 5 m. As noted earlier, molecular data have consistently supported a relationship to *Gavialis* (Fig. 9.45B), but anatomical and paleontological evidence places *Tomistoma* among Crocodyloidea (Brochu 2003a). Basal tomistomines such as *Maroccosuchus*, from the Paleogene (Eocene) of Morocco (Fig. 9.46), closely resemble early crocodyloids in features such as a rather broad snout and the enlarged fifth maxillary alveolus (Jouve et al. 2015) and support crocodyloid rather than gavialoid affinities for Tomistominae. Tomistomines were widely distributed across Eurasia, North and East Africa, and North America during the

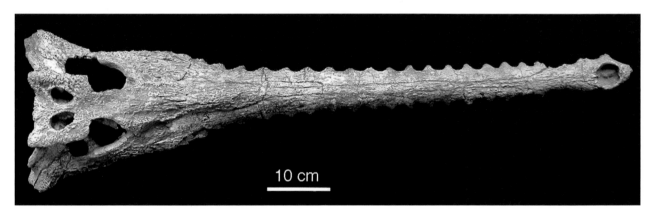

Figure 9.44. Cranium of the crocodylid *Euthecodon brumpti* in dorsal view. Note the prominent alveolar projections. Courtesy of Rodolfo Salas-Gismondi.

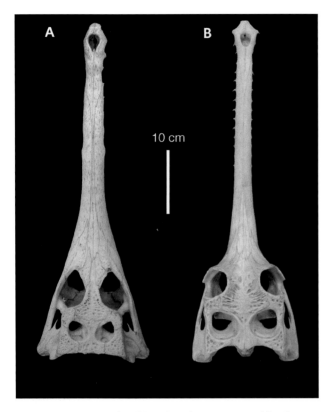

Figure 9.45. Crania of **A**, false gharial (*Tomistoma schlegelii*; Tomistominae), and **B**, gharial (*Gavialis gangeticus*; Gavialidae) in dorsal view. Courtesy of Rodolfo Salas-Gismondi.

Paleogene. Various extinct taxa are known from estuarine and coastal marine deposits and were probably capable of dispersing across the sea (Brochu 2003a).

Mekosuchinae is an ecologically diverse lineage of Crocodylidae that was endemic to Australia and neighboring islands in the Pacific Ocean (Salisbury and Willis 1996; Molnar et al. 2002). It ranged in time from the Paleogene (Eocene) to Quaternary (Pleistocene) in Australia and survived on several islands until after the arrival of the first human colonizers. Mekosuchines include large-bodied, *Crocodylus*-like predators such as the Paleogene (Oligocene) to Neogene (Miocene) *Baru*, small-sized, blunt-snouted forms such as the Neogene (Miocene) *Trilophosuchus*, and large-bodied, deep-snouted taxa with labiolingually flattened teeth such as the Paleogene (Oligocene) to Quaternary (Pleistocene) *Quinkana*. Synapomorphies for Mekosuchinae include the absence of an anterior process of the palatine and the prominent dorsal exposure of the supraoccipital on the skull deck (Salisbury and Willis 1996).

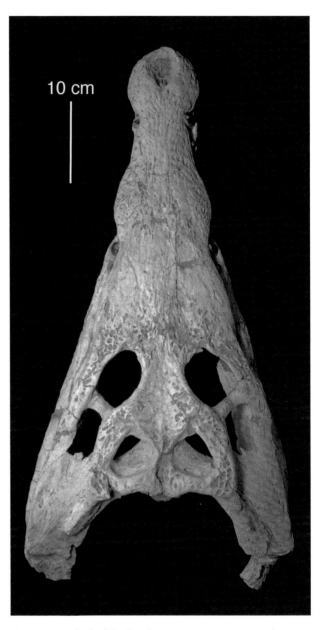

Figure 9.46. Skull of the basal tomistomine *Maroccosuchus zennaroi* in dorsal view. Courtesy of Rodolfo Salas-Gismondi.

EUSUCHIA: CROCODYLIA: GAVIALOIDEA

Gavialoidea comprises all crocodylians more closely related to *Gavialis gangeticus* than to *Alligator mississippiensis* or *Crocodylus niloticus* (Brochu 2003a). It is diagnosed by numerous synapomorphies such as a greatly elongated, narrow snout (Fig. 9.45B, 9.47B), a dorsally projecting external narial opening, a broad sheet of bone formed by the basisphenoid ventral to the basioccipital

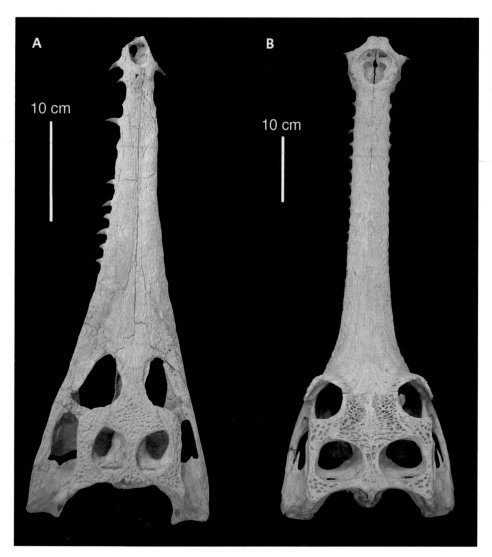

Figure 9.47. A, cranium of the gavialoid *Eosuchus lerichei* in dorsal view; **B**, cranium of an adult gharial (*Gavialis gangeticus*) in dorsal view. Courtesy of Rodolfo Salas-Gismondi.

(in adult individuals), homodont maxillary dentition, deep splenial symphysis, and the presence of a forked ventral (hypapohyseal) keel on the axis (Brochu 2004). Gavialoids include *Eothoracosaurus*, from the Late Cretaceous (Maastrichtian) of Mississippi and Tennessee (Brochu 2004), and *Eosuchus*, from the Paleogene (Paleocene) of Maryland, New Jersey, and Virginia, as well as France (Brochu 2006; Fig. 9.47A). The extant gharial (*Gavialis gangeticus*) inhabits rivers on the Indian subcontinent but retains salt-excreting glands on its tongue; the inside surfaces of its mouth are heavily keratinized, allowing prolonged exposure to saltwater (Grigg and Kirsh-

ner 2015; Fig. 9.47B). It can attain a total length of more than 6 m and subsists mainly on fish. *Gavialis* is first known from the Neogene (Miocene) of India. Its sister taxon Gryposuchinae (Vélez-Juarbe et al. 2007) is endemic to South America and the Caribbean. It ranged in time from the Paleogene (Oligocene) to Neogene (Miocene) and occurred in freshwater and nearshore marine settings. *Gryposuchus* attained a skull length of up to 1.4 m, placing it among the largest known crocodylians (Riff and Aguilera 2008). Vélez-Juarbe et al. (2007) hypothesized that gavialoids dispersed from Africa across the Atlantic Ocean during the Paleogene.

10   Avemetatarsalia

Bird-Line Archosaurs Excluding Dinosaurs

Gauthier (1986) proposed the name Ornithosuchia (from Greek *ornis*, bird, and *souchos*, crocodile) for the clade comprising all archosaurian reptiles that are more closely related to birds than to crocodylians. As the name Ornithosuchidae had already been used earlier for a pseudosuchian clade (Chapter 9), most authors did not adopt Gauthier's proposal. Gauthier also introduced Ornithodira (from Greek *ornis*, bird, and *dere*, neck) for a less inclusive clade comprising the most recent common ancestor of dinosaurs including birds and all descendants of that ancestor. Sereno (1991) slightly modified Gauthier's definition by adding *Scleromochlus*, an enigmatic small archosaur from the Late Triassic of Scotland. Benton (1999) placed *Scleromochlus* outside Ornithodira as defined by Gauthier and then coined Avemetatarsalia (from Latin *avis*, bird, Greek *meta*, near, and *tarsos*, flat of the foot) for the clade comprising *Scleromochlus* and Ornithodira sensu Gauthier (1986). Nesbitt et al. (2017) used Avemetatarsalia for a stem-based clade comprising Ornithodira and a new group Aphanosauria (Fig. 10.1).

Avemetatarsalia: *Scleromochlus*

Scleromochlus, from the Late Triassic of Scotland, attained a total length of about 18 cm (Benton 1999). The known specimens are preserved as mostly faint impressions in rather coarse-grained sandstone, and researchers continue to disagree on the interpretation of many anatomical details. Working with casts from the natural molds, Benton (1999) attempted a reconstruction of the skeleton of *Scleromochlus*. The skull is more or less triangular in plan view and twice as broad across the orbital region as it is deep. Broad parietals separate slit-like upper temporal fenestrae. The quadrate and quadratojugal are inclined posterodorsally. The forelimb of *Scleromochlus* is much shorter than the hind limb. The tibia is longer than the femur. Metatarsals I–IV are of equal length and tightly bundled. Benton (1999) interpreted *Scleromochlus* as a bipedal cursor that was possibly also capable of hopping.

Some authors (Huene 1914; Padian 1984) have hypothesized that *Scleromochlus* is closely related to Pterosauria based primarily on features of its limbs and overall body proportions. Sereno (1991) found some shared derived features for *Scleromochlus* and Pterosauria, such as the short scapula and the absence of a fourth trochanter on the femur but rejected other synapomorphies listed by Padian (1984). Benton (1999) found no evidence in support a closer relationship between *Scleromochlus* and Pterosauria and placed the former as the sister taxon to Ornithodira.

Avemetatarsalia: Aphanosauria

Nesbitt et al. (2017) defined Aphanosauria (from Greek *aphantos*, secret, and *sauros*, lizard) as the most inclusive clade containing *Teleocrater rhadinus* and *Yarasuchus deccanensis* but not *Passer domesticus* or *Crocodylus niloticus*. They listed several synapomorphies for this grouping including elongate cervical vertebrae with epipophyses and anteriorly overhanging neural spines that have rugose lateral edges at the apices; humerus with a large deltopectoral crest that extends for at least 35 percent of the length of the bone; femur with a straight, deep groove in its proximal portion and lacking an anteromedial tuber; and calcaneum with a tuber that is taller than broad. The best-known aphanosaur, *Teleocrater*, from the Middle Triassic (Anisian) of Tanzania (Fig. 10.2), attained a total length of about 1.8 m (Nesbitt et al. 2018). Its forelimbs are shorter than its hind limbs. The pelvis has a closed acetabulum. *Teleocrater* has certain features long considered diagnostic for more derived ornithodiran clades (e.g., extension of the supratemporal fossa onto the fron-

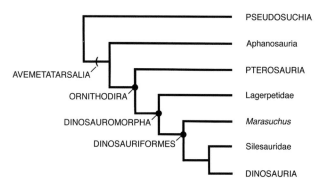

Figure 10.1. Phylogenetic hypothesis of the interrelationships of the major taxa of Avemetatarsalia. Dots denote node-based clades and parentheses denote stem-based clades. Based on Nesbitt (2011) and Nesbitt et al. (2017).

tal). It also has a well-developed "crocodile-normal" ankle although the calcaneal tuber is relatively smaller than in pseudosuchians. The ankles of *Teleocrater* and the closely related *Yarasuchus*, from the Middle Triassic (Anisian) of India, demonstrate that the crurotarsal ankle joint represents the plesiomorphic condition for Archosauria (Nesbitt et al. 2018). The microstructure of the limb bones of *Teleocrater* indicates sustained, elevated growth (and thus metabolic) rates as in other ornithodirans (Nesbitt et al. 2017).

Avemetatarsalia: Ornithodira

Nesbitt (2011) redefined Ornithodira as the least inclusive clade containing *Pterodactylus antiquus* and *Passer domesticus*. He listed several shared derived features including apices of neural spines on the cervical and dorsal vertebrae not expanded; second phalanx of manual digit II longer than the first phalanx; manual digits I-III with trenchant ungual phalanges; tibia longer than the femur; metatarsus compact, with metatarsals II-IV tightly bundled; and absence of osteoderms.

Even the earliest ornithodirans apparently had an upright stance and a parasagittal gait (Padian 1983a; Gauthier et al. 2011; Nesbitt 2011). This inference is based on a set of anatomical features including the head of the femur being distinctly offset from the shaft, the femoral shaft with a double (rather than sigmoidal) curvature that reflected anteroposterior excursion of the hind limb, the slender fibula (due to a lack of the torsion present in reptiles with a rotatory gait), the presence of a mesotarsal ankle joint, the elongation of the metatarsus (indicating habitual digitigrady, which is related to extended limb excursion common in cursorial animals with a parasagittal gait), and long pedal digits. The evolution of an obligate parasagittal gait appears to be linked to more rapid locomotion (Gauthier et al. 2011).

Figure 10.2. Reconstructed skeleton of the aphanosaurian *Teleocrater rhadinus*. Courtesy of Scott Hartman.

Avemetatarsalia: Ornithodira: Pterosauria

Pterosaurs (from Greek *pteron*, wing, and *sauros*, lizard) were the first group of amniotes to evolve active flapping flight (Wellnhofer 1978, 1991a; Padian 1983a, 1985; Chatterjee and Templin 2004; Unwin 2006; Witton 2013). They did so in a manner quite unlike that in birds or bats, the two other tetrapod clades that developed active flight. Unwin (2006) and Witton (2013) have published detailed accounts on pterosaurian paleobiology and diversity to which readers are referred for additional details.

Andres et al. (2014) defined Pterosauria (Fig. 10.3) as the most inclusive clade exhibiting the fourth metacarpal and digit hypertrophied to support the wing membrane synapomorphic with that in *Pterodactylus antiquus* (Fig. 10.4). Nesbitt (2011) listed a suite of shared derived features for this clade: anterodorsal process of the premaxilla is longer than the anteroposterior length of this bone; anterodorsal margin of the maxilla entering into the margin of the external naris; centra of cervical vertebrae 3 to 5 are longer than the centrum of a middorsal vertebra; postglenoid process of the coracoid long and expanded posteriorly; presence of a pteroid bone; length of manual digit IV equals 50 percent or more of the total length of the forelimb; anterior process of the ilium long but shorter than the posterior process; and metatarsal V with a dorsal eminence that is separated from the proximal end of the bone by a concave gap.

The flight apparatus and related structures of pterosaurs are unique among known tetrapods and thus merit more detailed consideration. Each wing was a multilayered membrane supported only by the greatly elongated

fourth digit of the manus (Figs. 10.4, 10.5). This membrane (brachiopatagium) is attached to the flank and extended back to the ankle. The wings could be folded when the animal was not flying. The robust, proportionately short humerus has a saddle-shaped proximal articular facet and a large deltopectoral crest for insertion of the substantial pectoralis muscle. Two prominent condyles on its distal end form part of the elbow joint, at which the forearm could be folded against the upper arm (Bennett 2001). The radius and ulna are straight and relatively long. The carpus consists of four elements: a proximal and a distal syncarpal (each formed through fusion of smaller carpal bones), a preaxial carpal, and the pteroid. The long and slender pteroid articulates on the side of the preaxial carpal (Bennett 2007a) and supported a small forewing (propatagium), which attached to the shoulder medially. It most likely projected medially rather than anteriorly (Palmer and Dyke 2010). Metacarpals I-III are long and slender, and metacarpal IV is much more robust than metacarpals I-III. The distal end of metacarpal IV forms a distinct roller joint for the wing finger, which could be folded back against the palm. The unguals on the free manual digits I-III are larger and more strongly curved than the pedal unguals and appear suitable for grasping. The joints between the, typically, four phalanges of manual digit IV were apparently largely inflexible, and thus motion of this finger was largely restricted to its articulation with metacarpal IV.

The scapula and coracoid are often fused into a sturdy bony bar around the shoulder joint. The distinctly saddle-shaped glenoid facet extends across the contact between the scapula and coracoid in more derived pterosaurs. It allowed the forelimb to move through an arc of 90 degrees when the limb was extended to the side (Padian 1983a; Bennett 2001). The glenoid facet continues onto the posterior surface of the scapulocoracoid so that the forelimb could be rotated toward the trunk to an angle of 10 degrees from the vertebral column. From this position, the wing could swing anteroposteriorly beneath the shoulder in most pterosaurs (Bennett 1997). The coracoid articulates with the large bony sternum. In some pterosaurs, the scapula forms a distal joint with the notarium, a structure formed by fusion of several dorsal vertebrae. This contact presumably helped brace the thoracic region against the forces generated by the flapping wings. The sternum was aligned anterodorsally in life,

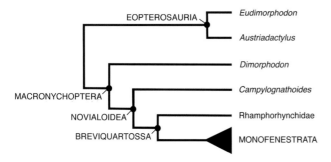

Figure 10.3. Phylogenetic hypothesis of the interrelationships of Pterosauria. Dots denote node-based clades. Based on Andres et al. (2014).

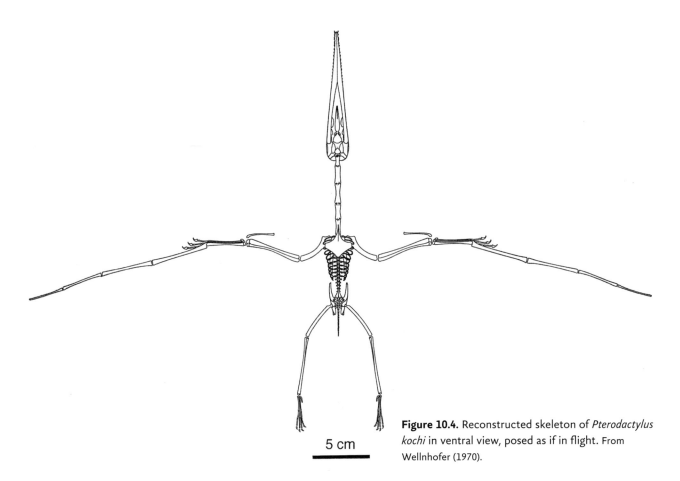

5 cm

Figure 10.4. Reconstructed skeleton of *Pterodactylus kochi* in ventral view, posed as if in flight. From Wellnhofer (1970).

as in birds (Claessens et al. 2009). It bears a distinct ventromedial crest (cristospina), which extends anteriorly from the articulation with the coracoid and served as the site of origin for the pectoralis and other flight muscles. Unlike in birds, where the two major muscles are used to raise and lower the wing, pterosaurs employed groups of muscles for this purpose (Bennett 2003; Chatterjee and Templin 2004; Witton 2013). Based on wind-tunnel tests with models, Palmer (2011) interpreted the wings of pterosaurs as adapted to low-speed flying and suitable for controlled, low-speed landing as well as for soaring on thermals and slopes. He noted that the wings would have been vulnerable to strong or turbulent winds in flight and on the ground.

Some exquisitely preserved fossils provide insight into the complex internal architecture of the wing membrane in pterosaurs (Frey et al. 2003). The most prominent component of the membrane is a layer of thin wing fibers (actinofibrils), which are embedded below the epidermis and dermis (Fig. 10.5). It is still not clear how these fibers attached to the wing bones. Bennett (2000) argued that the actinofibrils probably could not be compressed or stretched. Thus, he posited that these wing fibers could have kept the brachiopatagium stretched out from front to back while allowing it to be folded parallel to the direction of the fibers. Below the layer of actinofibrils, a network of muscle fibers extended perpendicular to the actinofibrils. This would have allowed pterosaurs to control tension in the wing membrane and thus the lift generated by the wings. Curving the membrane would have increased lift, whereas flattening it would have decreased lift. Finally, a layer of interconnected chambers and passages likely formed a network of blood vessels that extended from the shoulder region to the tip of the wing close to the ventral surface of the brachiopatagium (Frey et al. 2003). This vascular network could have aided in shedding excess heat generated by muscular activity during flight.

In the pelvic girdle, the ischium and pubis are fused in many pterosaurs. The acetabulum forms a hemispherical socket on the lateral surface of the pelvis. Paired pre-

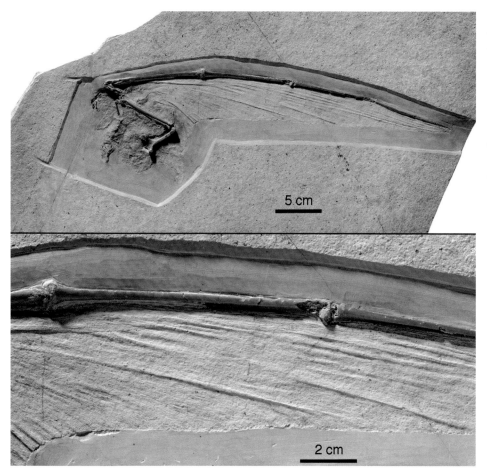

Figure 10.5. Isolated forelimb of *Rhamphorhynchus muensteri* (known as the "Zittel wing") with exceptionally preserved impression of the flight membrane (with detail magnified in the illustration below to show actinofibrils). Courtesy of Bayerische Staatssammlung für Paläontologie und Geologie.

pubic bones articulate with the anteroventral surfaces of the pubes and contact the gastralia anteriorly. Claessens et al. (2009) argued that these elements and gastralia could have moved up and down during breathing, thereby changing the volume of the abdominal cavity. The femur has a long, slightly curved, and tubular shaft, and its proximal head is set off from the latter at an acute angle. It has considerable mobility at the hip joint and could be moved laterally away from the body or brought close to the body during locomotion on the ground (but see Padian 2017). The distal end of the femur bears two large condyles. The long tibia and the proximal tarsal bones tend to fuse to form a tibiotarsus during growth (Padian 2017). The fibula is usually shorter than the tibiotarsus, and both its ends become fused to the latter. In the pes, the slender metatarsals I–IV are bundled together. Metatarsal V is slender and mobile in early pterosaurs but short and more robust in more derived taxa. Mobility at the joints between the phalanges of the long and slender

pedal digits was apparently limited (J. M. Clark et al. 1998). Webbing between the toes is evident on some exquisitely preserved skeletons and on tracks attributed to pterosaurs. In basal pterosaurs, a membrane (cruropatagium) extended between the hind feet but apparently did not attach to the tail (Unwin 2006). In more derived taxa, this membrane was divided into two smaller ones that extended from the base of the tail to pedal digit V. This change allowed the hind limbs to move independently of each other. Many early pterosaurs have a long tail with a few free anterior caudals followed by a succession of vertebrae connected by an interlacing network of thin, greatly elongated pre- and postzygapophyses and similarly elongated hemal processes (Wellnhofer 1975; Padian 2008).

There now exists a general consensus that at least the more derived pterosaurs could move on all fours while on the ground (Wellnhofer 1988; Unwin 2006; Costa et al. 2014; but see Padian 1983a). Pterosaurs likely employed an upright gait with the vertebral column held at a steep angle (Chatterjee and Templin 2004). They probably placed their hands and feet flat on the ground. More derived pterosaurs could swing their arms anteroposteriorly by flexing and extending them at the elbow joints as well as elevating and depressing the humerus. Many appear well adapted for climbing, based on the size and curvature of the unguals and the elongation of the penultimate phalanges in the free manual digits and in the pedal ones. Smaller forms could have become airborne by leaping off cliff faces or trees. Witton and Habib (2010) hypothesized that larger pterosaurs could have launched themselves from a quadrupedal pose by vaulting their hind limbs upward and pushing off the ground with their forelimbs, similar to the launch mode used by present-day vampire bats.

The postcranial skeleton of pterosaurs was highly pneumatized, to an even greater extent than in birds (Claessens et al. 2009; E. G. Martin and Palmer 2013). The centra of the dorsal vertebrae bear deep lateral recesses for the accommodation of air sacs, which formed as outgrowths of the lungs. Pterosaurs probably had an efficient flow-through respiratory system comparable to that in birds (Bonde and Christiansen 2003; Claessens et al. 2009), in which the air sacs forced oxygen-rich air into the solid lungs in one direction and oxygen-depleted air was moved through other parts of the respiratory system and then exhaled. Such a system is critically important because active flight requires substantial oxygen consumption. In many derived pterosaurs, the cervical vertebrae as well as the bones of the shoulder girdle and limbs are pneumatized. The bones of the wing finger are almost completely hollow except for delicate bony struts spanning the internal cavities and supporting the relatively thin bony walls. This structural change facilitated enlargement of the bones and would have increased their resistance to bending and twisting (Witton and Habib 2010; E. G. Martin and Palmer 2013). An overall reduction in body mass due to pneumatic modification would have resulted in improved flight performance.

Pterosaurs have proportionately large but lightly built skulls. The individual cranial and mandibular bones are often indistinguishably fused to each other in adult specimens. The configuration of the cranial openings is highly variable among pterosaurs. An external mandibular fenestra has been reported in the Early Jurassic *Dimorphodon* (Nesbitt and Hone 2010). The dentition is variably developed. In pterosaurs with teeth, the enamel is restricted to the apices of the crowns. Replacement teeth erupt behind functional teeth rather than below or lingual to them as in other archosaurs. Many pterosaurs lack teeth altogether, and their jawbones were covered by keratinous rhamphothecae in life.

Natural infillings of the cranial cavity as well as CT scans of well-preserved skulls have permitted reconstruction of the brains in several pterosaurs (Witmer et al. 2003; Codorniú et al. 2016). Although the brains of pterosaurs are smaller than those of birds relative to body mass, they have various bird-like features including the expansion of the cerebrum and cerebellum, which displaced the large optic lobes ventrolaterally, small olfactory lobes, and greatly enlarged flocculi. The last feature is consistent with the need for processing large amounts of sensory information from the wing and helping to coordinate eye, head, and neck movements that stabilized the gaze on a target while foraging (Witmer et al. 2003). Finally, the large semicircular canals of the inner ear are critical to maintaining equilibrium.

Some exceptionally preserved fossils of pterosaurs have revealed information about the structure of the skin. Frey et al. (2003) reported a prominent soft-tissue pad with small, non-overlapping scales in the ankle region of an unidentified Early Cretaceous pterosaur from Brazil. A number of well-preserved specimens of pterosaurs preserve millimeter-long, fiber-like structures (pycnofibers; Kellner et al. 2010) on the head (except on the jaws), neck, body, and the proximal portions of the limbs, sometimes forming extensive mats (Frey and Martill 1998; X. Wang et al. 2002). These fibers probably formed a fuzzy body covering that insulated the presumably endothermic pterosaurs against fluctuations in ambient temperatures. Pycnofibers differ in their structure from the filamentous epidermal structures present in various theropod dinosaurs (Chapters 11, 12), but it is conceivable that both types of integumentary covering shared a common developmental origin.

Traditionally, Pterosauria was divided into the long-tailed Rhamphorhynchoidea (from Greek *rhamphos*, beak, bill, and *rhynchos*, snout) and the short-tailed Pterodactyloidea (from Greek *pteron*, wing, and *daktylos*, finger)

(Romer 1956; Wellnhofer 1978). Phylogenetic analyses (Kellner 2003; Unwin 2003; Lü et al. 2010; Andres et al. 2010, 2014) have since established that Rhamphorhynchoidea is merely a grade containing pterosaurian taxa that are successively more closely related to Pterodactyloidea. In recent decades, a wealth of discoveries, especially from the Jurassic and Early Cretaceous of China, has demonstrated a previously unsuspected diversity of pterosaurs.

PTEROSAURIA: EOPTEROSAURIA

The oldest known pterosaurs are Late Triassic (Norian) in age. They already have a fully developed flight apparatus. Most fossils of Triassic pterosaurs have been recovered from marine strata in northern Italy (Dalla Vecchia 2013, 2014). Additional records are known from Arizona, Austria, and Greenland. Andres et al. (2014) united all known Triassic pterosaurs as Eopterosauria (from Greek *eos*, dawn, *pteron*, wing, and *sauros*, lizard), which is the least inclusive clade containing *Preondactylus buffarinii* and *Eudimorphodon ranzii*. Diagnostic synapomorphies include the presence of a heterodont dentition, triangular, sometimes bulbous, teeth, and an external narial opening that is distinctly longer than high.

Eudimorphodon, from the Norian of Italy, has small, closely spaced, and tricuspid or pentacuspid teeth in the upper and lower jaws (Wild 1978; Fig. 10.6). It attained a wingspan of about 1 m. *Austriadactylus*, from the Norian of Austria and Italy, also has multicuspid teeth, but its dentition differs from that of *Eudimorphodon*, and the up-

Figure 10.6. Skeleton of the eopterosaurian *Eudimorphodon ranzii*. From Wikipedia—CC BY 2.0.

per and lower teeth are distinctly different from each other. Its cranium bears a prominent dorsal bony crest (Dalla Vecchia et al. 2002). *Austriadactylus* and *Eudimorphodon* have long tails but lack the greatly elongated pre- and postzygapophyseal processes and chevrons present in the tails of certain more derived pterosaurs (Dalla Vecchia 2014).

PTEROSAURIA: MACRONYCHOPTERA

Macronychoptera (from Greek *makros*, long, *onyx*, claw, and *pteron*, wing) is the least inclusive clade comprising *Dimorphodon macronyx* and *Quetzalcoatlus northropi* (Unwin 2003; Andres et al. 2014). It is characterized by synapomorphies including the absence of an extension of the antorbital fossa onto the jugal; presence of a contact between the premaxilla and frontal; anterior end of the mandible prow-shaped; posterior tooth crowns without denticles; coracoid narrow, shaft-like; and enormously elongated pre- and postzygapophyses on the caudal vertebrae (Andres, pers. comm.).

Dimorphodon, from the Early Jurassic (Hettangian-Sinemurian) of England, has a deep skull with large cranial openings (Padian 1983b). The crowns of the teeth in the premaxilla and anterior portion of the dentary are fang-like, somewhat labiolingually flattened, and have finely serrated cutting edges. The upper teeth are larger than the closely spaced lower ones. *Dimorphodon* attained a wingspan of 1.4 m.

PTEROSAURIA: MACRONYCHOPTERA: NOVIALOIDEA

Novialoidea (from Latin *novus*, new, and *ala*, wing) is the least inclusive clade containing *Campylognathoides zitteli* and *Quetzalcoatlus northropi* (Kellner 2003; Andres et al. 2014).

Campylognathoides, from the Early Jurassic (Toarcian) of Germany, attained a wingspan of up to 1.8 m (Padian 2008). Its rather short but deep skull has large orbits and short, vertically extending teeth. The anterior tip of the mandible lacks teeth.

PTEROSAURIA: MACRONYCHOPTERA: NOVIALOIDEA: BREVIQUARTOSSA

Unwin (2003) united Rhamphorhynchidae and Pterodactyloidea in a clade Breviquartossa (from Latin *brevis*, short, *quartus*, fourth, and *os*, bone; in reference to the proportionately short metatarsal IV). Andres et al. (2014) defined this clade as containing *Rhamphorhynchus muensteri* and *Quetzalcoatlus northropi*. Synapomorphies for Breviquartossa include the fused mandibular symphysis, the concave ventral margin of the jugal, and manual unguals that are least twice as large as the pedal ones (Andres, pers. comm.)

Rhamphorhynchidae is the least inclusive clade containing *Rhamphorhynchus muensteri* and *Scaphognathus crassirostris* (Andres et al. 2014). Its diagnostic features are the low number of teeth (fewer than 11 in each jaw) and the shape of the deltopectoral crest, which has a distinctly constricted "neck" and an expanded distal portion. Rhamphorhynchinae (Andres et al. 2014) includes *Angustinaripterus*, from the Middle Jurassic of Sichuan (China) (He et al. 1983); *Dorygnathus*, from the Early Jurassic (Toarcian) of Germany (Padian 2008; Fig. 10.7); and *Rhamphorhynchus*, from the Late Jurassic (Tithonian) of Germany (Wellnhofer 1975; Fig. 10.8). It is distinguished by a number of synapomorphies including antorbital fenestra elongate; anterodorsally extending, edentulous tips of the upper and lower jaws (which supported keratinous extensions in life); quadratojugal forming the lower temporal bar; teeth strongly curved; and anterior teeth procumbent. *Rhamphorhynchus* attained a wingspan of nearly 2 m. Its tail ended in a slightly asymmetrical vane (Fig. 10.8), which was probably held vertically in life. *Scaphognathus*, from the Late Jurassic (Tithonian) of Germany (Fig. 10.9), shares with Rhamphorhynchinae

Figure 10.7. Skull and anterior cervical vertebrae of the rhamphorhynchid *Dorygnathus banthensis* in lateral view.

the dorsal position of the external naris and tall tooth crowns (Andres et al. 2010) but differs in having a more robust skull with a blunt snout and vertically projecting rather than procumbent teeth (Wellnhofer 1975).

PTEROSAURIA: MACRONYCHOPTERA: NOVIALOIDEA: BREVIQUARTOSSA: MONOFENESTRATA

The most diverse clade of Pterosauria, Monofenestrata (from Greek *monos*, one, and Latin *fenestra*, opening), is the most inclusive clade exhibiting confluent external naris and antorbital fenestra synapomorphic with that in *Pterodactylus antiquus* (Andres et al. 2014; Fig. 10.10). In addition to this large nasoantorbital fenestra, the clade is diagnosed by a long neck in which cervical vertebrae 3-7 are elongated (Lü et al. 2010). The skull is typically long and low, with the snout comprising more than 80 percent of the total length of the cranium. The quadrate extends anteroventrally, and the mandibular symphysis is short (equal to or less than 20 percent of the total mandibular length).

MONOFENESTRATA: DARWINOPTERA

Darwinopterus, from the Middle or Late Jurassic of Liaoning (China), represents a structural intermediate between "rhamphorhynchoid" and pterodactyloid pterosaurs (Lü et al. 2010; Fig. 10.11). It shares postcranial features of the former but has the proportionately long skull and neck characteristic of the latter. *Darwinopterus* attained a wingspan of 80 cm. One specimen preserves an egg close to the pelvis. It also differs from other skeletons referred to *Darwinopterus* in having a wider pelvic canal and the absence of a low bony cranial crest, establishing that these features are sexually dimorphic (Lü et al. 2011). Andres et al. (2014) united *Darwinopterus* with *Wukongopterus*, from the Middle or Late Jurassic of Liaoning (China) (X. Wang et al. 2009), and *Pterorhynchus*, from the Middle or Late Jurassic of Inner Mongolia (China) (Czerkas and Ji 2002), in Darwinoptera, the least inclusive clade containing *Darwinopterus modularis* and *Pterorhynchus wellnhoferi*.

MONOFENESTRATA: CAELICODRACONES

Caelicodracones (from Latin *caelicus*, heavenly, and *draco*, dragon) is the least inclusive clade containing

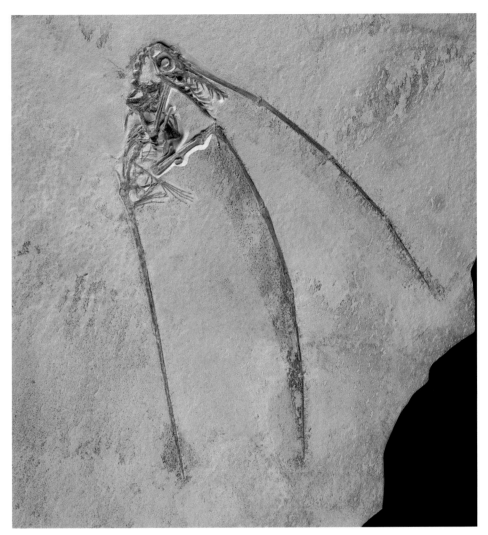

Figure 10.8. Skeleton of the rhamphorhynchid *Rhampho-rhynchus muensteri* with traces of both wing membranes and the rhomboidal caudal vane. No scale provided. Courtesy of the Royal Ontario Museum.

Anurognathus ammoni and *Quetzalcoatlus northropi* (Andres et al. 2014).

Anurognathidae includes *Anurognathus*, from the Late Jurassic (Tithonian) of Germany (Bennett 2007b; Fig. 10.12), and *Jeholopterus*, from the Early Cretaceous (Barremian) of Liaoning (China; X. Wang et al. 2002). The skull is deep and broader than long, with a short snout and large orbits. The jaws hold relatively few widely spaced teeth. The tail of *Anurognathus* has only up to 11 vertebrae. *Anurognathus* attained a wingspan of 50 cm, whereas *Jeholopterus* reached a wingspan of 90 cm (X. Wang et al. 2002). The wing metacarpal is unusually short relative to the humerus, the pteroid is short, and the wing finger includes only three phalanges (Unwin 2003). The short, broad wings and short tail suggest slow, maneuverable flight (Bennett 2007b). Anurognathids have been interpreted as insectivores. Long considered basal pterosaurs (Kellner 2003; Unwin 2003), Anurognathidae share derived features with Pterodactyloidea, especially the presence of a large nasoantorbital fenestra and a short tail comprising 15 or fewer caudal vertebrae (Andres et al. 2010).

MONOFENESTRATA: CAELICODRACONES: PTERODACTYLOIDEA

Pterodactyloidea (from Greek *pteron*, wing, and *daktylos*, finger) is the most inclusive clade exhibiting a metacarpus at least 80 percent as long as the humerus

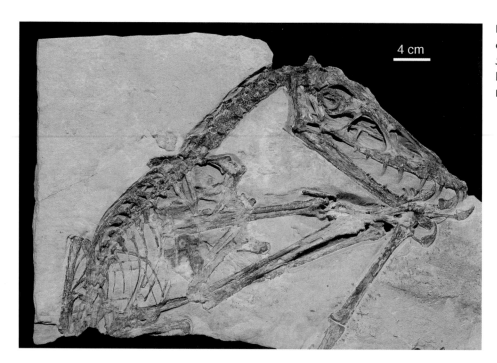

Figure 10.9. Partial skeleton of the rhamphorhynchid *Scaphognathus crassirostris* in lateral view. Courtesy of Oliver Rauhut.

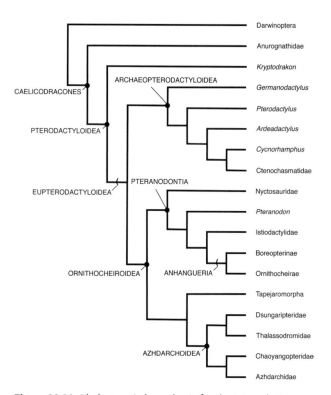

Figure 10.10. Phylogenetic hypothesis for the interrelationships of Monofenestrata. Dots denote node-based clades and parentheses denote stem-based clades. Based on Andres et al. (2014).

synapomorphic with *Pterodactylus antiquus* (Andres et al. 2014). It comprises the majority of Jurassic and Cretaceous macronychopterans. In addition to the elongation of the wing metacarpal, the subrectangular outline of that metacarpal in transverse section is diagnostic for this clade (Andres et al. 2014). In earlier pterosaurs, the length of the wing metacarpal equals at most 68 percent of the length of the humerus (Unwin 2003). Andres et al. (2014) argued that the elongation of the wing metacarpal allowed for greater variation in the wing configuration among pterodactyloid pterosaurs.

The oldest pterodactyloid reported to date is *Kryptodrakon*, from the Middle or Late Jurassic of Xinjiang (China; Andres et al. 2014). Although still poorly known, *Kryptodrakon* has the diagnostic elongation and cross-sectional outline of the wing metacarpal. Better-known members of Pterodactyloidea share additional derived features including the length of the mandibular symphysis being at least 36 percent of the total length of the mandible; jaw joint positioned below the center of the orbit; absence of elongated basipterygoid processes; length of the femur being almost equal to that of the humerus; and presence of at most one phalanx in pedal digit V (Andres, pers. comm.).

Figure 10.11. Skeleton of the darwinopteran *Darwinopterus robustodens*. Courtesy of the late Junchang Lü.

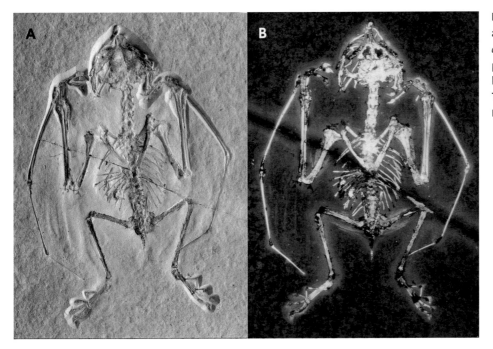

Figure 10.12. Skeleton of the anurognathid *Anurognathus ammoni* in dorsal view, photographed **A**, under natural light, and **B**, under UV light. Total length 9 cm. Courtesy of Helmut Tischlinger.

CAELICODRACONES: PTERODACTYLOIDEA: ARCHAEOPTERODACTYLOIDEA

Kellner (2003) proposed Archaeopterodactyloidea (from Greek *arkhaios*, ancient, *pteron*, wing, and *daktylos*, finger), which Andres et al. (2014) defined as the least inclusive clade containing *Ctenochasma elegans* and *Germanodacty-*

lus cristatus. This clade is characterized by a suite of synapomorphies including: squamosal rounded; quadrate posteriorly inclined relative to the ventral margin of the skull by 150 degrees; occiput facing ventrally; and atlas and axis not fused to each other. *Germanodactylus*, from the Late Jurassic (Tithonian) of Germany, has a low premaxillary crest that is set back relative to the anterior margin of the nasoantorbital fenestra and extends to the region of the orbit. It attained a wingspan of about

1 m (Wellnhofer 1970). *Pterodactylus*, from the Late Jurassic (Tithonian) of Germany, has a long, slightly tapered snout, large orbit, and an inflated, rounded postorbital portion of the cranium (Wellnhofer 1970; Fig. 10.13). It also reached a wingspan of about 1 m. The

Figure 10.13. Skeleton of the pterodactylid *Pterodactylus kochi*. Wingspan 46 cm. Courtesy of Bayerische Staatssammlung für Paläontologie und Geologie.

anterior portions of the jaws each hold about 40 conical teeth with labiolingually flattened crowns. *Pterodactylus*, *Ardeadactylus*, also from the Late Jurassic (Tithonian) of Germany (Bennett 2013a; Fig. 10.14), and *Cycnorhamphus*, from the Late Jurassic (Kimmeridgian-Tithonian) of France and Germany (Fabre 1981; Bennett 2013b), are successive sister taxa to Ctenochasmatidae, which is noteworthy for the extraordinary dental specializations. *Cycnorhamphus* has a posteriorly extended frontoparietal crest with a rounded posterior edge, and its teeth are confined to the anterior ends of the upper and lower jaws (Bennett 2013b).

Ctenochasmatidae is the least inclusive clade containing *Gnathosaurus sulcatus* and *Pterodaustro guinazui* (Andres et al. 2014). *Ctenochasma*, from the Late Jurassic (Tithonian) of France and Germany (Fig. 10.15), and *Pterodaustro*, from the Early Cretaceous (Aptian-Albian) of Argentina, share greatly elongated snouts (constituting more than 60 percent of total skull length) with highly modified dentitions. *Ctenochasma* has up to about 140 tightly packed needle-like teeth in each jaw that form a parallel-sided comb in plan view (Wellnhofer 1970; Bennett 2007c; Fig. 10.15). It attained a wingspan of at least 1.2 m. *Pterodaustro* has greatly elongated, curved jaws. The upper jaws hold tiny, nubbin-like teeth. By contrast, its lower jaws each have hundreds of tightly packed, long, and extraordinarily thin teeth (Chiappe and Chinsamy 1996). This comb-like structure superficially resembles the numerous keratinous lamellae in the beaks of extant flamingos and, like the latter, may have been used for straining small animals from water. *Pterodaustro* reached a wingspan of about 2.5 m.

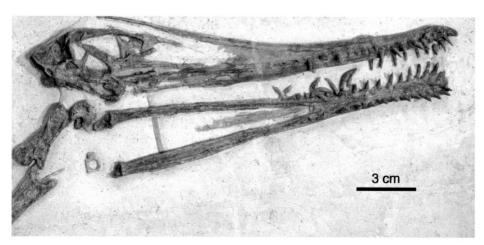

3 cm

Figure 10.14. Skull and anterior cervical vertebrae of the archaeopterodactyloid *Ardeadactylus longicollum*, with cranium in lateral view and mandible in dorsal view. Courtesy of Staatliches Museum für Naturkunde Stuttgart.

Figure 10.15. Skull of the ctenochasmatid *Ctenochasma elegans,* with cranium in lateral view and mandible in dorsal view. Note the specialized "tooth comb." Courtesy of Oliver Rauhut.

The subclade Gnathosaurinae, including *Gnathosaurus,* from the Late Jurassic (Tithonian) of Germany (Wellnhofer 1970), and *Plataleorhynchus,* from the Early Cretaceous (Berriasian) of England (Howse and A. R. Milner 1995). Each has well over 100 teeth that form spatulate rosettes at the tips of the jaws. *Gnathosaurus* has a bony crest on the premaxillae that can extend back to the region of the orbits. It attained a wingspan of 1.7 m.

CAELICODRACONES: PTERODACTYLOIDEA: EUPTERODACTYLOIDEA

Eupterodactyloidea (from Greek *eu,* true, *pteron,* wing, and *daktylos,* finger) is the least inclusive clade containing *Pteranodon longiceps* but not *Pterodactylus antiquus* (Andres et al. 2014). It is characterized by a number of shared derived features including the position of the jaw joint anterior to the center of the orbit; scapulocoracoid rotated laterally relative to the vertebral column; presence of a deep and short cristospina on the sternum; and shaft of the humerus straight (Andres, pers. comm.). It comprises *Haopterus,* from the Early Cretaceous (Barremian) of Liaoning (China) (X. Wang and Lü 2001), and Ornithocheiroidea.

PTERODACTYLOIDEA: EUPTERODACTYLOIDEA: ORNITHOCHEIROIDEA

Ornithocheiroidea (from Greek *ornis,* bird, and *cheir,* hand) is the least inclusive clade containing *Anhanguera blittersdorffi, Pteranodon longiceps, Dsungaripterus weii,* and *Quetzalcoatlus northropi* (Andres et al. 2014). It shares a number of diagnostic apomorphies including the presence of frontal and parietal crests; presence of a notarium (formed through fusion of the anterior dorsal vertebrae and for articular contact with the distal ends of the scapulae); humerus with a distinct ulnar crest; and articular contact only between the carpus and wing metacarpal (Andres, pers. comm.). Andres et al. (2014) divided Ornithocheiroidea into Pteranodontia and Azhdarchoidea.

PTERODACTYLOIDEA: EUPTERODACTYLOIDEA: ORNITHOCHEIROIDEA: PTERANODONTIA

Pteranodontia (from Greek *pteron,* wing, *an,* without, and *odous* (*odon*), tooth), is the least inclusive clade containing *Pteranodon longiceps* and *Nyctosaurus gracilis.* Synapomorphies for this clade include the extension of the nasoantorbital fossa onto the jugal; expanded posterior end of the scapula more or less oval; and sternum constricted behind the cristospina (Andres, pers. comm.).

ORNITHOCHEIROIDEA: PTERANODONTIA: NYCTOSAURIDAE

Nyctosauridae is the least inclusive clade containing *Nyctosaurus gracilis* and *Muzquizopteryx coahuilensis* (Andres et al. 2014). *Nyctosaurus,* from the Late Cretaceous (Coniacian) of Kansas, has long and edentulous jaws. Its broad scapulocoracoid does not contact the vertebral column. The forelimb is proportionately long. The humerus has a large, "hatchet-shaped" deltopectoral crest. The wing finger has only three phalanges. *Nyctosaurus* attained a wingspan of up to 3 m.

ORNITHOCHEIROIDEA: PTERANODONTIA: PTERANODONTOIDEA

Pteranodontoidea is the least inclusive clade containing *Anhanguera blittersdorffi* and *Pteranodon longiceps* (Andres et al. 2014). *Pteranodon*, from the Late Cretaceous (Coniacian) of Kansas, South Dakota, and Wyoming, has a posterodorsally projecting frontal crest that is sexually dimorphic in shape and size (Bennett 1992, 1994, 2001; Fig. 10.16). The jaws lack teeth and bore keratinous rhamphothecae in life. The upper jaws are longer than the lower ones. The mandibular symphysis extends for about two-thirds of the total length of the mandible (Bennett 1994). *Pteranodon* attained a wingspan of up to more than 7 m in presumed males and up to 4 m in presumed females (Bennett 2001). Its proportionately long tail consists of proximal caudal vertebrae with double centra and distal caudals forming a bony rod. Although *Pteranodon* is known from more than a thousand specimens, all apparently represent skeletally mature or nearly mature individuals. Bennett (2001) interpreted this as evidence that *Pteranodon* did not fly out to sea until it had reached maturity and that the young lived elsewhere. Adult individuals of *Pteranodon* subsisted on fish (Bennett 2001).

ORNITHOCHEIROIDEA: ORNITHOCHEIROMORPHA

Ornithocheiromorpha (from Greek *ornis*, bird, *cheir*, hand, and *morphe*, form) is the most inclusive clade containing *Ornithocheirus simus* but not *Pteranodon longiceps* (Andres et al. 2014). The still poorly known *Lonchodectes*, from the Early to Late Cretaceous (Berriasian-Turonian) of England, has slender jaws with raised alveolar margins and small teeth with slightly constricted bases (Unwin 2001).

Istiodactylidae is the least inclusive clade containing *Istiodactylus latidens* and *Nurhachius ignaciobritoi* (Andres et al. 2014). It is diagnosed by a relatively broad and short snout, a greatly enlarged nasoantorbital fenestra, and a posteriorly inclined, long posterior region of the cranium (Andres and Ji 2006; Witton 2012; Fig. 10.17). The teeth are closely spaced and have lancet-shaped crowns with distinct cutting edges. The tooth rows are restricted to the region anterior to the nasoantorbital fenestra. *Istio-*

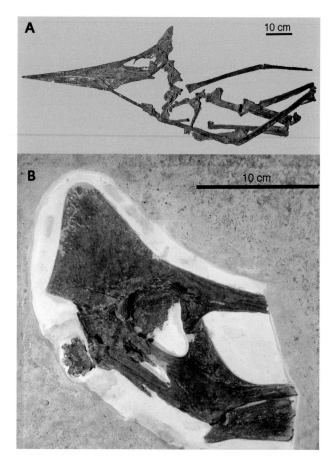

Figure 10.16. A, skull and partial postcranial skeleton of the pteranodontid *Pteranodon sternbergi*; **B,** partial skull of *Pteranodon longiceps.* Courtesy of Department of Paleobiology, National Museum of Natural History.

dactylus, from the Early Cretaceous (Aptian) of England and Liaoning (China), attained a wingspan of more than 4 m.

The pteranodontoid *Hamipterus*, from the Early Cretaceous of Xinjiang (China), is noteworthy for nesting in colonies and for the preservation of eggs, some of which contain embryos (X. Wang et al. 2014, 2017). It shows sexual dimorphism in the development of a distinct premaxillary crest.

Anhangueria is the most inclusive clade containing *Anhanguera blittersdorffi* but not *Istiodactylus latidens* (Andres et al. 2014). Boreopterinae is distinguished by a long snout with many long, slender teeth that extend well above or below the opposing jaws, especially near the tips of the jaws (Lü et al. 2006, 2010). *Boreopterus* and *Zhenyuanopterus* are known from the Early Cretaceous (Barremian) of Liaoning (China). The snout of the latter has a low bony crest extending over much of its length.

Ornithocheirae is particularly distinguished by the considerably elongated portion of the snout anterior to the nasoantorbital fenestra. The teeth in the anterior regions of the jaws are long. The forelimbs are proportionately long but the hind limbs are rather small. Some forms may have attained wingspans of at least 6 m, although wingspans between 4 and 5 m were more common (Martill and Unwin 2012). *Anhanguera* (Fig. 10.18) and *Tropeognathus*, from the Early Cretaceous (Albian) of Brazil, have prominent rounded median crests on the elongated portion of the snout and on the anterior portion of the mandible (Wellnhofer 1987, 1991b; Kellner and Tomida 2000).

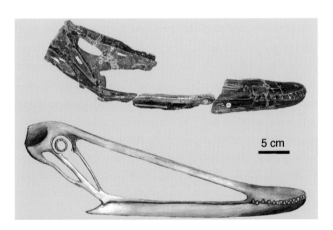

Figure 10.17. Partial skull of the istiodactylid *Istiodactylus latidens* in lateral view (with reconstruction). From Witton (2012)—CC BY 2.5.

PTERODACTYLOIDEA: EUPTERODACTYLOIDEA: ORNITHOCHEIROIDEA: AZHDARCHOIDEA

Azhdarchoidea (from Azhdar, a dragon-like creature in Persian mythology) is the least inclusive clade containing *Tapejara wellnhoferi* and *Quetzalcoatlus northropi* (Andres et al. 2014). Diagnostic features for this clade include the presence of a deep ventral flange on the coracoid; deltopectoral crest of the humerus tall, rectangular; and presence of a distinct proximal ridge on the ulnar crest of the humerus (Andres, pers. comm.). Azhdarchoidea is divided into Tapejaromorpha and Neoazhdarchia.

ORNITHOCHEIROIDEA: AZHDARCHOIDEA: TAPEJAROMORPHA

The most distinctive features of Tapejaromorpha are the short, deep cranium surmounted by a tall crest and the downturned tips of the jaws (Wellnhofer and Kellner 1991; Kellner 2004). *Tapejara*, from the Early Cretaceous (Albian) of Brazil (Fig. 10.19), attained a wingspan of 1.5 m, whereas the larger-sized *Tupandactylus*, also from the Early Cretaceous (Albian) of Brazil (Fig. 10.20), possibly reached a wingspan of 3 m. Crania of adult tapejarids bear two bony crests, one extending dorsally from the tip of the snout and the other extending back from the premaxilla over much of the cranium. Well-preserved specimens show that an enormous soft-tissue crest occupied the space between these bony crests and was anchored in fibrous bone along the edges of the former (Fig. 10.20). In addition, the skull also has a parietal crest. The cranial crests of tapejarids

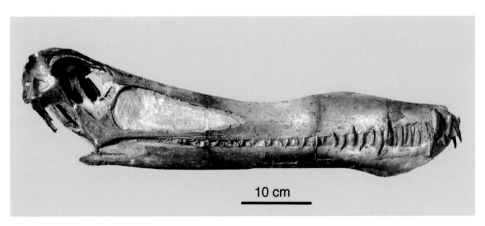

Figure 10.18. Skull of the anhanguerid *Anhanguera piscator* in lateral view. Courtesy of Andre Veldmeijer and Erno Endenburg.

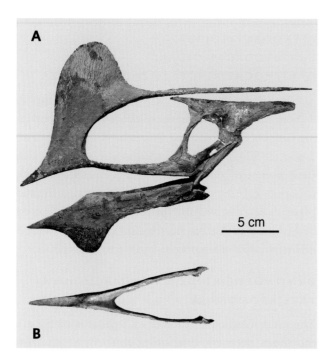

Figure 10.19. A, partial skull in lateral view, and **B**, mandible in dorsal view of the tapejarid *Tapejara wellnhoferi*. Courtesy of Andre Veldmeijer and Erno Endenburg.

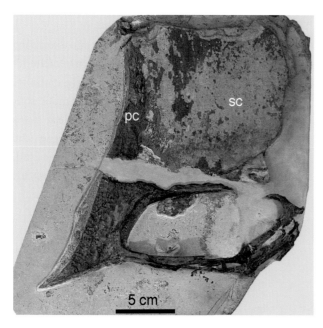

Figure 10.20. Cranium of the tapejarid *Tupandactylus navigans* in lateral view. The premaxillary crest (pc) supported a soft-tissue crest (sc), which is represented by an impression. Traces of a keratinous beak cover the anterior tip of the snout. Courtesy of David Martill.

changed in shape and size during growth, and presumably served primarily for display (Witton 2013). Tapejaridae differs from most other pterodactyloids in the absence of a notarium. Based on the structure of the jaw apparatus, Wellnhofer and Kellner (1991) suggested that tapejarids subsisted on fruits, seeds, and other plant matter.

ORNITHOCHEIROIDEA: AZHDARCHOIDEA: NEOAZHDARCHIA

Neoazhdarchia includes Dsungaripteromorpha and Neopterodactyloidea.

ORNITHOCHEIROIDEA: AZHDARCHOIDEA: NEOAZHDARCHIA: DSUNGARIPTEROMORPHA

Andres et al. (2014) divided Dsungaripteromorpha into Dsungaripteridae and Thalassodromidae.

Dsungaripteridae is distinguished by the presence of a short, tapered parietal crest and a median bony crest on the snout, the enclosure of the tooth bases by alveolar bone, and greatly expanded paroccipital processes (Fig. 10.21). The tips of the jaws lack teeth. *Dsungaripterus*, from the Early Cretaceous of Xinjiang (China), attained a wingspan of 3 m. The limb bones and vertebrae of dsungaripterids have relatively thick bony walls. Based on the stout jaws and blunt teeth, Unwin (2006) suggested that these pterosaurs possibly fed on hard-shelled prey such as mollusks.

Thalassodromidae is characterized by the presence of an enormous bony crest arising from the tip of the snout and extending posteriorly well beyond the occipital region. Thus, the cranium is deep above, but not below, the large nasoantorbital fenestra. In *Thalassodromeus*, from the Early Cretaceous (Albian) of Brazil, the premaxillae, frontal, parietal, and supraoccipital form the cranial crest, which bears vascular grooves, indicating a specialized covering in life (Kellner and Campos 2002; Fig. 10.22). The snout bears a distinct ventral ridge or protuberance, which fitted between the edges of the fused mandibular rami. Humphries et al. (2007) rejected the suggestion by Kellner and Campos (2002) that thalassodromids obtained food by plowing their lower jaws through water (skim feeding). These pterosaurs attained wingspans between 4 and 5 m.

Figure 10.21. Skull of the dsungaripterid *Dsungaripterus weii* in lateral view. Note the knob-like teeth and pincer-like anterior ends of the jaws. Courtesy of Stephen Godfrey.

Figure 10.22. Skull of the thalassodromid *Thalassodromeus sethi* in lateral view. The orbit is marked by "o." Composed by Mark Witton from images courtesy of Andre Veldmeijer and Erno Endenburg.

ORNITHOCHEIROIDEA: AZHDARCHOIDEA: NEOAZHDARCHIA: NEOPTERODACTYLOIDEA

Neopterodactyloidea includes Chaoyangopteridae and Azhdarchidae (Andres et al. 2014). It is especially distinguished by the structure of the midcervical vertebrae, which have prominent zygapophyses and low, ridge-like neural spines. They also share other derived features such as the scapula being longer than the coracoid.

Chaoyangopteridae is known only from the Early Cretaceous (Aptian) of Liaoning (China) (Lü and Ji 2005; Lü et al. 2008). The proportionately large skull has enormous nasoantorbital fenestrae and a short, posterodorsally projecting crest, which incorporates part of the premaxilla

(Lü et al. 2008). Chaoyangopterids attained wingspans ranging from 1.1 to 1.9 m (Witton 2013).

Azhdarchidae includes the largest known flying animals of all time, reaching wingspans of possibly up to 11 m (Witton and Habib 2010). It is characterized by the possession of typically greatly elongated, "tubular" cervical vertebrae with large, flared zygapophyses, neural spines that are greatly reduced or absent altogether, and prominent processes (postexapophyses) on the condyles of posterior articular surfaces of the centra (Averianov 2010; Fig. 10.23). However, some azhdarchid taxa have short, robust necks (Vremir et al. 2017). The skulls of *Quetzalcoatlus*, from the Late Cretaceous (Maastrichtian) of Texas, and *Zhejiangopterus*, from the Late Cretaceous (Campanian) of Zhejiang (China), are long and low, with an elongate, uncrested snout anterior to the nasoantorbital fenestra (Kellner and Langston 1996; Unwin and Lü 1997; Witton and Naish 2008). The posterior region of the cranium is low, and the orbit occupies a ventral position. The cranium of *Quetzalcoatlus* bears a short crest that extends over the posterior half of the nasoantorbital fenestra (Kellner and Langston 1996). The shallow mandibular symphysis forms up to 60 percent of the total mandibular length. Some smaller-sized azhdarchids, such as *Bakonydraco*, from the Late Cretaceous (Santonian) of Hungary, have proportionately shorter snouts (Ősi et al. 2005). The mode of life of azhdarchids has been the subject of much speculation (Witton 2013). They were apparently capable of soaring but probably not extended flapping flight (Witton and Habib 2010). Referred trackways with large hand- and footprints (*Haenamichnus*) from the Late Cretaceous of South Korea indicate that azhdarchids held their limbs directly beneath their bodies (Hwang et al. 2002; Witton and Naish 2008).

Dinosauromorpha

Dinosaurs have long excited human curiosity more than any other group of extinct animals. They include the largest land-dwelling tetrapods of all time and were astonishingly diverse, comparable only to Cenozoic mammals in that respect. There now exists a general consensus that birds, the most speciose group of extant land vertebrates, descended from small theropod dinosaurs. Thus, dinosaurs continue to be an enormous evolutionary success to the present day.

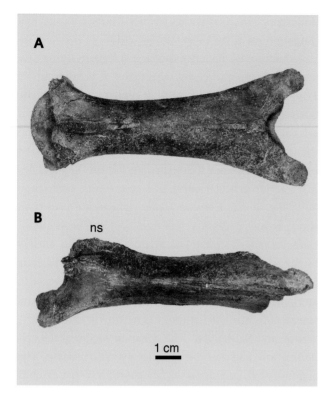

Figure 10.23. Cervical vertebra 6 of the azhdarchid *Azhdarcho longicollis* in **A**, dorsal, and **B**, lateral views. Abbreviation: ns, neural spine. Courtesy of Alexander Averianov.

Among Ornithodira, dinosaurs and their closest relatives form a clade Dinosauromorpha (derived from Greek *deinos*, fearfully great, *sauros*, lizard, and Greek *morphe*, form; Benton 1985). Sereno (2005) defined Dinosauromorpha as the most inclusive clade containing *Passer domesticus* but not *Pterodactylus antiquus*, *Ornithosuchus woodwardi*, or *Crocodylus niloticus*. It shares several derived features in the structure of the hind limb (Nesbitt 2011): length of the longest metatarsal more than 50 percent of that of the tibia; metatarsal V not "hooked," with its articular surface for the fourth distal tarsal extending more or less parallel to the long axis of the metatarsal shaft; and pedal digit V represented only by its tapered metatarsal. Dinosauromorpha dates back to the Middle Triassic (Anisian) (Nesbitt et al. 2010).

Dinosauromorpha: Lagerpetidae

Lagerpetidae comprises all taxa more closely related to *Lagerpeton chanarensis* than to *Alligator mississippiensis*, *Eudimorphodon ranzii*, *Marasuchus lilloensis*, *Silesaurus opo-*

lensis, *Triceratops horridus*, *Saltasaurus loricatus*, and *Passer domesticus* (Nesbitt et al. 2009b). It includes *Lagerpeton*, from the Late Triassic (Carnian) of Argentina (Sereno and Arcucci 1994a), and a subclade comprising *Dromomeron*, from the Late Triassic (Norian) of Argentina, New Mexico, and Texas (Irmis et al. 2007a; Nesbitt et al. 2009b; Martínez et al. 2016), and *Ixalerpeton*, from the Late Triassic (Carnian) of Brazil (Cabreira et al. 2016). These small-bodied, slender-limbed ornithodirans were clearly adapted for bipedal running. Unlike pseudosuchians, lagerpetids have mesotarsal ankle joints. *Ixalerpeton* and *Lagerpeton* attained total lengths of up to 70 cm. Diagnostic features for Lagerpetidae include the proximal head of the femur being hook-shaped in lateral view; proximal portion of the femur with a flat anterolateral surface of the; posterolateral portion of the tibial facet with a distinct process; and calcaneum with concave articular surface for the fibula (Nesbitt 2011). The transversely narrow pes of *Lagerpeton* is functionally two-toed (didactyl), with pedal digit IV being the longest and pedal digit I being the shortest and most slender. Based on the structure of the hind limb, the proportionately small pelvic girdle, and the anteriorly inclined neural spies on the posterior dorsal vertebrae, Sereno and Arcucci (1994a) reconstructed *Lagerpeton* as an agile runner that was also possibly capable of hopping. The frontals and parietals of *Ixalerpeton* are transversely broader than in most basal dinosaurs, and the supratemporal fossa does not extend onto the frontal (Cabreira et al. 2016).

Brusatte et al. (2011) attributed certain types of tracks from the Early to Middle Triassic (Olenekian-Anisian) of Poland to dinosauromorphs similar to *Lagerpeton*, but other researchers (e.g., Padian 2013) have rejected this interpretation.

Dinosauromorpha: Dinosauriformes

Novas (1992) defined Dinosauriformes (from Greek *deinos*, fearfully great, and *sauros* lizard, and Latin *forma*, form) as comprising the most recent common ancestor of *Marasuchus* and Dinosauria and all descendants of that ancestor. Nesbitt (2011) listed a suite of synapomorphies for this clade including the pubis being longer than the ischium; narrow contact between the pubis and ischium; anterior trochanter of the femur forming a shelf proximal to the fourth trochanter; lateral surface of the distal portion of the tibia with a proximodistal groove; and astragalus with an anterior ascending flange.

DINOSAUROMORPHA: DINOSAURIFORMES: *MARASUCHUS*

Bonaparte (1975) first proposed a close relationship between *Marasuchus* (formerly known as "*Lagosuchus*"), from the Late Triassic (Carnian) of Argentina, and dinosaurs. Phylogenetic analyses have consistently supported this hypothesis (Sereno 1991a; Sereno and Arcucci 1994b; Nesbitt 2011; Bittencourt et al. 2015). *Marasuchus* shares various apomorphies with dinosaurs such as the (in lateral view) parallelogram-shaped cervical centra, presence of an acetabular antitrochanter, and extension of the proximal articular surface of the femur under the femoral head. It also has a mesotarsal ankle joint. *Marasuchus*

attained a total length of about 50 cm and has short forelimbs and a long tail (Fig. 10.24A). Diagnostic features for this taxon include the anterodorsally inclined neural spines on presacral vertebrae 6–9, the midcaudal vertebrae being twice as long as the anterior caudals, and the transversely concave distal pubic blade (Sereno and Arcucci 1994b). Its cranial structure is still poorly known.

DINOSAUROMORPHA: DINOSAURIFORMES: *SALTOPUS*

Saltopus, from the Late Triassic of Scotland, is known only from a single, poorly preserved skeleton (Benton and Walker 2011). It attained a total length of possibly up to 1 m (of which the tail comprises more than half). Benton and Walker (2011) assigned *Saltopus* to Dinosauriformes based on the presence of an anterior trochanter,

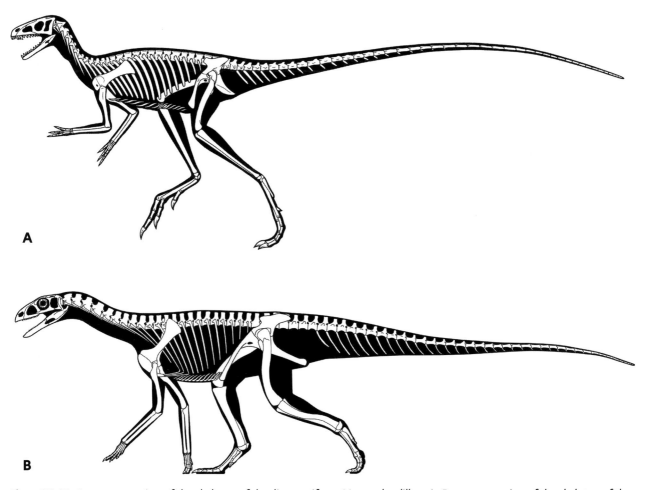

Figure 10.24. A, reconstruction of the skeleton of the dinosauriform *Marasuchus lilloensis*; **B,** reconstruction of the skeleton of the silesaurid *Asilisaurus kongwe*. A, courtesy of Paul Sereno and Carol Abraczinkas; B, courtesy of Scott Hartman.

the presence of a trochanteric shelf on the femur, and the "waisted" sacral ribs.

DINOSAUROMORPHA: DINOSAURIFORMES: SILESAURIDAE

Silesauridae is the most inclusive clade containing *Silesaurus opolensis* but not *Passer domesticus*, *Triceratops horridus*, and *Alligator mississippiensis* (Nesbitt et al. 2010). It ranged in time from the Middle to the Late Triassic (Anisian-Norian) and is known from Argentina, Brazil, Morocco, New Mexico and Texas, Poland, Tanzania, and Zambia (Nesbitt et al. 2010; Peecook et al. 2013). Dzik and Sulej (2007) argued that *Silesaurus*, from the Late Triassic (Norian) of Poland (Dzik 2003), and its relatives were basal ornithischian dinosaurs. However, Nesbitt et al. (2010) noted the absence of dinosaurian features such as the anterior extension of the supratemporal fossa onto the frontal and the long deltopectoral crest on the humerus. Silesaurids have proportionately long forelimbs, which indicate at least facultatively quadrupedal locomotion (Fig. 10.24B). An unexpected feature in at least two known silesaurid taxa is the presence of a "crocodile-normal" ankle, as in pseudosuchians and aphanosaurians but unlike the mesotarsal ankle in other derived ornithodirans (Nesbitt et al. 2017). Diagnostic cranial synapomorphies of Silesauridae include the short, more or less triangular tooth crowns with denticulated cutting edges and the edentulous anterior end of the dentary, which tapers to a point and possibly supported a small beak in life. Both features suggest omnivorous or herbivorous habits (Nesbitt et al. 2010). *Silesaurus* attained a total length of 2.3 m.

DINOSAUROMORPHA: DINOSAURIFORMES: *NYASASAURUS*

Nyasasaurus, known only from a partial humerus and some vertebrae from the Middle Triassic (Anisian) of Tanzania (Nesbitt et al. 2012), represents either a derived early dinosauriform or the oldest known dinosaur. If it is a dinosaur, it would greatly extend the known temporal range of that clade. The humerus of *Nyasasaurus* has a long, laterally deflected deltopectoral crest. Its bone microstructure indicates a sustained, elevated growth rate (Nesbitt et al. 2012).

11   Dinosauria I: Saurischia

Owen (1842) first established Dinosauria (from Greek *deinos*, "fearfully great" in the meaning used by Owen, and *sauros*, lizard) for the reception of three taxa of Mesozoic reptiles—*Hylaeosaurus*, *Iguanodon*, and *Megalosaurus*—that clearly differed in their skeletal structure from all other extinct and extant reptiles then known. In retrospect, this was a remarkable insight, because Owen was working with very incomplete specimens. Seeley (1887) distinguished two major groups among dinosaurs, Saurischia (from Greek *sauros*, lizard, and *ischion*, hip) and Ornithischia (from Greek *ornis*, bird, and *ischion*, hip), based primarily on differences in the structure of the pelvic girdle. However, he did not see any particularly close relationship between these two groups and rejected Owen's concept of Dinosauria. Starting with Huene (1914), Seeley's views were generally adopted until Bakker and Galton (1974) built a compelling new case for the monophyly of Dinosauria. The latter authors listed numerous synapomorphies in support of this hypothesis and explicitly included birds, which are a clade of derived theropod dinosaurs and thus are "living dinosaurs." Ostrom (1976a) found numerous anatomical features in support of the dinosaurian affinities of birds. Gauthier's (1986) pioneering phylogenetic analysis further supported the monophyly of Dinosauria including birds. Since then every major analysis has confirmed this hypothesis. Seeley's names Saurischia and Ornithischia remain in general use for the two principal clades of Dinosauria.

Padian and May (1993) first defined Dinosauria as comprising the most recent common ancestor of *Triceratops* and birds and all descendants of that ancestor. Sereno (2005) modified this definition by adding *Triceratops horridus* and *Passer domesticus* (house sparrow) as specifiers.

The fossil record for basal dinosauriforms and early dinosaurs is still inadequate, although new discoveries in the last few years will substantially improve it. Much additional and better-preserved material is needed to clarify the phylogenetic utility of many potentially informative characters, especially cranial features. Brusatte et al. (2010b) and Nesbitt (2011) listed a number of synapomorphies for Dinosauria: supratemporal fossa extending onto the dorsal surface of the frontal anteriorly; postaxial cervical vertebrae with accessory processes on the postzygapophyses (epipophyses); apex of the large deltopectoral crest is typically situated at more than 30 percent of the length of the humerus; femur with an asymmetrical fourth trochanter; articular facet for the fibula on the astragalus occupying less than 30 percent of the transverse width of that bone; and calcaneum bearing a concave facet for articulation with the fibula. Some of these features are also present in other ornithodirans. Furthermore, the phylogenetic significance of other skeletal features frequently cited as diagnostic for Dinosauria, such as the absence of the postfrontal, the number of sacral vertebrae, and the presence of an open acetabulum, remains uncertain because of considerable variation observed among basal dinosaurs and their dinosauriform relatives (Langer and Benton 2006; Brusatte et al. 2010b).

Dinosauromorphs including early dinosaurs had upright limb posture and were digitigrade. The presence of a supra-acetabular crest and the, typically, open acetabulum would have restricted femoral abduction compared to other archosaurs (J. R. Hutchinson and Gatesy 2000). The forelimbs were reduced in length relative to the hind limbs, and basal dinosaurs were predominantly bipedal. Various clades of more derived dinosaurs became secondarily quadrupedal.

The oldest known dinosaurs date from the Late Triassic (Carnian) (Martinez et al. 2011). Given that several dinosaurian clades were already present at that point in time, the initial diversification of Dinosauria occurred earlier. To date, however, none of the reports of stratigraphically older dinosaurian records, either in the form of bones or tracks, has withstood critical scrutiny (Irmis et al. 2007b; Brusatte et al. 2010b; Langer et al. 2010). The sole potential exception is *Nyasasaurus* (Chapter 10).

For detailed information on the anatomy, diversity, and distribution of the individual clades of dinosaurs excluding birds, interested readers should consult the authoritative reference work edited by Weishampel et al. (2004). Fastovsky and Weishampel (2016) and Lucas (2016) published introductory textbooks on this subject. Brusatte (2012) provided an excellent overview of the biology of dinosaurs.

Dinosauria: Saurischia

Saurischia is the most inclusive clade containing *Passer domesticus* and *Saltasaurus loricatus* but not *Triceratops horridus* (Sereno 2005). Seeley (1887) originally used the triradiate shape of the pelvis as the principal diagnostic feature for Saurischia, but this feature is plesiomorphic and thus not phylogenetically informative. Nesbitt (2011) provided a list of synapomorphies for Saurischia: subnarial foramen positioned on the suture between the premaxilla and maxilla; lacrimal overhanging the posterodorsal portion of the antorbital fenestra laterally; neural arches of cervical vertebrae 6-9 bearing epipophyses; posterior cervical and/or dorsal vertebrae forming accessory articulations with a hyposphene and a hypantrum between successive vertebrae; absence of a fifth distal carpal; shaft of metacarpal IV much narrower than those of metacarpals I-III; phalanx 1 of manual digit I

longest non-ungual phalanx of the manus; distal end of the femur bearing a deep groove between the lateral condyle and the tibiofibular crest; and astragalus with an elliptical, rimmed fossa behind the anterior ascending process on its proximal surface.

Saurischia is divided into Theropoda and Sauropodomorpha (Gauthier 1986). The phylogenetic relationships of the oldest known saurischians, especially Herrerasauridae and *Eoraptor*, remain contentious. Gauthier (1986) placed Herrerasauridae outside Saurischia and Ornithischia, whereas Novas (1994), Sereno (1994), Rauhut (2003), and Nesbitt (2011) found Herrerasauridae and *Eoraptor* as basal theropods. Langer and Benton (2006) included Herrerasauridae and *Eoraptor* in Saurischia, but placed them outside a clade comprising Theropoda and Sauropodomorpha. Subsequently, Martinez et al. (2011), Martínez et al. (2013c), and Sereno et al. (2013) reinterpreted *Eoraptor* as a basal sauropodomorph. Cabreira et al. (2016) placed Herrerasauridae and several other taxa previously considered basal theropods outside the clade comprising Sauropodomorpha and Theropoda. Baron et al. (2017a) presented a novel hypothesis that abolishes Seeley's division of Dinosauria. They grouped Herrerasauridae with Sauropodomorpha and hypothesized a separate clade comprising Theropoda and Ornithischia, for which they reintroduced Huxley's (1870) name Ornithoscelida (from Greek *ornis*, bird, and *skelos*, leg, thigh). This diversity of competing phylogenetic hypotheses underscores our still inadequate understanding of the fossil record of early dinosaurs.

Saurischia: Theropoda

Theropod dinosaurs other than birds are commonly referred to as "predatory dinosaurs" based on the structure of the dentition and other skeletal features, which indicate that the majority of taxa were faunivorous. However, several clades independently evolved features indicative of omnivory or herbivory (Zanno and Makovicky 2011). Unlike more derived sauropodomorphs and many ornithischians, theropods apparently were obligate bipeds throughout their evolutionary history. The pes is at least functionally tridactyl. Theropods have often very extensive pneumaticity of the skeleton, which has been interpreted as evidence for the presence of air sac-driven lung ventilation in these dinosaurs (Wedel 2009). Based on the structure of the axial skeleton, even early theropods already had a dorsally immobilized lung (Brocklehurst et al. 2008). As do some sauropodomorphs, various theropods other than birds have gaps in the pneumatization of the vertebral column, which indicate that the diverticula producing the pneumatic features originated from more than one source. This further supports the hypothesis of air sacs in these dinosaurs (Brocklehurst et al. 2018).

Sereno (2005) defined Theropoda (from Greek *therion*, wild beast, and *pous*, foot) as the most inclusive clade containing *Passer domesticus* but not *Saltasaurus loricatus*. Nesbitt (2011) listed a suite of synapomorphies for this clade: prezygapophyses on the distal caudal vertebrae extending for one-quarter or more of the length of the adjacent caudals; length of the humerus less than 60 percent that of the femur; proximal ends of the metacarpals abutting and not overlapping each other; distal end of the pubis forming a mediolaterally narrow "boot"-like expansion; and lateral edge of the lateral condyle on the proximal end of the tibia squared off rather than rounded.

Herrerasauridae is the most inclusive clade containing *Herrerasaurus ischigualastensis* but not *Passer domesticus* (Sereno 2005). It comprises *Herrerasaurus* and *Sanjuansaurus*, from the Late Triassic (Carnian) of Argentina (Novas 1994; Sereno 1994; Sereno and Novas 1994; Alcober and Martinez 2010), and *Staurikosaurus*, from the Late Triassic (Carnian-Norian) of Brazil (Bittencourt and Kellner 2009). *Herrerasaurus* (Fig. 11.1) attained a skull length of up to 30 cm and a total length of up to 4 m. Its snout is proportionately long. The antorbital fossa is anteroposteriorly narrow and apparently contains a small (promaxillary) fenestra (Sereno 2007). The lower jaw has mobile contacts between the dentary and postdentary bones similar to those in more derived theropods but with a different configuration of elements (Sereno 1994). *Herrerasaurus* has short forelimbs. Its manus has three functional digits (I-III) bearing large, trenchant unguals. Manual digit IV consists of a short metacarpal and a single phalanx, whereas manual digit V is reduced to a bony

Figure 11.1. A, skull of the basal theropod *Herrerasaurus ischigualastensis* in lateral view; **B**, reconstruction of the skeleton of *Herrerasaurus ischigualastensis*. B, courtesy of Paul Sereno and Carol Abraczinkas.

nubbin. The pelvis has an open acetabulum and a posteroventrally inclined pubis with a prominent distal "boot." The sacrum of *Herrerasaurus* has three vertebrae, but that of *Staurikosaurus* has only two, much like in more basal dinosauriforms. The pes of *Herrerasaurus* is functionally tridactyl. Pedal digit V has only a single phalanx.

The phylogenetic position of *Guaibasaurus*, from the Late Triassic (Norian) of Brazil (Bonaparte et al. 1999), remains unresolved. Bonaparte et al. (1999) interpreted it as close to the ancestry of all saurischians. Ezcurra (2010) considered *Guaibasaurus* a basal sauropodomorph, but Martínez et al. (2013c) found it as the sister taxon to derived theropods (Neotheropoda), and Cabreira et al. (2016) considered it a basal saurischian. The humerus of *Guaibasaurus* is shorter and more robust than that in basal theropods. Manual digit III is shorter than manual digit II, similar to the condition in more derived saurischians. The ilium lacks a brevis fossa, and the acetabulum has a complete bony wall.

Eodromaeus, from the Late Triassic (Carnian) of Argentina, attained a total length of 1.2 m (Martinez et al. 2011).

Figure 11.2. Skull of the basal theropod *Daemonosaurus chauliodus* in lateral view.

It retains teeth on the pterygoid. The posterior cervicals have pneumatic openings (pleurocoels) in the centra. The sacrum has three vertebrae. The ilium has an arched brevis fossa. The pubis ends distally in a small "foot."

Tawa, from the Late Triassic (Norian) of New Mexico, is more closely related to more derived theropods based on the presence of a short subnarial gap in the upper tooth row, the presence of rimmed pleurocoels on the anterior cervical vertebrae, the length of the manus reaching 40 percent or more of the length of the humerus, the absence of manual digit V, and the small size of the calcaneum (Nesbitt et al. 2009b). It attained a total length of about 2 m. The possibly related *Daemonosaurus*, from the Late Triassic (Rhaetian) of New Mexico (Sues et al. 2011), has a short and deep skull with procumbent premaxillary and anterior dentary teeth (Fig. 11.2).

SAURISCHIA: THEROPODA: NEOTHEROPODA

Neotheropoda (from Greek *neos*, new, *therion*, wild beast, and *pous*, foot) comprises the most recent common ancestor of *Coelophysis bauri* and *Passer domesticus* and all descendants of that ancestor (Sereno 2005). Shared derived features for this clade include the presence of three tympanic recesses in the braincase; presacral vertebrae with pneumatic openings behind the parapophyses; sacrum consisting of at least five vertebrae; presence of an olecranon process on the ulna; absence of manual digit V; and pedal digit V represented only by its metatarsal (Carrano et al. 2012).

SAURISCHIA: THEROPODA: NEOTHEROPODA: COELOPHYSOIDEA

Coelophysoidea (from Greek *koilos*, hollow, and *physis*, form) is the most inclusive clade containing *Coelophysis*

bauri but not *Carnotaurus sastrei*, *Ceratosaurus nasicornis*, and *Passer domesticus* (Sereno 2005). It comprises a variety of small- to medium-sized theropods of Late Triassic (Norian-Rhaetian) and Early Jurassic age. Diagnostic synapomorphies for this clade include the presence of a diastema in the upper tooth row below the external naris; dorsoventrally deep, slightly upturned anterior end of the dentary; elongate cervical vertebrae; the presence of long and thin epipophyses on the cervical vertebrae; and the fusion of the pelvic bones (Carrano et al. 2012). *Coelophysis* (including "*Syntarsus*"), from the Late Triassic (Rhaetian) of New Mexico (Colbert 1989; Fig. 11.3) and the Early Jurassic (Hettangian-Sinemurian) of South Africa and Zimbabwe, attained a total length of up to 3 m. The elongate skull is dorsoventrally low. *Coelophysis* is the oldest known theropod with a furcula (Nesbitt et al. 2009d). This bone formed through median fusion of the clavicles and is present in most theropods including most birds.

SAURISCHIA: THEROPODA: NEOTHEROPODA: AVEROSTRA

Allain et al. (2012) defined Averostra (from Latin *avis*, bird, and *rostrum*, snout) as comprising the most recent common ancestor of *Ceratosaurus nasicornis* and *Passer domesticus* and all descendants of that ancestor. Carrano et al. (2012) listed a suite of diagnostic synapomorphies for this clade: preacetabular portion of the ilium forms a lobe-like anteroventral extension; ischium with an expanded, triangular distal end; pubes enclosing a slit-like opening proximal to their expanded distal ends; and the presence of a fossa on the medial surface of the proximal end of the fibula occupying a central position. Averostra is divided into Ceratosauria and Tetanurae.

Originally grouped with Coelophysoidea, various medium- to large-sized basal neotheropods including *Dilophosaurus*, from the Early Jurassic (Sinemurian-Toarcian) of Arizona (Welles 1984), are now considered more closely related to averostrans than to *Coelophysis* based on derived features such as the presence of a promaxillary fenestra and the lower number of maxillary teeth (N. D. Smith et al. 2007). The cranium of *Dilophosaurus* bears a pair of tall, thin dorsal crests formed by the lacrimal, nasal, and premaxilla (Welles 1984). Similar cranial crests are also present in other large-sized basal neotheropods but likely evolved independently (Brusatte et al. 2010b; Gates et al. 2016). They possibly

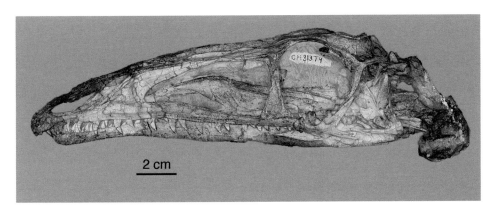

Figure 11.3. Skull and anterior cervical of the coelophysid *Coelophysis bauri* in lateral view. Courtesy of Robert Reisz and Diane Scott.

served for display. *Dilophosaurus* attained a total length of about 7 m.

SAURISCHIA: THEROPODA: NEOTHEROPODA: AVEROSTRA: CERATOSAURIA

Ceratosauria (from Greek *keras*, horn, and *sauros*, lizard) was originally conceived as a group comprising all basal neotheropods (e.g., Gauthier 1986). Later authors segregated Coelophysoidea from Ceratosauria. Wilson et al. (2003) offered a revised definition of Ceratosauria as the most inclusive clade containing *Ceratosaurus nasicornis* but not *Passer domesticus*. Diagnostic features for this clade include the fusion of the cervical ribs to the vertebrae in adult individuals; far-laterally projecting parapophyses on the dorsal vertebrae; sacral centra constricted at midlength; the presence of a continuous ridge extending from the supraacetabular crest to the brevis shelf of the ilium; and astragalus and calcaneum fused to each other in adult individuals (Carrano et al. 2012). *Saltriavenator*, from the Early Jurassic (Sinemurian) of Italy (Dal Sasso et al., 2018), is the oldest representative of this clade. The geologically youngest ceratosaurians are Late Cretaceous (Maastrichtian) in age.

Ceratosaurus, from the Late Jurassic (Kimmeridgian-Tithonian) of Colorado, Utah, and Wyoming, as well as Portugal, attained a total length of at least 6 m (Gilmore 1920; Madsen and Welles 2000; Malafaia et al. 2015). It is especially characterized by the presence of a mediolaterally narrow horn core on the fused nasals.

Abelisauroidea (named for Roberto Abel, an Argentine museum director, and Greek *sauros*, lizard) is the least inclusive clade containing *Carnotaurus sastrei* and *Noasaurus leali* (Sereno 2005). It is mainly known from the Cretaceous of the Southern Hemisphere, although a few forms have been reported from Europe, and recent discoveries extend the temporal range of this clade back into the Jurassic. Diagnostic apomorphies for Abelisauroidea include the large external mandibular fenestra; presence of a peg-and-socket articulation between the ilium and ischium; fusion of the ascending process of the astragalus to the fibula; and the presence of double grooves along the lateral and medial sides of the pedal ungual phalanges (Carrano et al. 2012).

Among Abelisauroidea, Wilson et al. (2003) defined Noasauridae as the most inclusive clade containing *Noasaurus leali* but not *Carnotaurus sastrei*. It is divided into two subclades, Noasaurinae and Elaphrosaurinae. The best-known noasaurine, *Masiakasaurus*, from the Late Cretaceous (Maastrichtian) of Madagascar (Carrano et al. 2011), has a relatively long and low skull. The anterior end of the dentary is deflected ventrally. The anterior dentary teeth are procumbent. The pubes form a distinct distal "boot." *Masiakasaurus* attained a total length of about 1.8 m. Elaphrosaurinae includes *Limusaurus*, from the Late Jurassic (Oxfordian) of Xinjiang, China (Xu et al. 2009), which is noteworthy for its proportionately small skull and greatly reduced forelimbs (Xu et al. 2009; Fig. 11.4). *Limusaurus* attained a total length of 1.7 m. Manual digit I is represented only by its greatly reduced metacarpal, and metacarpal IV is rather slender. The distal elements of the hind limbs are considerably elongated, indicating excellent cursorial abilities. Juveniles of *Limusaurus* have teeth in the upper and lower jaws but these teeth were completely lost and replaced by a beak in adults (S. Wang et al. 2016). More mature individuals acquired gastroliths. These features suggest that *Limusaurus* was omnivorous or even herbivorous. *Elaphrosaurus*, from the Late Jurassic (Kimmeridgian-Tithonian) of Tanzania (Janensch 1925), is closely related to *Limusaurus*

Figure 11.4. Nearly complete skeleton of the noasaurid *Limusaurus inextricabilis* in lateral view. Courtesy of Jim Clark.

(Rauhut and Carrano 2016). It attained an estimated total length of about 6 m.

Abelisauridae is characterized by an anteroposteriorly short but dorsoventrally deep skull with a thick skull roof and pronounced sculpturing on the facial bones (Bonaparte et al. 1990; Sampson and Krause 2007). The forelimbs, particularly the ulna and radius, are unusually short, even shorter than in tyrannosaurid theropods, but the hind limbs are robust. Abelisauridae is defined as the most inclusive clade containing *Carnotaurus sastrei* but not *Noasaurus leali* (Wilson et al. 2003). It is best known from the Late Cretaceous of Argentina, India, and Madagascar and includes the dominant dinosaurian predators in Gondwana during this time interval. The latter attained total lengths from about 5 to at least 10 m. *Majungasaurus*, from the Late Cretaceous (Maastrichtian) of Madagascar (Sampson and Krause 2007), has fused frontals that form a dorsomedian "dome" (Fig. 11.5). *Carnotaurus*, from the Late Cretaceous (Maastrichtian) of Argentina (Bonaparte et al. 1990), has a pair of distinct dorsolateral horns on the frontals. Pol and Rauhut (2012) interpreted *Eoabelisaurus*, from the Middle Jurassic (Aalenian-Bajocian) of Argentina, as a basal abelisaurid with incipient reduction of its forelimbs.

SAURISCHIA: THEROPODA: NEOTHEROPODA: AVEROSTRA: TETANURAE

Most neotheropods belong to Tetanurae (from Greek *tetanos*, stiff, and *oura*, tail), which is the most inclusive clade containing *Passer domesticus* but not *Ceratosaurus nasicornis* (Allain et al. 2012). Tetanurans have tightly articulated caudal vertebrae, especially in the distal portion of the tail, which presumably served as a dynamic stabilizer during bipedal locomotion (Gatesy and Dial

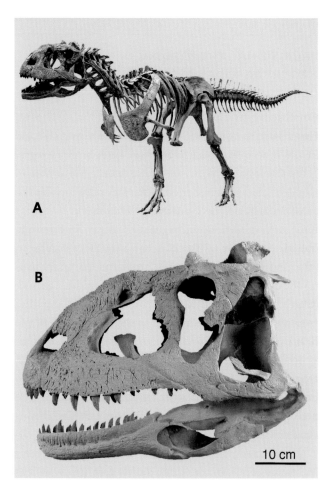

Figure 11.5. A, reconstructed skeleton, and **B**, skull in lateral view of the abelisaurid *Majungasaurus crenatissimus*. Total length c. 6.2 m. Courtesy of David Krause.

1996). Carrano et al. (2012) listed various synapomorphies for Tetanurae including the posterior termination of the maxillary tooth row anterior to the orbit; the participation of the nasal in the antorbital fossa; humerus with a large internal tuberosity and prominent deltopec-

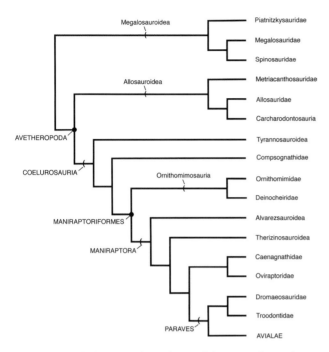

Figure 11.7. Skull of *Monolophosaurus jiangi* in lateral view. Length of cranium 80 cm. Courtesy of Philip Currie.

Figure 11.6. Phylogenetic hypothesis of the interrelationships of Tetanurae. Dots denote node-based clades and parentheses denote stem-based clades. Based on Carrano et al. (2012), A. H. Turner et al. (2012), and Brusatte et al. (2014).

toral crest; and the presence of a notch ventral to the obturator process on the ischium. Tetanurae is divided into Megalosauroidea and Avetheropoda (Fig. 11.6).

Monolophosaurus, from the Middle Jurassic of Xinjiang (China) (Zhao and Currie 1994), has been associated with various derived tetanurans. Phylogenetic analyses by Brusatte et al. (2010b) and Carrano et al. (2012) recovered a more basal position for this taxon. The cranium bears a prominent median crest with a straight dorsal edge, formed by the premaxillae, nasals, prefrontals, lacrimals, and frontals (Fig. 11.7). Each nasal has two large, equal-sized openings. *Monolophosaurus* attained a total length of about 5 m.

THEROPODA: NEOTHEROPODA: AVEROSTRA: TETANURAE: MEGALOSAUROIDEA

Megalosauroidea (from Greek *megas*, large, and *sauros*, lizard) is the most inclusive clade containing *Megalosaurus bucklandii* but not *Passer domesticus* and shares only a few derived features, such as the presence of a maxillary fossa and the U-shaped outline of the ventral process of the postorbital in transverse section (Carrano et al. 2012). *Megalosaurus*, from the Middle Jurassic (Bathonian)

of England, was the first named nonavian dinosaur (Buckland 1824), but it was described in detail and properly diagnosed only quite recently (Benson 2010). Historically, this genus name was indiscriminately applied to teeth and bones of mostly unidentifiable large-bodied theropods worldwide. Carrano et al. (2012) divided Megalosauroidea into Piatnitzkysauridae, Megalosauridae, and Spinosauridae.

Piatnitzkysauridae is the most inclusive clade containing *Piatnitzkysaurus floresi* but not *Megalosaurus bucklandii*. It is characterized especially by the presence of two rows of lateral foramina on the maxilla, striated interdental plates, and anteriorly inclined neural spines on the dorsal vertebrae (Carrano et al. 2012). Its best-known representative is *Piatnitzkysaurus*, from the Middle Jurassic (Bajocian-Callovian) of Argentina (Bonaparte 1986). *Marshosaurus*, from the Late Jurassic (Kimmeridgian-Tithonian) of Colorado and Utah, is referable to this clade (Carrano et al. 2012).

Megalosauridae is the most inclusive clade containing *Megalosaurus bucklandii* but not *Allosaurus fragilis*, *Spinosaurus aegyptiacus*, and *Passer domesticus*. Aside from *Megalosaurus* itself, it includes a variety of Jurassic and Early Cretaceous taxa such as *Eustreptospondylus*, from the Middle Jurassic (Callovian) of England (Sadleir et al. 2008); *Torvosaurus*, from the Late Jurassic (Kimmeridgian-Tithonian) of Colorado, Utah, and Wyoming, (Britt 1991) and Portugal (Hendrickx and Mateus 2014); and *Wiehenvenator*, from the Middle Jurassic (Callovian) of Germany (Rauhut et al. 2016). Megalosaurids have dorsoventrally low skulls with long snouts. *Torvosaurus* attained a total length of up to 10 m (Hendrickx and Mateus 2014; Fig. 11.8). A possible megalosaurid, *Sciurumimus,* from the Late Jurassic (Kimmeridgian) of Germany, is known from an exquisitely preserved skeleton of a juvenile individual with filamentous structures covering much of its body (Rauhut et al. 2012b; Fig. 11.9). It provides the first evidence that feather-like epidermal covering was present in basal tetanurans.

Sereno (2005) defined Spinosauridae as the most inclusive clade containing *Spinosaurus aegyptiacus* but not *Torvosaurus tanneri, Allosaurus fragilis,* and *Passer domesticus.* This clade is distinguished by greatly elongated snouts with sigmoidal alveolar margins and with coni-

cal rather than labiolingually flattened tooth crowns, often enormously elongated neural spines on the dorsal vertebrae, and a greatly enlarged ungual on manual digit I (Charig and Milner 1997; Sues et al. 2002; Dal Sasso et al. 2005; Allain et al. 2012; Fig. 11.10). *Spinosaurus,* from the Late Cretaceous (Cenomanian) of Egypt and Morocco, possibly reached a total length of more than 13 m and has up to 1.8-meter-tall dorsal neural spines (Stromer 1915). Ibrahim et al. (2014) reconstructed *Spinosaurus* as a giant semiaquatic predator based on features of the hind limbs and the solid limb bones, which lack medullary cavities. Features of the snout, gut contents, and the oxygen-isotope composition of the tooth enamel indi-

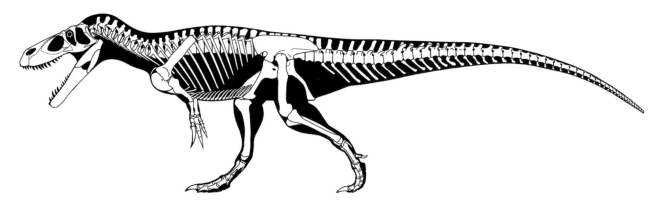

Figure 11.8. Skeleton of the megalosaurid *Torvosaurus tanneri.* Courtesy of Scott Hartman.

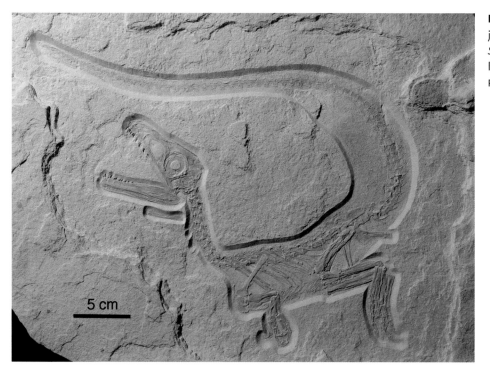

Figure 11.9. Skeleton of a juvenile of the megalosauroid *Sciurumimus albersdoerferi* in lateral view. Courtesy of Oliver Rauhut.

5 cm

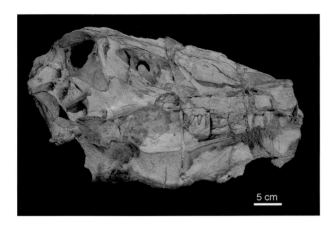

Figure 11.10. Skull of the spinosaurid *Irritator challengeri,* with cranium in lateral view. Courtesy of Diane Scott.

cate that fish formed a staple of spinosaurid diets (Charig and Milner 1997; Rayfield et al. 2007; Amiot et al. 2010).

THEROPODA: NEOTHEROPODA: AVEROSTRA: TETANURAE: AVETHEROPODA

Avetheropoda (from Latin *avis,* bird, and Greek *therion,* wild beast, and *pous,* foot) comprises the most recent common ancestor of *Allosaurus fragilis* and *Passer domesticus* and all descendants of that ancestor (Allain et al. 2012). Carrano et al. (2012) listed a suite of diagnostic features for this clade including the distinct curvature of the chevrons; the presence of a ridge on the medial surface of the ilium adjacent to the preacetabular notch; large, oval obturator foramen of the pubis; presence of a triangular, flange-like accessory trochanter of the femur; and a deep notch on the proximal end of metatarsal III. Avetheropoda is divided into Allosauroidea and Coelurosauria.

AVEROSTRA: TETANURAE: AVETHEROPODA: ALLOSAUROIDEA

Allosauroidea (from Greek *allos,* other, and *sauros,* lizard) is the most inclusive clade containing *Allosaurus fragilis* but not *Passer domesticus* (Sereno 2005). Carrano et al. (2012) divided it into Metriacanthosauridae, Allosauridae, and Carcharodontosauria.

Metriacanthosauridae is defined as the most inclusive clade containing *Metriacanthosaurus parkeri* but not *Allosaurus fragilis, Carcharodontosaurus saharicus,* and *Passer domesticus* (Sereno 2005). It includes *Sinraptor,* from the Late Jurassic (Oxfordian) of Xinjiang (China; Currie and

Zhao 1994), and *Metriacanthosaurus,* from the Late Jurassic (Oxfordian) of England (Walker 1964). Metriacanthosauridae is characterized by a number of synapomorphies including the presence of a lateral flange of the squamosal, which conceals the proximal head of the quadrate in lateral view; a longitudinal groove containing the dorsal row of neurovascular foramina on the lateral surface of the dentary; and the fusion of the distal ends of the ischia (Carrano et al. 2012). *Sinraptor* attained a total length of nearly 8 m.

Allosauridae is the most inclusive clade containing *Allosaurus fragilis* but not *Sinraptor dongi, Carcharodontosaurus saharicus,* and *Passer domesticus* (Sereno 2005). The best-known representative is *Allosaurus,* from the Late Jurassic (Kimmeridgian-Tithonian) of Colorado, Montana, New Mexico, Oklahoma, South Dakota, Utah, and Wyoming (Gilmore 1920; Madsen 1976; Fig. 11.11) and Portugal (Mateus et al. 2006). It has a prominent, mediolaterally compressed posterodorsal "horn" on the lacrimal, a small external mandibular fenestra, and an antarticular bone in the lower jaw (Madsen 1976; Carrano et al. 2012; Fig. 11.11B). *Allosaurus* reached a total length of at least 10 m.

Carcharodontosauria (from Greek *karkaros,* sharp-pointed, *odous* (*odon*), tooth, and *sauros,* lizard) is the most inclusive clade containing *Carcharodontosaurus saharicus* and *Neovenator salerii* but not *Allosaurus fragilis* or *Sinraptor dongi* (Benson et al. 2010). Diagnostic features for this clade include the pneumatization of the quadrate, the presence of two anterior pleurocoels on the cervical vertebrae, and the dorsomedially directed head of the femur (Carrano et al. 2012). Carcharodontosauria includes Neovenatoridae and Carcharodontosauridae.

Neovenatoridae is the most inclusive clade containing *Neovenator salerii* but not *Carcharodontosaurus saharicus, Allosaurus fragilis,* or *Sinraptor dongi* (Benson et al. 2010). It is characterized by a number of synapomorphies including the presence of large pneumatic foramina on and internal spaces in the ilium; a prominent medial shelf adjacent to acetabular notch of the ilium; and a prominent, ventrally curving anterolateral process on the lateral condyle of the tibia (Carrano et al. 2012). In addition to *Neovenator,* from the Early Cretaceous (Barremian) of England (Brusatte et al. 2008), it possibly includes Megaraptora, a group of several theropod taxa with elongate forelimbs and large manual unguals from the Early Cretaceous of Australia,

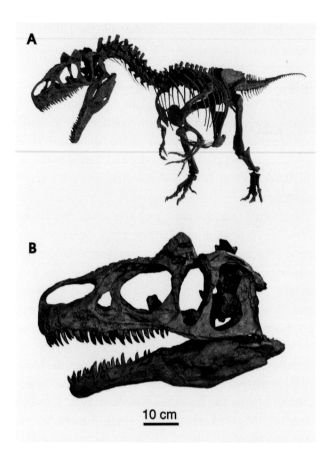

Figure 11.11. **A**, reconstructed skeleton, and **B**, skull in lateral view of the allosaurid *Allosaurus fragilis*. Total length 7.9 m. Courtesy of Black Hills Institute of Geological Research.

Japan, and Utah, and the Late Cretaceous of Argentina and possibly Inner Mongolia (China) (Carrano et al. 2012; Zanno and Makovicky 2013; Coria and Currie 2016).

Carcharodontosauridae is the most inclusive clade containing *Carcharodontosaurus saharicus* but not *Neovenator salerii*, *Allosaurus fragilis*, or *Sinraptor dongi* (Benson et al. 2010). Diagnostic synapomorphies for this clade include dorsal vertebrae with neural spines that are at least twice as tall as the corresponding centra and the peg-and-socket contact between the ilium and ischium (Carrano et al. 2012). *Acrocanthosaurus*, from the Early Cretaceous (Aptian-Albian) of Oklahoma, Texas, and Wyoming, has tall neural spines along much of its vertebral column and attained a total length of about 12 meters (Currie and Carpenter 2000; Eddy and Clarke 2011; Fig. 11.12). *Concavenator*, from the Early Cretaceous (Barremian) of Spain, has greatly elongated neural spines on the posterior two dorsal vertebrae and apparent quill

nodes for the attachment of feathers on the ulna (Ortega et al. 2010). Carcharodontosaurinae, with *Carcharodontosaurus*, from the Late Cretaceous (Cenomanian) of Algeria, Egypt, and Morocco (Stromer 1931; Sereno et al. 1996), and *Giganotosaurus*, from the Late Cretaceous (Cenomanian) of Argentina (Coria and Salgado 1995), is characterized by the distinct sculpturing on the maxilla and nasal, the presence of a lateral lamina of the nasal overhanging the antorbital fossa, and the extensively ossified interorbital septum and sphenethmoid. *Carcharodontosaurus* attained an estimated skull length of about 1.4 m (Carrano et al. 2012) and a total length of at least 12 m. Carcharodontosaurids were the dinosaurian apex predators in many communities of Cretaceous dinosaurs until the later part of the Late Cretaceous, when this ecological role was assumed by abelisaurids in the Southern Hemisphere and tyrannosauroids in the Northern Hemisphere (Brusatte et al. 2010c; Zanno and Makovicky 2013).

AVEROSTRA: TETANURAE: AVETHEROPODA: COELUROSAURIA

Coelurosauria (from Greek *koilos*, hollow, *oura*, tail, and *sauros*, lizard) is the most inclusive clade containing *Passer domesticus* but not *Allosaurus fragilis*, *Sinraptor dongi*, and *Carcharodontosaurus saharicus* (Sereno 2005). It dates back to at least the Middle Jurassic (Bathonian). A. H. Turner et al. (2012) listed a suite of synapomorphies for this clade: maxilla forming an extensive palatal shelf; prefrontal proportionately small; frontals forming a narrow anterior wedge between the nasals; quadratojugal shaped like a reversed L; crowns of the premaxillary teeth D-shaped in transverse section, with rounded labial and flat lingual sides; cervical and anterior dorsal vertebrae with amphiplatyan rather than opisthocoelous centra; a transverse groove or a fossa separating the ascending process of the astragalus from the condylar portion of that bone; and triangular obturator process confluent with the shaft of the ischium posteriorly.

The interrelationships of coelurosaurian theropods are still contentious. Brusatte et al. (2014) interpreted *Bicentenaria*, from the Late Cretaceous (Cenomanian) of Argentina (Novas et al. 2012), as the basalmost coelurosaur known. Their phylogenetic analysis found *Tugulusaurus*, from the Early Cretaceous of Xinjiang (China) (Rauhut and Xu 2005), and *Zuolong*, from the Late Jurassic (Oxford-

10 cm

Figure 11.12. Skull of the carcharodontosaurid *Acrocanthosaurus atokensis* in lateral view. Courtesy of Black Hills Institute of Geological Research.

ian) of Xinjiang (China) (Choiniere et al. 2010), as more derived than *Bicentenaria* and formed an unresolved polytomy with Tyrannosauroidea and Maniraptoriformes, which includes all remaining coelurosaurs.

TETANURAE: AVETHEROPODA: COELUROSAURIA: TYRANNOSAUROIDEA

Tyrannosauroidea (from Greek *tyrannos*, tyrant, and *sauros*, lizard) is the most inclusive clade containing *Tyrannosaurus rex* but not *Ornithomimus edmontonicus*, *Troodon formosus*, or *Velociraptor mongoliensis* (Sereno 2005). Diagnostic apomorphies for this clade include a long and slender medial process arising from the retroarticular process of the articular and the contribution of the scapula to the glenoid being markedly longer anteroposteriorly than that of the coracoid.

Brusatte et al. (2014) found *Coelurus* and *Tanycolagreus*, both from the Late Jurassic (Kimmeridgian-Tithonian) of Wyoming (Carpenter et al. 2005a,b), as a clade at the base of Tyrannosauroidea. These two taxa share apomorphies such as the division of the proximal articular surface of the ulna by a median ridge, the presence of a horizontal ridge on the anterior surface of the femoral head and neck, and the position of the medial ridge of the tibia at approximately the posteromedial corner of the distal end of this bone. *Tanycolagreus* attained a total length of about 3 m.

More derived tyrannosauroids are characterized by a suite of synapomorphies (Brusatte et al. 2014): anterior margin of the premaxilla vertical; presence of a deep foramen or fossa at the base of the nasal process of the premaxilla; premaxillary teeth smaller than the anterior maxillary teeth, with crowns that are U-shaped in transverse section; lacrimal bearing a small, conical dorsal projection; retroarticular process greatly reduced; parapophyses on the anterior dorsal vertebrae much enlarged; distal "boot" formed by the pubes markedly extended anteriorly; and proximal head of the femur dorsally inclined, with a deep trochanteric fossa on its posterior surface.

Tyrannosauroid theropods show an evolutionary trend toward increase in body size and robustness. Brusatte et al. (2010c) reviewed the paleobiology of these extensively studied predatory dinosaurs. Proceratosauridae includes the earliest known tyrannosauroids and is characterized by the long axis of the external naris being about as long as the long axis of the antorbital fenestra and the presence of a deep, extensive fossa ventral to the external naris (Rauhut et al. 2010; Brusatte et al. 2014). The proceratosaurid, *Guanlong*, from the Late Jurassic (Oxfordian) of Xinjiang (China), attained a total length of about 3 m. Its snout bears a tall, highly pneumatized crest formed by the fused nasals (Xu et al. 2006; Fig. 11.13). The length of the forelimb in *Guanlong* is only slightly less than 60 percent of that of the hind limb, proportionately much longer than in more derived tyrannosauroids. Its manus is tridactyl. *Dilong,* from the Early Cretaceous of Liaoning (China), is noteworthy for its possession of filamentous integumentary covering (Xu et al. 2004). It attained a total length of 1.6 m. Taxa such as *Xiongguanlong,* from the Early Cretaceous (Aptian-Albian) of Gansu (China) (D. Li et al. 2010); *Appalachiosaurus,* from the Late Cretaceous (Campanian) of Alabama (Carr et al. 2005); and *Timurlengia,* from the Late Cretaceous (Turonian) of Uzbekistan (Brusatte et al. 2016), represent an intermediate evolutionary grade of tyrannosauroids. *Appalachiosaurus* reached a total length of 6 m. *Yutyrannus,* from the Early Cretaceous (Barremian) of Liaoning (China), is noteworthy for the presence of extensive, dense integumentary covering of filamentous structures and apparently attained large body size (total length about 9 m) independent from the more derived Tyrannosauridae (Xu et al. 2012).

Tyrannosauridae is the least inclusive clade containing *Tyrannosaurus rex, Gorgosaurus libratus,* and *Albertosaurus sarcophagus* (Sereno 2005). It comprises the dinosaurian apex predators of the Late Cretaceous in western North America and East Asia (Brusatte et al. 2010c). Diagnostic cranial features of this clade include the distinct cornual process of the lacrimal, a pronounced lateral ridge on the postorbital ramus of the jugal at its contact with the postorbital, and the distinctly inflated jugal process of the ectopterygoid (Brusatte et al. 2014). The skull is proportionately larger and more robust than in more basal tyrannosauroids (Currie et al. 2003). The anterolaterally facing orbits suggest some degree of binocular vision (Stevens 2006). Large olfactory bulbs indicate a keen sense of smell (Zelenitsky et al. 2009). The tooth crowns are labiolingually thick (incrassate) rather than blade-like, especially in *Tyrannosaurus* and *Tarbosaurus.* Together with evidence from bite marks, their structure indicates unusually powerful biting (Bates and Falkingham 2012). The forelimbs are proportionately short. The manus has only two functional digits. Albertosaurinae, comprising *Albertosaurus,* from the Late Cretaceous (Maastrichtian) of Alberta, and *Gorgosaurus,* from the Late Cretaceous (Campanian) of Alberta), is closely related to the more derived Tyrannosaurinae (Carr 1999; Currie 2003; Currie et al. 2003). It differs from the latter in the proportionately longer hind limbs, the promaxillary fenestra of the maxilla being concealed in lateral view, the extensive dorsal expansion of the quadratojugal, and several other cranial apomorphies (Currie et al. 2003). Tyrannosaurinae is divided into two subclades. One includes *Tyrannosaurus,* from the Late Cretaceous (Maastrichtian) of Alberta, Colorado, Montana, New Mexico, North and South Dakota, Saskatchewan, Texas, and Wyoming (Brochu 2003b; Fig. 11.14), which attained a total length of nearly 13 m, and *Tarbosaurus,* from the Late Cretaceous (Maastrichtian) of China, Kazakhstan, and

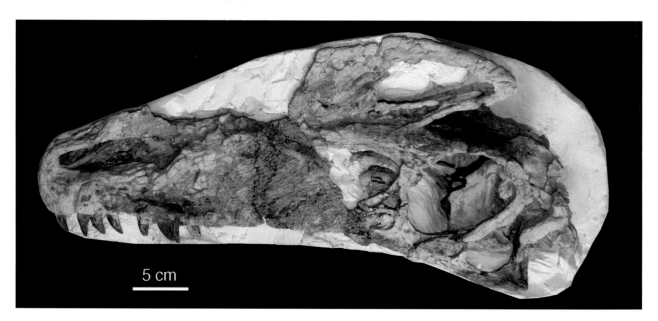

Figure 11.13. Cranium of the tyrannosauroid *Guanlong wucaii* in lateral view. Note the prominent dorsal crest. Courtesy of Jim Clark.

Mongolia (Maleev 1974; Hurum and Sabath 2003). It is characterized by a transversely broad postorbital portion of the cranium and the absence of a lacrimal "horn." The other subclade includes *Alioramus*, from the Late Cretaceous (Maastrichtian) of Mongolia (Brusatte et al. 2012), and *Qianzhousaurus*, from the Late Cretaceous (Maastrichtian) of Jiangxi (China) (Lü et al. 2014), and is distinguished by a long and dorsoventrally low skull, with the snout comprising at least two-thirds of the total skull length, and pronounced rugosities on the nasal (Lü et al. 2014). Alioramines are also smaller in overall size and have more gracile skeletons.

TETANURAE: AVETHEROPODA: COELUROSAURIA: MANIRAPTORIFORMES

Maniraptoriformes (from Latin *manus*, hand, *raptor*, robber, and *forma*, form) is the least inclusive clade containing *Passer domesticus* and *Ornithomimus edmontonicus*

A

B

Figure 11.14. **A**, reconstructed skeleton, and **B**, skull and cervical vertebrae and ribs of the tyrannosaurid *Tyrannosaurus rex*. Total length c. 12.2 m. Courtesy of Black Hills Institute of Geological Research.

20 cm

(Maryánska et al. 2002). Brusatte et al. (2014) noted that this clade is not well supported and the interrelationships of its various constituent taxa remain poorly resolved. Maniraptoriformes contains *Ornitholestes*, Compsognathidae, Ornithomimosauria, and Maniraptora.

COELUROSAURIA: MANIRAPTORIFORMES: *ORNITHOLESTES*

Ornitholestes, from the Late Jurassic (Kimmeridgian-Tithonian) of Wyoming, attained a total length of about 2 m (Carpenter et al. 2005b). Although this taxon is known from a fairly complete skeleton, its structure has yet to be documented in detail.

COELUROSAURIA: MANIRAPTORIFORMES: COMPSOGNATHIDAE

Compsognathidae is the most inclusive clade containing *Compsognathus longipes* but not *Passer domesticus* (Holtz et al. 2004). It comprises a number of mostly small-sized, lightly built forms from the Late Jurassic (Tithonian) of Germany and France (Figs. 11.15, 11.16) and the Early Cretaceous of Brazil, China, and Italy (Ostrom 1978; P.-j. Chen et al. 1998; Currie and P.-j. Chen 2001; Hwang et al. 2004; Göhlich et al. 2006; K. Peyer 2006; Chiappe and Göhlich 2010; Dal Sasso and Manganuco 2011). Compsognathidae is characterized by a suite of synapomorphies (Brusatte et al. 2014): the absence of a contact between the prefrontal and nasal; long and slender retroarticular process; cutting edges on the crowns of some maxillary and dentary teeth without serrations; anterior cervical centra extending posteriorly beyond their neural arches; slender cervical ribs; posterior dorsal vertebrae with fan-shaped neural spines; sacral vertebrae with fused neural spines; and pubis extending vertically and terminating in a distinct distal "boot" that is not extended anteriorly.

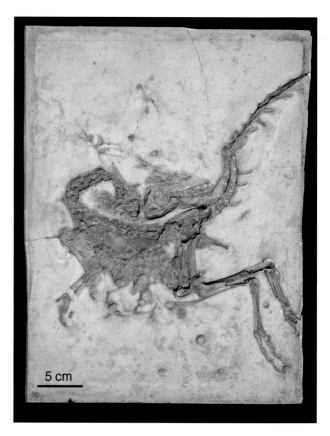

Figure 11.15. Skeleton of the compsognathid *Compsognathus longipes*. Courtesy of Bayerische Staatssammlung für Paläontologie und Geologie.

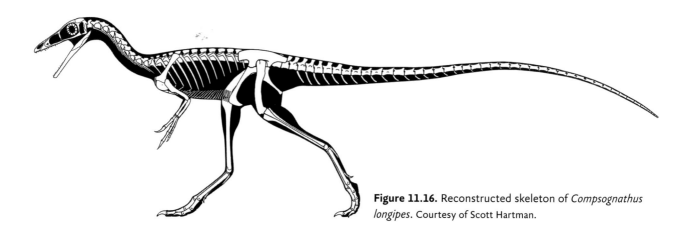

Figure 11.16. Reconstructed skeleton of *Compsognathus longipes*. Courtesy of Scott Hartman.

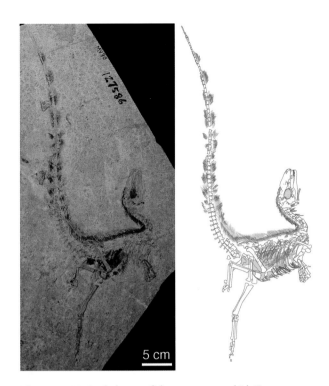

Figure 11.17. A, skeleton of the compsognathid *Sinosauropteryx prima* with "protofeathers"; **B**, outline drawing of this specimen with integumentary structures highlighted in brown. From Smithwick et al. (2017)—CC BY 4.0.

Compsognathids attained total lengths ranging from 1 to 2.5 m. At least *Sinosauropteryx*, from the Early Cretaceous (Aptian) of Liaoning (China), has a "fuzzy" integumentary covering with filamentous structures that range in length from 5 to 40 mm and apparently have short shafts with "barb" branches. It had a complex color pattern with a "bandit mask" over the eye, a body with a darker back and lighter underside, and a striped tail (Smithwick et al. 2017; Fig. 11.17); this pattern suggests life in an open habitat. Gastrointestinal contents of several compsognathid skeletons indicate that these dinosaurs preyed on a wide range of small vertebrates (Ostrom 1978; Dal Sasso and Manganuco 2011).

COELUROSAURIA: MANIRAPTORIFORMES: ORNITHOMIMOSAURIA

Ornithomimosauria (from Greek *ornis*, bird, *mimos*, actor, and *sauros*, lizard) is the most inclusive clade containing *Ornithomimus velox* but not *Allosaurus fragilis*, *Tyrannosaurus rex*, *Compsognathus longipes*, *Alvarezsaurus calvoi*, *Therizinosaurus cheloniformis*, *Deinonychus antirrho-*

pus, *Troodon formosus*, and *Passer domesticus* (Y. Lee et al. 2014). Brusatte et al. (2014) listed a number of diagnostic derived features for this clade: premaxilla with a long posterolateral process, which separates the maxilla from the nasal behind the external naris; surangular with dorsolateral flange for contact with the lateral condyle of the quadrate; length of metacarpal I about half that of metacarpal II; phalanx 1 of manual digit I longer than metacarpal II; paired flexor processes on the proximoventral surfaces of the first manual phalanges; manual unguals weakly curved, with small flexor tubercles; and proximomedial surface of the fibula bearing distinct oval fossa.

Nqwebasaurus, from the Early Cretaceous (Berriasian-Valanginian) of South Africa (Sereno 2017), and *Pelecanimimus*, from the Early Cretaceous (Barremian) of Spain (Pérez-Moreno et al. 1994), are the oldest undisputed ornithomimosaurs. *Pelecanimimus* attained a total length of 2.5 m. Its jaws hold some 220 small teeth. The tooth crowns are D-shaped in transverse section anteriorly and blade-like more posteriorly. *Pelecanimimus* had a small soft-tissue or keratinous crest at the back of the head as well as a throat pouch (Pérez-Moreno et al. 1994). More derived ornithomimosaurs lack premaxillary, maxillary, and (with a single exception) dentary teeth. The anterior end of the dentary is deflected ventrally.

Ornithomimidae is the most inclusive clade containing *Ornithomimus velox* but not *Deinocheirus mirificus* (Y. Lee et al. 2014). Ornithomimids resemble present-day ratite birds such as ostriches in their overall body proportions, especially the distally elongated, powerful hind limbs. The forelimbs are long and slender. Zelenitsky et al. (2012) and van der Reest et al. (2016) documented the presence of feathers on the forelimbs of adults and filamentous body covering on specimens of *Ornithomimus* from the Late Cretaceous (Campanian) of Alberta. Some ornithomimid skulls (Fig. 11.18) preserve remnants of a keratinous beak with vertical grooves along its lingual surface. Barrett (2005) argued that ornithomimids fed on plants. Kobayashi and Lü (2003) and Varricchio et al. (2008) inferred gregarious behavior for *Sinornithomimus*. *Ornithomimus* attained a total length of about 4 m, whereas *Gallimimus*, from the Late Cretaceous (Maastrichtian) of Mongolia, possibly reached 8 m (Osmólska et al. 1972).

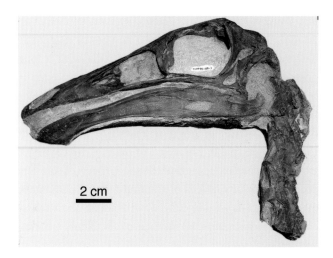

Figure 11.18. Skull and anterior neck vertebrae of the ornithomimid *Ornithomimus edmontonicus* in lateral view. Courtesy of Don Brinkman.

Deinocheiridae is the most inclusive clade containing *Deinocheirus mirificus* but not *Ornithomimus velox* (Y. Lee et al. 2014). *Deinocheirus*, from the Late Cretaceous (Maastrichtian) of Mongolia, attained a skull length of more than 1 m and a total length of 11 m. It has a long snout and a dorsoventrally deep mandible. In the postcranial skeleton, posterior dorsal and sacral neural spines are greatly elongated and vertebrae at the distal tip of the tail are fused. *Deinocheirus* has unusually long forelimbs and proportionately short hind limbs (Y. Lee et al. 2014). Gastrointestinal contents with numerous gastroliths and fish remains suggest that *Deinocheirus* was omnivorous (Y. Lee et al. 2014)

COELUROSAURIA: MANIRAPTORIFORMES: MANIRAPTORA

Maniraptora (from Latin *manus*, hand, and *raptor*, robber) is the most inclusive clade containing *Passer domesticus* but not *Ornithomimus edmontonicus* (Maryańska et al. 2002). Brusatte et al. (2014) listed numerous synapomorphies for this clade: maxilla broadly entering into the margin of the external naris; maxillary and dentary teeth without a constriction between the crown and root; frontals contacting the nasals along a transverse suture; parietals fused to each other in subadult and adult individuals; centra of the proximal caudal vertebrae boxlike; prezygapophyses on the distal caudals much reduced or absent altogether; lateral proximal carpal is triangular; well-developed semilunate carpal covering the proximal ends of the first and second metacarpals; medial portion of the furcula anteroposteriorly compressed; small notch separating the rectangular internal tuberosity from the head of the humerus; iliac blades diverging from each other posteriorly; well-defined brevis fossa on the ilium deeply concave and facing mainly ventrally and medially; and ischia not forming a median symphysis. Brusatte et al. (2014) divided Maniraptora into Alvarezsauroidea, Therizinosauroidea, Oviraptorosauria, and Paraves.

COELUROSAURIA: MANIRAPTORIFORMES: MANIRAPTORA: ALVAREZSAUROIDEA

Alvarezsauroidea (named for the Argentine historian Gregorio Alvarez and Greek *sauros*, lizard) comprises all taxa more closely related to *Alvarezsaurus calvoi* than to *Passer domesticus* (Choiniere et al. 2010). Brusatte et al. (2014) listed a suite of diagnostic derived features for this clade: the presence of a flat bony bar between the external nares; absence of basal tubera; retroarticular process with a long and slender medial process; epipophyses positioned anterior to the postzygapophyses on the cervical vertebrae; parapophyses are on distinct pedicels on the dorsal vertebrae and positioned at about the same horizontal level as the transverse processes on the posterior dorsals; ulna subequal in length to the humerus, with a robust olecranon process; metacarpal III about as long or slightly longer than metacarpal I; and manual digit I with a greatly enlarged ungual.

The oldest known alvarezsauroid is *Haplocheirus*, from the Late Jurassic (Oxfordian) of Xinjiang (China) (Choiniere et al. 2010). Unlike the more derived Alvarezsauridae, *Haplocheirus* has recurved tooth crowns with serrated cutting edges and manual digits II and III are only slightly reduced.

Alvarezsauridae is defined as the least inclusive clade containing *Alvarezsaurus calvoi* and *Mononykus olecranus* (Choiniere et al. 2010). It is known from the Late Cretaceous of Central and East Asia and the Americas (Novas 1997; Xu et al. 2013). *Mononykus*, from the Late Cretaceous (Maastrichtian) of Mongolia, has a sternum with a robust median keel (Perle et al. 1993). Its forelimb is unusually short and the bones of the carpus are fused into a single block of bone. The manus has a single, robust digit I terminating in a large ungual, which may have been used for digging. Manual digits II and III are greatly

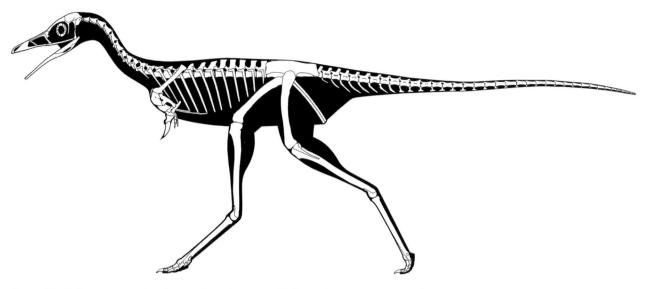

Figure 11.19. Reconstructed skeleton of the alvarezsaurid *Shuvuuia deserti*. Courtesy of Scott Hartman.

reduced. The pubis extends posteroventrally and tapers distally. The long hind limbs indicate excellent cursorial ability. The tail is proportionately short. *Mononykus* reached a length of about 1 m. At least *Shuvuuia*, from the Late Cretaceous (Campanian) of Mongolia (Fig. 11.19), had a filamentous body covering (Schweitzer et al. 1999).

COELUROSAURIA: MANIRAPTORIFORMES: MANIRAPTORA: THERIZINOSAURIA

Therizinosauria (from Greek *therizo*, reap, and *sauros*, lizard) is the most inclusive clade containing *Therizinosaurus cheloniformis* but not *Tyrannosaurus rex*, *Ornithomimus edmontonicus*, *Mononykus olecranus*, *Oviraptor philoceratops*, or *Troodon formosus* (Zanno 2010b). Zanno et al. (2009) distinguished *Falcarius*, from the Early Cretaceous (Aptian) of Utah, from a less inclusive grouping Therizinosauroidea, which includes *Beipiaosaurus inexpectus* and *Therizinosaurus cheloniformis*. The proportionately small teeth have a constriction between the denticulated crown and the root. In more derived taxa, the anterior portions of the jaws lack teeth and probably supported keratinous beaks (Lautenschlager et al. 2014). Diagnostic shared derived features of Therizinosauria include the presence of a pneumatic fossa (subotic recess) ventral to the fenestra ovalis; conjoined dentaries distinctly U-shaped in plan view; 12 or more cervical vertebrae; cervical centra with two pneumatic foramina on either side; all sacral centra with pneumatic openings; head of the humerus prominent and domed proximally; distal condyles of the humerus on the anterior surface of the bone; metacarpal I with a rectangular buttress on the ventrolateral aspect of its proximal surface; preacetabular portion of the iliac blade distinctly deflected laterally; anterior portion of the proximal surface of the fibula markedly broader mediolaterally than the posterior portion; short and slender ascending process of the astragalus covering only the lateral portion of the anterior surface of the tibia; metatarsals II–IV are not closely appressed to each other; and pes with four functional digits that appear to be more splayed than in other maniraptorans.

Therizinosaurs are readily distinguished from other theropod dinosaurs by the following combination of features: proportionately small skull; long neck; greatly elongated forelimbs ending in long, sometimes scythe-like unguals; broad abdominal and pelvic region; and less horizontal orientation of the vertebral column (Zanno 2010a,b; Fig. 11.20). The latter two suggest the possibility of a tripodal stance during feeding. Based on their dentition and overall body proportions, therizinosaurs were likely omnivorous or herbivorous.

Aside from a questionable record from strata of allegedly Early Jurassic age in Yunnan (China), therizinosauroids are known from the Early to Late Cretaceous of Central and East Asia and from the Late Cretaceous (Turonian) of New Mexico and Early to Late Cretaceous (Aptian-Turonian) of Utah (Zanno 2010b). *Beipiaosaurus* and *Jianchangosaurus*, from the Early Cretaceous

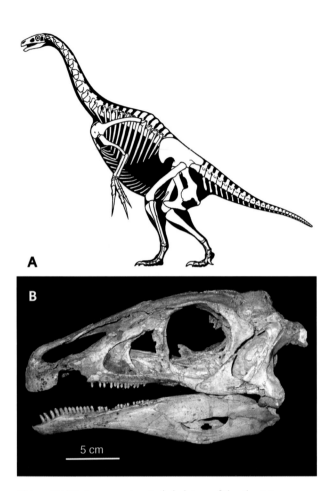

Figure 11.20. A, reconstructed skeleton of the therizinosauroid *Nothronychus graffami*; **B**, skull of the therizinosauroid *Erlikosaurus andrewsi* in lateral view. A, courtesy of Scott Hartman; B, courtesy of Stephan Lautenschlager.

(Barremian-Aptian) of Liaoning (China), have filamentous integumentary structures (Chuong et al. 2003; Xu et al. 2009; Pu et al. 2013). *Therizinosaurus*, from the Late Cretaceous (Maastrichtian) of Mongolia, has a hypertrophied deltopectoral crest on the humerus and up to 60 cm long, weakly curved manual unguals (Barsbold 1983).

COELUROSAURIA: MANIRAPTORIFORMES: MANIRAPTORA: OVIRAPTOROSAURIA

Oviraptorosauria (from Latin *ovum*, egg, Latin *raptor*, robber, and Greek *sauros*, lizard) is the most inclusive clade containing *Oviraptor philoceratops* but not *Passer domesticus* (Maryańska et al. 2002). Representatives of this clade have somewhat bird-like skulls, long forelimbs, and short tails that terminate in fused distal vertebrae (pygostyle) in some taxa. *Oviraptor*, from the Late Cretaceous

(Campanian) of Mongolia, was originally thought to have fed on the eggs of other dinosaurs because the first skeleton was discovered on top of a nest of eggs, then ascribed to the ceratopsian dinosaur *Protoceratops*. Later discoveries revealed that oviraptorids actually brooded their own eggs in a manner similar to that of many extant birds (J. M. Clark et al. 1999). Brusatte et al. (2014) listed a suite of diagnostic apomorphies for Oviraptorosauria: dentary more or less triangular in lateral view, distinctly increasing in dorsoventral height posteriorly; dorsal margin of dentary strongly convex in lateral view; dorsal margin of the symphyseal portion of the dentary downturned relative to the remainder of the dorsal margin; dentary teeth (if present) restricted to the anterior ends of the jaws; and glenoid facet of the articular positioned well below the alveolar margin of the dentary. Oviraptorosaurs were capable of fore-and-aft mandibular motion and probably were omnivorous or herbivorous. They ranged in total length from about 1 to 8 m.

The basal oviraptorosaur *Incisivosaurus*, from the Early Cretaceous (Barremian) of Liaoning (China), has incisor-like premaxillary teeth and teeth in the maxilla and dentary (Balanoff et al. 2009). *Caudipteryx*, from the Early Cretaceous (Aptian) of Liaoning, has only premaxillary teeth (Ji et al. 1998), and more derived oviraptorosaurs lack teeth altogether. *Caudipteryx* and the related *Protarchaeopteryx* have fully formed branching feathers on the forelimbs (remiges) and tail (rectrices) even though they were flightless (Ji et al. 1998; Chuong et al. 2003; Xu et al. 2010, 2014). *Avimimus*, from the Late Cretaceous (Campanian-Maastrichtian) of Mongolia, has edentulous jaws and small tooth-like bony projections on the premaxillae. Adult individuals of this taxon show extensive co-ossification in the limbs, with formation of a carpometacarpus, tibiotarsus, and tarsometatarsus (Kurzanov 1987; Funston et al. 2016).

Derived oviraptorosaurs form a clade Caenagnathoidea, which is the least inclusive clade containing *Oviraptor philoceratops* and *Caenagnathus collinsi* (Sereno 2005). In both Oviraptoridae and Caenagnathidae, at least some taxa have prominent bony cranial crests.

Oviraptoridae is defined as the least inclusive clade containing *Oviraptor philoceratops* but not *Caenagnathus collinsi* (Maryánska et al. 2002). It is known only from the Late Cretaceous of China and Mongolia. The cranium and mandible of oviraptorids such as *Citipati*, from the

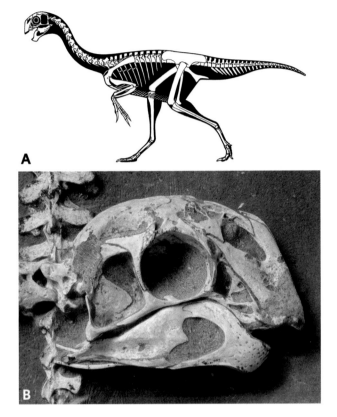

Figure 11.21. A, reconstructed skeleton of the oviraptorid *Khaan mckennai*; **B**, skull of the *Khaan mckennai* in lateral view. A, courtesy of Scott Hartman; B, courtesy of Mick Ellison.

Figure 11.22. Skull of the oviraptorid *Citipati osmolskae* in lateral view. Courtesy of Mick Ellison.

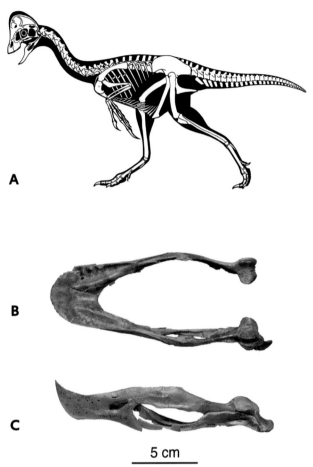

5 cm

Figure 11.23. A, reconstructed skeleton of the caenagnathid *Anzu wyliei*; **B-C**, mandible referred to the caenagnathid *Chirostenotes pergracilis* in **B**, dorsal, and **C**, lateral views. A, from Lamanna et al. (2014)—CC BY 2.5; B, courtesy of Greg Funston and Phil Currie.

Late Cretaceous (Campanian) of Mongolia, are dorso-ventrally deep but anteroposteriorly short, and the external naris extends posteriorly above the antorbital fenestra (Barsbold 1983; J. M. Clark et al. 2002; Figs. 11.21, 11.22).

Caenagnathidae is the most inclusive clade containing *Caenagnathus collinsi* but not *Oviraptor philoceratops* (Maryánska et al. 2002). It is known from the Late Cretaceous of western North America and Central and East Asia. *Anzu*, from the Late Cretaceous (Maastrichtian) of North and South Dakota, has a prominent cranial crest (Lamanna et al. 2014; Fig. 11.23A). The mandible is dorsoventrally shallow and has a long, undivided external mandibular fenestra (Currie et al. 1994; Funston and Currie 2014; Lamanna et al. 2014; Fig. 11.23B–C).

COELUROSAURIA: MANIRAPTORIFORMES: MANIRAPTORA: SCANSORIOPTERYGIDAE

Scansoriopterygidae is the least inclusive clade containing *Epidendrosaurus ningchengensis* and *Epidexipteryx hui* (F. Zhang et al. 2008). The known representatives of this clade are small: the only known subadult has a body length of about 25 cm. The short and deep skull has large, procumbent anterior teeth. *Epidexipteryx* has four long, ribbon-like tail feathers, but its limbs lack pennaceous plumage (F. Zhang et al. 2008). A diagnostic feature for Scansoriopterygidae is the slender and greatly elongated manual digit III (F. Zhang et al. 2008; Fig. 11.24). *Yi*, from the Middle or Late Jurassic (Callovian-Oxfordian) of Hebei (China), is distinguished by the possession of a distally tapering, slightly curved, rod-like element that is attached to the wrist and longer than the ulna. Xu et al. (2015) reconstructed this rod (which is composed of bone or mineralized cartilage) as supporting, along with the manual digits, a gliding membrane rather than a wing formed by feathers. *Yi* also has filamentous structures covering parts of its body. Agnolín and Novas (2013) and Brusatte et al. (2014) referred Scansoriopterygidae, from the Middle or Late Jurassic of China (F. Zhang et al. 2002, 2008; Xu et al. 2015), to Oviraptorosauria. However, other phylogenetic analyses left the relationships of these enigmatic theropods unresolved (A. H. Turner et al. 2012) or found them as basal paravians (Godefroit et al. 2013; Xu et al. 2015) or even basal avialians (F. Zhang et al. 2008).

COELUROSAURIA: MANIRAPTORIFORMES: MANIRAPTORA: PARAVES

The most diverse group of Maniraptora, Paraves (from Greek *para*, near, and Latin *avis*, bird), is the most inclusive clade containing *Passer domesticus* but not *Oviraptor philoceratops* (Holtz and Osmólska 2004). Paraves comprises Deinonychosauria (Colbert and Russell 1969) and Avialae (Gauthier 1986). However, some recent phylogenetic analyses (Godefroit et al. 2013; Foth et al. 2014) found various deinonychosaurian taxa as stem-birds, bringing the monophyly of Deinonychosauria into question.

MANIRAPTORA: PARAVES: DEINONYCHOSAURIA

Deinonychosauria (from Greek *deinos*, terror, *onyx*, claw and *sauros*, lizard) comprises the most recent common ancestor of *Velociraptor mongoliensis* and *Troodon formosus* and all descendants of that ancestor (A. H. Turner et al. 2012). A key diagnostic feature of this clade is the enlarged, sickle-shaped ungual on the hyperextendible pedal digit II (A. H. Turner et al. 2012).

MANIRAPTORA: PARAVES: DEINONYCHOSAURIA: DROMAEOSAURIDAE

Dromaeosauridae is the most inclusive clade containing *Dromaeosaurus albertensis* but not *Troodon formosus, Ornithomimus edmontonicus*, and *Passer domesticus* (Sereno 2005). Brusatte et al. (2014) listed a suite of synapomorphies for Dromaeosauridae: small anterior tympanic recess positioned dorsal to the crista interfenestralis;

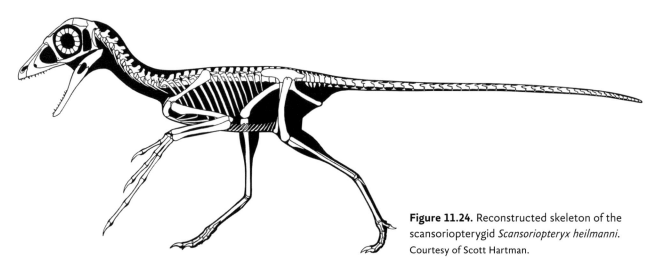

Figure 11.24. Reconstructed skeleton of the scansoriopterygid *Scansoriopteryx heilmanni*. Courtesy of Scott Hartman.

A

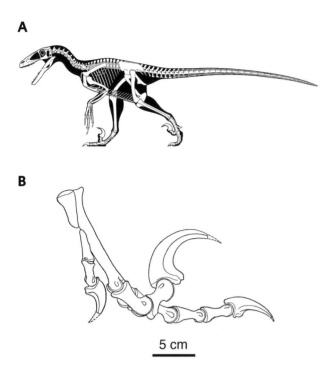

B

5 cm

Figure 11.25. A, reconstructed skeleton of the dromaeosaurid *Deinonychus antirrhopus*; **B**, left pes of *Deinonychus antirrhopus* in medial view. A, courtesy of Scott Hartman; B, modified from Ostrom (1969).

dorsal edge of the distal end of the paroccipital process twisted anterolaterally; distal end of the paroccipital process with a ventral flange; parapophyses situated on distinct pedicels on the posterior dorsal vertebrae; metatarsal II with a well-developed hinge joint facet (ginglymus) extending onto the extensor surface of the bone; penultimate and ungual phalanges of pedal digit II modified for hyperextension; and ungual of pedal digit II more strongly curved and much larger than that of pedal digit III.

Deinonychus, from the Early Cretaceous (Aptian-Albian) of Montana, Oklahoma, and Wyoming (Ostrom 1969, 1976b; D. L. Brinkman et al. 1998; Fig. 11.25A), attained a total length of up to 3 m. Its forelimb is proportionately long. The length of the manus is nearly half that of the forelimb. The carpus permitted a wide range of motion. The three long manual digits bear recurved, trenchant claws. The pes is functionally didactyl because the hyperextendible digit II bore a large ungual, which was kept off the ground during locomotion (Fig. 11.25B). A noteworthy feature of the tail is the presence of long bony anterior extensions of the prezygapophyses and chevrons

that extend forward and suggest that the tail served as a dynamic stabilizer during locomotion (Ostrom 1969).

Dromaeosauridae is definitely known from the Early Cretaceous (Barremian) to the end of the Cretaceous (Maastrichtian). This clade has been divided into Dromaeosaurinae, Halszkaraptorinae, Microraptorinae, Unenlagiinae, and Velociraptorinae.

Dromaeosaurinae includes *Dromaeosaurus*, from the Late Cretaceous (Campanian) of Alberta (Colbert and Russell 1969; Currie 1995), and *Utahraptor*, from the Early Cretaceous (Barremian) of Utah (Kirkland et al. 1993). The latter attained a total length of nearly 5 m.

Halszkaraptorinae includes *Halszkaraptor* and two other taxa of small-bodied dromaeosaurids, all from the Late Cretaceous of Mongolia (Cau et al. 2017). The premaxillae of *Halszkaraptor* form nearly one-third of the length of the dorsoventrally flattened snout, and each hold 11 teeth. The neck forms 50 percent of the snout-to-sacrum length. The ulna is flattened and has a sharp posterior margin. The ilium has a shelf-like supratrochanteric shelf. Numerous skeletal features suggest a semiaquatic mode of life and piscivory (Cau et al. 2017).

Microraptorinae includes *Microraptor,* from the Early Cretaceous (Aptian) of Liaoning (China) (Xu et al. 2003); *Sinornithosaurus*, from the Early Cretaceous (Aptian) of Liaoning (China) (Xu et al. 1999; Fig. 11.25); and *Hesperonychus*, from the Late Cretaceous (Campanian) of Alberta (Longrich and Currie 2009). Microraptorines range from 50 cm to 2 m in total length. All four limbs of *Microraptor* form wings with long pennaceous feathers, which are much like those in extant birds, and the feathers on the more distal parts of the limbs have asymmetrical vanes. The four wings appear more suitable for gliding than for powered flapping flight (Xu et al. 2003; Hone et al. 2010; Palmer 2014). *Microraptor* had dark, glossy plumage (Q. Li et al. 2012).

Unenlagiinae comprises a number of taxa from the Late Cretaceous of Argentina, of which *Buitreraptor* is the best-known representative (Agnolin and Novas 2013). *Buitreraptor* has a long and low skull, the length of which exceeds that of the femur (Makovicky et al. 2005). The tooth crowns lack serrated cutting edges; the crown is separated from the root by a constriction. The gracile manus has a greatly elongated digit II (Agnolin and Novas 2013).

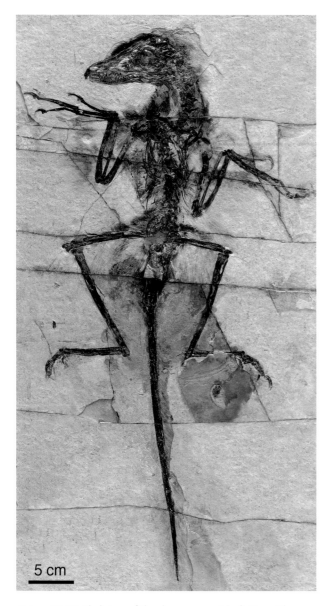

5 cm

Figure 11.26. Skeleton of the dromaeosaurid cf. *Sinornithosaurus* sp. with preserved epidermal covering of "protofeathers." Courtesy of Mick Ellison.

Figure 11.27. Skull of the dromaeosaurid *Tsaagan mangas* in lateral view. Length of skull 20.1 cm. Courtesy of Mick Ellison.

Velociraptorinae includes *Deinonychus* (Fig. 11.25) and several dromaeosaurid taxa from the Late Cretaceous of Asia (Figs. 11.26, 11.27) and Europe including *Velociraptor,* from the Late Cretaceous (Campanian) of Mongolia (Barsbold 1983; Norell and Makovicky 1997; Barsbold and Osmólska 1999; Fig. 11.27). A diagnostic feature of Velociraptorinae is the presence of pneumatic openings on all dorsal vertebrae (A. H. Turner et al. 2012). A skeleton of *Velociraptor mongoliensis* buried while fighting with an individual of the ceratopsian dinosaur *Protoceratops andrewsi* provides spectacular evidence for the predatory habits of dromaeosaurids (Barsbold 1974; Fig. 11.28). *Balaur,* from the Late Cretaceous (Maastrichtian) of Romania (Brusatte et al. 2013; Fig. 11.29), is noteworthy for the presence of enlarged unguals on pedal digits I and II. Cau et al. (2015) argued that *Balaur* is a basal avialian rather than a dromaeosaurid.

MANIRAPTORA: PARAVES: DEINONYCHOSAURIA: TROODONTIDAE AND ANCHIORNITHIDAE

Troodontidae is the most inclusive clade containing *Troodon formosus* but not *Velociraptor mongoliensis, Ornithomimus edmontonicus,* and *Passer domesticus* (Sereno 2005). It is diagnosed by a suite of synapomorphies (Brusatte et al. 2014; Fig. 11.30): maxilla with a distinct anterior process, which is set off by a concave step in the anterior margin of the element; internarial bar is flat in transverse section; quadrate distinctly inclined anteroventrally; dentary more or less triangular in lateral view, with lateral groove of neurovascular foramina along its middle and posterior portions; anterior teeth of the dentary smaller, more numerous, and more closely spaced than those in the middle of the tooth row; and proximal portion of metatarsal III strongly "pinched" between metatarsals II and IV or absent altogether.

Stenonychosaurus ("*Troodon*"), from the Late Cretaceous (Campanian) of Alaska, Alberta, Montana, North and South Dakota, and Wyoming, reached a total length of more than 2 m. It has the largest relative brain size recorded among dinosaurs excluding birds to date (Hopson 1979). Troodontids have plumage with large feathers on the fore- and hind limbs and frond-like tail feathering. At least one taxon, *Jianianhualong*, from the Early Cretaceous of Liaoning (China), has asymmetrical tail feathers (Xu et al. 2017).

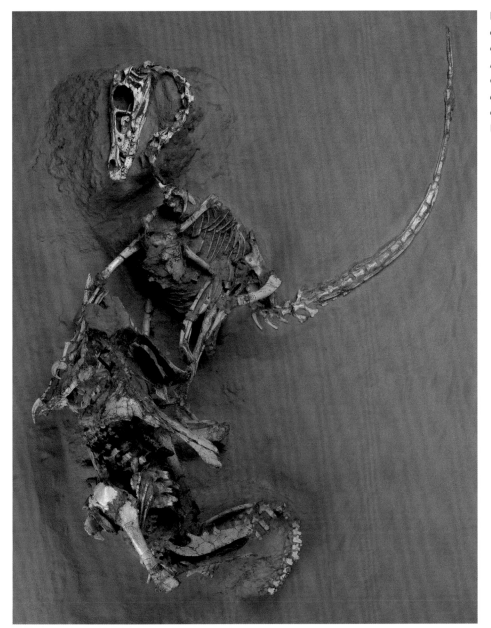

Figure 11.28. "The fighting dinosaurs"—skeletons of the dromaeosaurid *Velociraptor mongoliensis* and the protoceratopsid *Protoceratops andrewsi* buried together during predation. Courtesy of Mick Ellison.

Anchiornis, from the Late Jurassic (Oxfordian) of Liaoning (China), is particularly bird-like and apparently capable of at least gliding (Fig. 11.31). It attained a total length of only about 35 centimeters. The length of its forelimb is 80 percent that of the hind limb. Although initially referred to Troodontidae, several recent phylogenetic analyses recovered *Anchiornis* as an avialian (Godefroit et al. 2013; Foth et al. 2014; Foth and Rauhut 2017). Foth and Rauhut (2017) proposed a clade Anchiornithidae to which they also referred a fragmentary specimen previously assigned to *Archaeopteryx* from the Late Jurassic (Tithonian)

of Germany. Based on what the authors interpreted as fossilized melanosomes, Q. Li et al. (2010) reconstructed the plumage of *Anchiornis* as having a black base color, with white stripes across its wings as well as patches of black, gray, russet, and white over the head and neck. Longrich et al. (2012) observed that its wings have poorly differentiated, slender, and symmetrical feathers.

MANIRAPTORA: PARAVES: AVIALAE

Huxley (1868, 1870) first argued that birds evolved from dinosaurs. Although he marshaled compelling anatomical evidence in support of his hypothesis, other researchers thought that these similarities were either the result of convergent evolution or that birds and dinosaurs had a more remote common ancestor. Heil-

Figure 11.29. Articulated left crus and pes of the dromaeosaurid or basal avialian *Balaur bondoc* with two greatly enlarged pedal unguals. Courtesy of Mick Ellison.

mann (1926) dismissed Huxley's hypothesis because theropod dinosaurs were then (erroneously) thought to lack clavicles, which are generally interpreted as the precursor of the avian furcula. His view prevailed until Ostrom (1973, 1976a) and Bakker and Galton (1974) revived Huxley's idea. Gauthier (1986) reframed the Huxley-Ostrom hypothesis in explicitly phylogenetic terms and supported it with numerous synapomorphies. Starting in the 1990s, a spectacular array of discoveries representing an unexpected variety of bird-like dinosaurs and stem-birds, particularly from the Jurassic and Early Cretaceous of China, has beautifully confirmed the Huxley-Ostrom hypothesis (Chiappe 2007; Xu et al. 2014). Indeed, the transition from basal theropods to crown-group birds now ranks among the best-documented evolutionary transformations among vertebrates.

Gauthier (1986) distinguished crown-group birds (Aves) from the famous "Urvogel" (from German *Ur-*, original, and *Vogel*, bird) *Archaeopteryx*, from the Late Jurassic (Tithonian) of Germany (Wellnhofer 2009; Fig. 11.31). He defined Aves as comprising the most recent common ancestor of all present-day birds and all descendants of that ancestor. Gauthier and de Queiroz (2001) modified this definition of Aves as comprising the most recent common ancestor of ratites (specifier: ostrich, *Struthio camelus*), tinamous (specifier: great tinamou, *Tetrao major*), and Neognathae (specifier: Andean condor, *Vultur gryphus*), and all descendants of that ancestor. Although traditionally classified as "birds," *Archaeopteryx* and scores of other stem-birds discovered since Gauthier's study are not part of crown-group Aves. Gauthier

Figure 11.30. Skeleton of a juvenile of the troodontid *Mei long* in an apparent sleeping posture. Total length 53 cm. Courtesy of Mick Ellison.

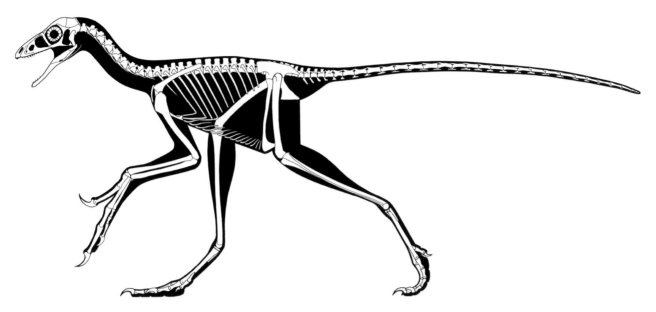

Figure 11.31. Reconstructed skeleton of the anchiornithid *Anchiornis huxleyi*. Courtesy of Scott Hartman.

(1986) united these forms with Aves in a clade Avialae (from Latin *avis*, bird, and *ala*, wing), which Maryańska et al. (2002) defined as the most inclusive clade containing *Passer domesticus* but not *Dromaeosaurus albertensis* or *Troodon formosus*. Brusatte et al. (2014) listed a number of synapomorphies for Avialae including the presence of asymmetrical, vaned feathers on the forelimb; extension of the posterior tympanic recess posterodorsal to the fenestra ovalis and confluent with this opening; 25 or fewer caudal vertebrae; and preacetabular process of the ilium markedly longer than the postacetabular one.

Archaeopteryx attained a total length of up to 50 cm and probably stood about 25 cm tall (Fig. 11.32). Some authors distinguished several genera of archaeopterygids (e.g., Elzanowski 2002), but most refer all known specimens to a single genus *Archaeopteryx* (Wellnhofer 2009). The jaws of *Archaeopteryx* hold small, widely spaced teeth with crowns that lack serrated cutting edges and are separated from the root by a constriction. The cervical vertebrae have openings that indicate the presence of an air-sac system. The shoulder girdle of *Archaeopteryx* has a long, narrow scapular blade and a short, more or less rectangular coracoid. The robust furcula is shaped like a boomerang. An ossified sternum is present in at least one specimen. The forelimbs are long, reaching the length of the hind limbs. The pelvic girdle has a more or less vertically extending pubis. Pedal digit I is short, extends posteriorly, and is positioned more proximally on the metatarsus.

Pedal digit III is the longest. Pennaceous feathers covered much of the neck, body and the proximal portions of the hind limbs and tail in *Archaeopteryx*. Eleven or 12 primary flight feathers (remiges) and at least 12 secondaries are attached to each ulna. The remiges are elongated, broad, and asymmetrical (Longrich et al. 2012). Relatively long covert feathers cover the bases of their quills. The long tail of *Archaeopteryx* has frond-like feathering with 16 or 17 pairs of feathers. Manning et al. (2013) inferred that the outer vanes and tips were probably colored black in life, whereas the inner vanes remained relatively unpigmented. *Archaeopteryx* was capable of moving on the ground as well as climbing trees (Wellnhofer 2009). Although it could probably fly, it was likely less maneuverable and agile than present-day flying birds.

Archaeopteryx has generally been interpreted as the basal avialian (Gauthier 1986; Wellnhofer 2009) although some recent studies (Godefroit et al. 2013; Foth et al. 2014; Xu and Pol 2014) have questioned that status. Recent discoveries of a considerable diversity of feathered paravians, especially from the Middle–Late Jurassic and Early Cretaceous of China, have demonstrated that various key "avian" features have a more complex phylogenetic distribution. Thus, the interrelationships of paravians in general and Avialae in particular will likely remain the subject of debate for some time to come.

One much-debated issue is the homology of the manual digits in crown-group birds and derived theropods

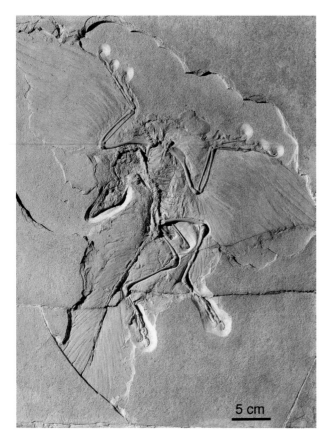

Figure 11.32. Skeleton with impressions of the plumage of the basal avialian *Archaeopteryx lithographica* (specimen in the Museum für Naturkunde, Berlin). Courtesy of Oliver Rauhut.

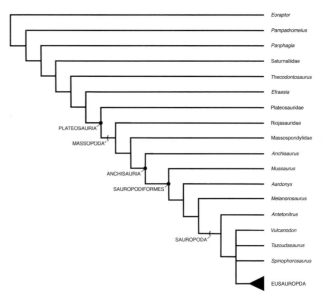

Figure 11.33. Phylogenetic hypothesis of the interrelationships of selected taxa of Sauropodomorpha. Dots denote node-based clades and parentheses denote stem-based clades. Based mainly on Otero et al. (2015).

closely related to the former (Bever et al. 2011; Xu et al. 2009, 2014). Traditionally, the manus of basal avialians was interpreted as comprising digits I, II, and III. Embryological studies on extant birds supported identification of the fingers as digits II, III, and IV. Wagner and Gauthier (1999) proposed the frame-shift hypothesis to resolve this conflict. According to this hypothesis, the embryonic condensations of mesenchyme that were originally associated with digits II, III, and IV enter the digits anterior (medial) to them. Thus, the condensations develop into digits I, II, and III. As noted above, the plesiomorphic condition for Theropoda is a manus with five digits, with greatly reduced digits IV and V along the palmar surface of the manus (e.g., *Herrerasaurus*; Sereno 1994).

For further details on the origin and early evolution of Avialae and Aves, interested readers should consult the detailed surveys by Chiappe (2007) and Mayr (2017). Xu et al. (2014) provided an overview concerning the evidence for the origin of birds.

Saurischia: Sauropodomorpha

Sereno (2007) defined Sauropodomorpha (from Greek *sauros*, lizard, *pous*, foot, and *morphe*, form) as the most inclusive clade containing *Saltasaurus loricatus* but not *Passer domesticus* or *Triceratops horridus*. Sauropodomorphs were predominantly herbivorous, although the earliest members of this clade were still carnivorous or omnivorous (Cabreira et al. 2016). Unlike many ornithischian dinosaurs (Chapter 12), they were not capable of oral processing of plant fodder but relied on lengthy retention of the food in a voluminous gastrointestinal tract to facilitate breakdown of the plant cellulose by microbial symbionts (Sander et al. 2011). Basal sauropodomorphs were small- to medium-sized and bipedal, but the larger-bodied, more derived forms were mostly quadrupedal.

Most of the Late Triassic and Early Jurassic sauropodomorph taxa were historically united in a group Prosauropoda (from Latin *pro*, before, and Sauropoda) (Huene 1932; Sereno 1999; Upchurch et al. 2007). However, most recent phylogenetic analyses have demonstrated that "prosauropods" constitute a paraphyletic assemblage of forms successively more closely related to Sauropoda (Fig. 11.33). These studies still differ considerably in the placement of individual taxa (Yates 2007a;

McPhee et al. 2015; Otero et al. 2015), which, at least in part, reflects apparently extensive homoplasy.

Otero et al. (2015) listed several synapomorphies for Sauropodomorpha: ratio of skull length to femur length less than 0.6; long axes of the tooth crowns not recurved distally; laminae on the neural arches of cervical vertebrae 4-8 weakly developed as low ridges; length of the humerus less than 55-65 percent of the length of the femur; and transverse width of the humerus more than 33 percent of the length of the humerus.

Cabreira et al. (2016) interpreted *Buriolestes*, from the Late Triassic (Carnian) of Brazil, as the basalmost sauropodomorph. R. T. Müller et al. (2018) posited that it possibly forms a distinct clade with other Carnian-age sauropodomorphs from South America. *Buriolestes* reached a total length of about 2.4 m and differs from more derived sauropodomorphs in the presence of a proportionately large skull (which is only slightly shorter than the femur) without enlarged external nares. The snout has a prominent subnarial gap. Its teeth have recurved crowns with serrated cutting edges, similar to those in theropods and indicating carnivorous habits. The ilium has a brevis fossa and a complete acetabular wall with a straight ventral margin.

Eoraptor, from the Late Triassic (Carnian) of Argentina, attained a total length of about 1 m (Sereno et al. 1993, 2013). Its cranium has a short snout, enlarged external nares, and large orbits (Fig. 11.34). The premaxillary and anterior maxillary teeth have more conical crowns, whereas the posterior maxillary teeth have recurved, blade-like crowns. The first dentary tooth is set back from the anterior end of the dentary. An unexpectedly plesiomorphic feature is the presence of two rows of small teeth on the pterygoid. The forelimb of *Eoraptor* measures almost half of the length of the hind limb. The radius and ulna are separated by a distinct gap, whereas the two bones are appressed in theropods. Manual digits I-III of *Eoraptor* bear non-trenchant claws, and manual digits IV and V are short. The proximal phalanx of manual digit I is twisted so that the tip of the ungual points medially, as in other sauropodomorphs (Fig. 11.35). The pelvis of *Eoraptor* has an open acetabulum, and the ilium has a distinct brevis fossa. Sereno et al. (1993) originally considered *Eoraptor* close to the most recent common ancestor of all dinosaurs but also listed

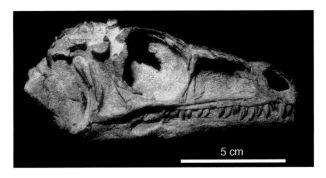

Figure 11.34. Skull of the basal sauropodomorph *Eoraptor lunensis* in lateral view. Courtesy of Sterling Nesbitt.

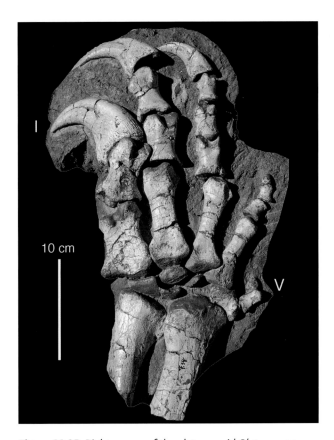

Figure 11.35. Right manus of the plateosaurid *Plateosaurus engelhardti* in dorsal view. Note the twisting of and greatly enlarged ungual on the first manual digit.

a number of features in support of a closer relationship to Theropoda. Subsequently, Martinez et al. (2011), Martínez et al. (2013c), and Sereno et al. (2013) reinterpreted *Eoraptor* as a basal sauropodomorph.

Martinez and Alcober (2009) described another early sauropodomorph, *Panphagia*, which coexisted with *Eoraptor*. It reached a total length of about 1.3 m. *Panphagia* has lanceolate tooth crowns with large, obliquely oriented serrations and a slight constriction between the crown and root. The teeth closer to the anterior end of the dentary have taller crowns than the more posterior ones. A longitudinal ridge extends on the lateral surface of the dentary below the tooth row. The external naris is large. The limb bones of *Panphagia* are hollow, as in *Eoraptor*.

The skull of *Pampadromaeus*, from the Late Triassic (Carnian) of Brazil, is longer than two-thirds of the length of the femur and closely resembles that of *Eoraptor* (Cabreira et al. 2011). The anterior teeth in the dentary and premaxilla lack distinct serrations, but the remaining teeth have leaf-shaped crowns with large marginal denticles. The acetabulum of *Pampadromaeus* has a complete bony wall. The femur has a well-developed trochanteric shelf.

Martínez et al. (2013c) interpreted *Panphagia*, *Eoraptor*, and *Pampadromaeus* as successively more closely related to more derived sauropodomorphs. Cabreira et al. (2016) placed *Buriolestes* at the base of Sauropodomorpha, followed crownward by *Eoraptor*, *Pampadromaeus*, and *Panphagia*. The latter two already have tooth crowns with large, apically angled denticles and swollen bases, as well as imbricating arrangement of the crowns in the jaws. These features are shared with predominantly herbivorous extant iguanid lizards and suggest that plants formed at least part of the diet for these basal sauropodomorphs (Button et al. 2017).

Saturnalia, from the Late Triassic (Carnian) of Brazil, attained a length of about 2 m (Langer et al. 1999; Langer 2003; Martínez et al. 2013c). The length of its skull is less than half that of the femur. The forelimb is short but robust. The hind limb has a long tibia and metatarsus. Derived features shared by *Saturnalia* with other sauropodomorphs include the proportionately small size of its skull; the leaf-shaped, denticulated tooth crowns; and the transversely broad distal end of the humerus. The acetabulum of *Saturnalia* has a partial bony wall (Langer 2003). The femur has a distinct trochanteric shelf. Metatarsal V has a slender proximal end. *Chromogisaurus*, from the Late Triassic (Carnian) of Argentina, is closely related to *Saturnalia* (Ezcurra 2010; Martínez et al. 2013c).

Thecodontosaurus (Benton et al. 2000) and *Pantydraco* (Yates 2003a; Galton and Kermack 2010), from the Late Triassic (Norian-Rhaetian) of England, are more closely related to all other sauropodomorphs than any of the aforementioned South American taxa. The skull is proportionately small. The teeth have leaf-shaped crowns with large, apically angled serrations. The femur has an elongated, ridge-like lesser trochanter but lacks the trochanteric shelf present in more basal sauropodomorphs such as *Saturnalia*. *Thecodontosaurus* reached a total length of up to about 2 m.

Efraasia, from the Late Triassic (Norian) of Germany (Galton 1973; Yates 2003b), resembles more derived sauropodomorphs in its larger body size (attaining a total length of at least 5 m), a longer neck, the contact between metacarpal I and the second distal carpal, and the first distal carpal capping both of the latter bones. However, the manus is proportionately long (about 60 percent of the combined lengths of the humerus and radius), as in *Thecodontosaurus*. Furthermore, the prefrontal of *Efraasia* is small and lacks the posterior extension along the dorsal margin of the orbit present in more derived sauropodomorphs such as *Plateosaurus* (Yates 2003b).

SAURISCHIA: SAUROPODOMORPHA: PLATEOSAURIA

Plateosauria (from Greek *platys*, broad, and *sauros*, lizard) comprises the most recent common ancestor of *Plateosaurus* and *Massospondylus* and all descendants of that ancestor (Sereno 1998). Otero et al. (2015) characterized this clade by the length of the humerus being less than 55–65 percent of the length of the femur, and the posterior margin of the postacetabular process of the ilium having a rounded posterodorsal margin and pointed ventral corner.

Plateosauridae comprises all plateosaurians more closely related to *Plateosaurus* than to *Massospondylus* (Sereno 1998). It includes *Plateosaurus*, from the Late Triassic (Norian) of France, Germany, Greenland, and Switzerland (Huene 1926, 1932; Galton 2001; Yates 2003b; Prieto-Márquez and Norell 2011; Figs. 11.35, 11.36), and *Unaysaurus*, from the Late Triassic (Norian) of Brazil (Leal et al. 2004). Otero et al. (2015) listed various synapomorphies for Plateosauridae: medial margin of the supratemporal fossa with a projection at the frontal/postorbital-parietal suture; ventral floor of the braincase "bent" in lateral view, with the basipterygoid process and the cultriform process of the parasphenoid positioned below the level of the basal tubera and occipital

A

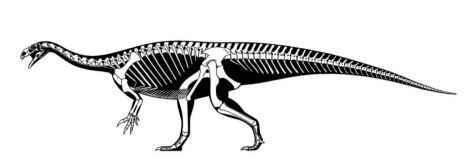

Figure 11.36. A, reconstructed skeleton of the plateosaurid *Plateosaurus engelhardti*; **B**, cranium of *Plateosaurus engelhardti*. A, courtesy of Scott Hartman; B, courtesy of Rainer Schoch.

B

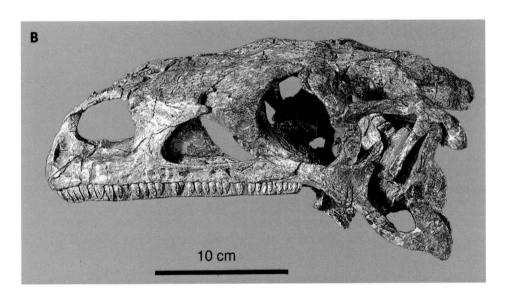

10 cm

condyle; symphyseal portion of the dentary distinctly deflected ventrally; jaw joint positioned below the level of the dentary tooth row; and centra of the proximal and middle caudal vertebrae with a longitudinal ventral groove. *Plateosaurus* attained a total length of up to 10 m. At least the adults of this taxon were bipedal (Bonnan and Senter 2007; Mallison 2010; Fig. 11.36A). *Thecodontosaurus, Efraasia*, and Plateosauria typically lack recurved teeth, and their lanceolate tooth dentary and maxillary tooth crowns overlap in the jaws and have large, apically oriented serrations (Fig. 11.36B). This, along with other cranial features, suggests that these sauropodomorphs had become more dependent on herbivory than more basal taxa (Button et al. 2017).

SAURISCHIA: SAUROPODOMORPHA: PLATEOSAURIA: MASSOPODA

Yates (2007a) defined Massopoda (from Greek *massos*, longer, and *pous*, foot) as comprising all sauropodomorphs

more closely related to *Saltasaurus loricatus* than to *Plateosaurus engelhardti*. Otero et al. (2015) provided a list of synapomorphies for this clade: antorbital fossa shorter anteroposteriorly than the orbit; anterior margin of the infratemporal fenestra extending under the posterior half of the orbit; metacarpal V nearly as broad as long, with a strongly convex proximal articular surface; length of the ungual of manual digit II less than 75 percent of that of the ungual of manual digit I; length of the hind limb exceeding that of the trunk; and fourth trochanter situated on the medial margin of the femur.

SAUROPODOMORPHA: PLATEOSAURIA: MASSOPODA: RIOJASAURIDAE

Among Massopoda, *Eucnemesaurus*, from the Late Triassic of South Africa, and *Riojasaurus*, from the Late Triassic (Norian) of Argentina, form a clade Riojasauridae, which is characterized by shared derived features in the structure of the femur such as a tall, crest-like lesser trochanter

Figure 11.37. Skull of the massospondylid *Massospondylus carinatus* in lateral view. Courtesy of Diane Scott.

5 cm

that is higher than broad in transverse section (Yates 2007a). *Riojasaurus* attained a total length of up to 11 m (Bonaparte 1972). The combined lengths of its humerus and radius equal 70 percent of the combined lengths of the femur and tibia.

SAUROPODOMORPHA: PLATEOSAURIA: MASSOPODA: MASSOSPONDYLIDAE

Massospondylidae is the most inclusive clade containing *Massospondylus carinatus* but not *Plateosaurus engelhardti* or *Saltasaurus loricatus* (Sereno 2007). Otero et al. (2015) listed a number of synapomorphies for this clade: dorsal surface of the snout with a depression behind the external nares; web of bone connecting the anterior and ventral rami of the lacrimal, concealing the posterodorsal corner of the antorbital fossa; length of at least cervical vertebra 4 or 5 exceeding four times the anterior height of the centrum; and femur with an asymmetrical fourth trochanter with a distinct distal corner and with a distal slope that is steeper than the proximal slope. Otero et al. (2015) distinguished two subclades among Massospondylidae, one with *Massospondylus*, from the Early Jurassic (Hettangian-Sinemurian) of Lesotho, South Africa, and Zimbabwe (Fig. 11.37), and the other including *Lufengosaurus*, from the Early Jurassic of Yunnan (China) (C.-c. Young 1947, 1951; Barrett et al. 2005a). *Massospondylus* attained a length of up to 6 m (Cooper 1981; Reisz et al. 2010; Chapelle and Choiniere 2018). It has a long neck and

tail and a slender trunk. Bonnan and Senter (2007) argued that adults of *Massospondylus* were bipedal, but tracks found near nests of this dinosaur indicate that the juveniles were quadrupedal (Reisz et al. 2010).

SAUROPODOMORPHA: PLATEOSAURIA: MASSOPODA: *ANCHISAURUS*

Anchisaurus, from the Early Jurassic (Pliensbachian) of Connecticut, was long placed with basal sauropodomorphs such as *Thecodontosaurus* based on its relatively small total length (2 to 3 m) and proportionally slender limbs (Huene 1932; Galton 1976). However, Yates (2004) noted that *Anchisaurus* shares a number of derived features with sauropods, such as the wrinkled enamel on the tooth crowns, the small size of the antorbital fossa, the dorsal position of the cultriform process of the parasphenoid, and the relatively distal position of the fourth trochanter on the femur. He defined a clade Anchisauria as comprising the most recent common ancestor of *Anchisaurus* and *Melanorosaurus* and all descendants of that ancestor.

SAUROPODOMORPHA: PLATEOSAURIA: MASSOPODA: SAUROPODIFORMES

Otero et al. (2015) placed *Anchisaurus* as the sister taxon to a clade Sauropodiformes, which Sereno (2007) defined as comprising the most recent common ancestor of *Mussaurus patagonicus* and *Saltasaurus loricatus* and all descendants of that ancestor. Sauropodiformes (from Greek

Figure 11.38. Reconstructed skeleton of the basal sauropodiform *Melanorosaurus readi*. Courtesy of Scott Hartman.

sauros, lizard, and *pous*, foot, and Latin *forma*, form) is diagnosed by several synapomorphies including the presence of a posterodistal tubercle on the radius; length of manual digit I greater than that of manual digit II; long axis of the femur weakly bent in lateral view with an offset of less than 10 degrees; and length of the ungual of pedal digit I being greater than that of all the nonterminal phalanges (Otero et al. 2015).

The basal sauropodiform *Mussaurus*, from the Late Triassic (Norian) of Argentina, was originally described on the basis of hatchlings but now is known from a growth series (Otero and Pol 2013). Adults possibly attained a total length of up to 3 m.

Aardonyx, from the Early Jurassic (Hettangian-Sinemurian) of South Africa (Yates et al. 2010), appears to have been predominantly bipedal, but features of the ulna and femur suggest the incipient development of a quadrupedal gait. Metatarsal I is rather robust compared to those of more basal sauropodomorphs and implies that the weight-bearing axis of the pes had already shifted to a more medial position (Yates et al. 2010).

Various large-bodied quadrupedal sauropodiforms from the Late Triassic and Early Jurassic have been interpreted as basal members of Sauropoda (Yates and Kitching 2003; Upchurch et al. 2007; Allain and Aquesbi 2008; Otero et al. 2015) or as successive sister taxa of that clade (McPhee et al. 2014, 2015; Y. Wang et al. 2017). One taxon, *Melanorosaurus*, from the Late Triassic of South Africa (Fig. 11.38), superficially resembles that of *Plateosaurus* in the proportions of its skull. Other cranial features, however, such as the inflection at the base of the dorsal process of the premaxilla, the formation of the posterior margin of the external naris by the maxilla, and the absence of the posterolateral process of the palatine, are derived features shared with Sauropoda (Yates 2007b).

Melanorosaurus also shares postcranial apomorphies with Sauropoda: sacrum composed of four or more vertebrae; ulna with a large anterolateral process on its proximal end, which forms a deep fossa for the reception of the proximal head of the radius; shift in the position of the radius to a more anteromedial position relative to the ulna; and femur with a straight shaft. The articular contact between the radius and ulna facilitated rotation of the manus so that its palm faced backward and thus the hand could swing in the direction of locomotion when the dinosaur moved quadrupedally (Bonnan and Yates 2007). *Melanorosaurus* reached a total length of about 8 m.

Antetonitrus, from the Late Triassic of South Africa (Yates and Kiching 2003; McPhee et al. 2014), has apparently long forelimbs that may have still been used for grasping. The shaft of its femur appears to have an elliptical rather than subcircular outline in transverse section, resembling the condition in sauropods. *Antetonitrus* attained a total length between 8 and 10 m (Yates and Kitching 2003).

SAUROPODIFORMES SAUROPODA

Sauropod dinosaurs ranged in time from the Late Triassic (Norian) to the Late Cretaceous (Maastrichtian) and are known from all continents except Antarctica. They include the largest land animals of all time, some attaining total lengths of more than 30 m and live weights of at least 70 tons.

Salgado et al. (1997) defined Sauropoda (from Greek *sauros*, lizard, and *pous*, foot) as the clade comprising the most recent common ancestor of *Vulcanodon karibaensis* and Eusauropoda (Fig. 11.39) and all descendants of that ancestor. Wilson (2002) provided a list of synapomorphies for Sauropoda: sacrum with four or more vertebrae; transverse processes of the anterior caudal vertebrae

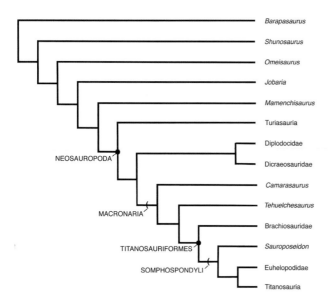

Figure 11.39. Phylogenetic hypothesis of the interrelationships of selected taxa of Sauropoda. Dots denote node-based clades and parentheses denote stem-based clades. Based mainly on Wilson and Sereno (1998) and D'Emic (2012).

deep and extending from the centrum to the neural arch; columnar limbs; humerus with a low deltopectoral crest; olecranon process of the ulna reduced or absent; metacarpals increasingly forming a tubular structure; ilium with low ischial peduncle; distal shaft of the ischium blade-like; femur with a crest- or ridge-like fourth trochanter; astragalus lacking a fossa at the base of its ascending process; third and fourth distal tarsals not ossified; proximal condyles of metatarsals I and V more or less equal in size to those of metatarsals II and IV; and ungual of pedal digit I longer than metatarsal I, deep, and mediolaterally narrow ("sickle-shaped").

Sauropods have have extensive pneumaticity of the vertebral column, especially the cervicals, which often have deep lateral recesses separated by narrow bony struts (Wedel 2005, 2009; Fig. 11.40). This pneumatization indicates the presence of a bird-like respiratory system and would have considerably reduced the weight of the long neck. Barrett and Upchurch (2007) and Sander et al. (2011) considered the often greatly elongated neck a key feature of the sauropod body plan. It would have considerably increased the range over which the animal could feed (feeding envelope) without having to move, and thus facilitated energetically more efficient foraging.

The proportionately small size of the head and rather simple teeth in sauropods rule out extensive oral food processing. Presumably these large-bodied herbivores relied on lengthy retention of plant fodder in an enormous gastrointestinal tract to facilitate fermentative breakdown by symbiotic microorganisms of the cellulose that would have constituted the bulk of their fodder (Sander et al. 2011). This would have allowed feeding even on high-fiber plant material with low nutritional quality. Increases in body size would have led to increased gut capacity. Sauropods have often been informally divided into "broad-crowned" and "narrow-crowned" groups based on the dimensions of the tooth crowns. The former, which include taxa like *Camarasaurus*, also have typically robust skulls and were capable of generating greater bite forces. Their teeth met in interdigitating occlusion. Button et al. (2017) inferred a diverse diet for this group. The second group, which includes *Diplodocus* and titanosaurs (although both differ in many detailed features), combines slender teeth with rather gracile mandibles and reduced cranial chambers for the adductor jaw muscles. They have high rates of tooth replacement, which suggest abrasive or gritty fodder (D'Emic et al. 2013). Button et al. (2017) argued that these sauropods probably employed little oral processing except for cropping.

Sauropods adopted a quadrupedal posture with robust, columnar limbs that facilitated weight support (Carrano 2005). The manus is relatively small and was held in a digitigrade posture. The metacarpals were nearly vertically aligned and tightly bound to form a tubular structure. By contrast, the pes had a spreading configuration with the metatarsals oriented more horizontally and supported by a large fleshy pad. Carrano (2005) noted that this represents a modification from the digitigrade condition with four weight-bearing pedal digits in basal sauropodomorphs and basal theropods.

Sauropod dinosaurs laid a fairly large number of eggs during each breeding season (Sander et al. 2011). The hatchlings apparently received little if any parental care. Thus, parental investment was low compared to that in extant large herbivorous mammals. Data from bone histology indicate that sauropods grew rapidly, increasing their body masses by up to 2 tons per year and attaining sexual maturity at ages between 15 and 30 years (Lehman and Woodward 2008; Griebeler et al. 2013). These growth rates suggest that sauropods had high basal metabolic rates (Sander et al. 2011).

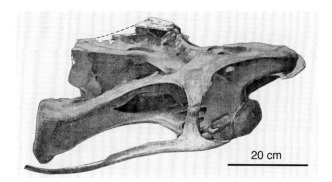

Figure 11.40. Cervical vertebra 5 (with articulated cervical rib) of the brachiosaurid *Giraffatitan brancai* in lateral view. Note the extensive pneumatization of the cervical (including the large pleurocoel). From Janensch (1914).

Hallett and Wedel (2016) have provided an excellent survey of the biology and evolution of sauropod dinosaurs to which interested readers are referred for further details.

Isanosaurus, from the Late Triassic (Norian) of Thailand (Buffetaut et al. 2000), is the oldest known sauropod, but little has been published about its skeletal structure to date. It has distinctly opisthocoelous cervical vertebrae, and the dorsal vertebrae bear tall neural spines.

Vulcanodon, from the Early Jurassic (Toarcian) of Zimbabwe (Cooper 1984), is a large-sized form with a femur length of 1.1 m. The femur is straight and distinctly flattened anteroposteriorly. Pedal digit V is enlarged relative to the condition in more basal sauropodomorphs. The combined lengths of humerus, ulna, and metacarpal III in *Vulcanodon* amount to 78 percent of the combined lengths of femur, tibia, and metatarsal III. The ulna has a pronounced anterolateral process and a deep fossa for the reception of the proximal head of the radius. What is known of the manus suggests that it was weight-bearing (Upchurch et al. 2007). This inference is supported by the closely related *Tazoudasaurus*, from the Early Jurassic (Toarcian) of Morocco, which was probably digitigrade and has a manus with a spreading configuration and a reduced number of phalanges (Allain and Aquesbi 2008). Allain and Aquesbi (2008) proposed a clade Vulcanodontidae for the reception of *Vulcanodon* and *Tazoudasaurus*, but other analyses (e.g., Remes et al. 2009) have not recovered this grouping. *Spinophorosaurus*, from the Jurassic of Niger (Remes et al. 2009), is closely related to but more derived than *Tazoudasaurus* and has spike-bearing osteoderms that were probably attached to the distal end of the tail.

SAUROPODA: EUSAUROPODA

Eusauropoda (from Greek *eu*, true, *sauros*, lizard, and *pous*, foot) comprises the most recent common ancestor of *Barapasaurus tagorei* and Neosauropoda and all descendants of this ancestor (Salgado et al. 1997; Fig. 11.39). Allain and Aquesbi (2008; see also Wilson 2002) listed a number of shared derived features for this clade: anteroposterior length of the frontal less than its transverse width; quadratojugal with a long anterior process; fossa present on the posterior surface of the quadrate; anterior portions of the tooth rows are U-shaped; manual digit II with a reduced number of phalanges; pubic symphysis S-shaped; tibia with a laterally projecting cnemial crest; length of metatarsal III is less than 25 percent of that of the tibia; metatarsus with a spreading configuration; nonterminal phalanges of the pedal digits short; pedal digits II and III with sickle-shaped unguals; and ungual on pedal digit IV small or absent. In many eusauropods, the upper and lower teeth came into occlusal contact, resulting in well-defined wear facets on the tooth crowns (Wilson and Sereno 1998).

The interrelationships of basal eusauropods are still poorly understood. Wilson (2002) interpreted *Shunosaurus*, from the Late Jurassic (Oxfordian) of Sichuan (China) (J. Wang et al. 2018), as a basal member of this clade. It attained a total length of 10 to 11 m (Y. Zhang 1988). The teeth of *Shunosaurus* have spatulate crowns with small apical denticles. The neck has 12 short opisthocoelous vertebrae. At the distal end of the tail, at least three large vertebrae and two dermal spines are fused into a tail club.

Barapasaurus, from the ?Early Jurassic of India (Jain et al. 1975; Bandyopadhyay et al. 2010), and *Patagosaurus*, from the Middle Jurassic (Aalenian-Bathonian) of Argentina (Bonaparte 1986), share various apomorphies with more derived eusauropods, such as the presence of complex bony laminae on the cervical and dorsal neural arches, the fusion of the distal ends of the sacral ribs (forming a sacricostal yoke; Wilson and Sereno 1998), a broad acromion process on the scapula, and laterally deflected pedal unguals (Wilson 2002). The better-known *Patagosaurus* also shares additional derived features with more derived eusauropods such as a depression surrounding the external naris, pneumatic excavations (pleurocoels) on the centra of the presacral vertebrae (Fig. 11.39), and a sacrum composed of five vertebrae (Bonaparte 1986).

Mamenchisauridae, with *Mamenchisaurus*, from the Late Jurassic (Oxfordian-Tithonian) of Sichuan and Xinjiang (China), and *Omeisaurus*, from the Late Jurassic (Oxfordian) of Sichuan, is characterized by a proportionately greatly elongated neck with at least 15 cervical vertebrae. In one species of *Mamenchisaurus*, the neck reached a length of 9.3 m (C.-c. Young and Zhao 1972). The middle cervical centra are more than four times as long as they are tall at their posterior articular surfaces and bear low neural arches (Upchurch 1998; Wilson 2002). *Mamenchisaurus* and *Omeisaurus* both attained total lengths of at least 15 m.

Cetiosaurus, from the Middle Jurassic (Bajocian-Bathonian) of England, has 13 rather short cervical vertebrae with undivided neural spines and simple pleurocoels and caudals with amphicoelous centra that lack pleurocoels (Upchurch and J. Martin 2002). The posterior cervical and anterior dorsal vertebrae have symmetrical and pyramid-shaped neural spines. Upchurch and J. Martin (2012) found *Cetiosaurus* outside Neosauropoda.

Royo-Torres et al. (2006) proposed a new clade of eusauropods, Turiasauria, for *Turiasaurus* and *Losillasaurus*, from the Jurassic-Cretaceous (Kimmeridgian-Berriasian) of Spain. *Zby*, from the Late Jurassic (Kimmeridgian) of Portugal (Mateus et al. 2014), and *Mierasaurus* and *Moabosaurus*, both from the Early Cretaceous (Berriasian-Aptian) of Utah (Royo-Torres et al. 2017), were also referred to Turiasauria. The heart-shaped, spatulate tooth crowns of turiasaurians bear small marginal denticles or lack them altogether. The proximal end of the radius is strongly flattened anteroposteriorly (Mateus 2014). *Turiasaurus* attained great size: one humerus is nearly 1.8 m long. Phylogenetic analyses by Royo-Torres et al. (2006, 2017) and Mateus et al. (2014) found Turiasauria outside of Neosauropoda.

Jobaria, from the Jurassic of Niger (Sereno et al. 1999), shares various apomorphies with more derived sauropods (Neosauropoda) including the absence of a contact between the jugal and ectopterygoid, metacarpals in broad proximal contact with each other and forming a U-shaped combined proximal articular surface, and a broad anterior end of the pelvis, with the distance between the preacetabular processes of the ilia greater than the anteroposterior length of these bones (Wilson 2002). The spatulate tooth crowns bear marginal denticles. *Jobaria* has only 12 cervicals. It retains clavicles and gastralia. *Jobaria* attained a total length of 18 m (Sereno et al. 1999).

SAUROPODA: EUSAUROPODA: NEOSAUROPODA

Neosauropoda (from Greek *neos*, new, *sauros*, lizard, and *pous*, foot) comprises the most recent common ancestor of Diplodocoidea (specifier: *Diplodocus*) and Macronaria (specifier: *Saltasaurus*) and all descendants of that ancestor (Wilson and Sereno 1998). Whitlock (2011) listed various diagnostic derived features for this clade: anterior rostral portion of the cranium with a preantorbital fenestra; anterior end of the lower temporal fenestra extending anterior to the anterior margin of the orbit; tooth crowns without mesial and distal denticles; chevrons lacking a proximal bony bridge; and carpus comprising two or fewer bones.

SAUROPODA: EUSAUROPODA: NEOSAUROPODA: DIPLODOCOIDEA

Diplodocoidea (from Greek *diploos*, twofold, and *dokos*, beam; in reference to the shape of the chevrons) comprises all neosauropods more closely related to *Diplodocus* than to *Saltasaurus* (Sereno 1998). D'Emic (2011) enumerated several synapomorphies for this clade such as tooth crowns slender, not overlapping, cervical ribs short and not overlapping more posterior cervical centra, and fibular facet of the astragalus posterolaterally facing. Wilson (2002) divided Diplodocoidea into Rebbachisauridae and a clade comprising Diplodocidae and Dicraeosauridae. Harris and Dodson (2004) named the latter Flagellicaudata (from Latin *flagellum*, whip, and *cauda*, tail) and defined it as comprising the most recent common ancestor of *Dicraeosaurus* and *Diplodocus* and all descendants of that ancestor. Whitlock (2011) added additional taxa of neosauropods to Diplodocoidea, but the placement of *Amazonsaurus*, from the Early Cretaceous (Aptian-Albian) of Brazil (Carvalho et al. 2003), remains uncertain (Wilson and Allain 2015). He listed a suite of synapomorphies for a less inclusive grouping comprising Rebbachisauridae and Flagellicaudata: premaxilla without a clear separation between the main body and the posteriorly directed ascending process; subnarial foramen not evident in lateral view; external nares positioned high up on the cranium between the orbits and facing dorsally; antorbital fenestra approximately equal in length to the greatest diameter of the orbit; quadrate strongly inclined posterodorsally; anterior margin of the snout squared or blunt; tooth rows restricted to the region anterior to the antorbital fenestra; four or more replacement teeth in each alveolus; anterior caudal vertebrae with wing-like transverse processes; mid-caudals at least twice as long as tall; and caudals forming the distal end of the tail at least five times longer than tall.

NEOSAUROPODA: DIPLODOCOIDEA: FLAGELLICAUDATA: DIPLODOCIDAE

Diplodocidae comprises all diplodocoids more closely related to *Diplodocus* than to *Dicraeosaurus* (Sereno 1998). Whitlock (2011) provided a number of diagnostic derived features for this clade: no contact between the quadratojugal and squamosal; lower jaw lacking a coronoid eminence; no direct crown-to-crown tooth occlusion; cervical centra with a longitudinal groove on the ventral surface; neural spines on the anterior caudal vertebrae rectangular in end view; distal portion of the tail

composed of more than 30 biconvex caudal vertebrae; and proximoventral corner of pedal phalanx I-1 drawn out and underlying metatarsal I. *Diplodocus* (including *Seismosaurus*), from the Late Jurassic (Kimmeridgian-Tithonian) of Colorado, New Mexico, Utah, and Wyoming (Hatcher 1901; Holland 1906; Berman and McIntosh 1978: Lucas et al. 2006), attained a total length of 25 to more than 30 m ("*Seismosaurus*") (Fig. 11.41). It has 13 to 15 elongate cervical vertebrae, and its long, whip-like tail includes up to 80 vertebrae. The alveolar edge of the maxilla of *Diplodocus* forms a thin sheet of bone that concealed the dorsal margin of a bony sheet formed by the dentary when the jaws were closed. The upper teeth are considerably larger than the lower ones. The slender, procumbent tooth crowns are elliptical in transverse section. *Diplodocus* is closely related to *Apatosaurus*, from the Late Jurassic (Kimmeridgian-Tithonian) of Colorado, Oklahoma, Utah, and Wyoming (Gilmore 1936; Berman and McIntosh 1978; Tschopp et al. 2015), which differs particularly in its more robust skeleton and relatively shorter cervical vertebrae. *Apatosaurus* attained a total length of about 23 m.

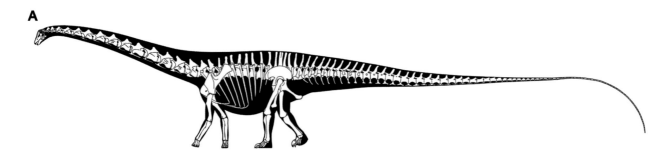

Figure 11.41. **A**, reconstructed skeleton of the diplodocid *Diplodocus carnegii*; **B**, skull of a diplodocid in lateral view. Length of skull 61.5 cm. A, courtesy of Scott Hartman; B, courtesy of Department of Paleobiology, National Museum of Natural History.

NEOSAUROPODA: DIPLODOCOIDEA: FLAGELLICAUDATA: DICRAEOSAURIDAE

Dicraeosauridae comprises all diplodocoids more closely related to *Dicraeosaurus* than to *Diplodocus* (Sereno 1998). This clade is characterized by several synapomorphies (Whitlock 2011): basal tubera transversely narrower than the occipital condyle; dentary with a lateral tuberosity near the symphysis; cervical centra with a longitudinal ridge on the ventral surface; neural spines of the posterior cervical and anterior dorsal vertebrae bifurcated and transversely narrow; dorsal centra lacking pneumatic openings; and humerus with a pronounced proximolateral corner. *Dicraeosaurus*, from the Late Jurassic (Kimmeridgian-Tithonian) of Tanzania (Janensch 1935-36, 1961), has a short neck with 12 cervical vertebrae. The tall neural spines of its middle and posterior cervicals extend anterodorsally. The cervical centra bear a distinct median keel on the anteroventral surface. *Dicraeosaurus* attained a total length of 12 m. The closely related *Amargasaurus*, from the Early Cretaceous of Argentina, has greatly elongated, deeply bifurcated neural spines on the more anterior cervical vertebrae and reached a total length of 10 m (Salgado and Bonaparte 1991). *Brachytrachelopan*, from the Late Jurassic (Tithonian) of Argentina, differs from other known dicraeosaurids in the presence of a proportionately very short neck (Rauhut et al. 2005).

NEOSAUROPODA: DIPLODOCOIDEA: REBBACHISAURIDAE

Rebbachisauridae comprises all diplodocoids more closely related to *Rebbachisaurus garasbae* than to *Diplodocus longus* (Salgado et al. 2004). It is diagnosed by the presence of a bony lamina between the diapophysis and neural spine on the anterior caudal vertebrae, the absence of a hyposphenal ridge on the anterior caudals, and cylindrical midcaudal centra with a flat ventral margin (Wilson and Allain 2015). Most members of this clade also have steeply dorsally inclined transverse processes on the dorsal vertebrae. *Rebbachisaurus*, from the Late Cretaceous (Cenomanian) of Morocco, has very tall neural arches (Wilson and Allain 2015). The distinctive, lightly built skull of *Nigersaurus*, from the Early Cretaceous (Aptian-Albian) of Niger (Sereno et al. 1999, 2007a), has a laterally expanded snout with transverse rows of slender-crowned

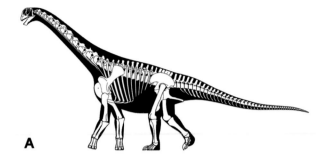

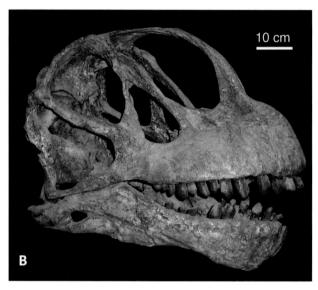

Figure 11.42. A, reconstructed skeleton of the camarasaurid *Camarasaurus lentus*; **B**, skull of *Camarasaurus lentus* in lateral view. A, courtesy of Scott Hartman; B, courtesy of Department of Paleobiology, National Museum of Natural History.

teeth, which are restricted to the anterior ends of the upper and lower jaws. Each individual alveolus contains a vertical column of as many as 10 teeth (Sereno et al. 2007a). *Nigersaurus* attained a total length of about 10 m.

NEOSAUROPODA: MACRONARIA

Macronaria (from Greek *makros*, long, and Latin *naris*, nostril) comprises all neosauropods more closely related to *Saltasaurus* than to *Diplodocus* (Wilson and Sereno 1998). Wilson (2002) listed several diagnostic synapomorphies for this clade such as the greatest diameter of the external naris exceeding the greatest diameter of the orbit, surangular forming the coronoid eminence of the lower jaw, 17 or fewer teeth in the dentary, and posterior dorsal vertebrae with opisthocoelous centra.

Camarasaurus, from the Late Jurassic (Kimmeridgian-Tithonian) of Colorado, New Mexico, Utah, and Wyoming (Osborn and Mook 1921; Gilmore 1925a; Madsen et al. 1995; McIntosh et al. 1996; Fig. 11.42), has mesiodistally broad, spatulate tooth crowns that lack denticles. Its skull is short and tall (Fig. 11.42B). The long axis of the lacrimal is aligned anterodorsally, the squamosal contacts the quadratojugal, and a groove extends anteroventrally from the surangular foramen to the ventral margin of the dentary (Wilson and Sereno 1998). The neck is proportionately short. The neural spines of middle and posterior cervical and anterior dorsal vertebrae are deeply divided. *Camarasaurus* reached a total length of up to 23 m.

Europasaurus, from the Late Jurassic (Kimmeridgian) of Germany (Sander et al. 2006), is distinguished by its small adult body size (with a total length of up to 6.2 m), which is probably an adaptation to life on an island. Its frontal has a very deep orbital rim, and there is no contact between the maxilla and quadratojugal. The cervical vertebrae have distinct pre- and postspinal laminae. The acromion on the scapula has a well-developed posterior projection. Carballido and Sander (2014) recovered this taxon as a sister taxon of Titanosauriformes.

NEOSAUROPODA: MACRONARIA: TITANOSAURIFORMES

Titanosauriformes (from Greek *titan*, a mythical giant, *sauros*, lizard, and Latin *forma*, form) comprises the most recent common ancestor of *Brachiosaurus* and *Saltasaurus* and all descendants of that ancestor (Salgado et al. 1997; Wilson and Sereno 1998). D'Emic (2012) listed a suite of shared derived features for this clade: tooth crowns not overlapping each other; presacral vertebrae with branching, centimeter- or decimeter-sized spaces that do not fully penetrate the centra; middle cervical centra more than three times as long as tall at their posterior articular surfaces; dorsal ribs containing pneumatic cavities; length of the humerus equal to 85 percent to 95 percent of the length of the femur; distal condyle of metacarpal I undivided and aligned perpendicular to the axis of the shaft; preacetabular process of the ilium semicircular in side view and flaring laterally at an angle of at least 45 degrees to the sagittal plane; pubis at least three times as long as the contact between the pubis and ischium; and ventrolateral surface of the ischium bearing a raised tubercle.

Tehuelchesaurus, from the Late Jurassic (Tithonian) of Argentina (Rich et al. 1999), shares with Titanosauri-

formes the presence of a lateral bulge on the shaft of the femur, the proximal third of which is deflected medially, and "plank-like" anterior dorsal ribs that are at least three times broader anteroposteriorly than thick transversely (D'Emic 2012).

NEOSAUROPODA: MACRONARIA: TITANOSAURIFORMES: BRACHIOSAURIDAE

Brachiosauridae comprises all titanosauriforms more closely related to *Brachiosaurus* than to *Saltasaurus* (Sereno 1998). It is characterized by various synapomorphies including the distance separating the supratemporal fenestrae being less than the long diameter of the fenestra; the presence of a triangular ventral projection on the anterior process of the quadratojugal; middle and posterior dorsal vertebrae with long, rod-like transverse processes; and ischium with a short pubic peduncle (D'Emic 2012). *Giraffatitan*, from the Late Jurassic (Kimmeridgian-Tithonian) of Tanzania (Janensch 1914, 1935–36, 1950, 1961; M. P. Taylor 2009), attained a total length of about 26 meters and a height of 12 meters with the long neck held nearly vertically (Fig. 11.43A). The tall skull has a proportionately long snout. A steeply arched bony bar separates the large external nares dorsally (Fig. 11.43B). The teeth are slightly spatulate; a few denticles are present at the apex of unworn tooth crowns. Although *Giraffatitan* has only 12 or 13 cervicals, the individual vertebrae are elongated. The centra of the posterior dorsal vertebrae are about twice as broad as tall. The forelimb of *Giraffatitan* is proportionately long. Its humerus bears a large deltopectoral crest. The pubis bears a distinct tubercle for the origin of the ambiens muscle. *Giraffatitan* is closely related to *Brachiosaurus*, from the Late Jurassic (Kimmeridgian-Tithonian) of Colorado and Utah (Riggs 1901).

NEOSAUROPODA: MACRONARIA: TITANOSAURIFORMES: SOMPHOSPONDYLI

Among Titanosauriformes, Wilson and Sereno (1998) defined a clade Somphospondyli (from Greek *somphos*, spongy, and *spondylos*, vertebra) for all titanosauriforms more closely related to *Saltasaurus* than to *Brachiosaurus*. Synapomorphies for this group include presacral vertebrae with minute spaces pervading the entire vertebra; presence of a prespinal lamina on the posterior cervical and anterior dorsal vertebrae; a medially beveled scapular

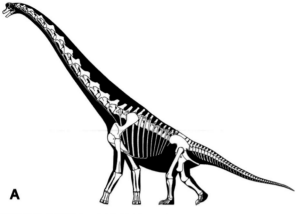

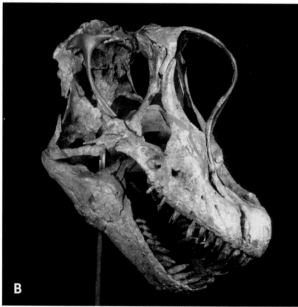

Figure 11.43. **A**, reconstructed skeleton of the brachiosaurid *Giraffatitan brancai*; **B**, reconstructed skull of *Giraffatitan brancai* in oblique lateral view. Length of skull 77 cm. A, courtesy of Scott Hartman.

glenoid; and fourth trochanter of the femur forming a subtle bulge (D'Emic 2012). *Sauroposeidon*, from the Early Cretaceous (Aptian-Albian) of Oklahoma, Texas, and Wyoming (Wedel et al. 2000), has enormously elongated middle cervical vertebrae, the largest one of which has a length of 1.4 m. D'Emic and Foreman (2012) interpreted it as a basal somphospondylian.

TITANOSAURIFORMES: SOMPHOSPONDYLI: EUHELOPODIDAE

Euhelopodidae comprises all neosauropods more closely related to *Euhelopus zdanskyi* than to *Neuquensaurus aus-*

tralis (D'Emic 2012). It is distinguished by the presence of bifid neural spines on the middle cervical vertebrae and the presence of a thick, more or less vertically extending bony lamina between the epipophysis and the prezygapophysis on the cervicals (D'Emic 2012). The best-known representative is *Euhelopus*, from the Early Cretaceous of Shandong (China) (Wiman 1929; Wilson and Upchurch 2009), which attained a total length of 15 m.

TITANOSAURIFORMES: SOMPHOSPONDYLI: TITANOSAURIA

Titanosauria (from Greek *titan*, a mythical giant, and *sauros*, lizard) comprises the most recent common ancestor of *Andesaurus delgadoi* and *Saltasaurus loricatus* and all descendants of that ancestor (Bonaparte and Coria 1993). D'Emic (2012) listed various diagnostic synapomorphies for this clade including centra of the anterior and middle caudal vertebrae bearing a longitudinal hollow on the ventral surface and a plate-like ischium without an emargination distal to its pubic peduncle. Titanosauria attained its greatest diversity in the Southern Hemisphere during the Late Cretaceous and includes the largest land animals of all time. The still incompletely known *Argentinosaurus*, from the Late Cretaceous (Cenomanian) of Argentina (Bonaparte and Coria 1993), has dorsal centra with diameters of up to 50 cm. One dorsal vertebra has a height of 1.59 m, and a referred femur has a minimum circumference of 1.18 m. *Argentinosaurus* probably attained a total length of more than 30 m and a live weight of as much as 70 tons (Mazzetta et al. 2004). Related taxa such as *Patagotitan*, from the Early Cretaceous (Albian) of Argentina, attained comparable size (Carballido et al. 2017). *Dreadnoughtus*, from the Late Cretaceous (Campanian-Maastrichtian) of Argentina (Lacovara et al. 2014), possibly reached similar dimensions when fully grown.

Titanosaurians differ from other neosauropods in various features of the appendicular skeleton (Carrano 2005). The fore- and hind limbs are short relative to the trunk region (Fig. 11.44A). The shoulder joint suggests increased mobility and flexibility of the forelimb relative to the condition in other sauropods. The distal articular surface of the humerus is long anteroposteriorly, indicating increased extension and flexion at the elbow joint. The manus is greatly shortened and has lost the carpus and phalanges (or replaced them by cartilage) (Fig. 11.44B). The

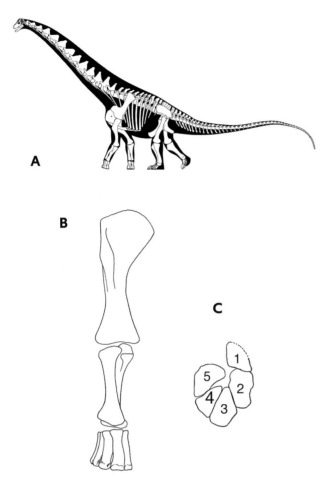

Figure 11.45. Skull of the lithostrotian *Sarmientosaurus musacchioi* in lateral view. Combined from illustrations in R. D. F. Martínez et al. (2016)—CC BY 2.5.

Figure 11.44. A, reconstructed skeleton of the lithostrotian *Alamosaurus sanjuanensis*; **B**, right forelimb of *Alamosaurus sanjuanensis* in anterior view. Length of humerus 1.36 m. **C**, proximal view of the articulated metacarpals. A, courtesy of Scott Hartman; B–C, modified from Gilmore (1946).

sacrum of titanosaurians is particularly wide transversely. The femur is canted laterally away from the body, and its distal articular surface is angled correspondingly. Its midshaft is distinctly more elliptical in transverse section. The femoral features indicate that the feet were planted far apart, creating "wide-gauge" trackways (Carrano 2005).

TITANOSAURIFORMES: SOMPHOSPONDYLI: TITANOSAURIA: LITHOSTROTIA

Lithostrotia (from Greek *lithostros*, inlaid with stones; in reference to the dermal armor) comprises the most recent common ancestor of *Malawisaurus dixeyi* and *Saltasaurus loricatus* and all descendants of that ancestor (Upchurch et al. 2004; Figs. 11.44, 11.45). D'Emic (2012) provided a suite of diagnostic derived features

for this clade: paroccipital process with a ventral, non-articulating process; tooth crowns slender; middle and posterior dorsal vertebrae with a single prespinal lamina; neural arches of the posterior dorsals without hyposphene-hypantrum accessory articulations; centra of the anterior caudal vertebrae procoelous; coracoid proximodistally long; length of the sternal plate more than 70 percent of that of the humerus; ischial margin of the acetabulum strongly embayed; and presence of often large osteoderms.

The basal lithostrotian *Malawisaurus*, from the Early Cretaceous (Aptian) of Malawi, has a short premaxilla, a ventrally curved dentary, cervical vertebrae with undivided pneumatic openings, and caudal vertebrae with short neural spines (Gomani 2005; D'Emic 2012).

Saltasauridae comprises the most recent common ancestor of *Opisthocoelicaudia* and *Saltasaurus* and all descendants of that ancestor (Sereno 1998). D'Emic (2012) listed a suite of synapomorphies for this clade including: 35 or fewer caudal vertebrae; anterior caudals with a tubercle on the dorsal margin of the prezygapophyses; deltopectoral crest of the humerus distally expanded; length of the humerus less than 80 percent of that of the femur; carpus unossified; and distal condyles of the femur dorsomedially beveled at 10 degrees. This clade includes *Alamosaurus*, from the Late Cretaceous (Maastrichtian) of

New Mexico, Texas, and Utah (Gilmore 1946; Fowler and R. M. Sullivan 2011; Fig. 11.44), and *Saltasaurus*, from the Late Cretaceous (Campanian-Maastrichtian) of Argentina (Bonaparte and Powell 1980). Based on recent discoveries, *Alamosaurus* apparently attain gigantic size comparable to that of the titanosaurian taxa from Argentina (Fowler and R. M. Sullivan 2011). By contrast, *Saltasaurus* reached a total length of 12 m. It has extensive dermal armor composed of large, oval osteoderms, some of which bear ridges or low spines, as well as smaller bony plates. *Opisthocoelicaudia*, from the Late Cretaceous (Maastrichtian) of Mongolia (Borsuk-Białynicka 1977), has distinctly opisthocoelous caudal vertebrae and co-ossified ischia and pubes. *Nemegtosaurus*, originally based only on a skull from the same formation and geographic area (Wilson 2005), is possibly congeneric with *Opisthocoelicaudia*, which is known only from postcranial remains. Its skull has a long snout with slender teeth. As in other lithostrotians (Fig. 11.45), the external nares are positioned high up on the dorsal surface of the cranium just anterior to the orbits. The basisphenoid contacts the quadrate laterally.

A Dinosaurian Enigma: *Chilesaurus*

Novas et al. (2015) reported an unusual new dinosaur, *Chilesaurus*, from the Late Jurassic (Tithonian) of Chile, and interpreted it as a basal tetanuran theropod. However, Baron and Barrett (2017) recovered it as a basal ornithischian dinosaur in their phylogenetic analysis. Until the skeletal structure of *Chilesaurus* has been documented in detail, its relationships among Dinosauria must remain uncertain. Attaining a total length of up to 3.2 m, *Chilesaurus* has slightly procumbent teeth with leaf-shaped crowns that bear small apical denticles. The deep and short premaxilla has a rugose anterior edge, which may have supported a small beak in life. The short and deep dentary of *Chilesaurus* has a downturned symphyseal portion. Only manual digits I and II are well developed. The pubis is rod-like and extends posteroventrally. The pes is tetradactyl with a large digit I.

Dinosauria II: Ornithischia

The second major clade of Dinosauria, Ornithischia, comprises all dinosaurs more closely related to *Triceratops horridus* than to *Passer domesticus* or *Saltasaurus loricatus* (Sereno 2005; Fig. 12.1). It likely dates back to the Late Triassic, but ornithischian dinosaurs became more common and diverse only in the Early Jurassic (Barrett et al. 2014). They subsequently diversified rapidly and included the dominant large-bodied herbivores in the Late Cretaceous continental ecosystems of the Northern Hemisphere. Along with saurischians other than birds, ornithischians became extinct at the end of the Cretaceous.

Seeley (1887) used the distinctive configuration of the pelvic girdle, with its postero-ventrally extending (retroverted) pubis (superficially resembling the condition in birds) and typically with an anterior (prepubic) process on the pubis, as the key diagnostic feature for Ornithischia (Fig. 12.2). Butler et al. (2008) and Nesbitt (2011) provided lists of synapomorphies for this clade: maxilla with a lateral (labial) emargination that is separated from the ventral margin of the antorbital fossa so that its tooth row is inset from the side of the snout (as is the dentary tooth row); a median predentary bone contacting, but usually is not firmly attached to, the dentaries and forming the mandibular symphysis; lower jaw with a dorsally expanded coronoid process; anterodorsal margin of coronoid process formed by the posterodorsal process of the dentary; short external mandibular fenestra; crowns of the maxillary and dentary teeth typically apicobasally short, more or less triangular in outline, and mesiodistally expanded above the roots; tooth crowns tallest at the middle of the tooth row; crowns of successive maxillary and dentary teeth overlapping to some extent labially or lingually; well-developed planar facets extending across multiple teeth in the maxilla and dentary; ilium with a long anterior (preacetabular) process, which is shorter than the posterior (postacetabular) process of the ilium; pubis extending posteroventrally parallel to the ischium and usually with a distinct prepubic process; anterior trochanter of the femur forming a steep margin with the shaft and separated from the latter by a distinct cleft; distal portion of the tibia with a posterolateral flange extending well behind the fibula; and calcaneum transversely narrow. Additional potential synapomorphies for Ornithischia include the presence of palpebral or supraorbital bones in the orbit (Maidment and Porro 2010) and the presence of few if any gastralia. Finally, many ornithischians have ossified spinal tendons, which may have stiffened the dorsal and caudal regions of the vertebral column (Organ 2006). Unlike saurischian dinosaurs, ornithischians lack pneumaticity of the axial skeleton and thus presumably air-sac-driven lung ventilation. However, Brocklehurst et al. (2018) noted that the structure of the ribcage indicates the lungs were dorsally immobilized in these dinosaurs.

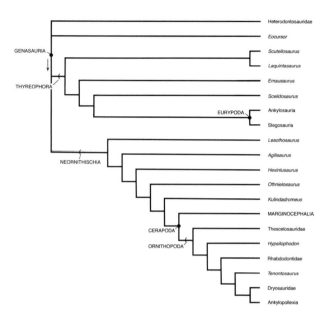

Figure 12.1. Phylogenetic hypothesis of the interrelationships of Ornithischia. Dots denote node-based clades and parentheses denote stem-based clades. Based on Butler et al. (2008).

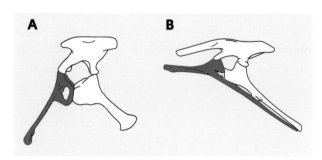

Figure 12.2. Pelvis (with pubis highlighted in blue) of **A**, *Plateosaurus engelhardti*, and **B**, *Thescelosaurus neglectus* in lateral view to illustrate the traditional distinction between Saurischia and Ornithischia. A, modified from Huene (1926) and B, modified from Romer (1927).

Ornithischians were predominantly if not exclusively herbivorous. Unlike sauropodomorph saurischians, however, they employed often extensive oral processing of food. Two ornithischian clades independently evolved complex dentitions for the breakdown of plant fodder. The predentary is loosely attached to the dentaries and, in many taxa, facilitated long-axis rotation of the latter (Nabavizadeh and Weishampel 2016). In life, a keratinous beak covered the premaxillae and the predentary, forming an effective cropping or gathering device. In some ornithischians, the beak is transversely narrow, whereas in others it is broad, suggesting differences in the mode of food gathering.

Pisanosaurus, from the Late Triassic (Norian) of Argentina, has long been interpreted as the oldest ornithischian dinosaur (e.g., Sereno 1991b; Irmis et al. 2007). It is known only from a poorly preserved partial skeleton. Agnolín and Rozadilla (2018) reinterpreted *Pisanosaurus* as a silesaurid, but it differs from other known silesaurids in features such as the shape of the dentary. Its phylogenetic position remains uncertain.

The oldest known undisputed ornithischian, *Eocursor*, possibly from the Late Triassic of South Africa, attained a total length of about 1 m (Butler 2010). Its teeth have apicobasally low, triangular crowns. The manus is proportionately large. The ilium of *Eocursor* has a long preacetabular process. The pubis has a short prepubic process. The distal portion of the hind limb is elongated.

Heterodontosauridae is the most inclusive clade containing *Heterodontosaurus tucki* but not *Parasaurolophus walkeri*, *Pachycephalosaurus wyomingensis*, *Triceratops horridus*, or *Ankylosaurus magniventris* (Sereno 2012). It ranges in time from the Early Jurassic (Hettangian-Sinemurian) to the Early Cretaceous (Berriasian) and is known from Argentina, Arizona, China, England, Lesotho, and South Africa (Norman et al. 2011; Sereno 2012). *Heterodontosaurus*, from the Early Jurassic (Hettangian-Sinemurian) of South Africa, attained a total length of slightly more than 1 m (Fig. 12.3A). Its premaxilla holds two small incisor-like teeth and a large caniniform tooth. The maxilla and dentary each have up to 12 tightly arranged, high-crowned "cheek" teeth with asymmetrical enamel covering (Fig. 12.3B). These teeth occluded with those in the opposing jaw, resulting in extensive wear. The lower jaw has a prominent coronoid process. The dentaries were capable of some rotation about their long axes to facilitate transverse chewing movements. Additional shared derived features of the skull of *Heterodontosaurus* and its close relatives include: posterior extension of the premaxilla to the contact between the prefrontal and lacrimal; presence of a prominent diastema and a lateral recess between the premaxilla and maxilla (for the reception of the caniniform tooth in the dentary); jugal with a laterally directed, horn-like process; and position of the jaw joint well below the plane of the occlusal contact between the maxillary and dentary teeth (Norman et al. 2011; Sereno 2012). The forelimb of *Heterodontosaurus* is proportionately long (Fig. 12.3A). The humerus bears a prominent deltopectoral crest. The proportionately large manus has long penultimate phalanges and trenchant unguals on digits I-III. The bones forming the distal portion of the hind limb are extensively co-ossified.

Tianyulong, from the Late Jurassic (Oxfordian; Y. Liu et al. 2012) of Liaoning (China), is noteworthy for the presence of long, unbranched filaments forming an integumentary covering below the neck and along the dorsal surfaces of the trunk and tail (Zheng et al. 2009). Although these structures resemble the "protofeathers" in coelurosaurian theropods (Chapter 11), it is uncertain whether they are homologous to the latter. *Fruitadens*,

from the Late Jurassic (Tithonian) of Colorado, is noteworthy for its small size, reaching an estimated total length of only 65 to 75 cm (Butler et al. 2010).

Dinosauria: Ornithischia: Genasauria

Genasauria (from Greek *genys*, cheek, and *sauros*, lizard) comprises the most recent common ancestor of *Ankylosaurus magniventris*, *Stegosaurus stenops*, *Parasaurolophus walkeri*, *Triceratops horridus*, and *Pachycephalosaurus wyomingensis*, and all descendants of that ancestor (Butler et al. 2008). Butler et al. (2008) listed a number of synapomorphies for this clade: dentary symphysis spout-shaped; crowns of the premaxillary teeth expanded above the roots; medial surfaces of the maxilla and dentary bear "special" foramina related to tooth replacement; length of the manus less than 40 percent the combined lengths of

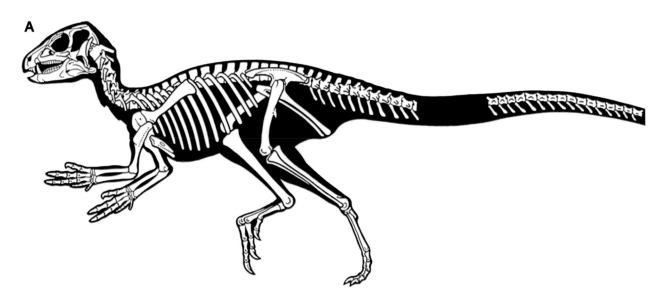

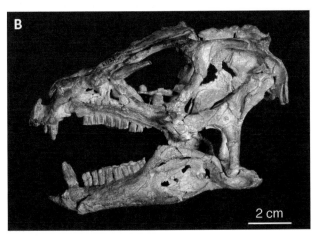

2 cm

Figure 12.3. A, reconstructed skeleton of the heterodontosaurid *Heterodontosaurus tucki*. Total length c. 1 m. **B**, skull of *Heterodontosaurus tucki* in lateral view. Note the caniniforms and the maxillary and posterior dentary "cheek" teeth. A, from Sereno (2012)—CC BY 3.0; B, courtesy of the late Alan Charig.

humerus and radius; penultimate phalanges of manual digits II and III shorter than first phalanges; and manual unguals not distinctly curved.

ORNITHISCHIA: GENASAURIA: THYREOPHORA

Thyreophora (from Greek *thyreos*, a large, four-cornered shield, and *phoro*, carry) comprises all genasaurs more closely related to *Ankylosaurus magniventris* than to *Parasaurolophus walkeri*, *Triceratops horridus*, and *Pachycephalosaurus wyomingensis* (Butler et al. 2008). This clade is especially characterized by the presence of extensive dermal armor, notably the presence of parasagittal rows of dorsal osteoderms. In the cranium, the anterior portion of the jugal is broader mediolaterally than deep dorsoventrally.

Diagnostic features for *Scutellosaurus*, from the Early Jurassic (Pliensbachian) of Arizona (Colbert 1981), include distinct medial flanges on the preacetabular process of the ilium and a long tail comprising at least 58 vertebrae (Butler et al. 2008). Attaining a total length of about 1.2 m, it was at least facultatively bipedal. *Laquintasaurus*, from the Early Jurassic (Hettangian) of Venezuela (Barrett et al. 2014), probably reached a total length of about 1 m and is distinguished by the shape of and ridging on its tooth crowns. No detailed anatomical account has been published to date, but Baron et al. (2017a,b) placed *Laquintasaurus* as the sister taxon of *Scutellosaurus*.

Emausaurus, from the Early Jurassic (Toarcian) of Germany (Haubold 1990), has a large triangular palpebral bone with a ridged, thick lateral margin. As in *Scutellosaurus* (Sereno 1986), the frontal still forms the dorsal margin of the orbit.

ORNITHISCHIA: GENASAURIA: THYREOPHORA: THYREOPHOROIDEA

Scelidosaurus, from the Early Jurassic (Sinemurian) of England (Owen 1863), forms a clade Thyreophoroidea with more derived thyreophorans. Diagnostic features for Thyreophoroidea are the formation of the dorsal rim of the orbit by a supraorbital bone and anterior caudal vertebrae with neural spines that are taller than the corresponding centra (Maidment et al. 2008). *Scelidosaurus* attained a total length of up to 4 m and has extensive dermal armor comprising multiple rows of keeled, more or less oval osteoderms. Adult individuals, at least, were likely quadrupedal (Maidment and Barrett 2014).

THYREOPHORA: THYREOPHOROIDEA: EURYPODA

Eurypoda (from Greek *eurys*, broad, and *pous*, foot) comprises the most recent common ancestor of *Stegosaurus stenops* and *Ankylosaurus magniventris* and all descendants of that ancestor (Butler et al. 2008). Shared derived features for this clade include the absence of the lateral ramus on the quadrate (so that the shaft and the pterygoid flange extend in a single plane), the length of the preacetabular process of the ilium being more than 50 percent of the total length of the bone, and a ridge-like fourth trochanter of the femur (Thompson et al. 2012).

THYREOPHORA: THYREOPHOROIDEA: EURYPODA: STEGOSAURIA

Stegosauria (from Greek *stegos*, roof, and *sauros*, lizard) comprises all eurypodans more closely related to *Stegosaurus stenops* than to *Ankylosaurus magniventris* (Butler et al. 2008). Maidment et al. (2008) listed various synapomorphies for this clade: two parasagittal rows of spine- or plate-like dorsal osteoderms extending dorsally from the neck to the distal end of the tail; humerus with a distinct tubercle for the triceps brachii muscle and an associated ridge posterolateral to its deltopectoral crest; and fourth trochanter of the femur indistinct or absent altogether in mature individuals. Some stegosaurs, such as *Kentrosaurus*, from the Late Jurassic (Kimmeridgian-Tithonian) of Tanzania (E. Hennig 1925), also bear a pair of shoulder spines. Stegosauria ranged in time from the Middle Jurassic to the Early Cretaceous and is known from Argentina, China, England, France, Portugal, South Africa, Tanzania, and the western United States.

Maidment et al. (2008) interpreted *Gigantspinosaurus*, from the Late Jurassic of Sichuan (China), as the basalmost stegosaur. All other stegosaurian taxa share the presence of a parallel-sided scapular blade and an expanded posterior end of the pubis. In *Stegosaurus*, from the Late Jurassic (Kimmeridgian-Tithonian) of Colorado, Utah, Wyoming, and Portugal, the anterior osteoderms are broad, more or less triangular plates that are arranged in two alternating rows along the neck and back, whereas

A

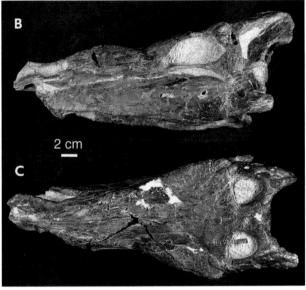

2 cm

B

C

Figure 12.4. A, reconstructed skeleton of the stegosaurian *Stegosaurus stenops* (throat armor omitted); **B–C**, skull of *Stegosaurus stenops* in **B**, lateral, and **C**, dorsal views. A, courtesy of Scott Hartman; B-C, courtesy of Department of Paleobiology, National Museum of Natural History.

the ones near the distal end of the tail form two pairs of posterolaterally directed, conical bony spikes (Gilmore 1914; Sereno and Dong 1992; Maidment et al. 2008, 2015; Fig. 12.4). The function of the large dorsal osteoderms has been the subject of much speculation. Although they presumably served primarily for species recognition (Padian and Horner 2011), the heavily vascularized large plates of *Stegosaurus* could also have facilitated passive heat loss (de Buffrénil et al. 1986: Farlow et al. 2010). The osteoderms were not attached to the vertebral column and were covered by an outer layer of keratin in life. The paired spines near the distal end of the tail in *Stegosaurus* could have been used for defense if the tail swung from side to side. *Stegosaurus* attained a total length of up to 9 m. *Miragaia*, from the Late Jurassic (Kimmeridgian-Tithonian) of Portugal, differs from other

known stegosaurs in the elongation of its neck, which comprises at least 17 vertebrae (Mateus et al. 2009). Stegosaurs were quadrupedal herbivores that likely foraged close to the ground.

THYREOPHORA: THYREOPHOROIDEA: EURYPODA: ANKYLOSAURIA

Ankylosauria (from Greek *ankylos*, curved, crooked, and *sauros*, lizard) comprises all eurypodans more closely related to *Ankylosaurus magniventris* than to *Stegosaurus stenops* (Butler et al. 2008). It ranged in time from the Middle Jurassic (Callovian) to the end of the Cretaceous (Maastrichtian) and is known from all continents including Antarctica. Ankylosauria is distinguished by substantial dorsal and lateral dermal armor formed by parasagittal rows of osteoderms (which were covered by keratin) and a highly modified cranial structure that involves fusion of the dermal ossifications to the underlying bones. Thompson et al. (2012) listed a number of synapomorphies for this clade: skull lacking upper temporal and external mandibular fenestrae; sutures between the cranial bones in adult individuals closed through remodeling of the cranial bones or through fusion of the dermal ossifications with the latter; quadratojugal and postorbital/squamosal each bearing a horn-like protuberance; osteoderm ttached to the mandibular ramus; posterior dorsal vertebrae fused into a bony rod; posterior hemal arches shaped like an inverted T in lateral view; and pubis contributing less than 25 percent of the acetabulum. Some ankylosaurs have complex nasal passageways, but the phylogenetic distribution of these features remains to be established (Witmer and Ridgely 2008). Ankylosaurs were heavily built, quadrupedal plant-eaters that foraged close to the ground. The broad, barrel-shaped trunk indicates a capacious gut that could have facilitated fermentative digestion of plant fodder. The proportionately small teeth have leaf-shaped crowns bearing large marginal denticles. Rybczynski and Vickaryous (2001) observed dental occlusion in the ankylosaurid *Euoplocephalus* and reconstructed jaw motion of this ankylosaur as involving backward motion of the mandible.

Ankylosauria has traditionally been divided into Nodosauridae (Figs. 12.5, 12.6) and Ankylosauridae (Figs. 12.7-12.9) (Coombs 1978). Most phylogenetic analyses have supported this classification (Vickaryous et al.

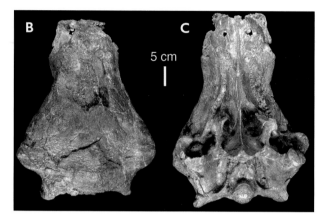

Figure 12.5. A, reconstructed skeleton of the nodosaurid *Edmontonia longiceps*; **B-C**, cranium of *Edmontonia rugosidens* in **B**, dorsal, and **C**, palatal views. A, courtesy of Scott Hartman; B-C, courtesy of Department of Paleobiology, National Museum of Natural History.

2004; Thompson et al. 2012; Arbour and Currie 2015a; C. M. Brown et al. 2017). The most conspicuous difference between Nodosauridae and Ankylosauridae is the presence of a tail club in the latter (Fig. 12.9). In ankylosaurids, the posterior caudal vertebrae (which are sometimes fused to each other) have elongated pre- and postzygapophyses and hemal arches and make up the "handle" for a large bony club at the distal end of the tail. This tail club is formed by a series of osteoderms that are fused to each other and the associated vertebrae and evolved after the development of the "handle" (Arbour and Currie 2015b). Presumably ankylosaurids employed lateral swings of the club-bearing tail for defense.

EURYPODA: ANKYLOSAURIA: NODOSAURIDAE

Nodosauridae comprises all ankylosaurs more closely related to *Panoplosaurus* than to *Ankylosaurus* (Sereno 1986). Aside from the absence of a tail club, it is characterized by the presence of a distinct notch or change in slope between the head of the femur and the greater trochanter (Thompson et al. 2012). Scheyer and Sander (2002) observed that nodosaurid osteoderms have a distinctive microstructure, with an outer layer (cortex) com-

Figure 12.6. Skull and dorsal armor of the nodosaurid *Borealopelta markmitchelli* in dorsal view. From C. M. Brown (2017)—CC BY 4.0.

posed of complex structural fibers arranged at an angle of 45 degrees to each other and overlying a thick zone of trabecular bone. Nodosaurids are known from Argentina, Antarctica, Canada, China, various regions of Europe, Japan, and the United States (Arbour and Currie 2015a; C. M. Brown et al. 2017). *Panoplosaurus*, from the Late Cretaceous (Campanian) of Alberta, and *Edmontonia*, from the Late Cretaceous (Campanian-Maastrichtian) of Alaska, Alberta, Montana, and South Dakota (Fig. 12.5), are representative taxa from western North America (Coombs 1978). Each attained a total length of up to 6 m.

EURYPODA: ANKYLOSAURIA: ANKYLOSAURIDAE

Ankylosauridae comprises all ankylosaurs closer to *Ankylosaurus* than to *Panoplosaurus* (Sereno 1986). The transversely broad cranium is distinguished by the lower

Figure 12.7. Cranium of the ankylosaurid *Ankylosaurus magniventris* in **A**, dorsal, and **B**, lateral views. From Arbour and Mallon (2017)—CC BY 4.0.

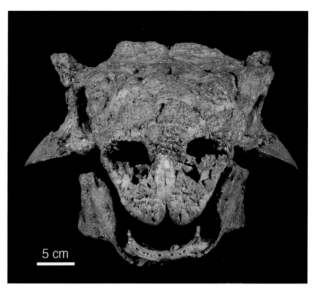

Figure 12.8. Skull of the ankylosaurid *Tarchia gigantea* in anterior view. Courtesy of Robert Sullivan.

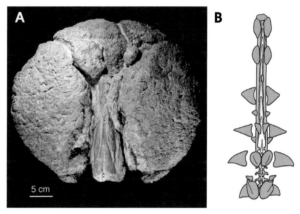

Figure 12.9. **A**, tail "club" with distal caudal vertebrae (part of the "handle") of the ankylosaurid *Anodontosaurus lambei* in dorsal view; **B**, simplified drawing of the tail in a specimen of the ankylosaurid cf. *Pinacosaurus* sp. Brown, osteoderms; green, ossified tendons. A, courtesy of Department of Paleobiology, National Museum of Natural History; B, modified from Arbour and Currie (2015b).

temporal fenestra being concealed in lateral view (Thompson et al. 2012). Ankylosaurid osteoderms are thinner than those of nodosaurids and the structural fibers in their outer layer are not as regularly organized (Scheyer and Sander 2002). Representatives of Ankylosauridae have been reported from Alberta, China, Mongolia, the western United States, and Uzbekistan. *Ankylosaurus*, from the Late Cretaceous (Maastrichtian) of Alberta, Montana, and Wyoming, reached a length of at least 8 m (Arbour and Mallon 2017; Fig. 12.7). *Pinacosaurus*, from the Late Cretaceous (Campanian) of Inner Mongolia (China) and Mongolia, and *Tarchia*, from the Late Cretaceous (Campanian-Maastrichtian) of Mongolia (Fig. 12.8),

are well-documented taxa from Asia (Maryańska 1977; Burns et al. 2011; Arbour and Currie 2015a).

The phylogenetic relationships of some ankylosaurian taxa remain uncertain. *Kunbarrasaurus*, from the Early Cretaceous (Albian) of Queensland (Australia) (Leahey et al. 2015), is probably a basal ankylosaurid (Vickaryous et al. 2004; Thompson et al. 2012; C. M. Brown et al. 2017). Blows (2015) proposed a third ankylosaurian group, Polacanthidae, but Thompson et al. (2012) found no support for this hypothesis in their phylogenetic analysis. *Polacanthus*, from the Early Cretaceous (Barremian-Aptian) of England and the Early Cretaceous (Barremian) of Spain, is distinguished by the presence over the pelvis and sacrum of a broad, flat shield composed of larger boss-like ossifications separated by smaller elements. The dermal armor along its trunk consists of dorsolaterally situated, conical osteoderms. *Polacanthus* attained a total length of 5 m (Blows 2015).

ORNITHISCHIA: GENASAURIA: NEORNITHISCHIA

The sister group of Thyreophora, Neornithischia (from Greek *neos*, new, *ornis*, bird, and *ischion*, hip) comprises all genasaurians more closely related to *Parasaurolophus walkeri* than to *Ankylosaurus magniventris* or *Stegosaurus stenops* (Butler et al. 2008). Character support for this clade is weak because of the scanty fossil record of early ornithischians. One widely accepted shared derived feature diagnostic for Neornithischia is the presence of a tab-shaped obturator process on the ischium (Butler et al. 2008; Boyd 2015; Baron et al. 2017a,b).

Basal neornithischians were bipedal, with hind-limb proportions indicating cursoriality, and attained total lengths ranging from 1 to 3 m. They typically have tapering snouts that bore keratinous beaks at the tips. The dentition includes five or fewer premaxillary teeth, and the maxillary and posterior dentary teeth have labiolingually flattened, leaf-shaped, and imbricating crowns (Norman 2015).

Lesothosaurus ("*Fabrosaurus*"), from the Early Jurassic (Hettangian-Sinemurian) of Lesotho and South Africa, was long considered the best-known basal ornithischian or ornithopod (Sereno 1991b; Butler 2005). Based on the presence of a distinct anteroposterior ridge on the lateral surface of the surangular, however, Butler et al. (2008) and Boyd (2015) found this taxon at the base of Thyreophora. Baron et al. (2017b) again considered *Lesothosaurus* the basal taxon of Neornithischia, although they also recovered it in an unresolved polytomy with Neornithischia and Thyreophora in some of their analyses. The maxillary and more posterior dentary teeth of *Lesothosaurus* (Fig. 12.10) have crowns that have a bulbous base and bear large mesial and distal denticles (Sereno 1991b). Steeply in-

clined wear facets indicate simple up-and-down (orthal) jaw motion for puncturing and crushing food. The maxilla has a slot for articulation with the lacrimal (Porro et al. 2015). The forelimbs of *Lesothosaurus* are much shorter than its hind limbs. Metatarsal V is apparently absent (Baron et al. 2017b). Knoll et al. (2010) and Baron et al. (2017b) argued that material interpreted by Butler (2005) as a distinct taxon *Stormbergia* probably represents larger, more mature individuals of *Lesothosaurus*, which would then have reached a total length of at least 2 m.

Based on recent phylogenetic analyses (Butler et al. 2008; Boyd 2015), various Jurassic and Cretaceous small-bodied neornithischians represent basal members of this clade rather than a distinct group Hypsilophodontidae as traditionally claimed (see below). These taxa include *Agilisaurus* and *Hexinlusaurus*, from the Middle Jurassic (Bathonian-Callovian) of (China) (Barrett et al. 2005b). Although more derived than *Lesothosaurus* in many skeletal features, both taxa lack key apomorphies of the more derived neornithischian clades such as the asymmetrical distribution of enamel on the maxillary and dentary teeth. They were small in size, with femoral lengths ranging from 15 to about 20 cm. *Othnielosaurus*, from the Late Jurassic (Kimmeridgian-Tithonian) of Colorado, Utah, and Wyoming (Galton 2007), attained a total length of about 1.5 m. The basal neornithischian *Ku-*

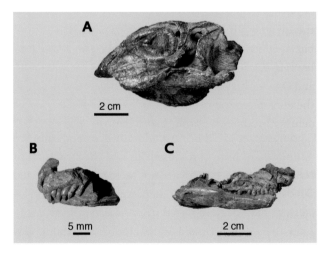

Figure 12.10. A, partial skull of the basal neornithischian *Lesothosaurus diagnosticus* in lateral view; **B**, left premaxilla of *Lesothosaurus diagnosticus* with conical, recurved tooth crowns in lateral view; **C**, right maxilla and dentary of *Lesothosaurus diagnosticus* with leaf-shaped tooth crowns. From Porro et al. (2015)—CC BY 4.0.

lindadromeus, from the Middle or Late Jurassic of Siberia (Russia), is known from skeletal remains that preserve various kinds of associated integumentary structures (Godefroit et al. 2014). The tail and the distal portions of the limbs were covered by various kinds of scales, whereas filaments were distributed around the head and trunk as well as on the back. The proximal portions of the limbs of *Kulindadromeus* bore groups of six or seven filaments arising from a basal plate and distinctly separated from each other. Finally, the proximal portion of the tibia is associated with clusters of six or seven ribbon-like structures that each comprise about 10 internal parallel filaments. The presence of filamentous integumentary structures in *Kulindadromeus*, along with similar features in the heterodontosaurid *Tianyulong*, raises the intriguing possibility that specialized epidermal covering was widespread among or perhaps even shared ancestrally by all early dinosaurs (Godefroit et al. 2014).

ORNITHISCHIA: GENASAURIA: NEORNITHISCHIA: CERAPODA

Cerapoda (combined from <u>Cera</u>topsia and Ornith<u>opoda</u>) comprises the most recent common ancestor of *Parasaurolophus walkeri* and *Triceratops horridus* and all descendants of that ancestor (Butler et al. 2008). Butler et al. (2008) listed as synapomorphies for this clade the length of postacetabular process being more than 35 percent of the total length of the ilium, and pedal digit I with robust metatarsal and the distal end of pedal phalanx I-1 extending beyond the distal end of metatarsal II. Cerapoda and *Othnielosaurus* share the asymmetrical distribution of the enamel on the maxillary and dentary teeth, but the dentary tooth crowns of *Othnielosaurus* lack distinct primary ridges.

NEORNITHISCHIA: CERAPODA: ORNITHOPODA

Ornithopoda comprises all genasaurians more closely related to *Parasaurolophus walkeri* than to *Triceratops horridus* (Butler et al. 2008). This clade is characterized by the presence of a depression on the boundary between the premaxilla and maxilla, narrow and elongate frontals, and a small opening positioned dorsally on the contact between the dentary and surangular. The definitions of Ornithopoda (from Greek *ornis*, bird, and Greek *pous*,

foot) by Butler et al. (2008) and Boyd (2015) differ considerably. Traditionally, Ornithopoda included various basal ornithischians such as Heterodontosauridae and *Lesothosaurus* and more derived taxa classified as Hypsilophodontidae and Iguanodontidae (which gave rise to Hadrosauridae) (e.g., Galton 1972). Hypsilophodontidae, which comprised a diversity of small- to medium-sized bipedal cerapodans, is now considered a grade rather than a clade (Butler et al. 2008; Boyd 2015). The phylogenetic analysis by Butler et al. (2008) found a number of "hypsilophodontid" taxa at the base of Ornithopoda. Boyd (2015) recognized a clade "Parksosauridae" as the sister group of Cerapoda. Earlier, C. M. Brown et al. (2013) had recovered the same clade, to which they applied the more widely used name Thescelosauridae but retained it among Ornithopoda.

NEORNITHISCHIA: CERAPODA: ORNITHOPODA: THESCELOSAURIDAE

Thescelosauridae comprises the most recent common ancestor of *Thescelosaurus neglectus* and *Orodromeus makelai* and all descendants of that ancestor (C. M. Brown et al. 2013). Synapomorphies for this clade include the length of the articular contact between the quadrate and quadratojugal being 25 to 50 percent of the length of the quadrate; fusion of the premaxillae; presence of a posterolateral recess on the premaxilla for an anterolateral boss on the maxilla; presence of a boss, horn, or sculpturing on the lateral surface of the jugal; and flat lateral surface of the greater trochanter of the femur (C. M. Brown et al. 2013; Fig. 12.11). Thescelosauridae can be divided into Orodrominae and Thescelosaurinae (see also Boyd [2015]). Orodrominae includes *Orodromeus*, from the Late Cretaceous (Campanian) of Montana (Horner and Weishampel 1988), and *Oryctodromeus*, from the Late Cretaceous (Cenomanian) of Idaho and Montana (Varricchio et al. 2007). It is characterized by the D-shaped outline of the shaft of the fibula in transverse section and the presence of a spine on the acromion process of the scapula (Boyd 2015). An adult and two juvenile skeletons of *Oryctodromeus* were discovered together in the expanded end chamber of a burrow—the first recorded instance of burrowing in dinosaurs other than birds (Varricchio et al. 2007). *Oryctodromeus* attained a total length of slightly more than 2 m. Boyd (2015) listed as shared derived features for Thescelosaurinae the

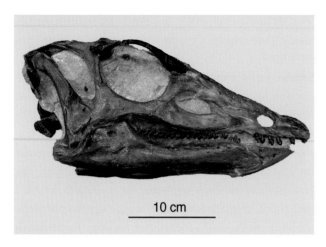

Figure 12.11. Skull of the thescelosaurid *Thescelosaurus neglectus* in lateral view. From Boyd (2014)—CC BY 4.0.

NEORNITHISCHIA: CERAPODA: ORNITHOPODA: *HYPSILOPHODON*

Hypsilophodon, definitely known only from the Early Cretaceous (Barremian) of England (Galton 1974; Fig. 12.12), is now considered more derived than other "hypsilophodontids" but has yet to be diagnosed by unique features (Butler et al. 2008). It reached a total length of 1.8 m. The limb proportions of *Hypsilophodon* suggest cursoriality.

NEORNITHISCHIA: CERAPODA: ORNITHOPODA: IGUANODONTIA

Iguanodontia (from the lizard genus *Iguana* and Greek *odous (odon)*, tooth) comprises all ornithopods more closely related to *Parasaurolophus walkeri* than to *Hypsilophodon foxii* or *Thescelosaurus neglectus* (Sereno 2005). Boyd (2015) listed as synapomorphies for Iguanodontia the position of the jugal wing of the quadrate well dorsal to the distal condyles of this element and the neural spines on the sacral vertebrae being at least twice as tall as the sacral centra. Using a different definition of the clade, Norman (2015) proposed several shared derived features for most or all iguanodontians: the lateral expansion of the premaxillae; absence of premaxillary teeth; presence of marginal denticles on the dentary and maxillary teeth; and the patterns of ridges on the dentary and maxillary tooth crowns.

ORNITHOPODA: IGUANODONTIA: *TENONTOSAURUS*

Tenontosaurus, from the Early Cretaceous (Aptian-Albian) of Montana, Oklahoma, Texas, and Wyoming (Ostrom 1970; Forster 1990; Winkler et al. 1997; Thomas 2015), is characterized by several derived features includ-

presence of two palpebral bones that are not fused to the dorsal margin of the orbit and a dorsally projecting process on the surangular anterior to the jaw joint (Fig. 12.11). In addition to *Thescelosaurus*, from the Late Cretaceous (Maastrichtian) of Alberta, Montana, North and South Dakota, Saskatchewan, and Wyoming (Boyd et al. 2009; Boyd 2014), and *Parksosaurus*, from the Late Cretaceous (Maastrichtian) of Alberta, this clade also comprises various Asian forms such as *Changchunsaurus*, from the Early or Late Cretaceous of Jilin (China) (Jin et al. 2010; Butler et al. 2011), and a subclade of South American taxa (Elasmaria; Calvo et al. 2007) including *Talenkauen*, from the Late Cretaceous (Maastrichtian) of Argentina (Novas et al. 2004). *Thescelosaurus* attained a total length of up to 4 m (Boyd et al. 2009). Its femur is longer than the tibia, unlike in Orodrominae and more basal ornithischians.

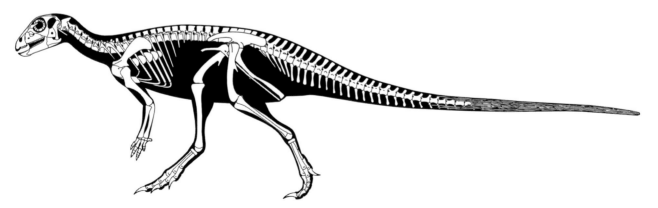

Figure 12.12. Reconstructed skeleton of the neornithischian *Hypsilophodon foxii*. Courtesy of Scott Hartman.

ing a dorsoventrally tall maxilla, large external naris, 12 cervical vertebrae, and a very long tail (with at least 59 caudal vertebrae) (Butler et al. 2008; Thomas 2015). Its dentary tooth crowns bear a prominent vertical ridge, but the maxillary tooth crowns lack such a ridge and have numerous smaller ridges. *Tenontosaurus* attained a total length of up to 8 m.

ORNITHOPODA: IGUANODONTIA: RHABDODONTIDAE

Rhabdodontidae comprises the most recent common ancestor of *Rhabdodon priscus* and *Zalmoxes robustus* and all descendants of that ancestor (Weishampel et al. 2003). This Late Cretaceous clade is now known from the Santonian of Hungary, the Campanian of Austria, the Campanian-Maastrichtian of France and Spain, and the Maastrichtian of Romania (Godefroit et al. 2017). Weishampel et al. (2003) characterized this clade by the presence of more than 12 sharp ridges on the lingual surface of each dentary tooth crown, the (in lateral view) straight to slightly convex dorsal margin of the ilium, and the (in anterior view) distinctly laterally bowed femur. *Zalmoxes* reached a total length of almost 3 m. Godefroit et al. (2017) interpreted the proportionately large, blade-like tooth crowns as suitable for feeding on tough, fibrous plant material.

McDonald et al. (2010) also placed *Muttaburrasaurus*, from the Early Cretaceous (Albian) of Australia (Bartholomai and Molnar 1981), with Rhabdodontidae, but the phylogenetic analysis by Boyd (2015) found it as a separate lineage of basal iguanodontians. Until a comprehensive anatomical study of this taxon is available, its phylogenetic position must remain unresolved. The cranium of *Muttaburrasaurus* has a distinctive inflated nasal region.

ORNITHOPODA: IGUANODONTIA: DRYOSAURIDAE

Dryosauridae comprises *Dryosaurus altus* and all taxa more closely related to it than to *Parasaurolophus walkeri* (Sereno 1998). It includes *Dryosaurus*, from the Late Jurassic (Kimmeridgian-Tithonian) of Colorado, Utah, and Wyoming (Gilmore 1925b), and *Dysalotosaurus*, from the Late Jurassic (Kimmeridgian) of Tanzania (Janensch 1955; Hübner and Rauhut 2010). Butler (2008) listed several synapomorphies for this clade including the broad brevis shelf on the ilium and the presence of a

deep pit at the base of the fourth trochanter on the femur. *Dryosaurus* reached a total length of up to 4 m. Dryosauridae is hypothesized as the sister taxon of Ankylopollexia.

ORNITHOPODA: IGUANODONTIA: ANKYLOPOLLEXIA

Ankylopollexia (from Greek *ankylos*, crooked, curved, and Latin *pollex*, thumb) comprises the most recent common ancestor of *Camptosaurus dispar* and *Parasaurolophus walkeri* and all descendants of that ancestor (Sereno 2005; see also McDonald [2011] and Norman [2015]). Norman (2015) listed a suite of shared derived features for Ankylopollexia including dentary tooth crowns broader than the maxillary tooth crowns; neural spines of the dorsal vertebrae taller dorsoventrally than long anteroposteriorly; metacarpal I short and fused with a block-like carpus; and short manual digit I with a spike-like ungual phalanx (Fig. 12.13B). Ankylopollexians differ from dryosaurids in their greater body size and more robust skeleton. They also have proportionately longer and deeper skulls with increasingly complex dentitions (Figs. 12.13A, 12.14A). The development of the co-ossified carpal block indicates that the manus could bear weight and is consistent with at least facultatively quadrupedal locomotion, possibly related to the increased body size in these ornithopods. The spike-like terminal phalanx on manual digit I possibly served as a defensive weapon (Norman 1980).

ORNITHOPODA: IGUANODONTIA: ANKYLOPOLLEXIA: *CAMPTOSAURUS*

Camptosaurus, from the Late Jurassic (Kimmeridgian-Tithonian) of Utah and Wyoming, is a basal ankylopollexian diagnosed by the triangular acromial process of the scapula, the straight dorsal margin and the shape of the postacetabular process of the ilium, and the rounded and expanded distal end of the ischium (McDonald 2011). It attained a total length of up to 7 m. Its limb proportions indicate that *Camptosaurus* was capable of both bipedal and quadrupedal locomotion.

ORNITHOPODA: IGUANODONTIA: ANKYLOPOLLEXIA: IGUANODONTOIDEA

Among Ankylopollexia, W. Wu and Godefroit (2012) defined Iguanodontoidea as comprising the most recent

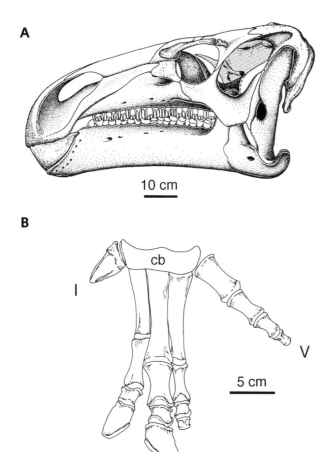

Figure 12.13. A, reconstructed skull of the iguanodontid *Iguanodon bernissartensis* in lateral view; **B**, left manus of *Mantellisaurus atherfieldensis*. Abbreviation: cb, carpal block. Roman numerals denote digits. From Norman (1980). With permission of the Royal Belgian Institute of Natural Sciences.

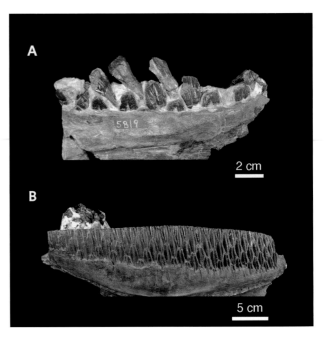

Figure 12.14. Dentitions of ankylopollexians. **A**, dentary of *Camptosaurus* in lingual view; **B**, left dentary of *Lambeosaurus* with dental battery in lingual view. Courtesy of Department of Paleobiology, National Museum of Natural History.

common ancestor of *Iguanodon* and *Parasaurolophus* and all descendants of that ancestor. They diagnosed it based on numerous shared derived features: exoccipitals excluding the supraoccipital from the dorsal margin of the foramen magnum; dorsal apex of the maxilla positioned posterior to the center of this bone; dentary with 18 to more than 30 tooth positions; tooth crowns with curved, mammillated cutting edges formed by marginal denticles; sternal elements hatchet-shaped; manual digits II and III with flattened, twisted, and hoof-like unguals; manual digit V long, with three or four phalanges; posterior process of the pubis shorter than the ischium; distal end of the prepubic process expanded; and pedal digit I absent. Iguanodontoidea includes a diversity of taxa that are successively more closely related to Euhadrosauria (see Norman [2014, 2015] for a different hypothesis of interrelationships).

Iguanodon, from the Early Cretaceous (Barremian) of Belgium, England, and Germany (Fig. 12.13A), reached a total length of up to 13 m. It was the second nonavian dinosaur to be scientifically documented and named (Mantell 1825; Dollo 1924) but a detailed anatomical study was published only much later (Norman 1980). The closely related *Mantellisaurus*, from the Early Cretaceous (Barremian) of Belgium, England, and possibly Germany, differs from *Iguanodon* especially in its smaller body size (with a total length of up to 7 m) and more gracile proportions (Norman 1986, 2012). Its proportionately longer and dorsoventrally less deep skull has a ventrally deflected snout that holds fewer teeth than in *Iguanodon*. *Ouranosaurus*, from the Early Cretaceous of Niger (Taquet 1976; Bertozzo et al. 2017), is distinguished by its greatly elongated neural spines on its dorsal, sacral, and caudal vertebrae and attained a total length of at least 8 m. Its snout is flattened dorsoventrally with posteriorly positioned external nares.

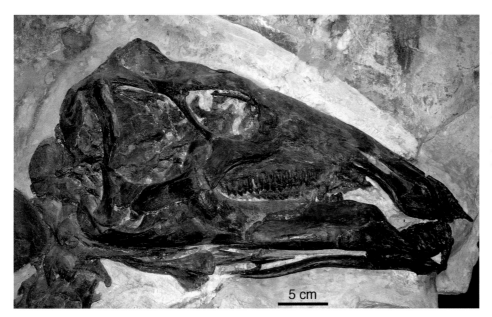

Figure 12.15. Skull and anterior cervical vertebrae of the basal hadrosauroid *Tethyshadros insularis* in lateral view. With permission of Ministero dei Beni e delle Attività Culturali e del Turismo–Soprintendenza per i Beni Archeologici del Friuli Venezia Giulia.

ANKYLOPOLLEXIA: IGUANODONTOIDEA: HADROSAUROIDEA

The largest and most derived representatives of Ornithopoda are the hadrosaurs, which include the first nonavian dinosaur reported from North America, *Hadrosaurus*, from the Late Cretaceous (Campanian) of New Jersey (Leidy 1858; Prieto-Márquez et al. 2006). Unfortunately, this taxon is known only from an incomplete skeleton that lacks diagnostic features, and this fact has led to nomenclatural complications (Prieto-Márquez 2010a). This discussion follows Sereno (1997) and W. Wu and Godefroit (2012) in defining Hadrosauroidea as comprising all iguanodontoideans more closely related to *Parasaurolophus* than to *Iguanodon*. W. Wu and Godefroit (2012) listed various shared derived features for this clade: antorbital fenestra not open on the lateral side of the cranium; basipterygoid processes extending ventrally well below the ventral margin of the occipital condyle; carpals well-ossified but not fused to each other; and ischium nearly straight in lateral view. Hadrosauroidea comprises hadrosaurs and various taxa from the Cretaceous of Central and East Asia and the western United States that are successively more closely related to Euhadrosauria. Basal hadrosauroids include *Eolambia*, from the Late Cretaceous (Cenomanian) of Utah (McDonald et al. 2012) and *Probactrosaurus*, from the Early Cretaceous of Inner Mongolia (China) (Norman 2002).

ANKYLOPOLLEXIA: IGUANODONTOIDEA: HADROSAUROIDEA: HADROSAURIDAE

W. Wu and Godefroit (2012) and Godefroit et al. (2012) recognized a clade Hadrosauridae that includes Euhadrosauria (Weishampel et al. 1993; corresponding to Hadrosauridae of traditional usage [e.g., Lull and Wright 1942]) and including several taxa closely related to (and sometimes included in) euhadrosaurians, such as *Tethyshadros*, from the Late Cretaceous (Maastrichtian) of Italy (Dalla Vecchia 2009; Fig. 12.15), and *Telmatosaurus*, from the Late Cretaceous (Maastrichtian) of Romania (Weishampel et al. 1993). Hadrosauridae as defined by Godefroit et al. (2012) is diagnosed by the absence of a sutural contact between the jugal and ectopterygoid, the absence of a surangular foramen, and the reduction of the marginal denticles on the tooth crowns.

HADROSAUROIDEA: HADROSAURIDAE: EUHADROSAURIA

Euhadrosauria shares a number of synapomorphies (W. Wu and Godefroit 2012): premaxilla with a "double-layered" labial margin, with a deep groove separating an external layer with bony projections from a somewhat setback palatal layer of thickened bone; anterior portion of the jugal dorsoventrally deep; distal end of the quadrate with a large, hemispherical lateral condyle; dentary with a moderately developed to long diastema

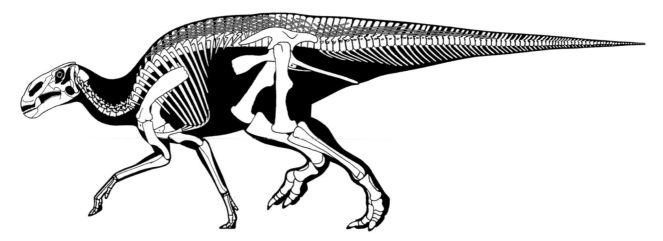

Figure 12.16. Reconstructed skeleton of the saurolophine *Brachylophosaurus canadensis*. Courtesy of Scott Hartman.

posterior to the predentary and forming most of the expanded apical portion of the coronoid process; dentary tooth crowns small, lanceolate, with a height/width ratio exceeding 3.1; dentary tooth crowns symmetrical, with the apex and primary ridge positioned on the midline and with at most weakly developed secondary ridges (Fig. 12.14B); and humerus with a proportionately long deltopectoral crest extending rather abruptly from the shaft, lending it a distinctive angular profile. Although euhadrosaurians probably could employ bipedal and quadrupedal locomotion, they predominantly used the latter (Maidment et al. 2012; Fig. 12.16).

Euhadrosauria is a diverse clade that represented one of the two major groups of dinosaurian megaherbivores in Late Cretaceous continental ecosystems in Asia, Europe, and North America (Horner et al. 2004; Prieto-Márquez 2010a,b). The vernacular term "hadrosaurs" ("duck-billed dinosaurs") is used here for this grouping. Although they are known from South America, hadrosaurs were not common on that continent and did not play a prominent ecological role as they did on the northern landmasses. The distinctive feeding apparatus of hadrosaurs may have been a key factor in their evolutionary diversification (Weishampel 1984). The transversely broad, edentulous anterior ends of the upper and lower jaws were covered by a keratinous beak and presumably served as a cropping device. The tooth batteries in the dentaries and maxillae formed large grinding and/or slicing surfaces (Lull and Wright 1942;

Ostrom 1961; Weishampel 1984; G. M. Erickson et al. 2012; LeBlanc et al. 2016, 2017). In each tooth battery, the small individual teeth are tightly packed together into vertical files or "tooth families" (Fig. 12.14B). Teeth of adjacent files interlock with each other. Each jaw holds as many as 60 of such tooth files, which are tightly packed together. The maxillary tooth crowns are covered by enamel only on the labial surface, whereas the dentary tooth crowns have enamel covering only on the lingual surface. Not only were all tooth families connected to the jaw by periodontal ligaments, but ligaments also held the individual teeth together (LeBlanc et al. 2016, 2017). Typically, each tooth has two to four replacement teeth stacked below it. LeBlanc et al. (2016) demonstrated that the pulp cavities of worn teeth became filled in with dentine, and these dead teeth continued to be part of the grinding surface until they were obliterated by wear.

Combining data from tooth wear and cranial structure, Nabavizadeh (2014) argued that hadrosaurs employed backwardly directed (palinal) jaw movement to shear fodder as well as medial rotation of the dentary teeth relative to the maxillary teeth. This would have facilitated simultaneous independent processing of food on both sides of the mouth. In order to account for transversal tooth wear, Norman and Weishampel (1985) postulated a mechanism (pleurokinesis) involving lateral rotation of the maxillae. Recent studies on potential cranial kinesis in dinosaurs (e.g., Holliday and Witmer 2008) have rendered this interpretation unlikely.

Hadrosaurs probably usually browsed within 2 m of the ground but could have extended their feeding range up to 5 m by adopting a bipedal stance (Mallon et al. 2013). Evidence from bonebeds, trackways, and nesting sites suggests that at least some taxa formed large, multigenerational herds (Brusatte 2012).

Prieto-Márquez (2010a) grouped all hadrosaurs other than *Hadrosaurus foulkii* together as Saurolophidae, which he divided into Saurolophinae and Lambeosaurinae. Prieto-Márquez et al. (2016) described a new hadrosaur, *Eotrachodon*, from the Late Cretaceous (Santonian) of Alabama, which they interpreted as a sister taxon to Saurolophidae but more derived than *Hadrosaurus*.

Saurolophinae (corresponding to the traditional "Hadrosaurinae" but excluding the problematical *Hadrosaurus*) and *Eotrachodon* share the presence of an extensive, well-defined depression that surrounds the large external naris and is divided into the circumnarial fossa proper and a smaller anterior recess (Prieto-Márquez and Wagner 2014; Prieto-Márquez et al. 2016; Fig. 12.17A). If present, cranial crests in saurolophines are formed by the nasals or frontals and are solid. By contrast, Lambeosaurinae is distinguished by the presence of prominent cranial crests that are formed by the premaxillae and nasals and enclose greatly elongated narial passages (Fig. 12.17B). These passages extend posterodorsally and then loop anteroventrally to pass back into the cranium dorsally in front of the orbits. Lambeosaurine crests show much variation within and among known taxa, and the shape of these features underwent considerable changes throughout ontogeny (Farke et al. 2013). Historically, differences in crest shape between growth stages were used to segregate juveniles and adults in different taxa (Dodson 1975). The cranial crests of lambeosaurines presumably served a variety of purposes (R. M. Sullivan and Williamson 1999), such as visual display for species recognition (Hopson 1975; Padian and Horner 2011) and possibly as a resonating device for vocalization (Wiman 1931; Weishampel 1981).

Prieto-Márquez (2010a) listed a number of shared derived features as diagnostic for Saurolophinae including the presence of a premaxillary foramen anterior and ventrolateral to the anterior margin of the external naris; ventral process of the predentary with a long indentation; a more or less triangular lower temporal fenes-

tra with a dorsal margin that is narrower than the ventral one; and pubic peduncle of the ischium extending parallel to the shaft of the ischium. *Saurolophus*, from the Late Cretaceous (Maastrichtian) of Alberta and Mongolia, attained a total length of at least 12 m and is distinguished by a prominent spike-like crest formed by the nasals (Maryańska and Osmólska 1981). In *Maiasaura*, from the Late Cretaceous (Campanian) of Montana, the low crest is formed by the nasals and does not contact the circumnarial depression. *Maiasaura* is famous for its nests of eggs and associated small juveniles that have been interpreted as evidence for some form of parental care (Horner and Makela 1979). *Edmontosaurus*, from the Late Cretaceous (Maastrichtian), of Colorado, Montana, North and South Dakota, Wyoming, and Alberta, has a transversely much expanded premaxillary beak that is as broad as the postorbital region across the jugals (Lambe 1920; Lull and Wright 1942; Campione and D. C. Evans 2011). Although it lacks a bony cranial crest, one particularly well-preserved specimen established that the head sported a soft-tissue structure similar to a cock's comb (P. R. Bell et al. 2014). *Edmontosaurus* reached a total length of up to 12 m.

Prieto-Márquez (2010a) hypothesized three synapomorphies as diagnostic for Lambeosaurinae: the absence of an anterodorsal process of the maxilla, so that the anterior end of this bone forms a ventrally sloping anterodorsal shelf that underlies the premaxilla; the dorsal process of the maxilla being taller dorsoventrally than broad anteroposteriorly, with a peaked and posteriorly inclined apex; and oval supratemporal fenestrae with their long axes extending anterolaterally. *Parasaurolophus*, from the Late Cretaceous (Campanian) of Alberta, New Mexico, and Utah, provides an example of particularly prominent crest development among lambeosaurines. Its premaxillary crest can extend posteriorly from the cranium for more than 1 m (R. M. Sullivan and Williamson 1999). *Parasaurolophus* attained a total length of at least 10 m. Adult *Corythosaurus*, from the Late Cretaceous (Campanian) of Alberta, have a (in lateral view) semicircular crest with its apex positioned above the orbits and its lateral surface formed mostly by the nasal (Ostrom 1961). The geologically slightly younger *Lambeosaurus*, from the same provenance, can be easily

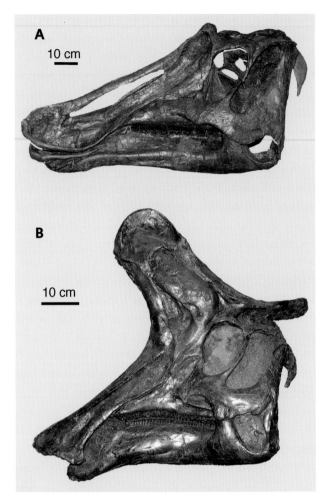

Figure 12.17. A, skull of the saurolophine *Prosaurolophus maximus* in lateral view; **B**, skull of the lambeosaurine *Lambeosaurus lambei* in lateral view. A, courtesy of Department of Paleobiology, National Museum of Natural History; B, courtesy of David Evans.

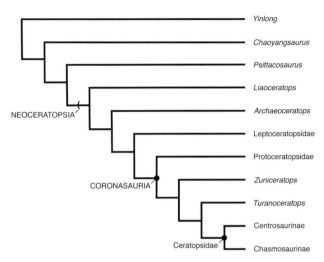

Figure 12.18. Phylogenetic hypothesis of the interrelationships of selected taxa of Ceratopsia. Dots denote node-based clades and parentheses denote stem-based clades. Based on Makovicky (2001) and Xu et al. (2006b).

distinguished from *Corythosaurus* by the shape of its crest, which is more or less hatchet-shaped in side view and has its apex positioned anterior to the orbits in adults (Ostrom 1961; Fig. 12.17B). The premaxilla forms most of the lateral surface of the crest.

Ornithischia: Genasauria: Marginocephalia

Marginocephalia (from Latin *margo*, edge, margin, and Greek *kephale*, head) comprises the most recent common ancestor of *Pachycephalosaurus wyomingensis* and *Triceratops horridus* and all descendants of that ancestor (Sereno 1986; Butler et al. 2008). It is characterized by a number of synapomorphies (Butler et al. 2008): presence of a pa-

rietosquamosal shelf posteriorly; premaxilla with three teeth; blade of the scapula at least nine times longer than broad at its narrowest point; ischium without a tab-shaped obturator process; and posterior process of the pubis short, extending for half or less the length of the ischium. The diagnostic parietosquamosal shelf became an elaborate bony frill in many members of one marginocephalian clade, Ceratopsia, whereas it is frequently absent because of substantial thickening of the skull roof in many representatives of the other clade, Pachycephalosauria. With the exception of a few poorly known basal taxa, all marginocephalians can be definitely assigned to either Ceratopsia or Pachycephalosauria.

ORNITHISCHIA: GENASAURIA: MARGINOCEPHALIA: CERATOPSIA

Ceratopsia (from Greek *keras*, horn, and *ops*, face) comprises all marginocephalians more closely related to *Triceratops horridus* than to *Pachycephalosaurus wyomingensis* (Sereno 2005; Fig. 12.18). Butler et al. (2008) listed a suite of diagnostic features for this clade: cranium transversely broad across the outwardly bowed jugals, lending it a distinctly triangular outline in plan view; presence of a median rostral bone anterior to the premaxilla; palatal region of the premaxilla horizontal or only slightly arched; presence of a fossa surrounding the external naris on the lateral surface of the premax-

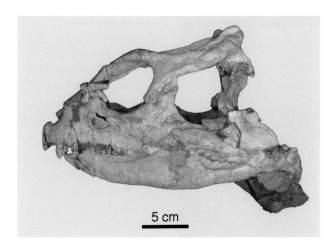

Figure 12.19. Skull of the basal ceratopsian *Yinlong downsi* in lateral view. Courtesy of Jim Clark.

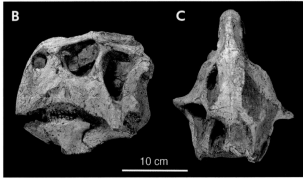

Figure 12.20. A, reconstructed skeleton of the psittacosaurid *Psittacosaurus neimongoliensis*; **B-C**, skull of *Psittacosaurus major* in **B**, lateral, and **C**, dorsal views. A, courtesy of Scott Hartman; B-C, from Sereno et al. (2007)—CC BY 2.0.

illa and separated from the ventral margin of this bone; presence of an eminence on the rim on the labial emargination of the maxilla near its contact with the jugal; anterior ramus of the jugal deeper than its posterior ramus; and lateral condyle of the quadrate larger than the medial one.

Yinlong, from the Late Jurassic (Oxfordian) of Xinjiang (China), is the basalmost ceratopsian known to date (Xu et al. 2006). The short snout of its proportionately large cranium (Fig. 12.19) has a small rostral bone. The premaxillary teeth of *Yinlong* are distinctly larger than the maxillary teeth. Its squamosals are posterolaterally expanded, reminiscent of the condition in pachycephalosaurs, but the parietals lack any expansion. The lateral surfaces of the jugal, postorbital, squamosal, and the more posterior region of the lower jaw, are rugose. *Yinlong* attained a total length of about 1 m.

Chaoyangsaurus, from the Late Jurassic or earliest Cretaceous of Liaoning (China) (X. Zhao et al. 1999), and *Xuanhuaceratops*, from the Late Jurassic of Hebei (China) (X. Zhao et al. 2006), are more derived than *Yinlong* in the presence of laterally arched jugals. X. Zhao et al. (2006) united these taxa as Chaoyangsauridae based on features of the quadrate and mandible.

Psittacosauridae is the most inclusive clade containing *Psittacosaurus mongoliensis* but not *Triceratops horridus* (Sereno 2005). The skull of *Psittacosaurus*, from the Early Cretaceous (Barremian-Albian) of China, Mongolia, and Russia (Sereno 2000, 2010; Averianov et al. 2006; Sereno

et al. 2007, 2010), has a deep and short snout (Fig. 12.20B–C). Both the antorbital fossa and fenestra are absent. The cranium is widest transversely across the jugals. Each jugal bears a lateral horn-like process below the orbit. The broad premaxilla is edentulous. The dentary and maxillary teeth of *Psittacosaurus* have mesiodistally broad crowns with a distinct but low primary ridge. They show low-angled shearing with different angles between the teeth along the tooth row. The tooth crowns have enamel around the occlusal surface of each crown, but the enamel is thicker along the leading cutting edge (G. M. Erickson et al. 2015). *Psittacosaurus* has proportionately long forelimbs (Fig. 12.20A). Q. Zhao et al. (2013) demonstrated that juvenile individuals were quadrupedal but that, starting at the age of about of four years, *Psittacosaurus* adopted bipedal locomotion. The manus is functionally tridactyl, with a short digit IV and without digit V. Sereno et al. (2010) argued that *Psittacosaurus* was adapted for feeding on high-fiber plant matter. Several skeletons preserve clusters of large gastroliths, documenting the presence of a gastric "mill" for additional processing of fodder. *Psittacosaurus* attained a total length of up to 2 m. Exceptionally preserved fossils of

Figure 12.21. Skeleton of *Psittacosaurus* sp. with soft-tissue preservation in ventral view. Note bristle-like structures on the tail and scalation on the body and limbs. Courtesy of Gerald Mayr.

this dinosaur from Liaoning (China) show that it sported a brush-like arrangement of possibly hollow, bristle- or quill-like structures extending along the dorsal surface of the tail (Mayr et al. 2002; Fig. 12.21). Much of the skin was covered by arrangements of large and small scales. Vinther et al. (2016) showed that *Psittacosaurus* had countershading, with a light underbelly and tail and a dark chin and jugal bosses. Based on this coloration pattern, they inferred that this dinosaur lived in closed habitats such as forests with a fairly dense canopy. Meng et al. (2004) reported on what appears to be a nest preserving the skeleton of an adult psittacosaur in undisturbed association with 34 post-hatchling juveniles, suggesting some form of parental care. Q. Zhao et al. (2014) described groups of juvenile psittacosaurs that apparently traveled together and were killed and buried by volcanic ash falls.

MARGINOCEPHALIA: CERATOPSIA: NEOCERATOPSIA

Neoceratopsia (from Greek *neos*, new, *keras*, horn, and *ops*, face) comprises all ceratopsians more closely related to *Triceratops* than to *Psittacosaurus* (Sereno 1998). Butler et al. (2008) provided a list of shared derived features for this clade: cranium attaining its greatest width posteriorly below the lower temporal fenestra; rostral bone with a well-developed ventrolateral process; width of the jugal-postorbital bar greater than that of the lower temporal fenestra; dorsal portion of the postorbital process of the jugal expanded posteriorly; postorbital triangular and plate-like; shaft of the quadrate reduced in anteroposte-

rior width and straight in lateral view; upper temporal fenestra long anteroposteriorly; parietosquamosal shelf extending posteriorly and forming a prominent bony frill; and primary ridges on maxillary and dentary tooth crowns offset, lending the crowns an asymmetrical appearance.

Neoceratopsia ranged in time from the Early Cretaceous (Barremian) to the end of the Cretaceous (Maastrichtian). The oldest known neoceratopsian, *Liaoceratops*, from the Early Cretaceous (Barremian) of Liaoning (China), has only a short frill formed primarily by the parietal with smaller lateral contributions from the squamosals (Xu et al. 2002). Its snout is narrow and beak-like, unlike in more basal ceratopsians. The jugal processes are posterior to the orbits. The tooth crowns of *Liaoceratops* are enameled on only one surface and form single cutting edges that sheared past one another during occlusion.

Archaeoceratops, from the Early Cretaceous (Aptian) of Gansu (China) (Dong and Azuma 1997), shares with more derived neoceratopsians an anteriorly keeled rostral bone that points ventrally, the presence of an epijugal ossification, a posterolaterally facing quadratojugal that is mediolaterally expanded and triangular in coronal section, a dorsally directed finger-like process on the lateral surface of the surangular, and tightly packed maxillary and dentary teeth (Butler et al. 2008). It has a short parietosquamosal frill and attained a total length of about 1 m.

Aquilops, from the Early Cretaceous (Albian) of Montana, is another basal neoceratopsian (Farke et al. 2014).

It has a strongly curved rostral bone with a median boss and an elongate, distinctly pointed antorbital fossa.

Leptoceratopsidae comprises all taxa more closely related to *Leptoceratops gracilis* than to *Triceratops horridus* (Makovicky 2001). It is known from the Late Cretaceous of western North America (Sternberg 1951; Chinnery 2004) and Central and East Asia (Sereno 2000; Xu et al. 2010). Leptoceratopsids are characterized by short and deep crania and robust mandibular rami with deeply convex ventral margins. Enamel is restricted to the lingual surface of the dentary teeth and the labial surface of the maxillary teeth, and wear facets extend at more or less the same angle, unlike in *Psittacosaurus* but resembling the condition in more derived ceratopsians (G. M. Erickson et al. 2015). *Leptoceratops*, from the Late Cretaceous (Maastrichtian) of Alberta and Wyoming, reached a total length of about 2 m.

MARGINOCEPHALIA: CERATOPSIA: NEOCERATOPSIA: CORONOSAURIA

Coronosauria (from Latin *corona*, crown, and Greek *sauros*, lizard) comprises the most recent common ancestor of *Protoceratops* and *Triceratops* and all descendants of that ancestor (Sereno 1986). This clade is distinguished by the presence of large, posterolaterally expanded parietosquamosal frills that are perforated by a pair of large openings in most taxa except *Triceratops*. Another diagnostic feature shared by most coronosaurians is the presence of a horn supported by the fused nasals dorsal or just posterior to the external nares. The first three cervical vertebrae are fused into a single structure (syncervical), presumably to support the proportionately large and heavy skull (Campione and Holmes 2006). The ungual phalanges of the pes are hoof-like.

NEOCERATOPSIA: CORONOSAURIA: PROTOCERATOPSIDAE

Makovicky (2001) listed various diagnostic synapomorphies for Protoceratopsidae such as the presence of an anterior prong on the quadratojugal, the bifurcated posterior portion of the splenial, and the anterior extension of the surangular on the lateral surface of the lower jaw. This clade includes *Protoceratops* and *Bagaceratops*, both from the Late Cretaceous (Campanian) of Mongolia (Brown and Schlaikjer 1940; Maryańska and Osmólska

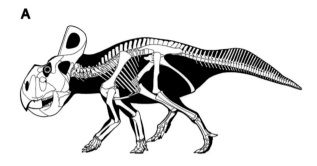

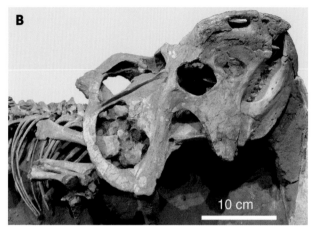

Figure 12.22. A, reconstructed skeleton of the protoceratopsid *Protoceratops andrewsi*; **B**, skull and anterior region of postcranial skeleton of *Protoceratops andrewsi* in oblique lateral view. A, courtesy of Scott Hartman; B, from Hone et al. (2014)—CC BY 4.0.

1975; Sereno 2000; Hone et al. 2014). *Protoceratops* is known from many well-preserved specimens, ranging from hatchlings to adults (Figs. 12.22, 12.23). Fastovsky et al. (1997) reported a nest of *Protoceratops* containing 15 aligned juvenile skeletons and inferred that this dinosaur had parental care. The nasals of *Protoceratops* are only slightly arched and have a bump rather than a bony horn core, unlike in *Bagaceratops*. Dodson (1976) inferred sexual dimorphism for a number of cranial features in *Protoceratops,* but Padian and Horner (2011) challenged this interpretation. *Protoceratops* reached a skull length of about 60 cm and a total length of up to 1.8 m. It was probably at least facultatively quadrupedal (Senter 2007). Makovicky (2001) found Protoceratopsidae as the sister taxon of Ceratopsidae plus *Zuniceratops*. Ösi et al. (2010) reported the discovery of *Ajkaceratops*, from the Late Cretaceous (Santonian) of Hungary, which they interpreted as a basal coronosaurian. However, the inferred

Figure 12.23. Skeleton of a juvenile of *Protoceratops andrewsi* in lateral view, showing an early ontogenetic stage of the parietosquamosal frill. Length as preserved c.14 cm. Courtesy of Mick Ellison.

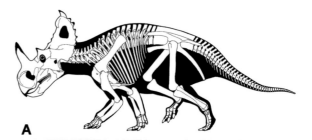

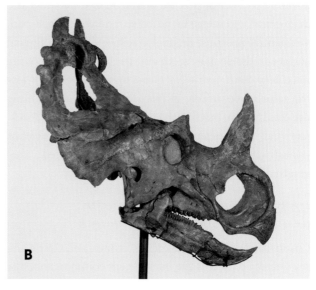

Figure 12.24. A, reconstructed skeleton of the centrosaurine *Centrosaurus apertus*; **B**, skull of *Centrosaurus apertus* in lateral view. Length of frill 58 cm. A, courtesy of Scott Hartman; B, courtesy of the Royal Ontario Museum.

rostral is fused to the premaxilla, and additional material is required to test this hypothesis.

NEOCERATOPSIA: CORONOSAURIA: CERATOPSIDAE

Ceratopsidae includes the majority of coronosaurians, primarily from the Late Cretaceous (Campanian-Maastrichtian) of western North America (Hatcher et al. 1907; Lull 1933; Dodson et al. 2004). Two poorly known taxa, *Zuniceratops*, from the Late Cretaceous (Turonian) of Utah (Wolfe and Kirkland 1998), and *Turanoceratops*, from the Late Cretaceous (Turonian) of Uzbekistan (Sues and Averianov 2009), can be grouped with Ceratopsidae as Ceratopsoidea (Sereno 1998). This clade is characterized by the presence of well-developed horn cores above the orbits, vertical tooth occlusion, and the curvature of the ischium (Makovicky 2001). Ceratopsidae comprises the most recent common ancestor of *Centrosaurus* and *Triceratops* and all descendants of that ancestor (Dodson et al. 2004). The posterolateral margins of the parietosquamosal frills in ceratopsids bear accessory ossifications (epoccipitals) that form wavy edges or are developed as marginal spines or hooks (Figs. 12.24, 12.25). The shapes of ceratopsid horns and frills changed considerably during ontogeny, and the taxon-specific configurations developed only in individuals approaching or reaching adulthood (Sampson et al. 1997). The exter-

nal nares are greatly enlarged and separated by a median bony lamina more anteriorly. The rostral and predentary bones form a transversely narrow beak with sharp cutting edges. The maxilla and dentary each have up to more than 35 tooth positions, each of which has a functional tooth as well as at least two replacement teeth stacked below. Unlike in hadrosaurs, each developing tooth appears to be nested within the pulp cavity of its predecessor. G. M. Erickson et al. (2015) and LeBlanc et al. (2017) reviewed the tissue composition of neoceratopsian teeth. The opposing batteries of maxillary and dentary teeth form steeply inclined occlusal surfaces that were well suited for shearing plant material. The lower jaw of ceratopsids has a tall coronoid process for insertion of powerful adductor jaw muscles. In the pelvis, the lateral surface of the ilium is distinctly everted and the ischium is strongly curved. The sacrum includes more than 10 pairs of sacral ribs. The forelimb and manus are large. Ceratopsids were obligate quadrupeds and foraged probably within 1-2 m of the ground (Mallon et al. 2013). Together with hadrosaurs, they formed communities of large-bodied herbivorous dinosaurs in the Late Cretaceous ecosystems of western North America. Bonebeds for some ceratopsids (e.g., *Centrosaurus*; Ryan et al. 2001) have been interpreted as evidence for large, multigenerational herds.

Ceratopsidae is divided into Centrosaurinae and Chasmosaurinae. Centrosaurinae comprises all ceratopsids more closely related to *Centrosaurus* than to *Triceratops* (Dodson et al. 2004). Diagnostic synapomorphies for this subclade include oral margin of premaxilla extending below the alveolar margin of that bone; length of postorbital horn cores less than 15 percent of the basal skull length; jugal with an infratemporal flange; squamosal much shorter than parietal; and six to eight parietal epoccipitals on either side (Dodson et al. 2004). *Centrosaurus* (Fig. 12.24) and *Styracosaurus*, both from the Late Cretaceous (Campanian) of Alberta, attained total lengths of up to about 6 m. The posterior margin of the frill in *Styracosaurus* bears four long, posterolaterally extending spikes on either side (Ryan et al. 2007). *Achelousaurus*, from the Late Cretaceous (Campanian) of Montana (Sampson 1995), and *Pachyrhinosaurus*, from the Late Cretaceous (Maastrichtian) of Alaska (Fiorillo and Tykoski 2012) and Alberta (Currie et al. 2008), differ from other centrosaurines in the presence of massive, rugose bony bosses on their snouts in place of nasal horn cores.

Chasmosaurinae is defined as all ceratopsids more closely related to *Triceratops* than to *Centrosaurus* (Dodson et al. 2004). Diagnostic derived features for this subclade include a large rostral, with a deeply concave posterior margin and hypertrophied dorsal and ventral processes; the presence of a premaxillary septum; presence of an interpremaxillary fossa; and triangular epoccipitals on the squamosal (Dodson et al. 2004; Fig. 12.25). The shape and size of the parietosquamosal frill differ considerably among chasmosaurines. The frill of *Triceratops*, from the Late Cretaceous (Maastrichtian) of Alberta, Colorado, Montana, North and South Dakota, Saskatchewan, and Wyoming, has a rounded posterolateral edge and lacks openings (Hatcher et al. 1907; Forster 1996a; Fig. 12.25B). The frill of *Chasmosaurus*, from the Late Cretaceous

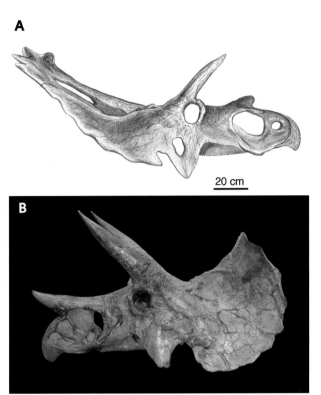

Figure 12.25. A, reconstructed cranium of the chasmosaurine *Anchiceratops ornatus* in lateral view; **B**, cranium of the chasmosaurine *Triceratops prorsus* in lateral view. Length of skull from rostral to the posterior edges of the quadrates 91 cm. A, digitally modified from D. A. Russell and Chamney (1967); B, from Longrich and Field (2012)—CC BY 2.5.

(Campanian) of Alberta, has large openings and a nearly straight or somewhat medially embayed posterior edge (Campbell et al. 2016). The facial horns of this taxon are rather short. The frill of *Kosmoceratops*, from the Late Cretaceous (Campanian) of Utah, is particularly ornate, with a parapet of 10 mostly forward curved accessory ossifications (Sampson et al. 2010). *Triceratops* is known from many specimens, which show considerable individual variation in features of the skull but likely represent only two species (Forster 1996b). It reached a total length of up to 9 m.

The biological significance of the frill and horns in ceratopsids has long been the subject of discussion. Differences in the structure of these features are pronounced among the numerous named ceratopsid taxa and are considered diagnostic in a clade that apparently has few differences in other skeletal features. The horns were

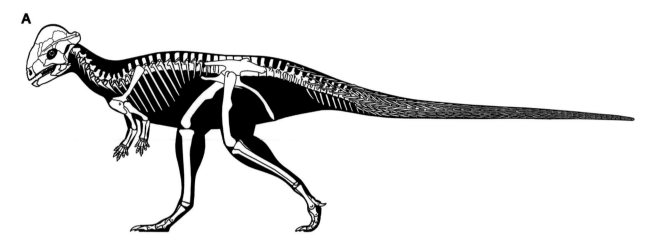

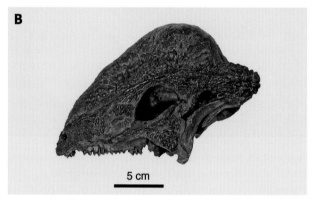

Figure 12.26. A, reconstructed skeleton of the pachycephalo-saurid *Stegoceras validum*; **B**, cranium of *Stegoceras validum* in lateral view. A, courtesy of Scott Hartman; B, courtesy of David Evans.

initially interpreted as defensive weapons against preda-tors. More recent studies (Farlow and Dodson 1975; Sampson et al. 1997; Sampson and Loewen 2010), how-ever, have highlighted their similarity to the horns in present-day bovid mammals, which are used primarily for intraspecific combat and competition for mates. The latter hypothesis gained additional support from the fact that the taxon-specific features of the horns became fully expressed only late during ontogeny (Sampson et al. 1997). However, Padian and Horner (2011) argued that there is little evidence for sexual dimorphism in the structure of the frill and horns among ceratopsids and that these features more likely played a role in species recognition.

The parietosquamosal frill in ceratopsids was long in-terpreted as providing an enlarged surface for the origin of adductor jaw muscles (e.g., Ostrom 1964). Although this hypothesis is plausible for more basal ceratopsians, the greatly expanded frills of coronosaurians more likely served primarily for visual display, and the jaw muscles

did not extend posteriorly much beyond the base of the frill (Makovicky and Norell 2006).

ORNITHISCHIA: GENASAURIA: MARGINOCEPHALIA: PACHYCEPHALOSAURIA

Pachycephalosauria (from Greek *pachys*, thick, *kephale*, head, and *sauros*, lizard) comprises all marginocephalians more closely related to *Pachycephalosaurus wyomingensis* than to *Triceratops horridus* (Sereno 2005). Superficially re-sembling ornithopods and basal neornithischians in many skeletal features, this clade was traditionally asso-ciated with the former until Sereno (1986) recognized a closer relationship to ceratopsians.

Wannanosaurus, from the Late Cretaceous of Anhui (China) (Hou 1977), has been considered the basalmost pachycephalosaur (e.g., Sereno 2000) but Butler and Q. Zhao (2009) questioned this assessment in view of the ju-venile nature of the only known specimen. *Wannanosau-rus* and other pachycephalosaurian taxa share a number of synapomorphies (Butler et al. 2008): two or three supraorbital bones integrated into the dorsal margin of the orbit; suture between the jugal and postorbital forming a short butt joint; postorbital-squamosal bar of the cranium broad and flat; postorbital and squamosal bearing a row of tubercles; frontal and parietal are thick dorsoventrally; presence of a single caniniform tooth in the dentary; length of the humerus less than 60 percent of that of the femur; humerus strongly bowed laterally along its length; preacetabular process of the ilium transversely expanded, forming a narrow dorsal shelf; pubic process of the ischium dorsoventrally compressed; and fourth trochanter of the femur forming a prominent ridge.

Goyocephale and *Homalocephale*, both from the Late Cretaceous (Campanian) of Mongolia (Maryańska and Osmólska 1974; Perle et al. 1982), have thick but flat skull roofs and large upper temporal fenestrae, unlike Pachycephalosauridae, which have distinctly domed skull roofs and often lack supratemporal fenestrae as adults. They share various derived features with Pachycephalosauridae (Butler et al. 2008): width of the jugal-postorbital bar greater than that of the lower temporal fenestra; postorbital and parietal forming a broad contact; the presence of a row of large tubercles along the posterior surface of the squamosal; preacetabular process of the ilium laterally deflected by more than 30 degrees and expanded transversely toward its anterior end; dorsal margin of the ilium with a subtriangular process extending medially; and tail surrounded by a distinctive double-layered lattice of ossified "tendons" (Fig. 12.26A), which represent ossifications (myorhabdoi) that formed in the myosepta between blocks of tail muscles (C. M. Brown and A. P. Russell 2012). The somewhat leaf-shaped dentary and maxillary tooth crowns of pachycephalosaurs closely resemble those of other ornithischians, and the transversely broad abdominal region is consistent with the presence of a capacious gut suitable for microbial fermentation of plant fodder. The disparity in size between the fore- and hind limbs in *Stegoceras* (Fig. 12.26A) indicates obligate bipedality.

In Pachycephalosauridae, the thickened frontals and parietals become fused to form a single dome (Fig. 12.26B). The dome consists of vertical columns of bone that grew outward during ontogeny. Juveniles have a flat skull roof that developed into a distinct dome in adults (Schott et al. 2011). The upper temporal fenestrae became closed late during ontogeny in taxa with highly domed skull roofs. In the latter, additional cranial bones became incorporated into the dome.

Stegoceras, from the Late Cretaceous (Campanian) of Alberta, Montana, and New Mexico, has a distinct parietosquamosal shelf that extends posterolaterally from the frontoparietal dome (Fig. 12.26B). *Pachycephalosaurus*, from the Late Cretaceous (Maastrichtian) of Montana, South Dakota, and Wyoming, developed a massive dome that posteriorly extends to the posterior edge of the skull roof and forms a steep posterior slope (B. Brown and Schlaikjer 1943; Sues and Galton 1987; Sereno 2000; R. M. Sullivan 2006). Recent studies (Snively and Theodor 2011; Peterson et al. 2013) have supported the hypothesis first elaborated by Galton (1971) that the cranial dome in pachycephalosaurs served in head- or flank-butting. This does not rule out that this structure, especially together with the squamosal nodes or horns, also served additional functions such as species recognition (e.g., Padian and Horner 2011). *Stegoceras* attained a total length of up to 2 m and *Pachycephalosaurus* possibly reached more than twice that length.

13   A Brief History of Reptiles

Following the discussion of reptilian diversity, this chapter presents a brief overview of the evolutionary history of this group.

The oldest known amniotes, from the Pennsylvanian (Bashkirian) of Nova Scotia (Canada), are the basal eureptile *Hylonomus* and the presumed basal synapsid *Protoclepsydrops*. They indicate that the two principal lineages of amniotes, Reptilia and Synapsida, had already diverged prior to this point in time. The earliest amniotes lived in the tropical forests of Euramerica while Gondwana experienced a major glaciation. Although the fossil record of reptiles and synapsids from the Pennsylvanian is still sparse, currently available data suggest that reptiles were represented by small-sized forms that subsisted on arthropods and other small animals. By contrast, synapsids were more diverse and, by the end of the Pennsylvanian, already included relatively large-sized carnivores and herbivores (Reisz and Fröbisch 2014; Modesto et al. 2015).

Deglaciation led to several intervals of marked warming during the early and middle Permian (Cisuralian-Guadalupian) (Montañez et al. 2007). Warmer, drier climates led to major changes in continental ecosystems. Basal synapsids ("pelycosaurs") and temnospondyl stem-amphibians dominated early Permian (Cisuralian) communities of continental tetrapods. Both carnivorous and herbivorous synapsids from the early Permian (Cisuralian) of the American Southwest show increases in body size; some herbivores attained much greater body sizes than the apex predators (Reisz and Fröbisch 2014).

Reptiles began to diversify during the early Permian (Cisuralian) (Sahney et al. 2010). The first herbivores appeared in several clades with otherwise faunivorous or omnivorous basal taxa. Captorhinid reptiles show an increase in body size in those taxa that have dentitions with multiple tooth rows suitable for feeding on high-fiber plant material (Hotton et al. 1997; Reisz and Sues 2000). However, other known early Permian reptiles attained total lengths of less than 1 m.

During the middle to late Permian (Guadalupian-Lopingian), more derived synapsids (therapsids) became the numerically dominant amniotes in the well-known assemblages of continental tetrapods from European Russia and from East and South Africa. Parareptiles and eureptiles also became much more diverse. One group of parareptiles, Pareiasauridae, included the first reptilian megaherbivores. Another clade, Mesosauridae, lived in an inland sea covering parts of southern Africa and South America during the early Permian (Cisuralian). Most known Permian eureptiles were small-sized and subsisted on arthropods and small vertebrates. Some may have been omnivorous or even herbivorous without exhibiting features that reflected changes in diet. In addition to the gliding weigeltisaurids, other small-sized reptiles probably also explored life in the trees. Based on extant small-bodied lizards, even primarily terrestrial species can readily move between the ground and trees and do not require specialized morphological traits for living and foraging in trees.

During the latter part of the Permian, two major biotic crises affected ecosystems on land and in the sea—one at the end of the middle Permian (Guadalupian) and the other the mass extinction at the end of the Permian (Lopingian). Synapsids were severely affected by the latter event, particularly the abundant herbivorous dicynodonts (J. Fröbisch 2008). By contrast, the effect of these extinction events on reptilian diversity is still poorly understood. Pareiasaurs disappeared at the end of the Permian, and most other parareptilian groups had already vanished earlier during this period. Only procolophonoids remained apparently unaffected by the end-Permian extinction (Modesto et al. 2001) and diversified during the Triassic. The known fossil record of eureptiles from the late Permian (Lopingian) is sparse. The last captorhinids date from the late Permian (Lopingian). Among diapsid reptiles, a variety of forms such as *Youngina*, weigeltisaurids, and aquatic taxa such as *Claudiosaurus* appear to have been restricted to the late Permian (Lopingian). The basal archosauromorph *Protorosaurus* and basal archosauriform *Archosaurus* indicate a still largely undocumented diversification of this major clade of diapsid reptiles that predated the end-Permian extinction event (Ezcurra et al. 2014; Ezcurra 2016; Pinheiro et al. 2016). The oldest undisputed lepidosauromorph is the Early Triassic *Paliguana* but the ghost lineage of this group extends well back into the Permian (Ezcurra et al. 2014).

The Triassic Period marks the beginning of the Mesozoic Era, which has long been known as the Age of Reptiles because of the abundance and diversity of reptiles living during this time interval (Sahney et al. 2010; Sues and Fraser 2010). "Greenhouse" conditions dominated global climates. It was also the period during which the supercontinent Pangaea attained its maximum extent, with few apparent major barriers to the dispersal of land-dwelling animals and plants.

At the beginning of the Triassic, synapsids and temnospondyl stem-amphibians dominated low-diversity communities of continental vertebrates. A few clades in both groups diversified again during this period, but the rapid evolutionary radiation of diapsid reptiles soon relegated them to fewer ecological niches. At the beginning of the Middle Triassic (Anisian), archosauromorphs had already attained considerable diversity (Nesbitt 2011; Nesbitt et al. 2015).

The Triassic was the heyday of archosauromorph diversification (Fig. 13.1). Along with archosaurs, several lineages of early archosauriforms included the oldest known reptilian macropredators (Erythrosuchidae) and a variety of probably semiaquatic forms (Nesbitt 2011). Among more basal archosauromorphs, several clades evolved craniodental features for omnivory or herbivory. One, rhynchosaurs, became especially abundant in

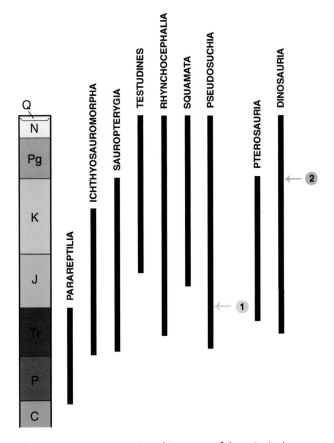

Figure 13.1. Known stratigraphic ranges of the principal reptilian clades. "1" marks the extinction of pseudosuchians other than crocodylomorphs and "2" the extinction of dinosaurs other than birds. Abbreviations for geological periods: C, Carboniferous; J, Jurassic; K, Cretaceous; N, Neogene; P, Permian; Pg, Paleogene; Q, Quaternary; Tr, Triassic.

Middle and early Late Triassic tetrapod communities of Gondwana. Archosauromorphs attained and maintained high growth rates (de Ricqlès et al. 2008), which may have provided them with a significant competitive advantage over other tetrapods.

Among archosaurs, pseudosuchians and avemetatarsalians first appeared in the fossil record at the beginning of the Middle Triassic (Nesbitt et al. 2010, 2017). Pseudosuchians were more common and diverse than avemetatarsalians during much of the Triassic but, with the exception of crocodylomorphs, most vanished at or before the end of period. Middle and Late Triassic pseudosuchians included a diversity of predominantly quadrupedal medium- to large-sized carnivores and herbivores. Early crocodylomorphs were terrestrial predators with limb proportions indicating cursoriality (Parrish 1987).

Avemetatarsalians had an upright stance and parasagittal gait, with movements of the femur restricted to a fore-and-aft plane. The structure of their girdles and limbs differs distinctly from that in pseudosuchians. Most early forms were bipedal and cursorial. Histological data from limb bones indicate that avemetatarsalians had high growth rates, which were particularly conspicuous in dinosaurs and pterosaurs. The latter groups attained rates comparable to those in extant birds and mammals (Padian et al. 2001). This indicates that these reptiles had high basal metabolic rates as well. The oldest known pterosaurs from the Late Triassic already have all the features for active flight (Wild 1978). The earliest undisputed dinosaurs date from the early Late Triassic. The simultaneous appearance of several dinosaurian clades at this point in time shows that the initial diversification of the group had occurred earlier. The earliest saurischian dinosaurs included both theropods and sauropodomorphs. The earliest sauropodomorphs were small- to medium-sized and apparently carnivorous or omnivorous, but large-bodied, predominantly herbivorous forms soon became common. Surprisingly, there is as yet no undisputed record of Late Triassic ornithischian dinosaurs, although most phylogenetic hypotheses predict their presence at that time.

The end-Permian extinction in the sea probably led to new ecological opportunities that triggered a great diversification of marine reptiles near the beginning of the Triassic. Early members of various clades of marine rep-

tiles lived in nearshore marine settings. Only ichthyosaurs and plesiosaurs fully adapted to life in the open sea by the end of the Triassic. Ichthyosaurs and most sauropterygians probably subsisted on cephalopods, fish, and marine reptiles. Some early ichthyosaurs and placodonts have dentitions suitable for feeding on hard-shelled invertebrates. Motani et al. (2015b) noted that the diversity of feeding modes among marine reptiles was highest during the Early Triassic and gradually decreased afterward.

The Triassic marks the first appearances of lepidosaurs and undisputed stem-turtles. Stem-squamates appeared in the fossil record in the early Middle Triassic (Anisian), and rhynchocephalians in the late Middle Triassic (Ladinian). In addition, other groups of small-bodied diapsid reptiles, such as the arboreal drepanosauromorphs and the gliding kuehneosaurids, apparently were restricted to the Triassic.

The end-Triassic extinction event again affected the diversity of marine and continental tetrapods. Among marine reptiles, only parvipelvian ichthyosaurs and plesiosaurian sauropterygians survived and then rapidly diversified during the Jurassic Period. Among continental tetrapods, pseudosuchians other than crocodylomorphs, procolophonid parareptiles, and other exclusively Triassic groups vanished near or at the end of the Triassic.

Dinosaurs apparently were largely unaffected by the end-Triassic faunal changes and rapidly diversified during the Jurassic. Sauropodomorph and theropod dinosaurs became not only more diverse than during the preceding period but also quickly increased in body size. However, one clade of theropods shows a progressive decrease in body size and gave rise to birds. Early Jurassic ornithischian dinosaurs include both neornithischians and the armored thyreophorans. Middle and Late Jurassic dinosaurian communities had a considerable diversity of carnivores and herbivores. The former included numerous small- to large-sized theropods, the latter especially including allosauroids. Among the dinosaurian herbivores, sauropods and ornithischians represent two major evolutionary radiations with different feeding mechanisms. Sauropods were bulk feeders that did not employ oral processing but relied on fermentative breakdown of plant cellulose by endosymbiotic microorganisms in their capacious digestive systems (Sander et al. 2011). They had already attained large body size early in the Jurassic, and some giant Late Jurassic sauropods such as *Giraffatitan* (with a height of 12 m) and *Diplodocus* (with a total length of up to 30 m) are among the largest terrestrial animals of all time. The elongation of the neck in sauropods facilitated more effective foraging and also feeding at higher levels above the ground, which other dinosaurian herbivores could not exploit. By contrast, Jurassic ornithischians were smaller in body size. Neornithischians evolved increasingly complex dentitions suitable for oral processing of plant food. Among thyreophorans, ankylosaurs and stegosaurs first appeared in the Middle Jurassic. At the Jurassic-Cretaceous transition, the low-browsing ankylosaurs and neornithischians began to dominate communities of herbivorous dinosaurs, while sauropods and stegosaurs declined in diversity (Butler et al. 2009).

The earliest crown-group turtles appeared in the Late Jurassic (Joyce 2007). Squamates have a known fossil record dating back to the Middle Jurassic, when scincomorphs, anguimorphs, and possibly gekkotans are first documented (S. E. Evans 1998, 2003). Crocodyliform archosaurs diversified rapidly during the Jurassic. During the Early Jurassic, they already included the marine thalattosuchians as well as various taxa presumably living in freshwater. Among pterosaurs, long-tailed basal forms ("rhamphorhynchoids") flourished until the Late Jurassic. The more derived pterodactyloids first appeared toward the end of the Middle Jurassic and diversified during the Late Jurassic and Early Cretaceous (Andres et al. 2014).

Ichthyosaurs and plesiosaurs flourished during the Jurassic Period, but their diversity decreased around the Jurassic-Cretaceous boundary (Fischer et al. 2012; Benson and Druckenmiller 2014). Ichthyosaurs vanished at the beginning of the Late Cretaceous, but plesiosaurs persisted to the end of the Cretaceous.

The Cretaceous Period witnessed profound changes in continental ecosystems (Lloyd et al. 2008). In particular, flowering plants (angiosperms) first appeared during the Early Cretaceous and already played a dominant ecological role toward the end of this period. Concurrently, gymnosperms declined in diversity. This floral change probably led to major diversifications among insects and various groups of vertebrates. Surprisingly, however, the

analysis by Butler et al. (2009) found no positive correlation between the diversity of dinosaurian herbivores and that of angiosperms. Dinosaurs continue to be represented by medium- to large-sized carnivores and herbivores, but significant changes in the composition of dinosaurian communities occurred during this period. The disintegration of Pangaea led to increased faunal provincialism. For example, allosauroid theropods remained the apex predators during the Early and mid-Cretaceous worldwide but, by the Late Cretaceous, they were replaced by tyrannosauroids in Laurasia and by abelisaurids in Gondwana. Similarly, ceratopsid marginocephalians and hadrosauroid ornithopods are the most abundant large-sized plant-eaters in Laurasia during the Late Cretaceous, while titanosaurian sauropods, which included the largest land animals of all time, were the principal megaherbivores in Gondwana. Ceratopsids and hadrosauroids had highly complex dentitions that allowed them to feed on a great variety of vegetation. Other neornithischians were much less common and diverse.

Cretaceous pterosaurs show a wide range of feeding adaptations and included azhdarchids, some of which were the largest flying animals of all time. Crocodyliform archosaurs attained considerable diversity during this period. Notosuchians, a predominantly Gondwanan clade, had diverse modes of life, from small-sized omnivores/herbivores to large-sized terrestrial predators (Pol et al. 2014). The earliest representatives of the three major lineages of extant crocodylians appeared in the fossil record during the Late Cretaceous (Brochu 2003). Squamates diversified rapidly during the Cretaceous, with most of the extant clades present by the end of this period (Vidal and Hedges 2009; Daza et al. 2016). Undisputed snakes first appeared in the mid-Cretaceous. Several groups of Cretaceous squamates adapted to marine life. One, mosasauroids, included the apex predators in the Late Cretaceous seas. Both major clades of crown-group turtles diversified during the Cretaceous. Cryptodiran turtles included a number of forms that adapted to life in the sea.

The end-Cretaceous extinction event led to a substantial reduction in the number of major reptilian clades (Fig. 13.1). On land, it led to the demise of nonavian dinosaurs, pterosaurs, and a variety of groups of squamates and crocodyliforms. In the sea, plesiosaurs and mosasaurs vanished. However, various reptilian groups survived this biotic crisis. Some possibly weathered the environmental changes at the end of the Cretaceous by sheltering in burrows or dens. Others were semiaquatic, and this mode of life may have conferred a measure of protection.

The Paleogene and Neogene witnessed the appearance and diversification of "modern" reptilian communities. Birds and mammals came to dominate continental tetrapod communities. Global climates changed from "greenhouse" to "ice-house" conditions at the transition from the Eocene to the Oligocene epoch (Zanazzi et al. 2007). As most reptiles rely on ambient temperatures to regulate their body temperatures, these climatic changes undoubtedly affected their diversity and geographic dis-tribution. Crocodylian diversity declined in Eurasia and North America from the Paleogene to the Neogene and again at the end of the latter (Markwick 1998). Rage (2013) related differences in the composition of assemblages of squamates in Europe during the Cenozoic to climatic changes. An important evolutionary event, commencing early in the Neogene, was the rapid diversification of colubroid snakes. Drying climates led to the spread of grassland habitats, which possibly favored these active predators over more basal snakes (Savitzky 1980).

The higher-level diversity of present-day reptiles other than birds is greatly diminished, with only three surviv-ing major clades—crocodylians, lepidosaurs (*Sphenodon* and squamates), and turtles. Of these, only squamates are ecologically diverse and remarkably speciose.

14 The Future of Reptiles

At the end of this survey of the diversity and evolution of reptiles, it is appropriate to take a look at their future. Like all animals and plants, reptiles are greatly affected by the current pervasive and increasingly rapid changes in Earth's ecosystems. The global human population has increased from fewer than two billion in the early 1900s to seven billion by 2013. This population explosion, with resulting pressures on natural resources and pollution of air, soil, and water, has already fundamentally altered the face of the planet. Fields and pastures make up more than half of Earth's land surface. The loss of natural habitats continues to accelerate as the human population keeps growing and standards of living are rising worldwide. Large-scale burning of fossil fuels releases vast quantities of carbon dioxide and other gases, leading to a buildup of these gases in and rapid warming of the atmosphere. In May of 2013, the concentration of carbon dioxide surpassed 400 parts per million (ppm) for the first time (and has done so every year since), up from 280 ppm during the pre-1800s (Ruddiman 2014). Recent studies project increases in average global temperatures of up to 4.5°C by the end of this century.

People often wonder why they should be concerned about the present losses of biodiversity. After all, extinction is a natural process that has operated without cease ever since life arose. Typically, however, it proceeds at a slow, fairly constant rate. Regardless of how well adapted a species may appear to be to a particular mode of life at any point in time, it will eventually vanish in a world of continuous change. This phenomenon is known as "background extinction."

At several points in the history of life on Earth, sudden, often dramatic increases in the rates of extinction have occurred. Large numbers of species perished over (geologically speaking) brief intervals of time, frequently leading to profound restructuring of ecosystems on land and in the sea. Such episodes are commonly referred to as "mass extinctions." Five severe biodiversity crises have occurred during the last 500 million years. Many biologists consider the present-day loss of biodiversity to be the sixth mass extinction. Such assessments may appear premature to some, but present-day environmental changes are leading to the rapidly accelerating loss of species. Pimm et al. (2014) estimated that current rates of extinction are about 1,000 times greater than the likely background rate of extinction.

The adverse and profound effects of humanity's activities on habitats and biodiversity around the world is no longer in dispute among scientists. However, the specific consequences of these changes for most animals and plants have yet to be fully understood. This poses an enormous challenge, because the available data indicate that

Figure 14.1. Gopher tortoise (*Gopherus polyphemus*). This ecologically important turtle is threatened by habitat destruction. From Wikipedia (photo by Tom Friedel)—CC BY 3.0.

the risk of extinction varies not only between different groups of organisms but even among different species within particular groups.

It was once widely but falsely believed that the extinction of reptiles would make little if any difference to ecosystems. In fact, even the loss of a single species of reptile can have significant ecological consequences. A good example is the gopher tortoise (*Gopherus polyphemus*; Fig. 14.1), which lives in the longleaf pine-wiregrass communities of the southeastern United States (A. J. Martin 2012). The tortoise excavates extensive tunnels, which maintain fairly constant temperature and humidity throughout the year and offer protection from periodic wildfires. These tunnels also provide shelter to many other species of animals, ranging from arthropods to rodents, including species found nowhere else. Furthermore, the turtle's tunneling churns the soil, which, in turn, alters the composition of communities of bacteria and fungi living in that soil. Thus, the disappearance of the gopher tortoise will result in the loss of many associated species.

Rising global temperatures are possibly the most important threat to reptiles. Climate change not only leads to shifts in their distribution, as species attempt to track suitable habitats, but also affects their lives more directly. Sinervo et al. (2010) observed that rising temperatures directly affect feeding and reproduction in lizards of the genus *Sceloporus*. Their surveys of populations of these lizards at 200 different sites in Mexico indicate that temperatures in those regions have changed too rapidly

to allow the lizards time to adapt. Higher temperatures are affecting the animals' ability to regulate their body temperatures, limiting the number of hours they can spend foraging and disturbing the development of embryos in those species that give birth to live young. Lizards appear to be far more susceptible to extinction caused by rising temperatures than was previously assumed. Many species already exist close to or right at the edge of their temperature tolerances. Thus, even relatively minor changes in average temperatures will lead to declines in population size or even extinction. In crocodylians and certain turtles, temperatures inside the nest during incubation of the eggs determine the male-female ratios of the hatchlings (Valenzuela 2004). High nest temperatures produce males in crocodylians and some squamates but females in turtles. Thus, increasing temperatures will alter sex ratios, which, in turn, will affect the long-term survival of reptilian populations. Changes in incubation temperature also influence aspects of the hatchling's biology, such as behavior and performance during locomotion (Deeming 2004; Booth 2006).

The destruction of natural habitats is another major cause of losses of reptilian biodiversity. Human activities particularly threaten the world's tropical rainforests, which harbor much of Earth's biodiversity. Although these forests together cover only 7 percent of Earth's land surface, biologists have estimated that they are home to at least half of all known species of animals and plants (Myers 1988). Almost half of the original rainforest cover had already vanished around the end of the twentieth century. In recent years, a few countries such as Costa Rica have made concerted efforts to set aside and protect large tracts of forest as biodiversity preserves, but much more needs to be done in this regard worldwide.

Habitat destruction is by no means confined to the tropical rainforests. For example, widespread losses of wetlands in the temperate regions have greatly affected natural habitats for many species of turtles and snakes. Even when habitats are not completely destroyed, they are often reduced to isolated small patches that are inadequate to ensure the long-term survival of many species (Fig. 14.2).

Species are affected not only directly by the loss or fragmentation of their habitats but also indirectly, by

Figure 14.2. Brown tree snake (*Boiga irregularis*). This colubrid snake was accidentally introduced to Guam where it extirpated most of the native forest-dwelling vertebrates. From Wikimedia (photo by Pavel Kirillov)—CC BY-SA 2.0.

Figure 14.3. Head of a young female tuatara (*Sphenodon punctatus*). This species was once widely distributed across New Zealand, but predation by invasive mammals arriving with human colonizers has limited its present geographic range to a few small islands. Courtesy of Paddy Ryan.

changes brought about by habitat alteration. For example, cutting down stands of trees leaves the ground bare. Soil temperature increases, as does the flow of water through the denuded soil, leading to nutrient leaching and increased soil erosion.

Invasive species affect indigenous reptiles as competitors, potential carriers of diseases and parasites (Kraus 2015). The arrival of the brown tree snake (*Boiga irregularis*; Fig. 14.2) on the island of Guam from either New Guinea or the Solomon Islands shortly after World War II resulted in dramatic declines among or extinction of native vertebrates. More recently, the Burmese python (*Python bivittatus*) was introduced to the Florida Everglades and has negatively affected indigenous reptiles and other vertebrates in this region. Over the millennia, mice and rats, domestic cats and dogs, and pigs traveling with human colonizers to oceanic islands have extirpated many native species of reptiles and other vertebrates. A well-known example is the tuatara (*Sphenodon punctatus*; Fig. 14.3), which once thrived in New Zealand but proved easy prey for the mammalian invaders. Today its natural range of distribution is reduced to a few small islands (Cree 2014).

Environmental pollution by fertilizers, herbicides, pesticides, radioactive waste, and toxic metals adversely affects many species of animals and plants directly and indirectly. Much research on this subject remains to be done for reptiles, especially because earlier studies have tended to focus on the immediate rather than the more long-term effects of chemical substances on species. Turtles and crocodylians have both environmental sex determination and large eggs, attributes that particularly expose these reptiles to endocrine-disrupting chemicals, which affect their sexual maturity and fertility (Crain and Guillette 1998; Gibbons et al. 2000). Some contaminants are not immediately lethal but indirectly affect physiological functions (e.g., Sasaki et al. 2016). Hopkins et al. (1999) reported that, for water snakes (*Nerodia fasciata*), consumption of prey containing metal contaminants in wetlands polluted by coal ash leads to elevated metabolic rates, which appear to reduce the energy available for growth and reproduction.

Diseases and parasites can severely affect species already weakened by other environmental changes. Among amphibians, infection by the chytrid fungus *Batrachochytrium dendrobatidis* has caused massive die-offs among populations of frogs in the Americas, Australia, and Europe in recent years (Fisher et al. 2009).

Some species of reptiles face an especially high risk of extinction because they are easily hunted, serve as food or as ingredients in folk medicine, or have skins sought for luxury trade goods. A well-known example is the decline in the numbers of freshwater turtles and tortoises in China, where there is great demand for their meat for soups and their shells for traditional medicine (Shi et al.

2008). Commercial turtle farming in China has not reduced the threat to wild tortoises and freshwater turtles, which some consumers prefer to farm-raised ones. Thus, the harvesting of wild populations of freshwater turtles and tortoises in China and neighboring countries continues unabated and will likely lead to their extinction in the not-too-distant future. In order to meet consumer demand, Chinese turtle farmers already import freshwater turtles from the United States.

It has become fashionable to talk about "sustainable development" of biological resources. This idea sounds appealing, but development of such harvesting strategies requires far more data on population densities and natural fluctuations in population size than are available for most species of animals and plants. Unless populations of particular species have been carefully monitored over long periods of time, it is difficult to assess the effects of short-term variations on their sizes and distributions. Such information is unavailable for the majority of known species of animals and plants, which have never been thoroughly studied in their natural habitat.

Larger-bodied species of reptiles often grow and mature at rates that are too slow to sustain continued harvesting of the most sought-after large individuals in the wild. In the case of crocodiles and alligators, captive breeding and carefully managed harvesting in several countries (Ross 1998) have reduced or even eliminated the threat to wild populations of some species (e.g., *Alligator mississippiensis*; Fig. 14.4). However, other species, such as the Philippine freshwater crocodile (*Crocodylus mindorensis*), remain critically endangered.

Turtles, lizards, and snakes have become popular worldwide as pets. Unfortunately, excessive collecting of certain species for the pet trade has already led to dramatic declines, or even local extirpation, of their wild populations. A well-documented example is the overharvesting of tortoises, first in Europe and then in Asia and Africa (Fig. 14.5). Although international efforts have been made to curb the trade in endangered wildlife, a vast global underground market for exotic species continues to flourish.

The best estimates of the 2013 Red List of Threatened Species, compiled by the International Union for the Conservation of Nature (IUCN), are that 13.2 percent of bird, 25 percent of mammal, and 41 percent of amphib-

Figure 14.4. American alligator (*Alligator mississippiensis*) in Florida. Once decimated, conservation efforts led to substantial increases in the size of populations and now the American alligator is no longer considered threatened. Courtesy of Harry Greene.

Figure 14.5. Indian star tortoise (*Geochelone elegans*). This and other species of tortoises are endangered by excessive collecting for the pet trade and by habitat loss. Courtesy of Division of Amphibians and Reptiles, National Museum of Natural History.

ian species are currently threatened with extinction (www.iucnredlist.org). How serious is the threat to reptiles? A 2013 study by a large international group of researchers (Böhm et al. 2013) quantitatively assessed the

Figure 14.6. A highly venomous Russell's viper (*Daboia russelli*) moving close to human habitation in India. Human overpopulation increasingly restricts natural habitats to the detriment of both people and reptiles. Courtesy of Wolfgang Wüster.

extinction risk for reptiles based on 1,500 randomly selected species from around the world. It found that 19 percent of the sampled species—nearly one in five—are threatened with extinction. Of these, 12 percent were classified as "critically endangered" (facing a particularly high risk of extinction in the wild). Three of the critically endangered species surveyed by Böhm and her team may have already been extinct in their native habitats at the time of publication for the report. Lovich et al. (2018) found that about 61 percent of the 356 cur-

rently recognized turtle species are threatened or already extinct, making turtles the most endangered group of vertebrates.

The survey by Böhm et al. (2013) confirmed that the risk of extinction is not evenly spread among reptiles. Freshwater turtles are at a particularly high risk, reflecting greater overall levels of threat to freshwater animals around the world. It estimated that 30 percent of species of freshwater reptiles, and 50 percent of species of freshwater turtles, are approaching extinction. Although the risk of extinction is lower for land-dwelling reptiles, small home ranges, specific ecological requirements, and limited mobility render these animals highly vulnerable to habitat changes. Ceballos et al. (2017) noted that biologists tend to focus on outright species losses and argued that often dramatic declines in populations, even among species not currently considered endangered, should be a much greater cause for concern.

Even under the most optimistic scenarios, most groups of animals and plants on land and in the sea—including reptiles—will incur substantial losses in species diversity in the decades ahead. It will require concerted national and international political efforts to slow the rates of climate change and human population growth in order to stem such losses. Individuals can contribute to this effort. They can support organizations dedicated to protecting biodiversity and working with governments to set up and maintain protected areas across the globe. At a time when a majority of people grow up and live in cities, parents and educators must make much more concerted efforts to introduce children to the natural world, highlighting both its riches and its vital importance to humanity's own survival.

Reptiles have flourished and survived three major extinction events during their 300-plus-million-year history but face an uncertain future. We share the world with them (Fig. 14.6). Our choices and actions will determine their fate and ours.

Glossary

Abduction   Movement of a limb or part of a limb away from the midline of the body.

Acetabulum   Socket on the pelvis for the reception of the **femur**.

Acrodont   Type of dentition in which the teeth are situated on top of and are fused to the jawbone.

Adduction   Movement of a limb or part of a limb toward the midline of the body.

Alveolus   Pocket-like structure. Example: tooth socket in jaw.

Amniota   Clade of tetrapods comprising reptiles, birds, mammals, and their close relatives. The eggs of amniotes are characterized by the presence of the **amnion**, an extraembryonic membrane that encloses a fluid-filled chamber surrounding the embryo.

Amphicoelous   Vertebral **centrum** that has concave anterior and posterior ends.

Amphibious   Living, or capable of living, both in water and on land.

Analog   Structures with similar function that have evolved completely independently. (Adjective: **analogous**.) Example: wings of birds and pterosaurs.

Anapsid   Type of cranium where the region behind the eye socket (orbit) is completely covered by bone.

Anterior   Pertaining to or directed toward the front end of the body.

Antorbital   Pertaining to the region of the skull in front of the orbit.

Apical   Directed toward or pertaining to the tip (apex) of a tooth crown.

Apomorphic   When different **character states** in two or more taxa are compared and one is thought to have evolved from the other, the former is interpreted as an apomorphic (or derived) character state. (Noun: **apomorphy**.)

Aquatic   Adapted for or pertaining to life in water.

Arboreal   Adapted for or pertaining to living and moving in trees and bushes.

Articulation   Point at which two bones contact each other in a joint.

Astragalus   Ankle bone that contacts the tibia and calcaneum to form the ankle joint. (Plural: **astragali**.)

Atlas   First **cervical** vertebra, connecting the skull to the vertebral column. (Plural: **atlantes**.)

Autapomorphy   Derived character state restricted to a particular taxon.

Axial   Pertaining to the vertebral column and associated structures.

Basal taxon   Taxon that diverged early in the evolutionary history of a **clade**.

Basicranium   Base of the **cranium**.

Beak   Sheaths of keratin covering the bony jaws. Also known as **rhamphotheca**.

Biostratigraphy   Correlation of sedimentary rocks by their fossil content.

Bipedal   Moving only on the hind limbs.

Bone   Skeleton-forming material produced by deposition of a form of calcium phosphate known as hydroxyapatite into a cartilaginous or fibrous matrix.

Calcaneum   Ankle bone that contacts the **astragalus** and **tibia** to form the ankle joint.

Carina   Cutting edge along the mesial and/or distal edge of a tooth crown.

Carnivorous   Subsisting predominantly or exclusively on animals.

Carpal   1. **(Adjective)** Pertaining to the wrist. 2. (Noun) A wrist bone.

Carpus   Set of bones or cartilages forming the wrist.

Caudal   1. (Adjective) Pertaining to or directed toward the tail. 2. (Noun) Short for caudal vertebra.

Centrum   Body of a **vertebra**, located below the spinal cord and surrounding (and partially replacing) the notochord.

Cervical   1. (Adjective) Of or pertaining to the neck. 2. (Noun) Short for cervical **vertebra**.

Character   Observable or measurable, heritable attribute of the structure or mode of life of a **taxon**.

Character state   Condition of a particular character found in a **taxon**. It can denote the presence or absence of that character or a specific expression of that character.

Choana   Internal opening of the nasal passage into the mouth cavity.

Clade   Group of organisms comprising their most recent common ancestor and all descendants of that ancestor and united by the shared possession of derived features (**synapomorphies**). Also known as a **monophyletic group** or **natural group**.

Cladogram   Branching diagram illustrating a hypothesis of relationships among taxa based on shared derived features. It does not have an implied time axis nor does it connote ancestor-descendant relationships.

Clavicle   Paired bone along the anterior edge of the **scapula**, joining it to the **interclavicle** or the **sternum**.

Cloaca   Common chamber into which the digestive, excretory, and reproductive systems discharge their contents. It opens to the outside through the vent.

Condyle   Rounded projection at the end of a bone that fits into a depression on another bone for movable contact between the two elements.

Convergence   Independent development of similar characters in only distantly related lineages of organisms. Also referred to as **homoplasy**. Example: overall body form in some fishes, certain ichthyosaurs, and small cetaceans.

Coracoid   Bone extending from the **scapula** to the **sternum**.

Coronoid process   Bony prominence on the lower jaw for insertion of the adductor jaw muscles.

Costal   Pertaining to the ribs.

Cranial   Pertaining to or directed toward the head.

Cranium   The skull proper. (Plural: **crania**.)

Crown group   Clade bracketed entirely by extant taxa but also including extinct close relatives of these taxa.

Crus   Distal portion of the hind limb (comprising **tibia** and **fibula**).

Cursorial   Pertaining to running.

Denticles   Small teeth or tooth-like projections.

Dentine   Bone-like hard tissue that forms much of the tooth and surrounds the pulp cavity and root canal.

Dentition   Set of teeth.

Dermal bone   Bone formed in the inner layer (dermis) of the skin and without a cartilaginous precursor.

Diaphysis   Shaft of a long bone.

Diapsida   Major clade of **amniotes**, typically characterized by the presence of two openings behind the orbit on either side of the cranium. However, many derived diapsids no longer retain one or both of these openings.

Digit   A finger or toe.

Digitigrade   Standing and walking on the digits with the hind part of the foot held off the ground.

Dimorphism   Condition of having two different forms or morphs.

Distal   1. For limbs, directed away from the body or attached end. 2. For teeth, toward the back of the tooth row.

Dorsal   1. (Adjective) Pertaining to the upper side of a body or body part. 2. (Noun) Short for dorsal **vertebra**.

Edentulous   Devoid of teeth.

Enamel   Shiny, heavily mineralized hard tissue covering the outer surface of a tooth crown.

Eon   Largest unit of geological time. Example: Phanerozoic Eon.

Epiphysis   Secondary center of bone-formation, especially at the joint ends of a long bone.

Era   Unit of geological time comprising two or more periods. Example: Mesozoic Era.

Extant   Existing in the present day, as opposed to **extinct**.

Extension   Motion in which the angle of a joint increases or straightens.

Facultative   Behavior that may or may not occur, as opposed to **obligate**.

Fauna   Group of animals living together in a particular region.

Femur   Proximal bone of the hind limb. (Plural: **femora**.)

Fenestra   Window-like opening in bones or between bones. (Plural: **fenestrae**.)

Fibula   Lateral bone of the crus. (Plural: **fibulae**.)

Flexion   Motion in which the angle of a joint decreases or becomes more acute.

Foramen   Small, typically rounded opening in a bony or cartilaginous structure. (Plural: **foramina**.)

Fossa   Depression or recess in a bony or cartilaginous structure. (Plural: **fossae**.)

Fossil   Remnant or trace of an ancient organism.

Fossorial   Adapted for or pertaining to digging.

Gastralia   Rod-like bones in the abdominal wall. (Singular: **gastralium**.)

Gastrolith   Stone in stomach used for breaking down food and/or as ballast.

Genus   Taxonomic unit ranking above species and denoting a group of closely related species. (Plural: **genera**.)

Ghost lineage Temporal range of a lineage predicted by phylogenetic analysis but not yet documented by the fossil record.

Glenoid Joint surface (1) between the shoulder girdle and **humerus** or (2) between the **cranium** and **mandible**.

Gondwana Name for the ancient southern landmass that encompassed much of present-day Africa, Antarctica, Australia, Madagascar, and South America as well as parts of Asia (especially India) and southern Europe.

Hallux First digit of the **pes**.

Hemal arch Structure formed by paired bony processes below a **caudal** centrum and enclosing blood vessels supplying the tail.

Herbivorous Subsisting predominantly or exclusively on plants.

Heterodont Type of dentition composed of teeth that differ in shape and/or size.

Homodont Type of dentition composed of teeth that all are similar in shape and/or size.

Homology Correspondence in origin, position or shape of a feature in different taxa due to shared ancestry. (Adjective: **homologous**.) Example: wings of birds and forelimbs of other reptiles.

Homoplasy Appearance of a similar feature in different taxa that is not due to common ancestry. (Adjective: **homoplastic**.) Also referred to as **convergence**.

Hyoid apparatus Set of bones, cartilages, and muscles supporting and moving the tongue.

Hypapophysis Ventral keel or ridge along the midline of a vertebral centrum.

Hyperdactyly Increase in the number of **digits**.

Hyperphalangy Increase in the number of **phalanges** in individual **digits**.

Humerus Proximal bone of the forelimb. (Plural: **humeri**.)

Ilium Dorsal bone of the pelvis, typically connecting it to the vertebral column. (Plural: **ilia**.)

Inner ear Sensory organ responsible for maintaining balance and equilibrium and receiving pressure waves that the brain can process as sound.

Intercentrum Typically small vertebral element situated between the vertebral bodies that are in contact with the ribs (**pleurocentra**).

Interclavicle Median bone on the ventral surface of the shoulder girdle that connects the **clavicles**.

Ischium Ventral bone of the pelvis behind the **pubis**. (Plural: **ischia**.)

Labial Of or directed toward the lip (or outside of the mouth).

Lateral Directed away from the midline of the body.

Laurasia Name for the ancient northern landmass that included present-day North America, much of Asia, and Europe.

Lineage Series of organisms representing a continuous line of descent.

Lingual Pertaining to or directed toward the tongue (or inside of the mouth).

Locomotion Active movement from one place to another.

Mandible Combined left and right lower jaws (**hemimandibles** or **mandibular rami**).

Mandibular Pertaining to the **mandible**.

Manus Hand or forefoot. (Plural: **manus**; adjective: **manual**.)

Maxilla Typically tooth-bearing principal bone of the upper jaw.

Meckel's groove Groove on the lingual side of the lower jaw left by Meckel's cartilage, the cartilaginous portion of the jaw in the embryo.

Medial Directed toward the midline of the body.

Mesial For teeth, directed toward the front of the tooth row.

Mesotarsal Ankle joint with fore-and-aft hinge motion between the main ankle bones and the rest of the foot.

Metacarpus Region of the hand between the wrist and the fingers.

Metatarsus Region of the foot between the ankle and the toes.

Middle ear Cavity and its bones/cartilages that transmit pressure waves to the inner ear.

Monophyletic Group of taxa that includes the last common ancestor of these taxa and all descendants of that ancestor. A strictly monophyletic taxon is known as a **clade**. (Noun: **monophyly**.)

Naris Opening in the **cranium** for passage to or from the nasal cavity. (Plural: **nares**.)

Neural arch Part of a **vertebra** located atop the **centrum**, surrounding the spinal cord and typically supporting a median bony process that extends dorsally (**neural spine**).

Neurocranium Part of the **cranium** enclosing the brain.

Occipital Pertaining to the back end of the skull.

Occiput Back end of the **cranium**.

Omnivorous Subsisting on both animal and plant food.

Ontogeny Pattern of changes in features of an organism from the formation of the zygote to the adult state.

Opisthocoelous Vertebral **centrum** that is convex at the anterior end and concave at the posterior end.

Orbit   Eye socket, containing the eyeball and associated blood vessels, muscles, and nerves. (Adjective: **orbital**.)

Ossicle   Small bone.

Osteoderm   Bone that develops within the dermis layer of the skin.

Osteosclerosis   Increase of the internal compactness of a bone without change in the external shape of the affected skeletal element.

Outgroup   Taxon used in phylogenetic analysis for comparisons, usually in order to determine whether a character state is **plesiomorphic** or **apomorphic** in a set of other taxa under consideration.

Pachyostosis   Increase in deposition of cortical bone, leading to change in the external shape of the affected skeletal element due to an increase in bone volume.

Palate   Roof of the mouth, which separates the nasal passage from the oral cavity. (Adjective: **palatal**.)

Pangaea   Name for the supercontinent that formed as a result of the collision of Gondwana with Laurasia during the late Paleozoic and disintegrated again during the Mesozoic.

Paraphyletic   Group of taxa that includes the last common ancestor of these taxa but excludes one or more descendants of that ancestor.

Parsimony   General scientific principle that the hypothesis that explains the data most simply should be accepted.

Pectoral   Pertaining to the shoulder and chest region.

Pelvic   Pertaining to the hip region (**pelvis**).

Period   Most commonly used unit of geological time, representing a subdivision of an **era**. Example: Cretaceous Period.

Pes   Hind foot. (Plural: **pedes**; adjective: **pedal**.)

Phalanx   Bone from a finger or toe. (Plural: **phalanges**; adjective: **phalangeal**.)

Phylogeny   Hypothesis concerning the genealogical relationships between groups of organisms or (most commonly) taxa.

Piscivorous   Subsisting exclusively or predominantly on fish.

Plantar   Pertaining to the sole of the **pes**.

Plantigrade   Standing or walking with the sole or palm more or less flat on the ground.

Plesiomorphic   When different character-states in two or more taxa are compared and one is thought to have evolved from the other, the latter is regarded as the plesiomorphic (or primitive) character state. (Noun: **plesiomorphy**.)

Pleurodont   Type of dentition in which the teeth are set in a groove in the jaw, with the outer (labial) wall of the groove higher than the inner (lingual) wall.

Posterior   Pertaining to or directed toward the rear end of a body.

Prehensile   Adapted for grasping or gripping.

Process   Structure that projects from the main body of a bone.

Procoelous   Vertebral centrum that is concave at the front and convex at the back.

Procumbent   For teeth, inclined forward.

Protraction   Movement away from the center of the body.

Proximal   Directed toward the body or the attached end.

Pubis   Ventral bone of the pelvis in front of the ischium. (Plural: **pubes**.)

Quadrupedal   Moving on four limbs.

Radius   Medial bone of the forearm. (Plural: **radii**.)

Retraction   Movement toward the center of the body.

Rostral   Pertaining to or directed toward the tip of the snout (**rostrum**).

Sacral   1. (Adjective) Pertaining to the sacrum. 2. (Noun) Short for sacral vertebra.

Sacrum   Portion of the vertebral column connected to the pelvic girdle.

Sagittal   Relating to an imaginary midline of a body, dividing the latter into symmetrical left and right halves.

Saltatory   Jumping on either two or four limbs.

Scapula   Shoulder blade. (Plural: **scapulae**.)

Scleral ring   Ring of bony platelets embedded in the sclera and surrounding the iris of the eyeball.

Sexual dimorphism   Differential development of a feature in males and females of the same species.

Sister taxa   Taxa that are each other's closest relatives.

Stem group   Artificial unresolved grouping of taxa at the base of a cladogram. Although related to the **crown group** these taxa lack the **apomorphic** features shared by the members of that crown group.

Sternum   Breastplate, along the midline below the thorax.

Suspensorium   Set of bones connecting the **cranium** with the **mandible**.

Suture   Joint between two bones, which abut, overlap, or form an interdigitating contact. The bones forming this joint are held together by fibrous connective tissue.

Symphysis   Fibrocartilaginous joint between two bones in the median (sagittal) plane of the body.

Synapomorphy   **Apomorphic** character state shared by two or more taxa. Synapomorphies characterize (**diagnose**) clades.

Synapsida   Major **clade** of **amniotes** comprising mammals and their close relatives, which is characterized by the presence of a single opening behind the orbit on either side of the cranium.

Systematics   Scientific study of how taxa are related.

Tarsus   Set of bones (**tarsals**) or cartilages forming the ankle.

Taxon   Group of organisms considered a unit within a hierarchical biological classification. A taxon is usually given a formal name and assigned a rank. (Plural: **taxa**.)

Taxonomy   Science of recognizing, describing, formally naming, and classifying organisms.

Temporal   1. Pertaining to the region of the **cranium** behind the orbit. 2. Pertaining to time.

Terrestrial   Adapted for or pertaining to life on solid ground.

Tethys   Name for the vast sea that separated **Gondwana** and **Laurasia** in the east.

Tetrapod   Member of the clade **Tetrapoda**, which comprises the last common ancestor of amphibians and amniotes and all descendants of that ancestor. Tetrapods typically have four limbs although many taxa have modified or even lost one or both pairs of limbs.

Thecodont   Type of dentition in which the teeth are attached by ligaments in distinct sockets (**alveoli**) in the jawbone.

Tibia   Median bone of the **crus**. (Plural: **tibiae**.)

Trackway   Set of aligned footprints left by a moving animal.

Tridactyl   Presence of three functional digits in each hand or foot.

Tympanum   Membrane (eardrum) covering the outer opening of the ear.

Ulna   Lateral bone of the forearm. (Plural: **ulnae**.)

Undulatory   Mode of locomotion in which the body moves through a succession of curves.

Ungual   **Phalanx** at the tip of a **digit**, which supports a claw in life.

Ventral   Pertaining to the underside of a body (venter) or body part.

Vertebra   One of the elements forming the backbone (**vertebral column**). (Plural: **vertebrae**.)

Vertebrata   Clade comprising the last common ancestor of lampreys and gnathostomes (jawed vertebrates) and all descendants of that ancestor.

Viviparous   Giving birth to live young that developed within the mother's body.

Zygapophyses   Paired bony processes on the **neural arch** of a **vertebra** that articulate with corresponding processes on the neural arches of successive vertebrae. The processes on the anterior end of the neural arch are the **prezygapophyses** and those on the posterior end of the arch are the **postzygapophyses**.

References

Adams, T. L. 2014. Small crocodyliform from the Lower Cretaceous (late Aptian) of central Texas and its systematic relationship to the evolution of Eusuchia. Journal of Paleontology 88: 1031-1049.

Agassiz, L. 1833-1844. Recherches sur les Poissons Fossiles. Imprimerie Petitpierre, Neuchâtel.

Agnolín, F. L., and F. E. Novas. 2013. Avian Ancestors. A Review of the Phylogenetic Relationships of the Theropods Unenlagiidae, Microraptoria, *Anchiornis* and Scansoriopterygidae. Springer, Dordrecht.

Agnolín, F. L., and S. Rozadilla. 2018. Phylogenetic reassessment of *Pisanosaurus mertii* Casamiquela, 1967, a basal dinosauriform from the Late Triassic of Argentina. Journal of Systematic Palaeontology 16: 853-879.

Aguilera, O. A., D. Riff, and J. Bocquentin-Villanueva. 2006. A new giant *Purussaurus* (Crocodyliformes, Alligatoridae) from the upper Miocene Urumaco Formation, Venezuela. Journal of Systematic Palaeontology 4: 221-232.

Albino, A. M., and S. Brizuela. 2014. An overview of the South American fossil squamates. The Anatomical Record 297: 349-368.

Alcober, O. A., and R. N. Martinez. 2010. A new herrerasaurid (Dinosauria, Saurischia) from the Upper Triassic Ischigualasto Formation of northwestern Argentina. ZooKeys 63: 55-81.

Alibardi, L. 2003. Adaptation to the land: the skin of reptiles in comparison to that of amphibians and endotherm amniotes. Journal of Experimental Zoology (Molecular and Developmental Evolution) 298B: 12-41.

Alifanov, V. R. 2000. The fossil record of Cretaceous lizards of Mongolia. Pp. 368-389 in M. J. Benton, M. A. Shishkin, D. M. Unwin, and E. N. Kurochkin (eds.), The Age of Dinosaurs in Russia and Mongolia. Cambridge University Press, Cambridge, UK.

Allain, R., and N. Aquesbi. 2008. Anatomy and phylogenetic relationships of *Tazoudasaurus naimi* (Dinosauria, Sauropoda) from the late Early Jurassic of Morocco. Geodiversitas 30: 345-424.

Allain, R., T. Xaisanavong, P. Richir, and B. Khentavong. 2012. The first definitive Asian spinosaurid (Dinosauria: Theropoda) from the Early Cretaceous of Laos. Naturwissenschaften 99: 369-377.

Amiot, R., et al. 2010. Oxygen isotope evidence for semiaquatic habits among spinosaurid theropods. Geology 38: 139-142. [16 co-authors]

Andrade, M. B., R. Edmonds, M. J. Benton, and R. Schouten. 2011. A new Berriasian species of *Goniopholis* (Mesoeucrocodylia, Neosuchia) from England, and a review of the genus. Zoological Journal of the Linnean Society 163: S66-S108.

Andres, B., J. M. Clark, and X. Xing. 2010. A new rhamphorhynchid pterosaur from the Upper Jurassic of Xinjiang, China, and the phylogenetic relationships of basal pterosaurs. Journal of Vertebrate Paleontology 30: 163-187.

Andres, B., J. M. Clark, and X. Xing. 2014. The earliest pterodactyloid and the origin of the group. Current Biology 24: 1011-1016.

Andres, B., and Q. Ji. 2006. A new species of *Istiodactylus* (Pterosauria, Pterodactyloidea) from the Lower Cretaceous of Liaoning, China. Journal of Vertebrate Paleontology 26: 70-78.

Andrews, C. W. 1906. A Descriptive Catalogue of the Tertiary Vertebrata of the Fayum, Egypt. Trustees of the British Museum, London.

Andrews, C. W. 1910. A Descriptive Catalogue of the Marine Reptiles of the Oxford Clay. Part 1. Trustees of the British Museum, London.

Andrews, C. W. 1913. A Descriptive Catalogue of the Marine Reptiles of the Oxford Clay. Part 2. Trustees of the British Museum, London.

Anquetin, J. 2010. The anatomy of the basal turtle *Eileanchelys waldmani* from the Middle Jurassic of the Isle of Skye, Scotland. Earth and Environmental Science Transactions of the Royal Society of Edinburgh 101: 67-96.

Anquetin, J. 2012. Reassessment of the phylogenetic interrelationships of basal turtles (Testudinata). Journal of Systematic Palaeontology 10: 3-45.

Anquetin, J., P. M. Barrett, M. E. H. Jones, S. Moore-Fay, and S. E. Evans. 2009. A new stem turtle from the Middle Jurassic of Scotland: new insights into the evolution and palaeoecology of basal turtles. Proceedings of the Royal Society B 276: 879-886.

Anquetin, J., C. Püntener, and J.-P. Billon-Bruyat. 2014. A taxonomic review of the Late Jurassic eucryptodiran turtles from the Jura Mountains (Switzerland and France). PeerJ 2: e369. doi:10.7717/peerj.369.

Anquetin, J., H. Tong, and J. Claude. 2017. A Jurassic stem pleurodire sheds light on the functional origin of neck retraction in turtles. Scientific Reports 7: 42376. doi:10.1038/srep42376.

Apesteguía, S., and M. E. H. Jones. 2012. A Late Cretaceous "tuatara" (Lepidosauria: Sphenodontidae) from South America. Cretaceous Research 34: 154-160.

Apesteguía, S., and F. E. Novas. 2003. Large Cretaceous sphenodontian from Patagonia provides insight into lepidosaur evolution in Gondwana. Nature 425: 609-612.

Apesteguía, S., and H. Zaher. 2006. A Cretaceous terrestrial snake with robust hindlimbs and a sacrum. Nature 440: 1037-1040.

Arbour, V. M., and P. J. Currie. 2015a. Systematics, phylogeny and palaeobiogeography of the ankylosaurid dinosaurs. Journal of Systematic Palaeontology 14: 385-444.

Arbour, V. M., and P. J. Currie. 2015b. Ankylosaurid dinosaur tail clubs evolved through stepwise acquisition of key features. Journal of Anatomy 227: 514-523.

Arbour, V. M., and J. C. Mallon. 2017. Unusual cranial and postcranial anatomy in the archetypal ankylosaur *Ankylosaurus magniventris*. Facets 2: 764-794. doi:10.1139/facets-2017-0063.

Arnold, E. N., O. Arribas, and S. Carranza. 2007. Systematics of the Palaearctic and Oriental lizard tribe Lacertini (Squamata: Lacertidae: Lacertinae), with descriptions of eight new genera. Zootaxa 1430: 1-86.

Auffenberg, W. 1974. Checklist of fossil land tortoises (Testudinidae). Bulletin of the Florida State Museum 18: 121-251.

Augé, M. L. 1988. Une nouvelle espèce de Lacertidae (Sauria, Lacertilia) de l'Oligocène français: *Lacerta filholi*. Place de cette espèce dans l'histoire des Lacertidae de l'Éocène supérieur au Miocène inférieur. Neues Jahrbuch für Geologie und Paläontologie Monatshefte 1988: 464-478.

Augé, M. L. 2005. Évolution des lézards du Paléogène en Europe. Mémoires du Muséum National d'Histoire Naturelle 192: 1-369.

Augé, M. L. 2012. Amphisbaenians from the European Eocene: a biogeographical review. Palaeobiodiversity and Palaeoenvironments 92: 425-443.

Augé, M. L., and J.-C. Rage. 2006. Herpetofaunas from the Upper Paleocene and Lower Eocene of Morocco. Annales de Paléontologie 92: 235-253.

Augé, M. L., and R. Smith. 1997. Les Agamidae (Reptilia, Squamata) du Paléogène d'Europe occidentale. Belgian Journal of Zoology 127: 123-138.

Aureliano, T., A. M. Ghilardi, E. Guilherme, J. P. Souza-Filho, M. Cavalcanti, and D. Riff. 2015. Morphometry, bite-force, and paleobiology of the late Miocene caiman *Purussaurus brasiliensis*. PLoS ONE 10(2): e0117944. doi:10.1371/journal.pone.0117944.

Autumn, K., Y. A. Liang, S. T. Hsieh, W. Zesch, W. P. Chan, T. W. Kenny, R. Fearing, and R. J. Full. 2000. Adhesive force of a single gecko foot-hair. Nature 405: 601-605.

Averianov, A. O. 2010. The osteology of *Azhdarcho lancicollis* Nessov, 1984 (Pterosauria, Azhdarchidae) from the Late Cretaceous of Uzbekistan. Proceedings of the Zoological Institute RAS 314: 264-317.

Averianov, A. O., A. V. Voronkevich, S. V. Lechchinskiy, and A. Fayngertz. 2006. A ceratopsian dinosaur *Psittacosaurus sibiricus* from the Early Cretaceous of west Siberia, Russia and its phylogenetic relationships. Journal of Systematic Palaeontology 4: 359-395.

Baczko, M. B. von, and M. D. Ezcurra. 2013. Ornithosuchidae: a group of Triassic archosaurs with a unique ankle joint. Pp. 187-202 in S. J. Nesbitt, J. B. Desojo, and R. B. Irmis (eds.), Anatomy, Phylogeny and Palaeobiology of Early Archosaurs and Their Kin. Geological Society, London, Special Publications 379.

Baczko, M. B. von, and M. D. Ezcurra. 2016. Taxonomy of the archosaur *Ornithosuchus*: reassessing *Ornithosuchus woodwardi* Newton, 1894 and *Dasygnathoides longidens* (Huxley 1877). Earth and Environmental Science Transactions of the Royal Society of Edinburgh 106: 199-205.

Bailon, S. 2000. Amphibiens et reptiles du Pliocène terminal d'Ahl al Oughlam (Casablanca, Maroc). Geodiversitas 22: 539-558.

Bakker, R. T., and P. M. Galton. 1974. Dinosaur monophyly and a new class of vertebrates. Nature 248: 168-172.

Balanoff, A., X. Xu, Y. Kobayashi, Y. Matsufune, and M. A. Norell. 2009. Cranial osteology of the theropod dinosaur

Incisivosaurus gauthieri (Theropoda: Oviraptorosauria). American Museum Novitates 3651: 1-35.

Bandyopadhyay, S., D. D. Gillette, S. Ray, and D. P. Sengupta. 2010. Osteology of *Barapasaurus tagorei* (Dinosauria: Sauropoda) from the Early Jurassic of India. Palaeontology 53: 533-569.

Barahona, F., S. E. Evans, J. A. Mateo, M. García-Marquez, and L. F. López-Jurado. 2000. Endemism, gigantism and extinction in island lizards: the genus *Gallotia* on the Canary Islands. Journal of Zoology, London 250: 373-388.

Barberena, M. C., N. M. B. Gomes, and L. M. P. Sanchotene. 1970. Osteologia craniana de *Tupinambis teguixin*. Escola de Geologia, Universidae Federal do Rio Grande do Sul, Publicaço Especial 21: 1-32.

Barbosa, J. A., A. W. A. Kellner, and M. S. S. Viana. 2008. New dyrosaurid crocodylomorph and evidences for faunal turnover at the K-P transition in Brazil. Proceedings of the Royal Society B 275: 1385-1391.

Bardet, N., N.-E. Jalil, F. de Lapparent de Broin, D. Germain, O. Lambert, and M. Amaghzaz. 2013. A giant chelonioid turtle from the Late Cretaceous of Morocco with a suction feeding apparatus unique among tetrapods. PLoS ONE 8(7): e63586. doi:10.1371/journal.pone.0063586.

Bardet, N., X. Pereda Suberbiola, M. Iarochène, M. Amalik, and B. Bouya. 2005. Durophagous Mosasauridae (Squamata) from the Upper Cretaceous phosphates of Morocco, with description of a new species of *Globidens*. Netherlands Journal of Geosciences 84: 167-175.

Barley, A. J., P. Q. Spinks, R. C. Thomson, and H. B. Shaffer. 2010. Fourteen nuclear genes provide phylogenetic resolution for difficult nodes in the turtle tree of life. Molecular Phylogenetics and Evolution 55: 1189-1194.

Baron, M. G., and P. M. Barrett. 2017. A dinosaur missing-link? *Chilesaurus* and the early evolution of ornithischian dinosaurs. Biology Letters 13: 20170220. doi:10.1098/rsbl.2017.0220.

Baron, M. G., D. B. Norman, and P. M. Barrett. 2017a. A new hypothesis of dinosaur relationships and early dinosaur evolution. Nature 543: 501-506.

Baron, M. G., D. B. Norman, and P. M. Barrett. 2017b. Postcranial anatomy of *Lesothosaurus diagnosticus* (Dinosauria: Ornithischia) from the Lower Jurassic of southern Africa: implications for basal ornithischian taxonomy and systematics. Zoological Journal of the Linnean Society 179: 125-168.

Barrett, P. M. 2005. The diet of ostrich dinosaurs (Theropoda: Ornithomimosauria). Palaeontology 48: 347-358.

Barrett, P. M., R. J. Butler, and F. Knoll. 2005b. Small-bodied ornithischian dinosaurs from the Middle Jurassic of Sichuan, China. Journal of Vertebrate Paleontology 25: 823-834.

Barrett, P. M., R. J. Butler, R. Mundil, T. M. Scheyer, R. B. Irmis, and M. R. Sánchez-Villagra. 2014. A palaeoequatorial ornithischian and new constraints on early dinosaur diversification. Proceedings of the Royal Society B 281: 20141147. doi:10.1098/rspb.2014.1147.

Barrett, P. M., and P. Upchurch. 2007. The evolution of feeding mechanisms in early sauropodomorph dinosaurs. Special Papers in Palaeontology 77: 91-112.

Barrett, P. M., P. Upchurch, and X.-l. Wang. 2005a. Cranial osteology of *Lufengosaurus huenei* Young (Dinosauria: Prosauropoda) from the Lower Jurassic of Yunnan, People's Republic of China. Journal of Vertebrate Paleontology 25: 806-822.

Barrett, P. M., and A. M. Yates. 2006. New information on the palate and lower jaw of *Massospondylus* (Dinosauria: Sauropodomorpha). Palaeontologia africana 41: 123-130.

Barsbold, R. 1974. [Dueling dinosaurs.] Priroda 1974(2): 81-83. [Russian]

Barsbold, R. 1983. [Carnivorous dinosaurs from the Cretaceous of Mongolia.] Sovmestnaya Sovyetskogo-Mongol'skaya Paleontologicheskaya Ekspeditsiya, Trudy 19: 1-117. [Russian]

Barsbold, R., and H. Osmólska. 1999. The skull of *Velociraptor* (Theropoda) from the Late Cretaceous of Mongolia. Acta Palaeontologica Polonica 44: 189-219.

Bartholomai, A., and R. E. Molnar. 1981. *Muttaburrasaurus*, a new iguanodontid (Ornithischia: Ornithopoda) dinosaur from the Lower Cretaceous of Queensland. Memoirs of the Queensland Museum 20: 319-349.

Bates, K. T., and P. L. Falkingham. 2012. Estimating maximum bite performance in *Tyrannosaurus rex* using multi-body dynamics. Biology Letters 8: 660-664.

Bauer, A. M., W. Böhme, and W. Weitschat. 2005. An Early Eocene gecko from Baltic amber and its implications for the evolution of gecko adhesion. Journal of Zoology, London 265: 327-332.

Bell, G. L., Jr., and M. J. Polcyn. 2005. *Dallasaurus turneri*, a new primitive mosasauroid from the Middle Turonian of Texas and comments on the phylogeny of Mosasauridae (Squamata). Netherlands Journal of Geosciences 84: 177-194.

Bell, P. R., F. Fanti, P. J. Currie, and V. M. Arbour. 2014. A mummified duck-billed dinosaur with a soft-tissue cock's comb. Current Biology 24: 70-75.

Bellairs, A. 1970. The Life of Reptiles. (Two volumes.) Universe Books, New York.

Bennett, S. C. 1993. The ontogeny of *Pteranodon* and other pterosaurs. Paleobiology 19: 92-106.

Bennett, S. C. 1994. Taxonomy and systematics of the Late Cretaceous pterosaur *Pteranodon* (Pterosauria, Pterodactyloidea). Occasional Papers of the Natural History Museum, The University of Kansas 169: 1-70.

Bennett, S. C. 1997. Terrestrial locomotion of pterosaurs: a reconstruction based on *Pteraichnus* trackways. Journal of Vertebrate Paleontology 17: 104-113.

Bennett, S. C. 2000. Pterosaur flight: the role of actinofibrils in wing function. Historical Biology 14: 255-284.

Bennett, S. C. 2001. Osteology and functional morphology of the Late Cretaceous pterosaur *Pteranodon*. Palaeontographica A 260: 1-153.

Bennett, S. C. 2003. Morphological evolution of the pectoral girdle of pterosaurs: myology and function. Pp. 191-215 in E. Buffetaut and J.-M. Mazin (eds.), Evolution and Palaeobiology of Pterosaurs. Geological Society, London, Special Publications 217.

Bennett, S. C. 2007a. Articulation and function of the pteroid bone in pterosaurs. Journal of Vertebrate Paleontology 27: 881-891.

Bennett, S. C. 2007b. A second specimen of the pterosaur *Anurognathus ammoni*. Paläontologische Zeitschrift 81: 376-398.

Bennett, S. C. 2007c. A review of the pterosaur *Ctenochasma*: taxonomy and ontogeny. Neues Jahrbuch für Geologie und Paläontologie Abhandlungen 245: 23-31.

Bennett, S. C. 2013a. New information on body size and cranial display structures of *Pterodactylus antiquus*, with a revision of the genus. Paläontologische Zeitschrift 87: 269-289.

Bennett, S. C. 2013b. The morphology and taxonomy of the pterosaur *Cycnorhamphus*. Neues Jahrbuch für Geologie und Paläontologie Abhandlungen 267: 23-41.

Benson, R. B. J. 2010. A description of *Megalosaurus bucklandii* (Dinosauria: Theropoda) from the Bathonian of the UK and the relationships of Middle Jurassic theropods. Zoological Journal of the Linnean Society 158: 882-935.

Benson, R. B. J., and P. S. Druckenmiller. 2014. Faunal turnover of marine tetrapods during the Jurassic-Cretaceous transition. Biological Reviews 89: 1-23.

Benson, R. B. J., M. Evans, and P. S. Druckenmiller. 2012. High diversity, low disparity and small body size in plesiosaurs (Reptilia, Sauropterygia) from the Triassic-Jurassic boundary. PLoS ONE 7(3): e31838. doi:10.1371/journal.pone.0031838.

Benson, R. B. J., M. Evans, A. S. Smith, J. Sassoon, S. Moore-Faye, H. F. Ketchum, and R. Forrest. 2013a. A giant pliosaurid skull from the Late Jurassic of England. PLoS ONE 8(5): e65989. doi:10.1371/journal.pone.0065989.

Benson, R. B. J., H. F. Ketchum, D. Naish, and L. E. Turner. 2013b. A new leptocleidid (Sauropterygia, Plesiosauria) from the Vectis Formation (early Barremian-early Aptian; Early Cretaceous) of the Isle of Wight and the evolution of Leptocleididae, a controversial clade. Journal of Systematic Palaeontology 11: 233-250.

Benton, M. J. 1983. The Triassic reptile *Hyperodapedon* from Elgin: functional morphology and relationships. Philosophical Transactions of the Royal Society of London B 302: 605-720.

Benton, M. J. 1985. Classification and phylogeny of the diapsid reptiles. Zoological Journal of the Linnean Society 84: 97-164.

Benton, M. J. 1990. The species of *Rhynchosaurus*, a rhynchosaur (Reptilia, Diapsida) from the Middle Triassic of England. Philosophical Transactions of the Royal Society of London B 328: 213-306.

Benton, M. J. 1999. *Scleromochlus taylori* and the origin of dinosaurs and pterosaurs. Philosophical Transactions of the Royal Society of London B 354: 1423-1446.

Benton, M. J. 2014. Vertebrate Palaeontology. Fourth Edition. Wiley-Blackwell, Oxford.

Benton, M. J., and J. M. Clark. 1988. Archosaur phylogeny and the relationships of the Crocodylia. Pp. 295-338 in M. J. Benton (ed.), The Phylogeny and Classification of the Tetrapods. Volume 1: Amphibians, Reptiles, Birds. Clarendon Press, Oxford.

Benton, M. J., P. C. J. Donoghue, R.J. Asher, M. Friedman, T. J. Near, and J. Vinther. 2015. Constraints on the timescale of animal evolutionary history. Palaeontologia Electronica 18.1.1FC.

Benton, M. J., L. Juul, G. W. Storrs, and P. M. Galton. 2000. Anatomy and systematics of the prosauropod dinosaur *Thecodontosaurus antiquus* from the Upper Triassic of southwest England. Journal of Vertebrate Paleontology 20: 77-108.

Benton, M. J., and A. D. Walker. 2002. *Erpetosuchus*, a crocodile-like basal archosaur from the Late Triassic of Elgin, Scotland. Zoological Journal of the Linnean Society 136: 25-47.

Benton, M. J., and A. D. Walker. 2011. *Saltopus*, a dinosauriform from the Upper Triassic of Scotland. Earth and

Environmental Science Transactions of the Royal Society of Edinburgh 101: 285-299.

Berg, D. E. 1966. Die Krokodile, insbesondere *Asiatosuchus* and aff. *Sebecus*?, aus dem Eozän von Messel bei Darmstadt/Hessen. Abhandlungen des Hessischen Landesamtes für Bodenforschung 52: 1-105.

Berkovitz, B., and P. Shellis. 2017. The Teeth of Non-Mammalian Vertebrates. Academic Press, London.

Berman, D. S. 1973. *Spathorhynchus fossorium*, a middle Eocene amphisbaenian (Reptilia) from Wyoming. Copeia 1973: 704-721.

Berman, D. S, R. A. Kissel, A. C. Henrici, S. S. Sumida, and T. Martens. 2004. A new diadectid (Diadectomorpha), *Orobates pabsti* from the Early Permian of central Germany. Bulletin of Carnegie Museum of Natural History 35: 1-37.

Berman, D. S, and J. S. McIntosh. 1978. Skull and relationships of the Upper Jurassic sauropod *Apatosaurus* (Reptilia, Saurischia). Bulletin of Carnegie Museum of Natural History 8: 1-35.

Berman, D. S, R. R. Reisz, and D. Scott. 2010. Redescription of the skull of *Limnoscelis paludis* Williston (Diadectomorpha: Limnoscelidae). New Mexico Museum of Natural History & Science Bulletin 49: 185-210.

Berman, D. S, R. R. Reisz, D. Scott, A. C. Henrici, and T. Martens. 2000. Early Permian bipedal reptile. Science 290: 969-972.

Berman, D. S, S. S. Sumida, and T. Martens. 1998. *Diadectes* (Diadectomorpha: Diadectidae) from the Early Permian of Germany, with description of a new species. Annals of Carnegie Museum 67: 53-93.

Bernard, A., C. Lécuyer, P. Vincent, R. Amiot, N. Bardet, E. Buffetaut, G. Cuny, F. Fourel, F. Martineau, J.-M. Mazin, and A. Prieur. 2010. Regulation of body temperature by some Mesozoic marine reptiles. Science 328: 1379-1382.

Bertozzo, F., F. M. Dalla Vecchia, and M. Fabbri. 2017. The Venice specimen of *Ouranosaurus nigeriensis* (Dinosauria, Ornithopoda). PeerJ 5: e3403. doi:10.7717/peerj.3403.

Bever, G. S., J. A. Gauthier, and G. P. Wagner. 2011. Finding the frame shift: digit loss, developmental variability, and the origin of the avian hand. Evolution & Development 13: 269-279.

Bever, G. S., T. R. Lyson, D. J. Field, and B.-A. S. Bhullar. 2015. Evolutionary origin of the turtle skull. Nature 525: 239-242.

Bever, G. S., and M. A. Norell. 2017. A new rhynchocephalian (Reptilia: Lepidosauria) from the Late Jurassic of Solnhofen (Germany) and the origin of the marine Pleurosauridae. Royal Society Open Science 4: 170570. doi:10.1098/rsos.170570.

Bhullar, B.-A. S. 2011. The power and utility of morphological characters in systematics: a fully resolved phylogeny of *Xenosaurus* and its fossil relatives (Squamata: Anguimorpha). Bulletin of the Museum of Comparative Zoology, Harvard University 160: 65-181.

Bhullar, B.-A. S., and G. S. Bever. 2009. An archosaur-like laterosphenoid in early turtles (Reptilia: Pantestudines). Breviora 518: 1-11.

Bhullar, B.-A. S., and K. T. Smith. 2008. Helodermatid lizard from the Miocene of Florida, the evolution of the dentary in Helodermatidae, and comments on dentary morphology in Varanoidea. Journal of Herpetology 42: 286-302.

Bickelmann, C., J. Müller, and R. R. Reisz. 2009. The enigmatic diapsid *Acerosodontosaurus piveteaui* (Reptilia: Neodiapsida) from the Upper Permian of Madagascar and the paraphyly of "younginiform" reptiles. Canadian Journal of Earth Sciences 46: 651-661.

Bittencourt, J. S., A. B. Arcucci, C. A. Marsicano, and M. C. Langer. 2015. Osteology of the Middle Triassic archosaur *Lewisuchus admixtus* Romer (Chañares Formation, Argentina), its inclusivity, and relationships amongst early dinosauromorphs. Journal of Systematic Palaeontology 13: 189-219.

Bittencourt, J. S., and A. W. A. Kellner. 2009. The anatomy and phylogenetic position of the Triassic dinosaur *Staurikosaurus pricei* Colbert, 1970. Zootaxa 209: 1-56.

Blows, W. T. 2015. British Polacanthid Dinosaurs. Siri Scientific Press, Manchester.

Böhm, M., et al. 2013. The conservation status of the world's reptiles. Biological Conservation 157: 372-385. [217 co-authors]

Böhme, W. 1988. Zur Genitalmorphologie der Sauria: Funktionelle und stammesgeschichtliche Aspekte. Bonner Zoologische Monographien 27: 1-176.

Böttcher, R. 1989. Über die Nahrung eines *Leptopterygius* (Ichthyosauria, Reptilia) aus dem süddeutschen Posidonienschiefer (Unterer Jura) mit Bemerkungen über den Magen der Ichthyosaurier. Stuttgarter Beiträge zur Naturkunde B 155: 1-19.

Böttcher, R. 1990. Neue Erkenntnisse über die Fortpflanzungsbiologie der Ichthyosaurier (Reptilia). Stuttgarter Beiträge zur Naturkunde B 164: 1-51.

Bojanus, L. H. 1819-1821. Anatomia Testudinis Europaeae. Zawadski, Vilna.

Bolet, A., M. Delfino, J. Fortuny, S. Almécija, J. M. Robles, and D. M. Alba. 2014. An amphisbaenian skull from the European Miocene and the evolution of Mediterranean worm lizards. PLoS ONE 9(6): e98082. doi:10.1371/journal.pone.0098082.

Bonaparte, J. F. 1972. Los tetrápodos del sector superior de la Formación Los Colorados, La Rioja, Argentina (Triásico Superior). Parte I. Opera Lilloana 22: 1-183.

Bonaparte, J. F. 1975. Nuevos materiales de *Lagosuchus talampayensis* Romer (Thecodontia-Pseudosuchia) y su significado en el origen de los Saurischia. Chañarense inferior, Triásico medio de Argentina. Acta Geológica Lilloana 13: 1-90.

Bonaparte, J. F. 1976. *Pisanosaurus mertii* Casamiquela and the origin of the Ornithischia. Journal of Paleontology 50: 808-820.

Bonaparte, J. F. 1984. Locomotion in rauisuchid thecodonts. Journal of Vertebrate Paleontology 3: 210-218.

Bonaparte, J. F. 1986. Les Dinosaures (Carnosaures, Allosauridés, Sauropodes, Cétiosauridés) du Jurassique Moyen de Cerro Cóndor (Chubut, Argentine). Annales de Paléontologie (Vertébrés-Invertébrés) 72: 247-289 and 325-386.

Bonaparte, J. F., and R. A. Coria. 1993. Un nuevo y gigantesco sauropodo titanosaurio de la Formación Rio Limay (Albiano-Cenomaniano) de la Provincia del Neuquen, Argentina. Ameghiniana 30: 271-282.

Bonaparte, J. F., J. Ferigolo, and A. M. Ribeiro. 1999. A new early Late Triassic dinosaur from Rio Grande do Sul State, Brazil. Pp. 89-109 in Y. Tomida, T. H. Rich, and P. Vickers-Rich (eds.), Proceedings of the Second Gondwanan Dinosaur Symposium. National Science Museum Monographs 15. National Science Museum, Tokyo.

Bonaparte, J. F., F. E. Novas, and R. A. Coria. 1990. *Carnotaurus sastrei* Bonaparte, the horned, lightly built carnosaur from the Middle Cretaceous of Patagonia. Contributions in Science, Natural History Museum of Los Angeles County 416: 1-41.

Bonaparte, J. F., and J. E. Powell. 1980. A continental assemblage of tetrapods from the Upper Cretaceous beds of El Brete, northwestern Argentina (Sauropoda-Coelurosauria-Carnosauria-Aves). Mémoires de la Société Géologique de France, Nouvelle Série 139: 19-28.

Bonde, N., and P. Christiansen. 2003. The detailed anatomy of *Rhamphorhynchus*: axial pneumaticity and its implications. Pp. 217-232 in E. Buffetaut and J.-M. Mazin (eds.), Evolution and Palaeobiology of Pterosaurs. Geological Society, London, Special Publications 217.

Bonnan, M. F., and P. Senter. 2007. Were the basal sauropodomorph dinosaurs *Plateosaurus* and *Massospondylus* habitual quadrupeds? Special Papers in Palaeontology 77: 139-155.

Bonnan, M. F., and A. M. Yates. 2007. A new description of the forelimb of the basal sauropodomorph *Melanorosaurus*: implications for the evolution of pronation, manus shape and quadrupedalism in sauropod dinosaurs. Special Papers in Palaeontology 77: 157-168.

Booth, D. T. 2006. Influence of incubation temperature on hatchling phenotype in reptiles. Physiological and Biochemical Zoology 79: 274-281.

Borsuk-Białynicka, M. 1977. A new camarasaurid sauropod *Opisthocoelicaudia skarzynskii* gen. n., sp. n. from the Upper Cretaceous of Mongolia. Palaeontologia Polonica 37: 5-64.

Borsuk-Białynicka, M. 1984. Anguimorphans and related lizards from the Late Cretaceous of the Gobi Desert, Mongolia. Palaeontologia Polonica 46: 5-105.

Borsuk-Białynicka, M., and S. E. Evans. 2009. A long-necked archosauromorph from the Early Triassic of Poland. Palaeontologia Polonica 65: 203-234.

Borsuk-Białynicka, M., M. Lubka, and W. Böhme. 1999. A lizard from Baltic amber (Eocene) and the ancestry of the crown group lacertids. Acta Palaeontologica Polonica 44: 349-382.

Borsuk-Białynicka, M., and S. M. Moody. 1984. Priscagaminae, a new subfamily of the Agamidae (Sauria) from the late Cretaceous of the Gobi Desert. Acta Palaeontologica Polonica 29: 51-81.

Bostrom, B. L., T. T. Jones, M. Hastings, and D. R. Jones. 2010. Behaviour and physiology: the thermal strategy of leatherback turtles. PLoS ONE 5(11): e13925. doi:10.1371/journal.pone.0013925.

Boy, J. A., and T. Martens. 1991. Ein neues captorhinomorphes Reptil aus dem thüringischen Rotliegend (Unter-Perm; Ost-Deutschland). Paläontologische Zeitschrift 65: 363-389.

Boyd, C. M. 2014. The cranial anatomy of the neornithischian dinosaur *Thescelosaurus neglectus*. PeerJ 2: e669. doi:10.7717/peerj.669.

Boyd, C. M. 2015. The systematic relationships and biogeographic history of ornithischian dinosaurs. PeerJ 3: e1523. doi:10.7717/peerj.1523.

Boyd, C. M., C. M. Brown, R. D. Scheetz, and J. A. Clarke. 2009. Taxonomic revision of the basal neornithischian taxa *Thescelosaurus* and *Bugenasaura*. Journal of Vertebrate Paleontology 29: 758-770.

Brainerd, E. L., and T. Owerkowicz. 2006. Functional morphology and evolution of aspiration breathing in tetrapods. Respiratory Physiology & Neurobiology 154: 73-88.

Brinkman, D. 1980. The hind limb step cycle of *Caiman sclerops* and the mechanics of the crocodile tarsus and metatarsus. Canadian Journal of Zoology 58: 2187-2200.

Brinkman, D. 1981. The hindlimb step cycle of *Iguana* and primitive reptiles. Journal of Zoology, London 181: 91-103.

Brinkman, D. B. 2005. Turtles: diversity, paleoecology, and distribution. Pp. 202-220 in P. J. Currie and E. B. Koppelhus (eds.), Dinosaur Provincial Park: A Spectacular Ancient Ecosystem Revealed. Indiana University Press, Bloomington.

Brinkman, D. L., R. L. Cifelli, and N. J. Czaplewski. 1998. First occurrence of *Deinonychus antirrhopus* (Dinosauria: Theropoda) from the Antlers Formation (Lower Cretaceous: Aptian-Albian) of Oklahoma. Oklahoma Geological Survey Bulletin 146: 1-27.

Brinkman, D. B., and X.-c. Wu. 1999. The skull of *Ordosemys*, an Early Cretaceous turtle from Inner Mongolia, People's Republic of China, and the interrelationships of Eucryptodira (Chelonia, Cryptodira). Paludicola 2: 134-147.

Britt, B. 1991. Theropods of Dry Mesa Quarry (Morrison Formation, Late Jurassic), Colorado, with emphasis on the osteology of *Torvosaurus tanneri*. Brigham Young University Geology Studies 37: 1-72.

Brochu, C. A. 1997. A review of *"Leidyosuchus"* (Crocodyliformes, Eusuchia) from the Cretaceous through Eocene of North America. Journal of Vertebrate Paleontology 17: 679-697.

Brochu, C. A. 1999. Phylogeny, systematics, and historical biogeography of Alligatoroidea. Society of Vertebrate Paleontology Memoir 6: 9-100.

Brochu, C. A. 2000. Phylogenetic relationships and divergence timing of *Crocodylus* based on morphology and the fossil record. Copeia 2000: 657-673.

Brochu, C. A. 2003a. Phylogenetic approaches toward crocodylian history. Annual Review of Earth and Planetary Sciences 31: 357-397.

Brochu, C. A. 2003b. Osteology of *Tyrannosaurus rex*: insights from a nearly complete skeleton and high-resolution computed tomographic analysis of the skull. Society of Vertebrate Paleontology Memoir 7: 1-138.

Brochu, C. A. 2004. A new Late Cretaceous gavialoid crocodylian from eastern North America and the phylogenetic relationships of *Thoracosaurus*. Journal of Vertebrate Paleontology 24: 610-633.

Brochu, C. A. 2006. Osteology and phylogenetic significance of *Eosuchus minor* (Marsh, 1870) new combination, a

longirostrine crocodylian from the late Paleocene of North America. Journal of Paleontology 80: 162-186.

Brochu, C. A. 2007a. Morphology, relationships, and biogeographical significance of an extinct horned crocodile (Crocodylia, Crocodylidae) from the Quaternary of Madagascar. Zoological Journal of the Linnean Society 150: 835-863.

Brochu, C. A. 2007b. Systematics and taxonomy of Eocene tomistomine crocodylians from Britain and northern Europe. Palaeontology 50: 917-928.

Brochu, C. A. 2010. A new alligatorid from the lower Eocene Green River Formation of Wyoming and the origin of caimans. Journal of Vertebrate Paleontology 30: 1109-1126.

Brochu, C. A. 2011. Phylogenetic relationships of *Necrosuchus ionensis* Simpson, 1937 and the early history of caimanines. Zoological Journal of the Linnean Society 163(Supplement 1): S228-S256.

Brochu, C. A. 2013. Phylogenetic relationships of Palaeogene ziphodont eusuchians and the status of *Pristichampsus* Gervais, 1853. Earth and Environmental Science Transactions of the Royal Society of Edinburgh 103: 521-550.

Brochu, C. A., and P. D. Gingerich. 2000. New tomistomine crocodylian from the middle Eocene (Bartonian) of Wadi Hitan, Fayum Province, Egypt. Contributions from the Museum of Paleontology, University of Michigan 30: 251-268.

Brochu, C. A., and G. W. Storrs. 2012. A giant crocodile from the Plio-Pleistocene of Kenya, the phylogenetic relationships of Neogene African crocodylines, and the antiquity of *Crocodylus* in Africa. Journal of Vertebrate Paleontology 32: 587-602.

Brocklehurst, R. J., E. R. Schachner, and W. I. Sellers. 2018. Vertebral morphometrics and lung structure in non-avian dinosaurs. Royal Society Open Science 5: 180983. doi:10.1098/rsos.180983.

Broili, F. 1912. Zur Osteologie des Schädels von *Placodus*. Palaeontographica 59: 147-155.

Brongniart, A. 1800. Essai d'une classification naturelle des Reptiles. Iere partie. Etablissement des ordres. Bulletin des Sciences de la Société Philomathique, Paris 2(35): 81-82.

Broom, R. 1914. A new thecodont reptile. Proceedings of the Zoological Society of London 1914: 1072-1077.

Broschinski, A. 1999. Ein Lacertilier (Scincomorpha, Paramacellodidae) aus dem Ober-Jura von Tendaguru (Tansania). Mitteilungen aus dem Museum für Naturkunde Berlin, Geowissenschaftliche Reihe 2: 155-158.

Brown, B., and E. M. Schlaikjer. 1940. The structure and relationships of *Protoceratops*. Annals of the New York Academy of Sciences 40: 133–266.

Brown, B., and E. M. Schlaikjer. 1943. A study of the troödont dinosaurs, with description of a new genus and four new species. Bulletin of the American Museum of Natural History 82: 115–150.

Brown, C. M. 2017. An exceptionally preserved armored dinosaur reveals the morphology and allometry of osteoderms and their horny epidermal coverings. PeerJ 5:e4066. doi:10.7717/peerj.4066.

Brown, C. M., D. C. Evans, M. J. Ryan, and A. P. Russell. 2013. New data on the diversity and abundance of small-bodied ornithopods (Dinosauria, Ornithischia) from the Belly River Group (Campanian) of Alberta. Journal of Vertebrate Paleontology 33: 495–520.

Brown, C. M., D. M. Henderson, J. Vinther, I. Fletcher, A. Sistiaga, J. Herrera, and R. E. Summons. 2017. An exceptionally preserved three-dimensional armored dinosaur reveals insights into coloration and Cretaceous predator-prey dynamics. Current Biology 27: 2514–2521.

Brown, C. M., and A. P. Russell. 2012. Homology and architecture of the caudal basket of Pachycephalosauria (Dinosauria: Ornithischia): the first occurrence of myorhabdoi in Tetrapoda. PLoS ONE 7(1): e30212. doi:10.1371/journal.pone.0030212.

Brown, D. S. 1981. The English Upper Jurassic Plesiosauroidea (Reptilia), and a review of the phylogeny and classification of the Plesiosauria. Bulletin of the British Museum (Natural History), Geology 35: 253–347.

Brusatte, S. L. 2012. Dinosaur Paleobiology. Wiley-Blackwell, Chichester.

Brusatte, S. L., A. Averianov, H.-D. Sues, A. Muir, and I. Butler. 2016. New tyrannosaur from the mid-Cretaceous of Uzbekistan clarifies evolution of giant body sizes and advanced senses in tyrant dinosaurs. Proceedings of the National Academy of Sciences of the United States of America 113: 3447–3452.

Brusatte, S. L., R. B. J. Benson, and S. Hutt. 2008. The osteology of *Neovenator salerii* (Dinosauria: Theropoda) from the Wealden Group (Barremian) of the Isle of Wight. Palaeontographical Society Publication 631: 1–75.

Brusatte, S. L., M. J. Benton, J. B. Desojo, and M. C. Langer. 2010a. The higher-level phylogeny of Archosauria (Tetrapoda: Diapsida). Journal of Systematic Palaeontology 8: 3–47.

Brusatte, S. L., T. D. Carr, and M. A. Norell. 2012. The osteology of *Alioramus*, a gracile and long-snouted tyrannosaurid (Dinosauria: Theropoda) from the Late Cretaceous of Mongolia. Bulletin of the American Museum of Natural History 366: 1–197.

Brusatte, S. L., G. T. Lloyd, S. C. Wang, and M. A. Norell. 2014. Gradual assembly of avian body plan culminated in rapid rates of evolution across the dinosaur-bird transition. Current Biology 24: 2386–2392.

Brusatte, S. L., S. J. Nesbitt, R. B. Irmis, R. J. Butler, M. J. Benton, and M. A. Norell. 2010b. The origin and early radiation of dinosaurs. Earth-Science Reviews 101: 68–100.

Brusatte, S. L., G. Niedźwiedzki, and R. J. Butler. 2011. Footprints pull origin and diversification of dinosaur stem lineage deep into Early Triassic. Proceedings of the Royal Society B 278: 1107–1113.

Brusatte, S. L., M. A. Norell, T. D. Carr, G. M. Erickson, J. R. Hutchinson, A. M. Balanoff, G. S. Bever, J. N. Choiniere, P. J. Makovicky, and X. Xu. 2010c. Tyrannosaur paleobiology: new research on ancient exemplar organisms. Science 329: 1481–1485.

Brusatte, S. L., M. Vremir, Z. Csiki-Sava, A. H. Turner, A. Watanabe, G. M. Erickson, and M. A. Norell. 2013. The osteology of *Balaur bondoc*, an island-dwelling dromaeosaurid (Dinosauria: Theropoda) from the Late Cretaceous of Romania. Bulletin of the American Museum of Natural History 374: 1–100.

Buckland, W. 1824. Notice on the Megalosaurus or great Fossil Lizard of Stonesfield. Transactions of the Geological Society of London, ser. 2, 1(2): 390–396.

Buffetaut, E. 1982. Radiation évolutive, paléoécologie et biogéographie des Crocodiliens Mésosuchiens. Mémoires de la Société Géologique de France, Nouvelle Série 142: 1–88.

Buffetaut, E., V. Suteethorn, G. Cuny, H. Tong, J. Le Loeuff, S. Khansubha, and S. Jongautchariyakul. 2000. The earliest known sauropod dinosaur. Nature 407: 72–74.

Bulanov, V. V., and A. G. Sennikov. 2010. New data on the morphology of Permian gliding weigeltisaurid reptiles of eastern Europe. Paleontological Journal 44: 682–694.

Bulanov, V. V., and A. G. Sennikov. 2015. New data on the morphology of the Late Permian gliding reptile *Coelurosauravus elivensis* Piveteau. Paleontological Journal 49: 413–423.

Burbrink, F. T. 2005. Inferring the phylogenetic position of *Boa constrictor* among the Boinae. Molecular Phylogenetics and Evolution 34: 167–180.

Burbrink, F. T., and B. I. Crother. 2011. Evolution and taxonomy of snakes. Pp. 19–53 in R. D. Aldridge and

D. M. Sever (eds.), Reproductive Biology and Phylogeny of Snakes. CRC Press, Boca Raton.

Burgin, C. J., J. P. Colella, P. L. Kahn, and N. S. Upham. 2018. How many species of mammals are there? Journal of Mammalogy 99: 1-14.

Burke, A. C. 1989. Development of the turtle carapace: implications for the evolution of a novel Bauplan. Journal of Morphology 199: 363-378.

Burns, M. E., P. J. Currie, R. L. Sissons, and V. M. Arbour. 2011. Juvenile specimens of *Pinacosaurus grangeri* Gilmore, 1933 (Ornithischia: Ankylosauria) from the Late Cretaceous of China, with comments on the specific taxonomy of *Pinacosaurus*. Cretaceous Research 32: 174-186.

Busbey, A. B., III, and C. E. Gow. 1984. A new protosuchian crocodile from the Upper Triassic Elliot Formation of South Africa. Palaeontologia africana 25: 127-149.

Buscalioni, A. D. 2017. The Gobiosuchidae in the early evolution of Crocodyliformes. Journal of Vertebrate Paleontology 37: e1324459. doi:10.1080/02724634.2017.132 4459.

Butler, R. J. 2010. The anatomy of the basal ornithischian dinosaur *Eocursor parvus* from the lower Elliot Formation (Late Triassic) of South Africa. Zoological Journal of the Linnean Society 160: 648-684.

Butler, R. J., P. M. Barrett, P. Kenrick, and M. G. Penn. 2009. Diversity patterns amongst herbivorous dinosaurs and plants during the Cretaceous: implications for hypotheses of dinosaur/angiosperm co-evolution. Journal of Evolutionary Biology 22: 446-459.

Butler, R. J., S. L. Brusatte, M. Reich, S. J. Nesbitt, R. R. Schoch, and J. J. Hornung. 2011a. The sail-backed reptile *Ctenosauriscus* from the latest Early Triassic of Germany and the timing and biogeography of the early archosaur radiation. PLoS ONE 6(10): e25693. doi:10.1371/journal. pone.0025963.

Butler, R. J., M. D. Ezcurra, F. C. Montefeltro, A. Samathi, and G. Sobral. 2015. A new species of basal rhynchosaur (Diapsida: Archosauromorpha) from the early Middle Triassic of South Africa, and the early evolution of Rhynchosauria. Zoological Journal of the Linnean Society 174: 571-588.

Butler, R. J., P. M. Galton, L. B. Porro, L. M. Chiappe, D. M. Henderson, and G. M. Erickson. 2010. Lower limits of ornithischian dinosaur body size inferred from a new Upper Jurassic heterodontosaurid from North America. Proceedings of the Royal Society B 277: 375-381.

Butler, R. J., L. Jin, J. Chen, and P. Godefroit. 2011b. The postcranial osteology and phylogenetic position of the small ornithischian dinosaur *Changchunsaurus parvus* from the Quantou Formation (Cretaceous: Aptian-Cenomanian) of Jilin Province, north-eastern China. Palaeontology 54: 667-683.

Butler, R. J., C. Sullivan, M. D. Ezcurra, J. Liu, A. Lecuona, and R. B. Sookias. 2014. New clade of enigmatic early archosaurs yields insights into early pseudosuchian phylogeny and the biogeography of the archosaur radiation. BMC Evolutionary Biology 14: 128. doi:10.1186/1471-2148-14-128.

Butler, R. J., P. Upchurch, and D. B. Norman. 2008. The phylogeny of the ornithischian dinosaurs. Journal of Systematic Palaeontology 6: 1-40.

Butler, R. J., and Q. Zhao. 2009. The small-bodied ornithischian dinosaurs *Micropachycephalosaurus hongtuyanensis* and *Wannanosaurus yansiensis* from the Late Cretaceous of China. Cretaceous Research 30: 63-77.

Button, D. J., P. M. Barrett, and E. J. Rayfield. 2017. Craniodental functional evolution in sauropodomorph dinosaurs. Paleobiology 43: 435-462.

Bystrov, A. P. 1957. [Past, Present and Future of Man.] Medgiz, Leningrad. [Russian]

Cabreira, S. F., et al.. 2016. A unique Late Triassic dinosauromorph assemblage reveals dinosaur ancestral anatomy and diet. Current Biology 26: 3090-3095. [12 co-authors]

Cabreira, S. F., C. L. Schultz, J. S. Bittencourt, M. B. Soares, D. C. Fortier, L. R. Silva, and M. C. Langer. 2011. New stem-sauropodomorph (Dinosauria, Saurischia) from the Triassic of Brazil. Naturwissenschaften 98: 1035-1040.

Cadena, E. 2015. A global phylogeny of Pelomedusoides turtles with new material of *Neochelys franzeni* Schleich, 1993 (Testudines, Podocnemididae) from the middle Eocene, Messel Pit, of Germany. PeerJ 3: e1221. doi:10.7717/peerj.1221.

Cadena, E. A., J. I. Bloch, and C. A. Jaramillo. 2015. New bothremydid turtle (Testudines, Pleurodira) from the Paleocene of northeastern Colombia. Journal of Paleontology 86: 688-698.

Cadena, E., and W. G. Joyce. 2015. A review of the fossil record of turtles of the clades *Platychelyidae* and *Dortokidae*. Bulletin of the Peabody Museum of Natural History, Yale University 56: 3-20.

Cadena, E. A., and J. F. Parham. 2015. Oldest known marine turtle? A new protostegid from the Lower Cretaceous of Colombia. PaleoBios 32: 1-42.

Caldwell, M. W. 1996. Ontogeny and phylogeny of the mesopodial skeleton in mosasauroid reptiles. Zoological Journal of the Linnean Society 116: 407-436.

Caldwell, M. W. 1997a. Limb ossification patterns of the ichthyosaur *Stenopterygius*, and a discussion of the proximal tarsal row of ichthyosaurs and other neodiapsid reptiles. Zoological Journal of the Linnean Society 120: 1-25.

Caldwell, M. W. 1997b. Limb osteology and ossification patterns in *Cryptoclidus* (Reptilia: Plesiosauroidea) with a review of sauropterygian limbs. Journal of Vertebrate Paleontology 17: 295-307.

Caldwell, M. W. 1999. Squamate phylogeny and the relationships of snakes and mosasauroids. Zoological Journal of the Linnean Society 125: 115-147.

Caldwell, M. W. 2000. An aquatic squamate reptile from the English Chalk, *Dolichosaurus longicollis* Owen, 1850. Journal of Vertebrate Paleontology 20: 720-735.

Caldwell, M. W. 2006. A new species of *Pontosaurus* (Squamata, Pythonomorpha) from the Upper Cretaceous of Lebanon and a phylogenetic analysis of Pythonomorpha. Memorie della Società Italiana di Scienze Naturali e del Museo Civico di Storia Naturale di Milano 34: 1-42.

Caldwell, M. W. 2012. A challenge to categories: "What, if anything, is a mosasaur?" Bulletin de la Société Géologique de France 183: 7-33.

Caldwell, M. W., and A. Albino. 2002. Exceptionally preserved skeletons of the Cretaceous snake *Dinilysia patagonica* Woodward, 1901. Journal of Vertebrate Paleontology 22: 861-866.

Caldwell, M. W., R. L. Carroll, and H. Kaiser. 1995. The pectoral girdle and forelimb of *Carsosaurus marchesetti* (Aigialosauridae), with a preliminary phylogenetic analysis of mosasauroids and varanoids. Journal of Vertebrate Paleontology 15: 516-531.

Caldwell, M. W., and M. S. Y. Lee. 1997. A snake with legs from the marine Cretaceous of the Middle East. Nature 386: 705-709.

Caldwell, M. W., R. L. Nydam, A. Palci, and S. Apesteguía. 2015. The oldest known snakes from the Middle Jurassic-Lower Cretaceous provide insights on snake evolution. Nature Communications 6: 5996. doi: 10.1038/ncomms6996.

Caldwell, M. W., and A. Palci. 2007. A new basal mosasauroid from the Cenomanian (U. Cretaceous) of Slovenia with a review of mosasauroid phylogeny and evolution. Journal of Vertebrate Paleontology 27: 863-880.

Calvo, J. O., J. D. Porfiri, and F. E. Novas. 2007. Discovery of a new ornithopod dinosaur from the Portezuelo Formation (Upper Cretaceous), Neuquén, Patagonia, Argentina. Arquivos do Museu Nacional, Rio de Janeiro 65: 471-483.

Camp, C. L. 1923. Classification of the lizards. Bulletin of the American Museum of Natural History 48: 289-480.

Camp, C. L. 1930. A study of the phytosaurs with description of new material from western North America. Memoirs of the University of California 10: 1-161.

Camp, C. L. 1942. California mosasaurs. Memoirs of the University of California 13: 1-68.

Camp, C. L. 1945. *Prolacerta* and the protorosaurian reptiles. American Journal of Science 243: 17-32 and 84-101.

Camp, C. L. 1980. Large ichthyosaurs from the Upper Triassic of Nevada. Palaeontographica A 170: 139-200.

Campbell, J. A., M. J. Ryan, R. B. Holmes, and C. J. Schröder-Adams. 2016. A re-evaluation of the chasmosaurine ceratopsid genus *Chasmosaurus* (Dinosauria: Ornithischia) from the Upper Cretaceous (Campanian) Dinosaur Park Formation of western Canada. PLoS ONE 11(1): e0145805. doi:10.1371/journal.pone.0145805.

Campione, N. E., and D. C. Evans. 2011. Cranial growth and variation in edmontosaurs (Dinosauria: Hadrosauridae): implications for latest Cretaceous megaherbivore diversity in North America. PLoS ONE 6(9): e25186. doi:10.1371/journal.pone.0025186.

Campione, N. E., and R. B. Holmes. 2006. The anatomy and homologies of the ceratopsid syncervical. Journal of Vertebrate Paleontology 26: 1014-1017.

Carballido, J. L., D. Pol, A. Otero, I. A. Cerda, L. Salgado, A. C. Garrido, J. Ramezzani, N. Cúneo, and M. J. Krause. 2017. A new giant titanosaur sheds light on body mass evolution among sauropod dinosaurs. Proceedings of the Royal Society B 284: 20171219. doi:10.1098/rspb.2017.1219.

Carballido, J. L., and P. M. Sander. 2014. Postcranial axial skeleton of *Europasaurus holgeri* (Dinosauria, Sauropoda) from the Upper Jurassic of Germany: implications for sauropod ontogeny and phylogenetic relationships of basal Macronaria, Journal of Systematic Palaeontology 12: 335-387.

Carpenter, K., C. Miles, and K. Cloward. 2005a. New small theropod from the Upper Jurassic Morrison Formation of Wyoming. Pp. 23-48 in K. Carpenter (ed.), Carnivorous Dinosaurs. Indiana University Press, Bloomington.

Carpenter, K., C. Miles, J. H. Ostrom, and K. Cloward. 2005b. Redescription of the small maniraptoran theropods

Ornitholestes and *Coelurus* from the Upper Jurassic Morrison Formation of Wyoming. Pp. 49-71 in K. Carpenter (ed.), Carnivorous Dinosaurs. Indiana University Press, Bloomington.

Carr, T. D. 1999. Craniofacial ontogeny in Tyrannosauridae (Dinosauria, Coelurosauria). Journal of Vertebrate Paleontology 19: 497-520.

Carr, T. D., T. E. Williamson, and D. R. Schwimmer. 2005. A new genus and species of tyrannosaurid from the Late Cretaceous (middle Campanian) Demopolis Formation of Alabama. Journal of Vertebrate Paleontology 25: 119-143.

Carrano, M. T. 2005. The evolution of sauropod locomotion: morphological diversity of a secondarily quadrupedal radiation. Pp. 229-251 in K. A. Curry Rogers and J. A. Wilson (eds.), The Sauropods: Evolution and Paleobiology. University of California Press, Berkeley.

Carrano, M. T., R. B. J. Benson, and S. D. Sampson. 2012. The phylogeny of Tetanurae (Dinosauria: Theropoda). Journal of Systematic Palaeontology 10: 211-300.

Carrano, M. T., M. A. Loewen, and J. J. W. Sertich. 2011. New materials of *Masiakasaurus knopfleri* Sampson, Carrano, and Forster, 2001, and implications for the morphology of Noasauridae (Theropoda: Ceratosauria). Smithsonian Contributions to Paleobiology 95: 1-53.

Carrano, M. T., and S. D. Sampson. 2008. The phylogeny of Ceratosauria (Dinosauria: Theropoda). Journal of Systematic Palaeontology 6: 183-236.

Carroll, R. L. 1964. The earliest reptiles. Journal of the Linnean Society (Zoology) 45: 61-83.

Carroll, R. L. 1969a. A Middle Pennsylvanian captorhinomorph, and the interrelationships of primitive reptiles. Journal of Paleontology 43: 151-170.

Carroll, R. L. 1969b. Problems of the origin of reptiles. Biological Reviews 44: 393-432.

Carroll, R. L. 1975. Permo-Triassic "lizards" from the Karroo. Palaeontologia africana 18: 71-87.

Carroll, R. L. 1977. The origin of lizards. Pp. 359-396 in S. M. Andrews, R. S. Miles, and A. D. Walker (eds.), Problems in Vertebrate Evolution. Academic Press, London.

Carroll, R. L. 1980. The hyomandibular as a supporting element in the skull of primitive tetrapods. Pp. 293-317 in A. L. Panchen (ed.), The Terrestrial Environment and the Origin of Land Vertebrates. Academic Press, London.

Carroll, R. L. 1981. Plesiosaur ancestors from the Upper Permian of Madagascar. Philosophical Transactions of the Royal Society of London B 293: 315-383.

Carroll, R. L. 1985. A pleurosaur from the Lower Jurassic and the taxonomic position of the Sphenodontida. Palaeontographica A 189: 1-28.

Carroll, R. L. 1988. Vertebrate Paleontology and Evolution. W. H. Freeman and Company, New York.

Carroll, R. L. 2009. The Rise of Amphibians: 365 Million Years of Evolution. Johns Hopkins University Press, Baltimore.

Carroll, R. L., and D. Baird. 1972. Carboniferous stem-reptiles of the Family Romeriidae. Bulletin of the Museum of Comparative Zoology, Harvard University 143: 321-364.

Carroll, R. L., and Z. Dong. 1991. *Hupehsuchus*, an enigmatic aquatic reptile from the Triassic of China, and the problem of establishing relationships. Philosophical Transactions of the Royal Society of London B 331: 131-153.

Carroll, R. L., and P. Gaskill. 1985. The nothosaur *Pachypleurosaurus* and the origin of plesiosaurs. Philosophical Transactions of the Royal Society of London B 309: 343-393.

Carroll, R. L., and W. Lindsay. 1985. The cranial anatomy of the primitive reptile *Procolophon*. Canadian Journal of Earth Sciences 22: 1571-1587.

Carroll, R. L., and R. Wild. 1994. Marine members of the Sphenodontia. Pp. 70-83 in N. C. Fraser and H.-D. Sues (eds.), In the Shadow of the Dinosaurs: Early Mesozoic Tetrapods. Cambridge University Press, Cambridge, UK.

Carvalho, I. S. 1994. *Candidodon*: um crocodilo com heterodontia (Notosuchia, Cretáceo Inferior—Brasil). Anais da Academia Brasileira de Ciências 60: 437-446.

Carvalho, I. S., L. S. Avilla, and L. Salgado. 2003. *Amazonsaurus maranhensis* gen. et sp. nov. (Sauropoda, Diplodocoidea) from the Lower Cretaceous (Aptian-Albian) of Brazil. Cretaceous Research 24: 697-713.

Carvalho, I. S., A. C. A. Campos, and P. H. Nobre. 2005. *Baurusuchus salgadoensis*, a new Crocodylomorpha from the Bauru Basin (Upper Cretaceous), Brazil. Gondwana Research 8: 11-30.

Carvalho, I. S., L. C. B. Ribeiro, and L. S. Avilla. 2004. *Uberabasuchus terrificus* sp. nov., a new Crococodylomorpha from the Bauru Basin (Upper Cretaceous), Brazil. Gondwana Research 7: 975-1002.

Carvalho, I. S., V. P. A. Teixeira, M. L. F. Ferraz, L. C. B. Ribeiro, A. G. Martinelli, F. M. Neto, J. W. Sertich, G. C. Cunha, I. C. Cunha, and P. F. Ferraz. 2011. *Campinasuchus dinizi* gen. et sp. nov., a new Late Cretaceous baurusuchid (Crocodyliformes) from the Bauru Basin, Brazil. Zootaxa 2871: 19-42.

Case, E. C. 1911. A revision of the Cotylosauria of North America. Carnegie Institution of Washington Publication 145: 1-122.

Cau, A., V. Beyrand, D. F. A. E. Voeten, V. Fernandez, P. Tafforeau, K. Stein, R. Barsbold, K. Tsogtbaatar, P. J. Currie, and P. Godefroit. 2017. Synchrotron scanning reveals amphibious ecomorphology in a new clade of bird-like dinosaurs. Nature 552: 395-399.

Cau, A., T. Broughman, and D. Naish. 2015. The phylogenetic affinities of the bizarre Late Cretaceous Romanian theropod *Balaur bondoc* (Dinosauria, Maniraptora): dromaeosaurid or flightless bird? PeerJ 3: e1032. doi:10.7717/peerj.1032.

Ceballos, G., P. R. Ehrlich, and R. Dirzo. 2017. Biological annihilation via the ongoing sixth mass extinction signaled by vertebrate population losses and declines. Proceedings of the National Academy of Sciences of the United States of America 114: E6089-E6096.

Čerňanský, A. 2010. A revision of chamaeleonids from the Lower Miocene of the Czech Republic with description of a new species of *Chamaeleo* (Squamata, Chamaeleonidae). Geobios 43: 605-613.

Čerňanský, A. 2012. The oldest known European Neogene girdled lizard fauna (Squamata, Cordylidae), with comments on Early Miocene immigration of African taxa. Geodiversitas 34: 837-848.

Čerňanský, A., M. L. Augé, and J.-C. Rage. 2015. A complete mandible of a new amphisbaenian reptile (Squamata, Amphisbaenia) from the late middle Eocene (Bartonian, MP 16) of France. Journal of Vertebrate Paleontology 35: e902379. doi:10.1080/02724634.2014.902379.

Čerňanský, A., and K. T. Smith. 2017. Eolacertidae: a new extinct clade of lizards from the Paleogene; with comments on the origin of the dominant European reptile group—Lacertidae. Historical Biology 30: 994-1014.

Chapelle, K. E. J., and J. N. Choiniere. 2018, A revised cranial description of *Massospondylus carinatus* Owen (Dinosauria: Sauropodomorpha) based on computed tomographic scans and a review of cranial characters for basal Sauropodomorpha. PeerJ 6: e4224. doi:10.7717/peerj.4224.

Charig, A. J. 1976. Order Thecodontia Owen 1859. Pp. 7-10 in O. Kuhn (ed.), Handbuch der Paläoherpetologie, Teil 13: Thecodontia. Gustav Fischer Verlag, Stuttgart.

Charig, A. J., and C. Gans. 1990. Two new amphisbaenians from the Lower Miocene of Kenya. Bulletin of the British Museum (Natural History), Geology 46: 19-36.

Charig, A. J., and A. C. Milner. 1997. *Baryonyx walkeri*, a fish-eating dinosaur from the Wealden of Surrey. Bulletin of the Natural History Museum, Geology 53: 11-70.

Chatterjee, S. 1974. A rhynchosaur from the Upper Triassic Maleri Formation of India. Philosophical Transactions of the Royal Society of London B 267: 209-261.

Chatterjee, S. 1978. A primitive parasuchid (phytosaur) reptile from the Upper Triassic Maleri Formation of India. Palaeontology 21: 83-127.

Chatterjee, S. 1982. Phylogeny and classification of thecodontian reptiles. Nature 295: 317-320.

Chatterjee, S., and R. J. Templin. 2004. Posture, locomotion, and paleoecology of pterosaurs. Geological Society of America Special Paper 376: 1-64.

Chen, P.-j., Z.-m. Dong, and S.-n. Zhen. 1998. An exceptionally well-preserved theropod dinosaur from the Yixian Formation of China. Nature 391: 147-152.

Chen, X.-h., R. Motani, L. Cheng, D.-y. Jiang, and O. Rieppel. 2014a. A carapace-like bony 'body tube' in an Early Triassic marine reptile and the onset of marine tetrapod predation. PLoS ONE 9(4): e94396. doi:10.1371/journal.pone.0094396.

Chen, X.-h., R. Motani, L. Cheng, D.-y. Jiang, and O. Rieppel. 2014b. The enigmatic marine reptile *Nanchangosaurus* from the Lower Triassic of Hubei, China and the phylogenetic affinities of Hupehsuchia. PLoS ONE 9(7): e102361. doi:10.1371/journal.pone.0102361.

Chen, X.-h., R. Motani, L. Cheng, D.-y. Jiang, and O. Rieppel. 2015. A new specimen of Carroll's mystery hupehsuchian from the Lower Triassic of China. PLoS ONE 10(5): e0126024. doi:10.1371/journal.pone.0126024.

Cheng, L., X.-h. Chen, Q.-h. Shang, and X.-c. Wu. 2014. A new marine reptile from the Triassic of China, with a highly specialized feeding adaptation. Naturwissenschaften 101: 251-259.

Cheng, Y.-n., X.-c. Wu, and Q. Ji. 2004. Triassic marine reptiles gave birth to live young. Nature 432: 383-386.

Chiappe, L. M. 2007. Glorified Dinosaurs: The Origin and Early Evolution of Birds. Wiley-Liss, New York.

Chiappe, L. M., and A. Chinsamy. 1996. *Pterodaustro*'s true teeth. Nature 379: 212-213.

Chiappe, L. M., and U. B. Göhlich. Anatomy of *Juravenator starki* (Theropoda: Coelurosauria) from the Late Jurassic of Germany. Neues Jahrbuch für Geologie und Paläontologie Abhandlungen 258: 257-296.

Chiari, Y., V. Cahais, N. Galtier, and F. Delsuc. 2012. Phylogenomic analyses support the position of turtles as the sister group of birds and crocodiles (Archosauria). BMC Biology 10: 65. doi:10.1186/1741-7007-10-65.

Chinnery, B. 2004. Description of *Prenoceratops pieganensis* gen. et sp. nov. (Dinosauria: Neoceratopsia) from the Two Medicine Formation of Montana. Journal of Vertebrate Paleontology 24: 572-590.

Choiniere, J. N., X. Xu, J. M. Clark, C. A. Forster, Y. Guo, and F. Han. 2010. A basal alvarezsauroid theropod from the early Late Jurassic of Xinjiang, China. Science 327: 571-574.

Chun, L., D.-y. Jiang, L. Cheng., X.-c. Wu, and O. Rieppel. 2014. A new species of *Largocephalosaurus* (Diapsida: Saurosphargidae), with implications for the morphological diversity and phylogeny of the group. Geological Magazine 151: 100-120.

Cidade, G. M., A. Solórzano, A. D. Rincón, D. Riff, and A. S. Hsiou. 2017. A new *Mourasuchus* (Alligatoroidea, Caimaninae) from the late Miocene of Venezuela, the phylogeny of Caimaninae and considerations on the feeding habits of *Mourasuchus*. PeerJ 5: e3056. doi:10.7717/peerj.3056.

Cisneros, J. C. 2008. Phylogenetic relationships of procolophonid parareptiles with remarks on their geological record. Journal of Systematic Palaeontology 6: 345-366.

Cisneros, J. C., R. Damiani, C. Schultz, A. da Rosa, C. Schwanke, L. W. Neto, and P. L. P. Aurélio. 2004. A procolophonoid reptile with temporal fenestration from the Middle Triassic of Brazil. Proceedings of the Royal Society of London B 271: 1541-1546.

Cisneros, J. C., C. Marsicano, K. D. Angielczyk, R. M. H. Smith, M. Richter, J. Fröbisch, C. F. Kammerer, and R. W. Sadleir. 2015. New Permian fauna from tropical Gondwana. Nature Communications 6: 8676. doi:10.1038/ncomms9676.

Clack, J. A. 2012. Gaining Ground: The Origin and Evolution of Tetrapods. Second Edition. Indiana University Press, Bloomington.

Claessens, L. P. A. M., P. M. O'Connor, and D. M. Unwin. 2009. Respiratory evolution facilitated the origin of pterosaur flight and aerial gigantism. PLoS ONE 4(2): e4497. doi:10.1371/journal.pone.0004497.

Claessens, L. P. A. M., and M. K. Vickaryous. 2013. The evolution, development and skeletal identity of the crocodylian pelvis: revisiting a forgotten scientific debate. Journal of Morphology 273: 1185-1198.

Clark, J., and R. L. Carroll. 1973. Romeriid reptiles from the Lower Permian. Bulletin of the Museum of Comparative Zoology, Harvard University 144: 353-406.

Clark, J. M. 1986. Phylogenetic relationships of the crocodylomorph archosaurs. PhD dissertation, The University of Chicago, Chicago.

Clark, J. M. 1994. Patterns of evolution in Mesozoic Crocodyliformes. Pp. 84-97 in N. C. Fraser and H.-D. Sues (eds.), In the Shadow of the Dinosaurs: Early Mesozoic Tetrapods. Cambridge University Press, New York.

Clark, J. M. 2011. A new shartegosuchid crocodyliform from the Upper Jurassic Morrison Formation of western Colorado. Zoological Journal of the Linnean Society 163: S152-S172.

Clark, J. M., and R. Hernandez R. 1994. A new burrowing diapsid from the Jurassic La Boca Formation of Tamaulipas, Mexico. Journal of Vertebrate Paleontology 14: 180-195.

Clark, J. M., J. A. Hopson, R. Hernández R., D. E. Fastovsky, and M. Montellano. 1998. Foot posture in a primitive pterosaur. Nature 391: 886-889.

Clark, J. M., M. A. Norell, and T. B. Rowe. 2002. Cranial anatomy of *Citipati osmolskae* (Theropoda, Oviraptorosauria), and a reinterpretation of the holotype of *Oviraptor philoceratops*. American Museum Novitates 3364: 1-24.

Clark, J. M., and H.-D. Sues. 2002. Two new basal crocodylomorph archosaurs from the Lower Jurassic and the monophyly of the Sphenosuchia. Zoological Journal of the Linnean Society 136: 77-95.

Clark, J. M., X. Xu, C. A. Forster, and Y. Wang. 2004. A Middle Jurassic 'sphenosuchian' from China and the origin of the crocodylian skull. Nature 430: 1021-1024.

Clos, L. M. 1995. A new species of *Varanus* (Reptilia: Sauria) from the Miocene of Kenya. Journal of Vertebrate Paleontology 15: 254-267.

Cocude-Michel, M. 1963. Les Rhynchocéphales et les Sauriens des calcaires lithographiques (Jurassique supérieur) d'Europe Occidentale. Nouvelles Archives du Muséum d'Histoire Naturelle de Lyon 7: 1-187.

Codorniú, L., A. Paulina Carabajal, D. Pol, D. Unwin, and O. W. M. Rauhut. 2016. A Jurassic pterosaur from Patagonia and the origin of the pterodactyloid neurocranium. PeerJ 4: e2311. doi:10.7717/peerj.2311.

Codrea, V. A., M. Venczel, and A. Solomon. 2017. A new family of teiioid lizards from the Upper Cretaceous of Romania with notes on the evolutionary history of early teiioids. Zoological Journal of the Linnean Society 181: 385-399.

Cohn, M. J., and C. Tickle. 1999. Developmental basis of limblessness and axial patterning in snakes. Nature 399: 474-479.

Colbert, E. H. 1945. The Dinosaur Book. The Ruling Reptiles and Their Relatives. American Museum of Natural History, New York.

Colbert, E. H. 1970. The gliding Triassic reptile *Icarosaurus*. Bulletin of the American Museum of Natural History 143: 85-142.

Colbert, E. H. 1981. A primitive ornithischian dinosaur from the Kayenta Formation of Arizona. Museum of Northern Arizona Bulletin 53: 1-61.

Colbert, E. H. 1989. The Triassic dinosaur *Coelophysis*. Museum of Northern Arizona Bulletin 57: 1-160.

Colbert, E. H., and C. C. Mook. 1951. The ancestral crocodilian *Protosuchus*. Bulletin of the American Museum of Natural History 97: 147-182.

Colbert, E. H., and P. E. Olsen. 2001. A new and unusual aquatic reptile from the Lockatong Formation of New Jersey (Late Triassic, Newark Supergroup). American Museum Novitates 3334: 1-24.

Colbert, E. H., and D. A. Russell. 1969. The small Cretaceous dinosaur *Dromaeosaurus*. American Museum Novitates 2380: 1-49.

Conrad, J. L. 2006. An Eocene shinisaurid (Reptilia, Squamata) from Wyoming, U.S.A. Journal of Vertebrate Paleontology 26: 113-126.

Conrad, J. L. 2008. Phylogeny and systematics of Squamata (Reptilia) based on morphology. Bulletin of the American Museum of Natural History 310: 1-182.

Conrad, J. L. 2015. A new casquehead lizard (Reptilia, Corytophanidae) from North America. PLoS ONE 10(7): e0127900. doi:10.1371/journal.pone.0127900.

Conrad, J. L., J. J. Head, and M. T. Carrano. 2014. Unusual soft-tissue preservation of a crocodile lizard (Squamata, Shinisauria) from the Green River Formation (Eocene) and shinisaur relationships. The Anatomical Record 297: 545-559.

Conrad, J. L., K. Jenkins, T. Lehmann, F. K. Manthi, D. J. Peppe, S. Nightingale, A. Cossette, H. M. Dunsworth, W. E. H. Harcourt-Smith, and K. P. McNulty. 2013. New specimens of '*Crocodylus*' *pigotti* (Crocodylidae) from Rusinga Island, Kenya, and generic reallocation of the species. Journal of Vertebrate Paleontology 33: 629-646.

Conrad, J. L., and M. A. Norell. 2007. A complete Late Cretaceous iguanian (Squamata, Reptilia) from the Gobi and identification of a new iguanian clade. American Museum Novitates 3584: 1-47.

Conrad, J. L., O. Rieppel, and L. Grande. 2007. A Green River (Eocene) polychrotid (Squamata: Reptilia) and a re-examination of iguanian systematics. Journal of Paleontology 81: 1365-1373.

Coombs, W. P., Jr. 1978. The families of the ornithischian dinosaur order Ankylosauria. Palaeontology 21: 143-170.

Cooper, M. R. 1981. The prosauropod dinosaur *Massospondylus carinatus* Owen from Zimbabwe: its biology, mode of life and phylogenetic significance. Occasional Papers of the National Museums and Monuments of Rhodesia B 6(10): 689-840.

Cooper, M. R. 1984. A reassessment of *Vulcanodon karibaensis* Raath (Dinosauria: Saurischia) and the origin of sauropoda. Palaeontologia africana 25: 203-231.

Cope, E. D. 1869. On the reptilian orders Pythonomorpha and Streptosauria. Proceedings of the Boston Society of Natural History 12: 250-266.

Coria, R. A., and P. J. Currie. 2016. A new megaraptoran dinosaur (Dinosauria, Theropoda, Megaraptoridae) from the Late Cretaceous of Patagonia. PLoS ONE 11(7): e0157973. doi:10.1371/journal.pone.0157973.

Coria, R. A., and L. Salgado. 1995. A new giant carnivorous dinosaur from the Cretaceous of Patagonia. Nature 377: 224-226.

Costa, F. R., O. Rocha-Barbosa, and A. W. A. Kellner. 2014. A biomechanical approach on the optimal stance of *Anhanguera piscator* (Pterodactyloidea) and its implications for pterosaur gait on land. Historical Biology 26: 582-590.

Crain, D. A., and L. J. Guillette Jr. 1998. Reptiles as models of contaminant-induced endocrine disruption. Animal Reproduction Science 53: 77-86.

Crawford, N. G., B. C. Faircloth, J. E. McCormack, R. T. Brumfield, K. Winker, and T. C. Glenn. 2012. More than 1000 ultraconserved elements provide evidence that turtles are the sister group of archosaurs. Biology Letters 8: 783-786.

Crawford, N. G., J. F. Parham, A. B. Sellas, B. C. Faircloth, T. C. Glenn, T. J. Papenfuss, J. B. Henderson, M. H. Hansen, and W. B. Simison. 2015. A phylogenomic analysis of turtles. Molecular Phylogenetics and Evolution 83: 250-257.

Cree, A. 2014. Tuatara: Biology and Conservation of a Venerable Survivor. Canterbury University Press, Christchurch, New Zealand.

Cruickshank, A. R. I. 1972. The proterosuchian thecodonts. Pp. 89-119 in K. A. Joysey and T. S. Kemp (eds.), Studies in Vertebrate Evolution. Oliver and Boyd, Edinburgh.

Crumly, C. R. 1985. A hypothesis for the relationships of land tortoise genera (family Testudinidae). Studia Palaeochelonologica 1: 115-124.

Crush, P. J. 1984. A late Upper Triassic sphenosuchid crocodilian from Wales. Palaeontology 27: 131-157.

Csiki-Sava, Z., E. Buffetaut, A. Ösi, X. Pereda-Suberbiola, and S. L. Brusatte. 2015. Island life in the Cretaceous—faunal composition, biogeography, evolution, and extinction of land-living vertebrates on the Late Cretaceous European archipelago. ZooKeys 469: 1-161.

Cundall, D. L., and F. J. Irish. 2008. The snake skull. Pp. 349-692 in C. Gans, A. S. Gaunt, and K. Adler (eds.), Biology of the Reptilia. Volume 20: The Skull of Lepidosauria. Society for the Study of Amphibians and Reptiles, Ithaca, New York.

Cundall, D., V. Wallach, and D. A. Rossman. 1993. The systematic relationships of the snake *Anomochilus*. Zoological Journal of the Linnean Society 109: 275-299.

Currie, P. J. 1981. *Hovasaurus boulei*, an aquatic eosuchian from the Upper Permian of Madagascar. Palaeontologia africana 24: 99-168.

Currie, P. J. 1995. New information on the anatomy and relationships of *Dromaeosaurus albertensis* (Dinosauria: Theropoda). Journal of Vertebrate Paleontology 15: 576-591.

Currie, P. J. 2003. Cranial anatomy of tyrannosaurid dinosaurs from the Late Cretaceous of Alberta, Canada. Acta Palaeontologica Polonica 48: 191-226.

Currie, P. J., and K. Carpenter. 2000. A new specimen of *Acrocanthosaurus atokensis* (Theropoda, Dinosauria) from the Lower Cretaceous Antlers Formation (Lower Cretaceous, Aptian) of Oklahoma. Geodiversitas 22: 207-246.

Currie, P. J., and P.-j. Chen. 2001. Anatomy of *Sinosauropteryx prima* from Liaoning, northeastern China. Canadian Journal of Earth Sciences 38: 1705-1727.

Currie, P. J., S. J. Godfrey, and L. Nessov. 1994. New caenagnathid (Dinosauria: Theropoda) specimens from the Upper Cretaceous of North America and Asia. Canadian Journal of Earth Sciences 30: 2255-2272.

Currie, P. J., J. H. Hurum, and K. Sabath. 2003. Skull structure and evolution in tyrannosaurid dinosaurs. Acta Palaeontologica Polonica 48: 227-234.

Currie, P. J., W. Langston, Jr., and D. H. Tanke. 2008. A New Horned Dinosaur from an Upper Cretaceous Bone Bed in Alberta. NRC Research Press, Ottawa.

Currie, P. J., and X.-j. Zhao. 1994. A new carnosaur (Dinosauria, Theropoda) from the Jurassic of Xinjiang, People's Republic of China. Canadian Journal of Earth Sciences 30: 2037-2081.

Cuthbertson, R. S., A. P. Russell, and J. S. Anderson. 2013. Reinterpretation of the cranial morphology of *Utatsusaurus hataii* (Ichthyopterygia) Osawa Formation, Lower Triassic, Miyagi, Japan). Journal of Vertebrate Paleontology 33: 817-830.

Cuvier, G. 1825. Recherches sur les ossements fossiles, où l'on rétablit les caractères de plusieurs animaux dont les révolutions du globe ont détruit les espèces. Troisième Édition. Tome 5, Partie 2. G. Dufour et E. D'Ocagne, Paris.

Czerkas, S. A., and Q. Ji. 2002. A new rhamphorhynchoid with a headcrest and complex integumentary structures. Pp. 15-41 in S. J. Czerkas (ed.), Feathered Dinosaurs and the Origin of Flight. The Dinosaur Museum, Blanding, Utah.

Dalla Vecchia, F. M. 2006. A new sauropterygian reptile with plesiosaurian affinity from the Late Triassic of Italy. Rivista Italiana di Paleontologia e Stratigrafia 112: 207-225.

Dalla Vecchia, F. M. 2009. *Tethyshadros insularis*, a new hadrosauroid dinosaur (Ornithischia) from the Upper Cretaceous of Italy. Journal of Vertebrate Paleontology 29: 1100-1116.

Dalla Vecchia, F. M. 2013. Triassic pterosaurs. Pp. 119-155 in S. J. Nesbitt, J. B. Desojo, and R. B. Irmis (eds.), Anatomy, Phylogeny and Palaeobiology of Early Archosaurs and Their Kin. Geological Society, London, Special Publication 379.

Dalla Vecchia, F. M. 2014. Gli Pterosauri Triassici. Edizioni del Museo Friulano di Storia Naturale, Udine.

Dalla Vecchia, F. M., R. Wild, H. Hopf, and J. Reitner. 2002. A crested rhamphorhynchoid pterosaur from the Late Triassic of Austria. Journal of Vertebrate Paleontology 22: 196-199.

Dal Sasso, C., and S. Maganuco. 2011. *Scipionyx samniticus* (Theropoda: Compsognathidae) from the Lower Cretaceous of Italy: osteology, ontogenetic assessment, phylogeny, soft tissue anatomy, taphonomy and palaeobiology. Memorie della Società Italiana di Scienze Naturali e del Museo Civico di Storia Naturale di Milano 37(1): 1-281.

Dal Sasso, C., S. Maganuco, E. Buffetaut, and M. A. Mendez. 2005. New information on the skull of the enigmatic theropod *Spinosaurus*, with remarks on its size and affinities. Journal of Vertebrate Paleontology 25: 888-896.

Dal Sasso, C., S. Maganuco, and A. Cau. 2018. The oldest ceratosaurian (Dinosauria: Theropoda), from the Lower Jurassic of Italy, sheds light on the evolution of the three-fingered hand of birds. PeerJ 6: e5976. doi:10.7717/peerj.5976.

Dal Sasso, C., G. Pasini, G. Fleury, and S. Maganuco. 2017. *Razanandrongobe sakalavae*, a gigantic mesoeucrocodylian from the Middle Jurassic of Madagascar, is the oldest known notosuchian. PeerJ 5: e3481. doi:10.7717/peerj.3481.

Dal Sasso, C., and G. Pinna. 1997. *Aphanizocnemus libanensis* n. gen. n. sp., a new dolichosaur (Reptilia, Varanoidea) from the Upper Cretaceous of Lebanon. Palaeontologia Lombarda 7: 1-31.

Danilov, I. G., and J. F. Parham. 2006. A redescription of '*Plesiochelys*' *tatsuensis* from the Late Jurassic of China, with comments on the antiquity of the crown clade Cryptodira. Journal of Vertebrate Paleontology 26: 573-580.

Danilov, I. G., and V. B. Sukhanov. 2001. New data on the lindholmemydid turtle *Lindholmemys* from the Late Cretaceous of Mongolia. Acta Palaeontologica Polonica 46: 125-131.

Danilov, I. G., and N. S. Vitek. 2013. Soft-shelled turtles (Trionychidae) from the Bissekty Formation (Late Cretaceous: late Turonian) of Uzbekistan: shell-based taxa. Cretaceous Research 41: 55-64.

Darwin, C. 1859. On the Origin of Species by Means of Natural Selection, or the Preservation of Favoured Races in the Struggle for Life. John Murray, London.

Davydov, V. I., D. Korn, and M. D. Schmitz. 2012. The Carboniferous Period. Pp. 603-651 in F. M. Gradstein, J. G. Ogg, M. D. Schmitz, and G. M. Ogg (eds.), The Geologic Time Scale. Elsevier, Amsterdam.

Daza, J. D., V. Abdala, J. S. Arias, D. García-López, and P. Ortiz. 2012. Cladistic analysis of Iguania and a fossil lizard from the late Pliocene of northwestern Argentina. Journal of Herpetology 46: 104-119.

Daza, J. D., and A. M. Bauer. 2012. A new amber-embedded sphaerodactyl gecko from Hispaniola, with comments on morphological synapomorphies of the Sphaerodactylidae. Breviora 529: 1-28.

Daza, J. D., A. M. Bauer, and E. Snively. 2013. *Gobekko cretacicus* (Reptilia: Squamata) and its bearing on the interpretation of gekkotan affinities. Zoological Journal of the Linnean Society 167: 430-448.

Daza, J. D., A. M. Bauer, and E. D. Snively. 2014. On the fossil record of the Gekkota. The Anatomical Record 297: 433-462.

Daza, J. D., E. L. Stanley, P. Wagner, A. M. Bauer, and D. A. Grimaldi. 2016. Mid-Cretaceous amber fossils illuminate the past diversity of tropical lizards. Science Advances 2: e1501080. doi:10.1126/sciadv.1501080.

de Blainville, H. M. D. 1816. Prodrome d'une distribution systematique du règne animal. Bulletin des Sciences de la Société Philomatique, Paris, Série 3, 3: 105-124.

deBraga, M., and R. L. Carroll. 1993. The origin of mosasaurs as a model of macroevolutionary patterns and processes. Evolutionary Biology 27: 245-322.

deBraga, M., and R. R. Reisz. 1995. A new diapsid reptile from the uppermost Carboniferous (Stephanian) of Kansas. Palaeontology 38: 199-212.

deBraga, M., and R. R. Reisz. 1996. The Early Permian reptile *Acleistorhinus pteroticus* and its phylogenetic position. Journal of Vertebrate Paleontology 16: 384-395.

deBraga, M., and O. Rieppel. 1997. Reptile phylogeny and the interrelationships of turtles. Zoological Journal of the Linnean Society 120: 281-354.

de Buffrénil, V., J. O. Farlow, and A. de Ricqlès. 1986. Growth and function of *Stegosaurus* plates: evidence from bone histology. Paleobiology 6: 208-232.

Deeming, D. C. 2004. Post-hatching phenotypic effects of incubation in reptiles. Pp. 211-228 in D. C. Deeming (ed.), Reptilian Incubation Environment, Evolution and Behaviour. Nottingham University Press, Nottingham.

de la Fuente, M. S., F. de Lapparent de Broin, and T. M. Bianco. 2001. The oldest and first nearly complete skeleton of a chelid, of the *Hydromedusa* sub-group (Chelidae, Pleurodira), from the Upper Cretaceous of Patagonia. Bulletin de la Société Géologique de France 172: 237-244.

de Lapparent de Broin, F., N. Bardet, M. Amaghzaz, and S. Meslouh. 2014. A strange new chelonioid turtle from the Latest Cretaceous Phosphates of Morocco. Comptes Rendus Palevol 13: 87-96.

Delfino, M., and T. Smith. 2009. A reassessment of the morphology and taxonomic status of '*Crocodylus*' *depressifrons* Blainville, 1855 (Crocodylia, Crocodyloidea) based on the Early Eocene remains from Belgium. Zoological Journal of the Linnean Society 156: 140-167.

DeMar, D. G., Jr., J. L. Conrad, J. J. Head, D. J. Varricchio, and G. P. Wilson. 2017. A new Late Cretaceous iguanomorph from North America and the origin of New World Pleurodonta (Squamata, Iguania). Proceedings of the Royal Society B 284: 20161902. doi:10.1098/rspb.2016.1902.

D'Emic, M. D. 2012. Early evolution of titanosauriform sauropod dinosaurs. Zoological Journal of the Linnean Society 166: 624-671.

D'Emic, M. D., and B. Z. Foreman. 2012. The beginning of the sauropod hiatus in North America: insights from the Cloverly Formation of Wyoming. Journal of Vertebrate Paleontology 32: 883-902.

D'Emic, M. D., J. A. Whitlock, K. M. Smith, D. C. Fisher, and J. A. Wilson. 2013. Evolution of high tooth replacement rates in sauropod dinosaurs. PLoS ONE 8:e69235. doi: 10.1371/journal.pone.0069235.

de Ricqlès, A., and J. R. Bolt. 1983. Jaw growth and tooth replacement in *Captorhinus aguti* (Reptilia, Captorhinomorpha): a morphological and histological analysis. Journal of Vertebrate Paleontology 3: 7-24.

de Ricqlès, A., and V. de Buffrénil. 2001. Bone histology, heterochronies and the return of tetrapods to life in water: where are we? Pp. 289-310 in J.-M. Mazin and V. de Buffrénil (eds.), Secondary Adaptations of Tetrapods to Life in Water. Verlag Dr. Friedrich Pfeil, Munich.

de Ricqlès, A., K. Padian, F. Knoll, and J. R. Horner. 2008. On the origin of high growth rates in archosaurs and their ancient relatives: complementary histological studies on Triassic archosauriforms and the problem of a "phylogenetic signal" in bone histology. Annales de Paléontologie 94: 57-76.

de Ricqlès, A., and P. Taquet. 1982. La faune de vertébrés du Permien supérieur du Niger. I. Le captorhinomorphe *Moradisaurus grandis* (Reptilia, Cotylosauria). Annales de Paléontologie (Vertébrés-Invertébrés) 68: 33-106.

Desojo, J. B., A. B. Heckert, J. W. Martz, W. G. Parker, R. R. Schoch, B. J. Small, and T. Sulej. 2013. Aetosauria: a clade of armoured pseudosuchians from the Upper Triassic continental beds. Pp. 203-239 in S. J. Nesbitt, J. B. Desojo, and R. B. Irmis (eds.), Anatomy, Phylogeny and Palaeobiology of Early Archosaurs and Their Kin. Geological Society, London, Special Publications 379.

Dilkes, D. W. 1995. The rhynchosaur *Howesia browni* from the Lower Triassic of South Africa. Palaeontology 38: 665-685.

Dilkes, D. W. 1998. The Early Triassic rhynchosaur *Mesosuchus browni* and the interrelationships of basal archosauromorph reptiles. Philosophical Transactions of the Royal Society of London B 353: 501-541.

Dodick, J. T., and S. P. Modesto. 1995. The cranial anatomy of the captorhinid reptile *Labidosaurikos meachami* from the Lower Permian of Oklahoma. Palaeontology 38: 687-711.

Dodson, P. 1975. Taxonomic implications of relative growth and sexual dimorphism in lambeosaurine dinosaurs. Systematic Zoology 24: 37-54.

Dodson, P. 1976. Quantitative aspects of relative growth and sexual dimorphism in *Protoceratops*. Journal of Paleontology 50: 929-940.

Dodson, P., C. A. Forster, and S. D. Sampson. 2004. Ceratopsidae. Pp. 494-513 in D. B. Weishampel, P. Dodson, and H. Osmólska (eds.), The Dinosauria. Second Edition. University of California Press, Berkeley.

Dollo, L. 1924. Le centenaire des Iguanodons (1822-1922). Philosophical Transactions of the Royal Society of London B 212: 67-78.

Dong, Z.-m., and Y. Azuma. 1997. On a primitive neoceratopsian from the early Cretaceous of China. Pp. 68-89 in Z.-m. Dong (ed.), Sino-Japanese Silk Road Dinosaur Expedition. China Ocean Press, Beijing.

Drevermann, F. 1933. Die Placodontier. 3. Das Skelett von *Placodus gigas* Agassiz im Senckenberg-Museum. Abhandlungen der Senckenbergischen Naturforschenden Gesellschaft 38: 319-364.

Druckenmiller, P. S., and A. P. Russell. 2008. A phylogeny of Plesiosauria (Sauropterygia) and its bearing on the systematic status of *Leptocleidus* Andrews, 1922. Zootaxa 1863: 1-120.

Dufeau, D. L., and L. M. Witmer. 2015. Ontogeny of the middle-ear air-sinus system in *Alligator mississippiensis* (Archosauria: Crocodylia). PLoS ONE 10(9): e0137060. doi:10.1371/journal.pone.0137060.

Dutchak, A. R., and M. W. Caldwell. 2006. Redescription of *Aigialosaurus dalmaticus* Kramberger, 1892, a Cenomanian mosasauroid lizard from Hvar Island, Croatia. Canadian Journal of Earth Sciences 43: 1821-1834.

Dutchak, A. R., and M. W. Caldwell. 2009. A redescription of *Aigialosaurus* (=*Opetiosaurus*) *bucchichi* (Kornhuber, 1901) (Squamata: Aigialosauridae) with comments on mosasauroid systematics. Journal of Vertebrate Paleontology 29: 437-452.

Dutuit, J.-M. 1972. Découverte d'un Dinosaure ornithischien dans le Trias supérieur de l'Atlas occidental marocain. Comptes Rendus de l'Académie des Sciences à Paris D 275: 2841-2844.

Dzik, J. 2003. A beaked herbivorous archosaur with dinosaur affinities from the early Late Triassic of Poland. Journal of Vertebrate Paleontology 23: 556-574.

Dzik, J., and T. Sulej. 2007. A review of the early Late Triassic Krasiejów biota from Silesia, Poland. Palaeontologia Polonica 64: 3-27.

Dzik, J., and T. Sulej. 2016. An early Late Triassic long-necked reptile with a bony pectoral shield and gracile appendages. Acta Palaeontologica Polonica 61: 805-823.

Eckstut, M. E., D. M. Sever, M. E. White, and B. I. Crother. 2009. Phylogenetic analysis of sperm storage in female squamates. Pp. 185-218 in L. T. Dahnof (ed.), Animal Reproduction: New Research Developments. Nova Science Publishers, Hauppauge, NY.

Eddy, D. R., and J. A. Clarke. 2011. New information on the cranial anatomy of *Acrocanthosaurus atokensis* and its implications for the phylogeny of Allosauroidea (Dinosauria: Theropoda). PLoS ONE 6(3): e17932. doi:10.1371/journal.pone.0017932.

Edmund, A. G. 1969. Dentition. Pp. 117-200 in C. Gans, A. d'A. Bellairs, and T. S. Parsons (eds.), Biology of the Reptilia. Volume 1: Morphology A. Academic Press, London.

Eernisse, D. J., and A. G. Kluge. 1993. Taxonomic congruence versus total evidence, and amniote phylogeny inferred from fossils, molecules, and morphology. Molecular Biology and Evolution 10: 1170-1195.

Elzanowski, A. 2002. Archaeopterygidae (Upper Jurassic of Germany). Pp. 129-159 in L. M. Chiappe and L. M. Witmer (eds.), Mesozoic Birds: Above the Heads of Dinosaurs. University of California Press, Berkeley.

Erickson, B. R. 1972. The lepidosaurian reptile *Champsosaurus* in North America. Monographs of the Science Museum of Minnesota 1: 1-91.

Erickson, B. R. 1973. A new chelydrid turtle *Protochelydra zangerli* from the late Paleocene of North Dakota. Scientific Publications of the Science Museum of Minnesota 2(2): 1-16.

Erickson, B. R. 1987. *Simoedosaurus dakotensis*, n. sp., a lepidosaurian reptile (Diapsida: Choristodera) from the Paleocene of North America. Journal of Vertebrate Paleontology 7: 237-251.

Erickson, G. M., and C. A. Brochu. 2006. How the 'terror crocodile' grew so big. Nature 398: 205-206.

Erickson, G. M., B. A. Krick, M. Hamilton, G. R. Bourne, M. A. Norell, E. Lilleodden, and W. G. Sawyer. 2012. Complex dental structure and wear biomechanics in hadrosaurid dinosaurs. Science 338: 98-101.

Erickson, G. M., M. A. Sidebottom, D. I. Kay, K. T. Turner, N. Ip, M. A. Norell, W. G. Sawyer, and B. A. Krick. 2015. Wear biomechanics in the slicing dentition of the giant horned dinosaur *Triceratops*. Science Advances 1: e1500055. doi:10.1126/sciadv.1500055.

Estes, R. 1962. A fossil gerrhosaur from the Miocene of Kenya (Reptilia: Cordylidae). Breviora 158: 1-10.

Estes, R. 1983. Sauria terrestria, Amphisbaenia. Pp. 1-249 in P. Wellnhofer (ed.), Handbuch der Paläoherpetologie, Part 10A. Gustav Fischer Verlag, Stuttgart.

Estes, R., K. de Queiroz, and J. Gauthier. 1988. Phylogenetic relationships within Squamata. Pp. 119-281 in R. Estes and G. Pregill (eds.), Phylogenetic Relationships of the Lizard Families: Essays Commemorating Charles L. Camp. Stanford University Press, Stanford.

Estes, R., T. H. Frazzetta, and E. E. Williams. 1970. Studies on the fossil snake *Dinilysia patagonica* Woodward: Part I. Cranial morphology. Bulletin of the Museum of Comparative Zoology, Harvard University 140: 25-73.

Evans, M. 1999. A new reconstruction of the skull of the Callovian elasmosaurid plesiosaur *Muraenosaurus leedsi* Seeley. Mercian Geologist 14: 191-196.

Evans, S. E. 1980. The skull of a new eosuchian reptile from the Lower Jurassic of South Wales. Zoological Journal of the Linnean Society 70: 203-264.

Evans, S. E. 1981. The postcranial skeleton of the Lower Jurassic eosuchian *Gephyrosaurus bridensis*. Zoological Journal of the Linnean Society 73: 81-116.

Evans, S. E. 1990. The skull of *Cteniogenys*, a choristodere from the Middle Jurassic of Oxfordshire. Zoological Journal of the Linnean Society 99: 205-237.

Evans, S. E. 1991. A new lizard-like reptile (Diapsida: Lepidosauromorpha) from the Middle Jurassic of England. Zoological Journal of the Linnean Society 103: 391-412.

Evans, S. E. 1998. Crown group lizards (Reptilia: Squamata) from the Middle Jurassic of the British Isles. Palaeontographica A 250: 123-154.

Evans, S. E. 2003. At the feet of the dinosaurs: the early history and radiation of lizards. Biological Reviews 78: 513-551.

Evans, S. E. 2008. The skull of lizards and tuataras. Pp. 1-347 in C. Gans, A. S. Gaunt, and K. Adler (eds.), Biology of the Reptilia. Volume 20: The Skull of Lepidosauria. Society for the Study of Amphibians and Reptiles, Ithaca, New York.

Evans, S. E. 2009. An early kuehneosaurid reptile (Reptilia: Diapsida) from the Early Triassic of Poland. Palaeontologia Polonica 65: 145-178.

Evans, S. E., and M. Borsuk-Białynicka. 2009. A small lepidosauromorph reptile from the Early Triassic of Poland. Palaeontologia Polonica 65: 179-202.

Evans, S. E., and D. C. Chure. 1998. Paramacellodid lizard skulls from the Jurassic Morrison Formation at Dinosaur National Monument, Utah. Journal of Vertebrate Paleontology 18: 99-114.

Evans, S. E., and H. Haubold. 1987. A review of the Upper Permian genera *Coelurosauravus*, *Weigeltisaurus* and *Gracilisaurus* (Reptilia: Diapsida). Zoological Journal of the Linnean Society 90: 275-303.

Evans, S. E., and M. K. Hecht. 1993. A history of an extinct reptilian clade, the Choristodera: longevity, Lazarus-taxa, and the fossil record. Evolutionary Biology 27: 323-338.

Evans, S. E., and M. Manabe. 1999. Early Cretaceous lizards from the Okurodani Formation of Japan. Geobios 32: 889-899.

Evans, S. E., and Y. Wang. 2005. The early Cretaceous lizard *Dalinghosaurus* from China. Acta Palaeontologica Polonica 50: 725-742.

Everhart, M. J. 2004. Plesiosaurs as the food of mosasaurs: new data on the stomach contents of a *Tylosaurus proriger* (Squamata; Mosasauridae) from the Niobrara Formation of western Kansas. The Mosasaur 7: 41-46.

Ewer, R. F. 1965. The anatomy of the thecodont reptile *Euparkeria capensis* Broom. Philosophical Transactions of the Royal Society of London B 248: 379-435.

Ezcurra, M. D. 2010. A new early dinosaur (Saurischia: Sauropodomorpha) from the Late Triassic of Argentina: a reassessment of dinosaur origin and phylogeny. Journal of Systematic Palaeontology 8: 371-425.

Ezcurra, M. D. 2016. The phylogenetic relationships of basal archosauromorphs, with an emphasis on the systematics of proterosuchian archosauriforms. PeerJ 4: e1778. doi:10.7717/peerj.1778.

Ezcurra, M. D., and R. J. Butler. 2015. Taxonomy of the proterosuchid archosauriforms (Diapsida: Archosauromorpha) from the earliest Triassic of South Africa, and implications for the early archosauriform radiation. Palaeontology 58: 141-170.

Ezcurra, M. D., R. J. Butler, and D. J. Gower. 2013. 'Proterosuchia': the origin and early history of Archosauriformes. Pp. 9-33 in S. J. Nesbitt, J. B. Desojo, and R. B. Irmis (eds.), Anatomy, Phylogeny and Palaeobiology of Early Archosaurs and Their Kin. Geological Society, London, Special Publications 379.

Ezcurra, M. D., L. E. Fiorelli, A. G. Martinelli, S. Rocher, M. B. von Baczko, M. Ezpeleta, J. R. A. Taborda, E. M. Hechenleitner, M. J. Trotteyn, and J. B. Desojo. 2017. Deep faunistic turnovers preceded the rise of dinosaurs in southwestern Pangaea. Nature: Ecology and Evolution 1: 1477-1483.

Ezcurra, M. D., T. M. Scheyer, and R. J. Butler. 2014. The origin and early evolution of Sauria: reassessing the Permian saurian fossil record and the timing of the crocodile-lizard divergence. PLoS ONE 9(2): e89165. doi:10.1371/journal.ponc.0089165.

Fabre, J. 1981. Les Rhynchocéphales et les Ptérosaures à crête pariétale du Kiméridgien supérieur-Berriasien d'Europe occidentale. Éditions de la Fondation Singer-Polignac, Paris.

Falcon-Lang, H. J., M. R. Gibling, and M. Grey. 2010. Classic localities explained 4. Joggins, Nova Scotia. Geology Today 26: 108-114.

Falconnet, J. 2012. First evidence of a bolosaurid parareptile in France (latest Carboniferous-earliest Permian of the Autun basin) and the spatiotemporal distribution of the Bolosauridae. Bulletin de la Société Géologique de France 183: 495-508.

Farke, A., D. J. Chok, A. Herrero, B. Scolieri, and S. Werning. 2013. Ontogeny in the tube-crested dinosaur *Parasaurolophus* (Hadrosauridae) and heterochrony in hadrosaurids. PeerJ 1: e182. doi:10.7717/peerj.182.

Farke, A., W. D. Maxwell, R. L. Cifelli, and M. J. Wedel. 2014. A ceratopsian dinosaur from the Lower Cretaceous of western North America, and the biogeography of Neoceratopsia. PLoS ONE 9(12): e112055. doi:10.1371/journal.pone.0112055.

Farlow, J. O., and P. Dodson. 1975. The behavioral significance of frill and horn morphology in ceratopsian dinosaurs. Evolution 29: 353-361.

Farlow, J. O., S. Hayashi, and G. J. Tattersall. 2010. Internal vascularity of the dermal plates of *Stegosaurus* (Ornithischia, Thyreophora). Swiss Journal of Geosciences 103: 173-185.

Farmer, C. G., and D. R. Carrier. 2000. Ventilation and gas exchange during treadmill locomotion in the American alligator (*Alligator mississippiensis*). Journal of Experimental Biology 203: 1671-1678.

Fastovsky, D. E., D. Badamgarav, H. Ishimoto, M. Watabe, and D. B. Weishampel. 1997. The paleoenvironments of Tugrikin-Shireh (Gobi Desert, Mongolia) and aspects of the taphonomy and paleoecology of *Protoceratops* (Dinosauria: Ornithischia). Palaios 12: 59-70.

Fastovsky, D. E., and D. B. Weishampel. 2016. Dinosaurs: A Concise Natural History. Third Edition. Cambridge University Press, New York.

Feldman, C. R., and J. F. Parham. 2002. Molecular phylogenetics of emydine turtles: taxonomic revision and the evolution of shell kinesis. Molecular Phylogenetics and Evolution 22: 388-398.

Felsenstein, J. 1978. Cases in which parsimony or compatibility methods will be positively misleading. Systematic Zoology 27: 401-410.

Fernández, M., and Z. Gasparini. 2008. Salt glands in the Jurassic metriorhynchid *Geosaurus*: implications for the evolution of osmoregulation in Mesozoic marine crocodyliforms. Naturwissenschaften 95: 79-84.

Fernandez, V., E. Buffetaut, V. Suteethorn, J.-C. Rage, P. Tafforeau, and M. Kundrát. 2015. Evidence of egg diversity in squamate evolution from Cretaceous anguimorph embryos. PLoS ONE 10(7): e0128610. doi:10.1371/journal.pone.0128610.

Ferreira, G. S., A. D. Rincón, A. Solórzano, and M. C. Langer. 2016. Review of the fossil matamata turtles: earliest well-dated record and hypotheses on the origin of their present geographical distribution. The Science of Nature 103: 28. doi: 10.1007/s00114-016-1355-2.

Field, D. J., J. A. Gauthier, B. L. King, D. Pisani, T. R. Lyson, and K. J. Peterson. 2014. Toward consilience in reptile phylogeny: miRNAs support an archosaur, not lepidosaur, affinity for turtles. Evolution and Development 16: 189-196.

Field, D. J., A. LeBlanc, A. Gau, and A. D. Behlke. 2015. Pelagic neonatal fossils support viviparity and precocial life history of Cretaceous mosasaurs. Palaeontology 58: 401-407.

Fiorelli, L., and J. O. Calvo. 2008. New remains of *Notosuchus terrestris* Woodward, 1896 (Crocodyliformes: Mesoeucrocodylia) from the Late Cretaceous of Neuquén, Patagonia, Argentina. Arquivos do Museu Nacional, Rio de Janeiro 66: 83-124.

Fiorillo, A. R., and R. S. Tykoski. 2012. A new Maastrichtian species of the centrosaurine ceratopsid *Pachyrhinosaurus* from the North Slope of Alaska. Acta Palaeontologica Polonica 57: 561-573.

Fischer, V., R. M. Appleby, D. Naish, J. Liston, J. B. Riding, S. Brindley, and P. Godefroit. 2013. A basal thunnosaurian from Iraq reveals disparate phylogenetic origins for Cretaceous ichthyosaurs. Biology Letters 9: 20130021. doi:10.1098/rsbl.2013.0021.

Fischer, V., M. S. Arkhangelsky, D. Naish, I. M. Stenshin, G. N. Uspensky, and P. Godefroit. 2014. *Simbirskiasaurus* and *Pervushovisaurus* reassessed: implications for the taxonomy and cranial osteology of Cretaceous platyp-

terygiine ichthyosaurs. Zoological Journal of the Linnean Society 171: 822-841.

Fischer, V., M. S. Arkhangelsky, I. M. Stenshin, G. N. Uspensky, N. G. Zverkov, and R. B. J. Benson. 2015. Peculiar macrophagous adaptations in a new Cretaceous pliosaurid. Royal Society Open Science 2: 150552. doi:10.1098/rsos.150552.

Fischer, V., R. B. J. Benson, N. G. Zverkov, L. C. Soul, M. S. Arkhangelsky, O. Lambert, I. M. Stenshin, G. N. Uspensky, and P. S. Druckenmiller. 2017. Plasticity and convergence in the evolution of short-necked plesiosaurs. Current Biology 27: 1667-1676.

Fischer, V., M. W. Maisch, D. Naish, R. Kosma, J. Liston, U. Joger, F. J. Krüger, J. P. Pérez, J. Tainsh, and R. M. Appleby. 2012. New ophthalmosaurid ichthyosaurs from the European Lower Cretaceous demonstrate extensive ichthyosaur survival across the Jurassic-Cretaceous boundary. PLoS ONE 7(1): e29234. doi:10.1371/journal.pone.0029234.

Fisher, M. C., T. W. J. Garner, and S. F. Walker. 2009. Global emergence of *Batrachochytrium dendrobatidis* and the amphibian chytridiomycosis in space, time, and host. Annual Review of Microbiology 63: 291-310.

Fleischle, C. V., T. Wintrich, and P. M. Sander. 2018. Quantitative histological models suggest endothermy in plesiosaurs. PeerJ 6: e4955. doi:10.7717/peerj.4955.

Flynn, J. J., S. J. Nesbitt, J. M. Parrish, L. Ranivoharimanana, and A. R. Wyss. 2010. A new species of *Azendohsaurus* (Diapsida: Archosauromorpha) from the Triassic Isalo Group of southwestern Madagascar: cranium and mandible. Palaeontology 53: 669-688.

Foffa, D., A. R. Cuff, J. Sassoon, E. J. Rayfield, M. N. Mavrogordato, and M. J. Benton. 2014. Functional anatomy and feeding biomechanics of a giant Upper Jurassic pliosaur (Reptilia: Sauropterygia) from Weymouth Bay, Dorset, UK. Journal of Anatomy 225: 209-219.

Folie, A., R. Smith, and T. Smith. 2013. New amphisbaenian lizards from the Early Paleogene of Europe and their implications for the early evolution of modern amphisbaenians. Geologica Belgica 16: 227-235.

Ford, D. P., and R. B. J. Benson. 2018. A redescription of *Orovenator mayorum* (Sauropsida, Diapsida) using high-resolution μCT, and the consequences for early amniote phylogeny. Papers in Palaeontology doi:10.1002/spp2.1236.

Forster, C. A. 1990. The postcranial skeleton of the ornithopod dinosaur *Tenontosaurus tilletti*. Journal of Vertebrate Paleontology 10: 273-294.

Forster, C. A. 1996a. New information on the skull of *Triceratops*. Journal of Vertebrate Paleontology 16: 246-258.

Forster, C. A. 1996b. Species resolution in *Triceratops*: morphometric and cladistic approaches. Journal of Vertebrate Paleontology 16: 259-270.

Foth, C., and O. W. M. Rauhut. 2017. Re-evaluation of the Haarlem *Archaeopteryx* and the radiation of maniraptoran theropod dinosaurs. BMC Evolutionary Biology 17: 236. doi:10.1186/s12862-017-1076-y.

Foth, C., H. Tischlinger, and O. W. M. Rauhut. 2014. New specimen of *Archaeopteryx* provides insights into the evolution of pennaceous feathers. Nature 511: 79-82.

Fowler, D. W., and R. M. Sullivan. 2011. The first giant titanosaurian sauropod from the Upper Cretaceous of North America. Acta Palaeontologica Polonica 56: 685-690.

Fox, R. C., and M. C. Bowman. 1966. Osteology and relationships of *Captorhinus aguti* (Cope) (Reptilia: Captorhinomorpha). The University of Kansas Paleontological Contributions, Vertebrata 11: 1-79.

Fraser, N. C. 1982. A new rhynchocephalian from the British Upper Trias. Palaeontology 25: 709-725.

Fraser, N. C. 1988. The osteology and relationships of *Clevosaurus* (Reptilia: Sphenodontida). Philosophical Transactions of the Royal Society of London B 321: 125-178.

Frazzetta, T. H. 1962. A functional consideration of cranial kinesis in lizards. Journal of Morphology 111: 287-320.

Frazzetta, T. H. 1968. Adaptive problems and possibilities in the temporal fenestration of tetrapod skulls. Journal of Morphology 125: 145-157.

Frazzetta, T. H. 1983. Adaptation and function of cranial kinesis in reptiles: a time-motion analysis of feeding in alligator lizards. Pp. 222-244 in A. G. J. Rhodin and K. Miyata (eds.), Advances in Herpetology and Evolutionary Biology: Essays in Honor of Ernest E. Williams. Museum of Comparative Zoology, Harvard University, Cambridge, Massachusetts.

Frey, E. 1988. Das Tragsystem der Krokodile—eine biomechanische und phylogenetische Analyse. Stuttgarter Beiträge zur Naturkunde A 426: 1-60.

Frey, E., and D. M. Martill. 1998. Soft tissue preservation in a specimen of *Pterodactylus kochi* (Wagner) from the Upper Jurassic of Germany. Neues Jahrbuch für Geologie und Paläontologie Abhandlungen 210: 421-441.

Frey, E., H.-D. Sues, and W. Munk. 1997. Gliding mechanism in the Late Permian reptile *Coelurosauravus*. Science 275: 1450-1452.

Frey, E., H. Tischlinger, M.-C. Buchy, and D. M. Martill. 2003. New specimens of Pterosauria (Reptilia) with soft parts, with implications for pterosaurian anatomy and locomotion. Pp. 233-266 in E. Buffetaut and J.-M. Mazin (eds.), Evolution and Palaeobiology of Pterosaurs. Geological Society, London, Special Publications 217.

Fröbisch, J. 2008. Global taxonomic diversity of anomodonts (Tetrapoda, Therapsida) and the terrestrial rock record across the Permian-Triassic boundary. PLoS ONE 3(11): e3733. doi:10.1371/journal.pone.0003733.

Fröbisch, N. B., J. Fröbisch, P. M. Sander, L. Schmitz, and O. Rieppel. 2013. Macropredatory ichthyosaur from the Middle Triassic and the origin of modern trophic networks. Proceedings of the National Academy of Sciences of the United States of America 110: 1393-1397.

Fröbisch, N. B., P. M. Sander, and O. Rieppel. 2006. A new species of *Cymbospondylus* (Diapsida, Ichthyosauria) from ther Middle Triassic of Nevada and a re-evaluation of the skull osteology of the genus. Zoological Journal of the Linnean Society 147: 515-538.

Frost, D. R., and R. Etheridge. 1989. A phylogenetic analysis and taxonomy of iguanian lizards (Reptilia: Squamata). Miscellaneous Publications, Museum of Natural History, The University of Kansas 81: 1-65.

Frost, D. R., R. Etheridge, D. Janies, and T. A. Titus 2001. Total evidence, sequence alignment, evolution of polychrotid lizards, and a reclassification of the Iguania (Squamata: Iguania). American Museum Novitates 3343: 1-38.

Fry, B. G., et al. 2006. Early evolution of the venom system in lizards and snakes. Nature 439: 584-588. [13 co-authors]

Fry, B. G., et al. 2009. A central role for venom in predation by *Varanus komodoensis* (Komodo Dragon) and the extinct *Varanus* (*Megalania*) *priscus*. Proceedings of the National Academy of Sciences of the United States of America 106: 8969-8974. [28 co-authors]

Funston, G. F., and P. J. Currie. 2014. A previously undescribed caenagnathid mandible from the late Campanian of Alberta, and insights into the diet of *Chirostenotes pergracilis* (Dinosauria: Oviraptorosauria). Canadian Journal of Earth Sciences 51: 156-165.

Funston, G. F., P. J. Currie, D. A. Eberth, M. J. Ryan, C. Tsogtbataar, B. Demchig, and N. R. Longrich. 2016. The first oviraptorosaur (Dinosauria: Theropoda) bonebed: evidence of gregarious behavior in a maniraptoran theropod. Scientific Reports 6: 35782. doi:10.1038/srep35782.

Gaffney, E. S. 1972. The systematics of the North American family Baenidae (Reptilia, Cryptodira). Bulletin of the American Museum of Natural History 147: 241-320.

Gaffney, E. S. 1975. A phylogeny and classification of the higher categories of turtles. Bulletin of the American Museum of Natural History 155: 387-436.

Gaffney, E. S. 1979. Comparative cranial morphology of recent and fossil turtles. Bulletin of the American Museum of Natural History 164: 65-376.

Gaffney, E. S. 1980. Phylogenetic relationships of the major groups of amniotes. Pp. 593-610 in A. L. Panchen (ed.), The Terrestrial Environment and the Origin of Land Vertebrates. Academic Press, London.

Gaffney, E. S. 1983. The cranial morphology of the extinct horned turtle, *Meiolania platyceps*, from the Pleistocene of Lord Howe Island, Australia. Bulletin of the American Museum of Natural History 175: 326-479.

Gaffney, E. S. 1990. The comparative osteology of the Triassic turtle *Proganochelys*. Bulletin of the American Museum of Natural History 194: 1-263.

Gaffney, E. S. 1996. The postcranial morphology of *Meiolania platyceps* and a review of the Meiolaniidae. Bulletin of the American Museum of Natural History 229: 1-165.

Gaffney, E. S., J. H. Hutchinson, F. A. Jenkins, Jr., and L. J. Meeker. 1987. Modern turtle origins: the oldest known cryptodire. Science 237: 289-291.

Gaffney, E. S., and F. A. Jenkins Jr. 2010. The cranial morphology of *Kayentachelys*, an Early Jurassic cryptodire, and the early history of turtles. Acta Zoologica (Stockholm) 91: 335-368.

Gaffney, E. S., and J. W. Kitching. 1994. The most ancient African turtle. Nature 369: 55-58.

Gaffney, E. S., and P. A. Meylan. 1988. A phylogeny of turtles. Pp. 157-219 in M. J. Benton (ed.), The Phylogeny and Classification of the Tetrapoda. Volume 1: Amphibians, Reptiles, Birds. Clarendon Press, Oxford.

Gaffney, E. S., and P. A. Meylan. 1992. The Transylvanian turtle, *Kallokibotion*, a primitive cryptodire of Cretaceous age. American Museum Novitates 3040: 1-37.

Gaffney, E. S., P. A. Meylan, R. C. Wood, E. Simons, and D. de A. Campos. 2011. Evolution of the side-necked turtles: the family Podocnemididae. Bulletin of the American Museum of Natural History 350: 1-237.

Gaffney, E. S., H. Tong, and P. A. Meylan. 2006. Evolution of the side-necked turtles: the families Bothremydidae, Euraxemydidae, and Araripemydidae. Bulletin of the American Museum of Natural History 300: 1-698.

Galton, P. M. 1971. A primitive dome-headed dinosaur (Ornithischia: Pachycephalosauridae) from the Lower Cretaceous of England, and the function of the dome in pachycephalosaurids. Journal of Paleontology 45: 40-47.

Galton, P. M. 1972. Classification and evolution of ornithopod dinosaurs. Nature 239: 464-466.

Galton, P. M. 1973. On the anatomy and relationships of *Efraasia diagnostica* (Huene) n. gen., a prosauropod dinosaur (Reptilia: Saurischia) from the Upper Triassic of Germany. Paläontologische Zeitschrift 47: 229-255.

Galton, P. M. 1974. The ornithischian dinosaur *Hypsilophodon* from the Wealden of the Isle of Wight. Bulletin of the British Museum (Natural History), Geology 25: 1-152.

Galton, P. M. 1976. Prosauropod dinosaurs (Reptilia: Saurischia) from North America. Postilla 169: 1-98.

Galton, P. M. 2001. Prosauropod dinosaurs from the Upper Triassic of Germany. Pp. 25-92 in Actas de las Jornadas Internacionales sobre Paleontología de Dinosaurios y su Etorno, Salas de los Infantes, Septiembre de 1999. Colectivo Arqueológico-Paleontológico de Salas, C.A.S., Burgos.

Galton, P. M. 2007. Teeth of ornithischian dinosaurs (mostly Ornithopoda) from the Morrison Formation (Upper Jurassic) of the western United States. Pp. 17-47 in K. Carpenter (ed.), Horns and Beaks: Ceratopsian and Ornithopod Dinosaurs. Indiana University Press, Bloomington.

Galton, P. M., and D. Kermack. 2010. The anatomy of *Pantydraco caducus*, a very basal sauropodomorph dinosaur from the Rhaetian (Upper Triassic) of South Wales, UK. Revue de Paléobiologie 29: 341-404.

Gamble, T., A. M. Bauer, E. Greenbaum, and T. R. Jackman. 2008. Out of the blue: a novel, trans-Atlantic clade of geckos (Gekkota, Squamata). Zoologica Scripta 37: 355-366.

Gans, C. 1974. Biomechanics: An Approach to Vertebrate Biology. J. B. Lippincott, Philadelphia.

Gans, C. 1978. The characteristics and affinities of the Amphisbaenia. Transactions of the Zoological Society of London 34: 347-416.

Gans, C. 1983. Is *Sphenodon punctatus* a maladapted relic? Pp. 613-620 in A. G. J. Rhodin and K. Miyata (eds.), Advances in Herpetology and Evolutionary Biology. Museum of Comparative Zoology, Harvard University, Cambridge, Massachusetts.

Gans, C., and B. Clark. 1976. Studies on the ventilation of *Caiman crocodilus* (Crcodilia Reptilia). Respiration Physiology 26: 285-301.

Gans, C., and R. Montero. 2008. An atlas of amphisbaenian skull anatomy. Pp. 621-738 in C. Gans, A. S. Gaunt, and K. Adler (eds.), Biology of the Reptilia. Volume 21: The Skull and Appendicular Locomotor Apparatus of Lepidosauria. Society for the Study of Amphibians and Reptiles, Ithaca, New York.

Gao, K., S. E. Evans, Q. Ji, M. Norell, and S. Ji. 2000. Exceptional fossil material of a semi-aquatic reptile from China: the resolution of an enigma. Journal of Vertebrate Paleontology 20: 417-421.

Gao, K., and R. C. Fox. 1996. Taxonomy and evolution of Late Cretaceous lizards (Reptilia: Squamata) from western Canada. Bulletin of Carnegie Museum of Natural History 33: 1-107.

Gao, K.-q., and R. C. Fox. 1998. New choristoderes (Reptilia: Diapsida) from the Upper Cretaceous and Paleocene, Alberta and Saskatchewan, Canada, and phylogenetic relationships of Choristodera. Zoological Journal of the Linnean Society 124: 303-353.

Gao, K.-q., and R. C. Fox. 2005. A new choristodere (Reptilia: Diapsida) from the Lower Cretaceous of western Liaoning Province, China, and the phylogenetic relationships of Monjurosuchidae. Zoological Journal of the Linnean Society 145: 427-444.

Gao, K.-q., and D. T. Ksepka. 2008. Osteology and taxonomic revision of *Hyphalosaurus* (Diapsida: Choristodera) from the Lower Cretaceous of Liaoning, China. Journal of Anatomy 212: 747-768.

Gao, K.-q., and M. A. Norell. 1998. Taxonomic review of *Carusia* (Reptilia: Squamata) from the Late Cretaceous of the Gobi Desert and phylogenetic relationships of anguimorphan lizards. American Museum Novitates 3230: 1-51.

Gao, K.-q., and M. A. Norell. 2000. Taxonomic composition and systematics of Late Cretaceous lizard assemblages from Ukhaa Tolgod and adjacent localities, Mongolian Gobi Desert. Bulletin of the American Museum of Natural History 249: 1-118.

Gardiner, B. G. 1982. Tetrapod classification. Zoological Journal of the Linnean Society 74: 207-232.

Gardner, J. D., and R. L. Cifelli. 1999. A primitive snake from the Cretaceous of Utah. Special Papers in Palaeontology 60: 87-100.

Gardner, J. D., A. P. Russell, and D. B. Brinkman. 1995. Systematics and taxonomy of softshelled turtles (Family Trionychidae) from the Judith River Group (mid-Campanian) of North America. Canadian Journal of Earth Sciences 32: 631-643.

Gardner, N. M., C. M. Holliday, and F. R. O'Keefe. 2010. The braincase of *Youngina capensis* (Reptilia, Diapsida): new insights from high-resolution CT scanning of the holotype. Palaeontologia Electronica 13(3): 19A. http://palaeo-electronica.org/2010_3/217/index.html.

Gasc, J. P. 1974. L'interprétation fonctionelle de l'appareil musculo-squelettique de l'axe vertébral chez les serpents (Reptilia). Mémoires du Muséum National d'Histoire Naturelle, Série A: Zoologie 83: 1-182.

Gates, T. A., C. Organ, and L. Zanno. 2016. Bony cranial ornamentation linked to rapid evolution of gigantic theropod dinosaurs. Nature Communications 7: 12931. doi:10.1038/ncomms12931.

Gatesy, S. M., and K. P. Dial. 1996. Locomotor modules and the evolution of avian flight. Evolution 50: 331-340.

Gauthier, J. A. 1980. *Anniella* (Sauria: Anguidae) from the Miocene of California. PaleoBios 31: 1-7.

Gauthier, J. A. 1982. Fossil Xenosauridae and Anguidae from the Lower Eocene Wasatch Formation, southcentral Wyoming, and a revision of the Anguioidea. Contributions to Geology, University of Wyoming 21: 7-54.

Gauthier, J. A. 1986. Saurischian monophyly and the origin of birds. Pp. 1-55 in K. Padian (ed.), The Origin of Birds and the Evolution of Flight. Memoirs of the California Academy of Sciences 8.

Gauthier, J. A. 1994. The diversification of amniotes. Pp. 129-159 in D. R. Prothero and R. M. Schoch (convenors), Major Features of Vertebrate Evolution. The Paleontological Society, Short Courses in Paleontology 7. University of Tennessee, Knoxville.

Gauthier, J. A., and K. de Queiroz. 1990. Phylogeny as a central principle in taxonomy: phylogenetic definitions of taxon names. Systematic Zoology 39: 307-322.

Gauthier, J., and K. de Queiroz. 2001. Feathered dinosaurs, flying dinosaurs, crown dinosaurs, and the name "Aves." Pp. 7-41 in J. Gauthier and L. F. Gall (eds.), New Perspectives on the Origin and Early Evolution of Birds: Proceedings of the International Symposium in Honor of John H. Ostrom. Peabody Museum of Natural History, Yale University, New Haven, Connecticut.

Gauthier, J., R. Estes, and K. de Queiroz. 1988c. A phylogenetic analysis of Lepidosauromorpha. Pp. 15-98 in R. Estes and G. Pregill (eds.), Phylogenetic Relationships of the Lizard Families: Essays Commemorating Charles L. Camp. Stanford University Press, Stanford.

Gauthier, J. A., M. Kearney, J. A. Maisano, O. Rieppel, and A. D. B. Behlke. 2012. Assembling the squamate Tree of Life: perspectives from the phenotype and the fossil record. Bulletin of the Peabody Museum of Natural History, Yale University 53: 3-308.

Gauthier, J., A. G. Kluge, and T. Rowe. 1988a. Amniote phylogeny and the importance of fossils. Cladistics 4: 105-209.

Gauthier, J. A., A. G. Kluge, and T. Rowe. 1988b. The early evolution of the Amniota. Pp. 103-155 in M. J. Benton (ed.), The Phylogeny and Classification of Tetrapods. Volume 1: Amphibians, Reptiles, Birds. Clarendon Press, Oxford.

Gauthier, J. A., S. J. Nesbitt, E. R. Schachner, G. S. Bever, and W. G. Joyce. 2011. The bipedal stem crocodilian *Poposaurus gracilis*: inferring function in fossils and innovation in archosaur locomotion. Bulletin of the Peabody Museum of Natural History, Yale University 52: 107-126.

Gauthier, J., and K. Padian. 1985. Phylogenetic, functional, and aerodynamic analyses of the origin of birds and their flight. Pp. 185-197 in M. K. Hecht, J. H. Ostrom, G. Viohl, and P. Wellnhofer (eds.), The Beginnings of Birds: Proceedings of the International *Archaeopteryx* Conference, Eichstätt 1984. Freunde des Jura-Museums, Eichstätt.

Gentry, A. D. 2016. New material of the Late Cretaceous marine turtle *Ctenochelys acris* Zangerl, 1953 and a phylogenetic reassessment of the 'toxochelyid'-grade taxa. Journal of Systematic Palaeontology 15: 675-696.

Gervais, P. 1872. Ostéologie du *Sphargis luth*. Nouvelles Archives du Muséum d'Histoire Naturelle 8: 199-228.

Gibbons, J. W., et al. 2000. The global decline of reptiles, déjà vu amphibians. BioScience 50: 653-665. [11 co-authors]

Gilbert, S. F., G. Bender, E. Betters, M. Yin, and J. A. Cebra-Thomas. 2007. The contribution of neural crest cells to the nuchal bone and plastron of the turtle shell. Integrative and Comparative Biology 47: 401-408.

Gilbert, S. F., G. A. Loredo, A. Brukman, and A. C. Burke. 2001. Morphogenesis of the turtle shell: the development of a novel structure in turtle evolution. Evolution and Development 3: 47-58.

Gilmore, C. W. 1914. Osteology of the armored Dinosauria in the United States National Museum, with special reference to the genus *Stegosaurus*. Bulletin of the United States National Museum 89: 1-143.

Gilmore, C. W. 1920. Osteology of the carnivorous Dinosauria in the United States National Museum, with special reference to the genera *Antrodemus* [*Allosaurus*] and *Ceratosaurus*. Bulletin of the United States National Museum 110: 1-154.

Gilmore, C. W. 1925a. A nearly complete articulated skeleton of *Camarasaurus*, a saurischian dinosaur from the Dinosaur National Monument. Memoirs of the Carnegie Museum 10: 347-384.

Gilmore, C. W. 1925b. Osteology of ornithopodous dinosaurs from the Dinosaur National Monument, Utah. *Camptosaurus medius, Dryosaurus altus, Laosaurus gracilis*. Memoirs of the Carnegie Museum 10: 385-409.

Gilmore, C. W. 1936. Osteology of *Apatosaurus*, with special reference to specimens in the Carnegie Museum. Memoirs of the Carnegie Museum 11: 175-300.

Gilmore, C. W. 1942. Osteology of *Polyglyphanodon*, an Upper Cretaceous lizard from Utah. Proceedings of the United States National Museum 92: 229-265.

Gilmore, C. W. 1946. Reptilian fauna of the North Horn Formation of central Utah. United States Geological Survey Professional Paper 210-C: 29-51.

Giugliano, L. G., R. G. Collevatti, and G. R. Colli. 2007. Molecular dating and phylogenetic relationships among Teiidae (Squamata) inferred by molecular and morphological data. Molecular Phylogenetics and Evolution 45: 168-179.

Gleeson, T. T. 1979. Foraging and transport costs in the Galapagos marine iguana, *Amblyrhynchus cristatus*. Physiological Zoology 52: 549-557.

Godefroit, P., A. Cau, D. Hu, F. Escuillié, W. Wu, and G. Dyke. 2013. A Jurassic avialian dinosaur from China resolves the early phylogenetic history of birds. Nature 498: 359-362.

Godefroit, P., F. Escuillié, Y. L. Bolotsky, and P. Lauters. 2012. A new basal hadrosauroid dinosaur from the Upper Cretaceous of Kazakhstan. Pp. 335-358 in P. Godefroit (ed.), Bernissart Dinosaurs and Early Cretaceous Terrestrial Ecosystems. Indiana University Press, Bloomington.

Godefroit, P., S. M. Snitsa, D. Dhouailly, Y. L. Bolotsky, A. V. Sizov, M. E. McNamara, M. J. Benton, and P. Spagna. 2014. A Jurassic ornithischian dinosaur from Siberia with feathers and scales. Science 345: 451-455.

Göhlich, U. B., and L. M. Chiappe. 2006. A new carnivorous dinosaur from the Late Jurassic Solnhofen Archipelago. Nature 440: 329-332.

Gomani, E. M. 2005. Sauropod dinosaurs from the Early Cretaceous of Malawi, Africa. Palaeontologia Electronica 8(1): 27A.

Gómez, R. O., A. M. Báez, and G. W. Rougier. 2008. An anilioid snake from the Upper Cretaceous of northern Patagonia. Cretaceous Research 29: 481-488.

Gorniak, G. C., H. I. Rosenberg, and C. Gans. 1982. Mastication in the tuatara, *Sphenodon punctatus* (Reptilia: Rhynchocephalia): structure and activity of the motor system. Journal of Morphology 171: 321-353.

Gottmann-Quesada, A., and P. M. Sander 2009. A redescription of the early archosauromorph *Protorosaurus speneri* Meyer, 1832, and its phylogenetic relationships. Palaeontographica A 287: 123-220.

Gow, C. E. 1972. The osteology and relationships of the Millerettidae (Reptilia: Cotylosauria). Journal of Zoology, London 167: 219-264.

Gow, C. E. 1975. The morphology and relationships of *Youngina capensis* Broom and *Prolacerta broomi* Parrington. Palaeontologia africana 18: 89-131.

Gow, C. E. 1997. A note on the postcranial skeleton of *Milleretta* (Amniota: Parareptilia). Palaeontologia africana 34: 55-57.

Gower, D. J. 1999. The cranial and mandibular osteology of a new rauisuchian archosaur from the Middle Triassic of southern Germany. Stuttgarter Beiträge zur Naturkunde B 280: 1-49.

Gower, D. J. 2003. Osteology of the early archosaurian reptile *Erythrosuchus africanus*. Annals of the South African Museum 110: 1-84.

Gower, D. J., N. Vidal, J. N. Spinks, and C. J. McCarthy. 2005. The phylogenetic position of Anomochilidae (Reptilia: Serpentes): first evidence from DNA sequences. Journal of Zoological Systematics and Evolutionary Research 43: 315-320.

Gozzi, E., and S. Renesto. 2003. A complete specimen of *Mystriosuchus* (Reptilia, Phytosauria) from the Norian (Late Triassic) of Lombardy (Northern Italy). Rivista Italiana di Paleontologia e Stratigrafia 109: 475-498.

Gray, J. E. 1825. A synopsis of the genera of reptiles and Amphibia, with a description of some new species. Annals of Philosophy, ser. 2, 10: 193-217.

Gray, J. E. 1855. Catalogue of Shield Reptiles in the Collection of the British Museum. Part I. Testudinata (Tortoises). British Museum, London.

Greene, H. W. 1997. Snakes: The Evolution of Mystery in Nature. University of California Press, Berkeley.

Greer, A. E. 1989. The Biology and Evolution of Australian Lizards. Surrey Beatty and Sons, Chipping Norton, NSW.

Gregory, J. T. 1945. Osteology and relationships of *Trilophosaurus*. University of Texas Special Publication 4401: 273-359.

Griebeler, E., N. Klein, and P. M. Sander. 2013. Aging, maturation and growth of sauropodomorph dinosaurs as deduced from growth curves using long bone histological data: an assessment of methodological constraints and solutions. PLoS ONE 8(6): e67012. doi:10.1371/journal.pone.0067012.

Grigg, G., and D. Kirshner. 2015. Biology and Evolution of Crocodylians. Comstock Publishing Associates, Ithaca, New York.

Guillon, J.-M., L. Guéry, V. Hulin, and M. Girondot. 2012. A large phylogeny of turtles (Testudinata) using molecular data. Contributions to Zoology 81: 147-158.

Günther, A. 1867. Contribution to the anatomy of *Hatteria* (*Rhynchocephalus*, Owen). Philosophical Transactions of the Royal Society of London 157: 595-629.

Günther, A. 1877. The Gigantic Land Tortoises (Living and Extinct) in the Collection of the British Museum. British Museum, London.

Haeckel, E. 1866. Generelle Morphologie der Organismen. Allgemeine Grundzüge der organischen Formen-Wissenschaft, mechanisch begründet durch die von Charles Darwin reformirte Descendenz-Theorie. (Two volumes.) Verlag von Georg Reimer, Berlin.

Haines, R. W. 1969. Epiphyses and sesamoids. Pp. 81-115 in C. Gans, A. d'A. Bellairs, and T. S. Parsons (eds.), Biology of the Reptilia. Volume 1: Morphology A. Academic Press, London.

Hallett, M., and M. J. Wedel. 2016. The Sauropod Dinosaurs: Life in the Age of Giants. Johns Hopkins University Press, Baltimore.

Han, D., K. Zhou, and A. M. Bauer. 2004. Phylogenetic relationships among gekkotan lizards inferred from *C-mos* nuclear DNA sequences and a new classification of Gekkota. Biological Journal of the Linnean Society 83: 353-368.

Hargreaves, A., M. T. Swain, D. W. Logan, and J. Mulley. 2014. Testing the Toxicofera: comparative reptile transcriptomics casts doubt on the single, early evolution of the reptile venom system. Toxicon 92: 140-156.

Harrell, T. L., Jr., A. Pérez-Huerta, and C. A. Suarez. 2016. Endothermic mosasaurs? Thermoregulation of Late Cretaceous mosasaurs (Reptilia, Squamata) indicated by stable oxygen isotopes in fossil bioapatite in comparison with coeval marine fish and pelagic seabirds. Palaeontology 59: 351-363.

Harris, J. D., and P. Dodson. 2004. A new diplodocoid sauropod dinosaur from the Upper Jurassic Morrison Formation of Montana, USA. Acta Palaeontologica Polonica 49: 197-210.

Harris, J. M., and R. L. Carroll. 1977. *Kenyasaurus*, a new eosuchian reptile from the Early Triassic of Kenya. Journal of Paleontology 51: 139-149.

Harshman, J., C. J. Huddleston, J. P. Bollback, T. J. Parsons, and M. J. Braun. 2003. True and false gharials: a nuclear gene phylogeny of Crocodylia. Systematic Biology 52: 386-402.

Hastings, A. K., J. I. Bloch, E. A. Cadena, and C. A. Jaramillo. 2010. A new small short-snouted dyrosaurid (Crocodylomorpha, Mesoeucrocodylia) from the Paleocene of northeastern Colombia. Journal of Vertebrate Paleontology 30: 139-162.

Hastings, A. K., J. I. Bloch, and C. A. Jaramillo. 2011. A new longirostrine dyrosaurid (Crocodylomorpha, Mesoeucrocodylia) from the Paleocene of north-eastern Colombia: biogeographic and behavioural implications for New-World Dyrosauridae. Palaeontology 54: 1095-1116.

Hatcher, J. B. 1901. *Diplodocus* (Marsh), its osteology, taxonomy and probable habits, with a restoration of the skeleton. Memoirs of the Carnegie Museum 1: 1-63.

Hatcher, J. B., O. C. Marsh, and R. S. Lull. 1907. The Ceratopsia. Memoirs of the United States Geological Survey 49: 1-157.

Haubold, H. 1990. Ein neuer Dinosaurier (Ornithischia, Thyreophora) aus dem Unteren Jura des nördlichen Mitteleuropa. Revue de Paléobiologie 9: 149-177.

Hay, J. M., S. D. Sarre, D. M. Lambert, F. W. Allendorf, and C. H. Daugherty. 2010. Genetic diversity and taxonomy: a reassessment of species designation in tuatara (*Sphenodon*: Reptilia). Conservation Genetics 11: 1063-1081.

Hay, O. P. 1908. The fossil turtles of North America. Carnegie Institution of Washington Publication 75: 1-568.

He, X., D. Yang, and C. Su. 1983. [A new pterosaur from the Middle Jurassic of Dashanpu, Zigong, Sichuan.] Journal of the Chengdu College of Geology, Supplement 1: 27-33. [Chinese]

Head, J. J. 2005. Snakes from the Siwalik Group (Miocene of Pakistan): systematics and relationship to environmental change. Palaeontologia Electronica 8(1): 18A. http://palaeo-electronica.org/paleo/2005_1/head18/issue1_05.htm.

Head, J. J. 2015. Fossil calibration dates for molecular phylogenetic analysis of snakes 1: Serpentes, Alethinophidia, Boidae, Pythonidae. Palaeontologia Electronica 18(1): 6FC. www.palaeo-electronica.org/content/fc-6.

Head, J. J., J. I. Bloch, A. K. Hastings, J. R. Bourque, E. A. Cadena, F. A. Herrera, P. D. Polly, and C. A. Jaramillo. 2009. Giant boid snake from the Palaeocene neotropics reveals hotter past equatorial temperatures. Nature 457: 715-717.

Head, J. J., G. F. Gunnell, P. A. Holroyd, J. H. Hutchinson, and R. L. Ciochon. 2013. Giant lizards occupied herbivorous mammalian ecospace during the Paleogene greenhouse in Southeast Asia. Proceedings of the Royal Society B 280: 20130665. doi:10.1098/rspb.2013.0665.

Head, J. J., P. A. Holroyd, J. H. Hutchinson, and R. L. Ciochon. 2005. First report of snakes (Serpentes) from the late middle Eocene Pondaung Formation, Myanmar. Journal of Vertebrate Paleontology 25: 246-250.

Head, J. J., D. M. Mohabey, and J. A. Wilson. 2007. *Acrochordus* Hornstedt (Serpentes, Caenophidia) from the Miocene of Gujarat, Western India: temporal constraints on dispersal of a derived snake. Journal of Vertebrate Paleontology 27: 720-723.

Head, J. J., and P. D. Polly. 2015. Evolution of the snake body form reveals homoplasy in amniote *Hox* gene function. Nature 520: 86-89.

Head, J. J., S. M. Raza, and P. D. Gingerich. 1999. *Drazinderetes tethyensis*, a new large trionychid (Reptilia: Testudines) from the marine Eocene Drazinda Formation of the Sulaiman Range, Punjab (Pakistan). Contributions from the Museum of Paleontology, The University of Michigan 30: 199-214.

Heaton, M. J. 1979. The cranial anatomy of primitive captorhinid reptiles from the Pennsylvanian and Permian of Oklahoma and Texas. Oklahoma Geological Survey Bulletin 127: 1-84.

Heaton, M. J. 1980. The Cotylosauria: a reconsideration of a group of archaic tetrapods. Pp. 497-551 in A. L. Panchen (ed.), The Terrestrial Environment and the Origin of Land Vertebrates. Academic Press, London.

Heaton, M. J., and R. R. Reisz. 1980. A skeletal reconstruction of the Early Permian captorhinid reptile *Eocaptorhinus laticeps* (Williston). Journal of Paleontology 54: 136-143.

Heaton, M. J., and R. R. Reisz. 1986. Phylogenetic relationships of captorhinomorph reptiles. Canadian Journal of Earth Sciences 23: 402-418.

Hecht, M. K. 1992. A new choristodere (Reptilia, Diapsida) from the Oligocene of France: an example of the Lazarus effect. Geobios 25: 115-131.

Heckert, A. B. 2004. Late Triassic microvertebrates from the lower Chinle Group (Otischalkian-Adamanian: Carnian), southwestern U.S.A. New Mexico Museum of Natural History & Science Bulletin 27: 1-170.

Hedges, S. B. 2014. The high-level classification of skinks (Reptilia, Squamata, Scincomorpha). Zootaxa 3765: 317-338.

Hedges, S. B., and L. L. Poling. 1999. A molecular phylogeny of reptiles. Science 283: 998-1001.

Heilmann, G. 1926. The Origin of Birds. H. F. & G. Witherby, London.

Hembree, D. I. 2007. Phylogenetic revision of Rhineuridae (Reptilia: Squamata: Amphisbaenia) from the Eocene to Miocene of North America. The University of Kansas Paleontological Contributions 15: 1-20.

Hendrickx, C., and O. Mateus. 2014. *Torvosaurus gurneyi* n. sp., the largest terrestrial predator from Europe, and a proposed nomenclature of the maxilla anatomy in nonavian theropods. PLoS ONE 9(3): e88905. doi:10.1371/journal.pone.0088905.

Hennig, E. 1925. *Kentrurosaurus aethiopicus*. Die Stegosaurier-Funde vom Tendaguru, Deutsch-Ostafrika. Palaeontographica, Supplement 7, 1(1): 101-254.

Hennig, W. 1966. Phylogenetic Systematics. (Translated by D. D. Davis and R. Zangerl.) University of Illinois Press, Urbana.

Herrel, A., F. de Vree, V. Delheusy, and C. Gans. 1999. Cranial kinesis in gekkonid lizards. Journal of Experimental Biology 202: 3687-3698.

Hervet, S. 2006. The oldest European ptychogasterid turtle (Testudinoidea) from the lowermost Eocene amber locality of Le Quesnoy (France, Ypresian, MP7). Journal of Vertebrate Paleontology 26: 839-848.

Hill, R. V. 2005. Integration of morphological data sets for phylogenetic analysis of Amniota: the importance of integumentary characters and increased taxon sampling. Systematic Biology 54: 530-547.

Hirayama, R., and T. Chitoku. 1996. Family Dermochelyidae (superfamily Chelonioidea) from the Upper Cretaceous of North Japan. Transactions and Proceedings of the Palaeontological Society of Japan 184: 597-622.

Hirsch, K. F. 1979. The oldest vertebrate egg? Journal of Paleontology 53: 1068-1084.

Hoffstetter, R. 1957. Un saurien helodermatidé (*Eurheloderma gallicum* nov. gen. et sp.) dans la faune fossile des phosphorites du Quercy. Bulletin de la Société Géologique de France 7: 775-786.

Hoffstetter, R. 1967. Coup d'oeil sur les Sauriens (= Lacertiliens) des couches de Purbeck (Jurassique supérieur d'Angleterre). Résumé d'un Mémoire. Pp. 349-371 in Problèmes Actuels de Paléontologie (Evolution des Vertébrés). Colloques Internationaux du Centre National de la Recherche Scientifique, no. 163. Éditions du Centre National de la Recherche Scientifique, Paris.

Hoffstetter, R., and J.-C. Gasc. 1969. Vertebrae and ribs of modern reptiles. Pp. 201-310 in C. Gans, A. d'A. Bellairs, and T. S. Parsons (eds.), Biology of the Reptilia. Volume 1: Morphology A. Academic Press, London.

Hoffstetter, R., and J.-C. Rage. 1972. Les Erycinae fossiles de France (Serpentes, Boidae). Compréhension et histoire de la sous-famille. Annales de Paléontologie, Vertébrés 58: 81-124.

Hoffstetter, R., and J.-C. Rage. 1977. Le gisement de Vertébrés miocènes de La Venta (Colombie) et sa faune de serpents. Annales de Paléontologie, Vertébrés 63: 161-190.

Holland, W. J. 1906. The osteology of *Diplodocus* Marsh. With special reference to the restoration of the skeleton of *Diplodocus carnegiei* Hatcher, presented by Mr. Andrew Carnegie to the British Museum, May 12, 1905. Memoirs of the Carnegie Museum 2: 225-278.

Holliday, C. M., and N. M. Gardner. 2012. A new eusuchian crocodyliform with novel cranial integument and its significance for the origin and evolution of Crocodylia. PLoS ONE 7(1): e30471. doi:10.1371/journal.pone.0030471.

Holliday, C. M., and L. M. Witmer. 2008. Cranial kinesis in dinosaurs: intracranial joints, protractor muscles, and their significance for cranial evolution and function in diapsids. Journal of Vertebrate Paleontology 28: 1073-1088.

Holman, J. A. 1995. A new species of *Emydoidea* (Reptilia: Testudines) from the late Barstovian (medial Miocene) of Cherry County, Nebraska. Journal of Herpetology 29: 548-553.

Holmes, R. 1977. The osteology and musculature of the pectoral limb of small captorhinids. Journal of Morphology 152: 101-140.

Holmes, R. 2003. The hind limb of *Captorhinus aguti* and the step cycle of basal amniotes. Canadian Journal of Earth Sciences 40: 515-526.

Holroyd, P. A., and J. H. Hutchinson. 2002. Patterns of geographic variation in latest Cretaceous vertebrates: evidence from the turtle component. Pp. 177-190 in J. H. Hartman, K. R. Johnson, and D. J. Nichols (eds.), The Hell Creek Formation and Cretaceous-Tertiary Boundary in the Northern Great Plains: An Integrated Continental Record of the End of the Cretaceous. Geological Society of America Special Paper 361.

Holroyd, P. A., and J. F. Parham. 2003. The antiquity of African tortoises. Journal of Vertebrate Paleontology 23: 688-690.

Holtz, T. R., Jr., R. E. Molnar, and P. J. Currie. 2004. Basal Tetanurae. Pp. 71-110 in D. B. Weishampel, P. Dodson, and H. Osmólska (eds.), The Dinosauria. Second Edition. University of California Press, Berkeley.

Holtz, T. R., Jr., and H. Osmólska. 2004. Saurischia. Pp. 21-24 in D. B. Weishampel, P. Dodson, and H. Osmólska (eds.), The Dinosauria. Second Edition. University of California Press, Berkeley.

Hone, D. W. E., A. A. Farke, M. Watabe, S. Shigeru, and K. Tsogtbaatar. 2014. A new mass mortality of juvenile *Protoceratops* and size-segregated aggregation behaviour in juvenile non-avian dinosaurs. PLoS ONE 9(11): e113306. doi: 10.1371/journal.pone.0113306.

Hone, D. W. E., H. Tischlinger, X. Xu, and F. Zhang. 2010. The extent of the preserved feathers on the four-winged dinosaur *Microraptor gui* under ultraviolet light. PLoS ONE 5(2): e9223. doi:10.1371/journal.pone.0009223.

Hopkins, W. A., C. L. Rowe, and J. D. Congdon. 1999. Elevated trace element concentrations and standard metabolic rate in banded water snakes (*Nerodia fasciata*) exposed to coal combustion wastes. Environmental Toxicology and Chemistry 18: 1258-1263.

Hopson, J. A. 1975. The evolution of cranial display structures in hadrosaurian dinosaurs. Paleobiology 1: 21-43.

Hopson, J. A. 1979. Paleoneurology. Pp. 39-146 in R. G. Northcutt and P. Ulinski (eds.), The Biology of the Reptilia. Volume 9: Neurology A. Academic Press, New York.

Horner, J. R., and R. Makela. 1979. Nest of juveniles provides evidence of family structure among dinosaurs. Nature 282: 296-298.

Horner, J. R., and D. B. Weishampel. 1988. A comparative embryological study of two ornithischian dinosaurs. Nature 332: 256-257.

Horner, J. R., D. B. Weishampel, and C. A. Forster. 2004. Hadrosauridae. Pp. 438-463 in D. B. Weishampel, P. Dodson, and H. Osmólska (eds.), The Dinosauria. Second Edition. University of California Press, Berkeley.

Hotton, N. III, E. C. Olson, and R. Beerbower. 1997. Amniote origins and the discovery of herbivory. Pp. 207-264 in S. S. Sumida and K. L. M. Martin (eds.), Amniote Origins: Completing the Transition to Land. Academic Press, San Diego.

Hou, L.-h. 1977. [A new primitive Pachycephalosauria from Anhui, China.] Vertebrata PalAsiatica 15: 198-202. [Chinese]

Houssaye, A. 2009. "Pachyostosis" in aquatic amniotes: a review. Integrative Zoology 4: 325-340.

Houssaye, A. 2013a. Bone histology of aquatic reptiles: what does it tell us about secondary adaptation to an aquatic life? Biological Journal of the Linnean Society 108: 3-21.

Houssaye, A. 2013b. Palaeoecological and morphofunctional interpretation of bone mass increase: an example in Late Cretaceous shallow marine squamates. Biological Reviews 88: 117-138.

Houssaye, A., J.-C. Rage, N. Bardet, P. Vincent, M. Amaghzaz, and S. Meslouh. 2013. New highlights about the enigmatic marine snake *Palaeophis maghrebianus* (Palaeophiidae: Palaeophiinae) from the Ypresian (Lower Eocene) Phosphates of Morocco. Palaeontology 56: 647-661.

Houssaye, A., F. Xu, L. Helfen, V. de Buffrénil, T. Baumbach, and P. Tafforeau. 2011. Three-dimensional pelvis and limb anatomy of the Cenomanian hind-limbed snake *Eupodophis descouensi* (Squamata, Ophidia) revealed by synchrotron-radiation computed laminography. Journal of Vertebrate Paleontology 31: 2-7.

Howse, S. C. B., and A. R. Milner. 1995. The pterodactyloids from the Purbeck Limestone Formation of Dorset. Bulletin of the Natural History Museum, London (Geology) 51: 73-88.

Hsiang, A., D. J. Field, T. H. Webster, A. D. B. Behlke, M. B. Davis, R. A. Racicot, and J. A. Gauthier. 2015. The origin of snakes: revealing the ecology, behavior, and evolutionary history of early snakes using genomics, phenomics, and the fossil record. BMC Evolutionary Biology 15: 87. doi:10.1186/s12862-015-0358-5.

Hua, S., and V. de Buffrénil. 1996. Bone histology as a clue in the interpretation of functional adaptations in the Thalattosuchia (Reptilia, Crocodylia). Journal of Vertebrate Paleontology 16: 703-717.

Hübner, T. R., and O. W. M. Rauhut. 2010. A juvenile skull of *Dysalotosaurus lettowvorbecki* (Ornithischia: Iguanodontia), and implications for cranial ontogeny, phylogeny, and taxonomy in ornithopod dinosaurs. Zoological Journal of the Linnean Society 160: 366-396.

Huene, F. von. 1914. Beiträge zur Geschichte der Archosaurier. Geologische und Palaeontologische Abhandlungen, Neue Folge 13: 1-53.

Huene, F. von. 1926. Vollständige Osteologie eines Plateosauriden aus dem schwäbischen Keuper. Geologische und Palaeontologische Abhandlungen, Neue Folge 15: 139-179.

Huene, F. von. 1932. Die fossile Reptil-Ordnung Saurischia, ihre Entwicklung und Geschichte. Monographien zur Geologie und Palaeontologie 1(4): 1-361.

Huene, F. von. 1936. *Henodus chelyops*, ein neuer Placodontier. Palaeontographica A 84: 99-147.

Hugall, A. F., R. Foster, and M. S. Y. Lee. 2007. Calibration choice, rate smoothing, and the pattern of tetrapod diversification according to the long nuclear gene RAG-1. Systematic Biology 56: 543-563.

Humphries, S., R. H. C. Bonser, M. P. Witton, and D. M. Martill. 2007. Did pterosaurs feed by skimming? Physical modelling and anatomical evaluation of an unusual feeding method. PLoS Biology 5(8): e204. doi:10.1371/journal.pbio.0050204.

Hurum, J. H., and K. Sabath. 2003. Giant theropod dinosaurs from Asia and North America: skulls of *Tarbosaurus bataar* and *Tyrannosaurus rex* compared. Acta Palaeontologica Polonica 48: 161-190.

Hutchinson, J. H. 1991. Early Kinosterninae (Reptilia: Testudines) and their phylogenetic significance. Journal of Vertebrate Paleontology 11: 145-167.

Hutchinson, J. H. 1996. Testudines. Pp. 337-353 in D. R. Prothero and R. J. Emry (eds.), The Terrestrial Eocene-Oligocene Transition in North America. Cambridge University Press, New York.

Hutchinson, J. H. 2013. New turtles from the Paleogene of North America. Pp. 477-497 in D. B. Brinkman, P. A. Holroyd, and J. D. Gardner (eds.), Morphology and Evolution of Turtles. Springer, Dordrecht.

Hutchinson, J. H., M. J. Knell, and D. B. Brinkman. 2013. Turtles from the Kaiparowits Formation, Utah. Pp. 295-318 in A. L. Titus and M. A. Loewen (eds.), At the Top of the Grand Staircase: The Late Cretaceous of Southern Utah. Indiana University Press, Bloomington.

Hutchinson, J. R., and S. M. Gatesy. 2000. Adductors, abductors, and the evolution of archosaur locomotion. Paleobiology 26: 734-751.

Hutchinson, M. N. 1992. Origins of the Australian scincid lizards: a preliminary report on the skinks of Riversleigh. Beagle 9: 61-70.

Hutchinson, M. N. 1997. The first fossil pygopod (Squamata, Gekkota), and a review of mandibular variation in living species. Memoirs of the Queensland Museum 41: 355-366.

Hutson, J. D., and K. N. Hutson. 2015. Inferring the prevalence and function of finger hyperextension in Archosauria from finger-joint range of motion in the American alligator. Journal of Zoology 296: 189-199.

Huxley, T. H. 1868. On the animals which are most nearly intermediate between birds and reptiles. Annals and Magazine of Natural History, ser. 4, 2: 66-75.

Huxley, T. H. 1869. An Introduction to the Classification of Animals. John Churchill and Sons, London.

Huxley, T. H. 1870. Further evidence of the affinity between the dinosaurian reptiles and birds. Quarterly Journal of the Geological Society of London 26: 12-31.

Huxley, T. H. 1875. On *Stagonolepis Robertsoni*, and on the evolution of the Crocodilia. Quarterly Journal of the Geological Society of London 31: 423-438.

Hwang, K. G., M. Huh, M. G. Lockley, D. M. Unwin, and J. L. Wright. 2002. New pterosaur tracks (Pteraichnidae) from the Late Cretaceous Uhangri Formation, S. W. Korea. Geological Magazine 139: 421-435.

Ibrahim, N., P. C. Sereno, C. Dal Sasso, S. Maganuco, M. Fabri, D. M. Martill, S. Zouhri, N. Myhrvold, and D. A. Lurino. 2014. Semiaquatic adaptations in a giant predatory dinosaur. Science 345: 1613-1616.

Iori, F. V., T. S. Marinho, I. S. Carvalho, and A. C. A. Campos. 2013. Taxonomic reappraisal of the sphagesaurid crocodyliform *Sphagesaurus montealtensis* from the Late Cretaceous Adamantina Formation of São Paulo State, Brazil. Zootaxa 3686(2): 183-200.

Irmis, R. B., S. J. Nesbitt, K. Padian, N. D. Smith, A. H. Turner, D. Woody, and A. Downs. 2007a. A Late Triassic dinosauromorph assemblage from New Mexico and the rise of dinosaurs. Science 317: 358-361.

Irmis, R. B., S. J. Nesbitt, and H.-D. Sues. 2013. Early Crocodylomorpha. Pp. 275-302 in S. J. Nesbitt, J. B. Desojo, and R. B. Irmis (eds.), Anatomy, Phylogeny and Palaeobiology of Early Archosaurs and Their Kin. Geological Society, London, Special Publications 379.

Irmis, R. B., W. G. Parker, S. J. Nesbitt, and J. Liu. 2007b. Early ornithischian dinosaurs: the Triassic record. Historical Biology 19: 3-22.

Ivakhnenko, M. F. 1979. [Permian and Triassic procolophons of the Russian Platform.] Trudy Paleontologicheskogo Instituta Akademii Nauk SSSR 164: 1-80. [Russian]

Ivakhnenko, M. F. 1987. [Permian parareptiles of the USSR.] Trudy Paleontologicheskogo Instituta Akademii Nauk SSSR 233: 1-159. [Russian]

Ivanov, M., M. Ruta, J. Klembara, and M. Böhme. 2018. A new species of *Varanus* (Anguimorpha: Varanidae) from the early Miocene of the Czech Republic, and its relationships and palaeoecology. Journal of Systematic Palaeontology 16: 767-797.

Jain, S. L., T. S. Kutty, T. Roy-Chowdhury and S. Chatterjee. 1975. The sauropod dinosaur from the Lower Jurassic Kota Formation of India. Proceedings of the Royal Society of London A 188: 221-228.

Jalil, N.-E., and P. Janvier. 2005. Les pareiasaures (Amniota, Parareptilia) du Permien supérieur du Bassin d'Argana, Maroc. Geodiversitas 27: 35-132.

Janensch, W. 1914. Übersicht über die Wirbeltierfauna der Tendaguru-Schichten, nebst einer kurzen Charakterisierung der neu aufgeführten Arten von Sauropoden. Archiv für Biontologie 3: 80-110.

Janensch, W. 1925. Die Coelurosaurier und Theropoden der Tendaguru-Schichten Deutsch-Ostafrikas. Palaeontographica, Supplement 7, 1(1): 1-100.

Janensch, W. 1935-36. Die Schädel der Sauropoden Brachiosaurus, Barosaurus und Dicraeosaurus aus den Tendaguru-Schichten Deutsch-Ostafrikas. Palaeontographica, Supplement 7, 1(2): 147-298.

Janensch, W. 1950. Die Wirbelsäule von Brachiosaurus brancai. Die Skelettrekonstruktion von Brachiosaurus brancai. Palaeontographica, Supplement 7, 1(3): 27-103.

Janensch, W. 1955. Der Ornithopode Dysalotosaurus der Tendaguruschichten. Palaeontographica, Supplement 7, 1(3): 105-176.

Janensch, W. 1961. Die Gliedmaszen und Gliedmaszengürtel der Sauropoden der Tendaguru-Schichten. Palaeontographica, Supplement 7, 1(3): 177-235.

Jenkins, F. A., Jr. 1971. The postcranial skeleton of African cynodonts. Bulletin of the Peabody Museum of Natural History, Yale University 36: 1-216.

Ji, C., D,-y. Jiang, R. Motani, O. Rieppel, W.-c. Hao, and Z.-y. Sun. 2015. Phylogeny of the Ichthyopterygia incorporating recent discoveries from South China. Journal of Vertebrate Paleontology 36: e1025956. doi:10.1080/02724634.2015.1025956.

Ji, Q., P. J. Currie, M. A. Norell, and S.-a. Ji. 1998. Two feathered dinosaurs from northeastern China. Nature 393: 753-761.

Jiang, D.-y., R. Motani, A. Tintori, O. Rieppel, G.-b. Chen, J.-d. Huang, R. Zhang, Z.-y. Sun, and C. Ji. 2014. The Early Triassic eosauropterygian Majiashanosaurus discocoracoidis, gen. et sp. nov. (Reptilia: Sauropterygia), from Chaohu, Anhui Province, People's Republic of China. Journal of Vertebrate Paleontology 34: 1044-1052.

Jiang, D.-y., O. Rieppel, N. C. Fraser, R. Motani, W.-c. Hao, A. Tintori, Y.-l. Sun, and Z.-y. Sun. 2011. New informa-

tion on the protorosaurian reptile Macrocnemus fuyuanensis Li et al., 2007, from the Middle/Upper Triassic of Yunnan, China. Journal of Vertebrate Paleontology 31: 1230-1237.

Jiang, D.-y., et al. 2016. A large aberrant stem ichthyosauriform indicating early rise and demise of ichthyosauromorphs in the wake of the end-Permian extinction. Scientific Reports 6: 26232. doi:10.1038/srep26232. [11 co-authors]

Jin, L., J. Chen, S.-q. Zan, R. J. Butler, and P. Godefroit. 2010. Cranial anatomy of the small ornithischian dinosaur Changchunsaurus parvus from the Quantou Formation (Cretaceous: Aptian-Cenomanian) of Jilin Province, northeastern China. Journal of Vertebrate Paleontology 30: 196-214.

Johnson, M. M., M. T. Young, L. Steel, and Y. Lepage. 2015. Steneosaurus edwardsi (Thalattosuchia: Teleosauridae), the largest known crocodylomorph of the Middle Jurassic. Biological Journal of the Linnean Society 115: 911-918.

Jones, M. E. H. 2008. Skull shape and feeding strategy in Sphenodon and other Rhynchocephalia (Diapsida: Lepidosauria). Journal of Morphology 269: 945-966.

Jones, M. E. H., C. L. Anderson, C. A. Hipsley, J. Müller, S. E. Evans, and R. R. Schoch. 2013. Integration of molecules and new fossils supports a Triassic origin for Lepidosauria (lizards, snakes, and tuatara). BMC Evolutionary Biology 13: 208. doi:10.1186/1471-2148-13-208.

Jones, M. E. H., A. J. D. Tennyson, J. P. Worthy, S. E. Evans, and T. H. Worthy. 2009. A sphenodontine (Rhynchocephalia) from the Miocene of New Zealand and palaeobiogeography of the tuatara (Sphenodon). Proceedings of the Royal Society B 276: 1385-1390.

Jouve, S., B. Bouya, M. Amaghzaz, and S. Meslouh. 2015. Maroccosuchus zennaroi (Crocodylia: Tomistominae) from the Eocene of Morocco: phylogenetic and palaeobiogeographical implications of the basalmost tomistomine. Journal of Systematic Palaeontology 13: 421-445.

Jouve, S., M. Iarochène, B. Bouya, and M. Amaghzaz. 2006. A new species of Dyrosaurus (Crocodylomorpha, Dyrosauridae) from the early Eocene of Morocco: phylogenetic implications. Zoological Journal of the Linnean Society 148: 603-656.

Joyce, W. G. 2007. Phylogenetic relationships of Mesozoic turtles. Bulletin of the Peabody Museum of Natural History, Yale University 48: 3-102.

Joyce, W. G. 2014. A review of the fossil record of turtles of the clade Pan-Carettochelys. Bulletin of the Peabody Museum of Natural History, Yale University 55: 3-33.

Joyce, W. G. 2015. The origin of turtles: a paleontological perspective. Journal of Experimental Zoology (Molecular and Developmental Evolution) 324B: 181-193.

Joyce, W. G. 2017. A review of the fossil record of basal Mesozoic turtles. Bulletin of the Peabody Museum of Natural History, Yale University 58: 65-113.

Joyce, W. G., and J. R. Bourque. 2016. A review of the fossil turtles of the clade *Pan-Kinosternoidea*. Bulletin of the Peabody Museum of Natural History, Yale University 57: 57-95.

Joyce, W. G., and J. A. Gauthier. 2004. Paleoecology of Triassic stem turtles sheds new light on turtle origins. Proceedings of the Royal Society of London B 271: 1-5.

Joyce, W. G., and T. R. Lyson. 2010. *Pangshura tatrotia*, a new species of pond turtle (Testudinoidea) from the Pliocene Siwaliks of Pakistan. Journal of Systematic Palaeontology 8: 449-458.

Joyce, W. G., and T. R. Lyson. 2015. A review of the fossil record of turtles of the clade *Baenidae*. Bulletin of the Peabody Museum of Natural History, Yale University 56: 147-183.

Joyce, W. G., and M. A. Norell. 2005. *Zangerlia ukhaachelys*, new species, a nanhsiungchelyid turtle from the Late Late Cretaceous of Ukhaa Tolgod, Mongolia. American Museum Novitates 3481: 1-19.

Joyce, W. G., J. F. Parham, and J. A. Gauthier. 2004. Developing a protocol for the conversion of rank-based taxon names to phylogenetically defined clade names, as exemplified by turtles. Journal of Paleontology 78: 989-1013.

Joyce, W. G., J. F. Parham, T. R. Lyson, R. C. M. Warnock, and P. C. J. Donoghue. 2013b. A divergence dating analysis of turtles using fossil calibrations: an example of best practices. Journal of Paleontology 87: 612-634.

Joyce, W. G., R. R. Schoch, and T. R. Lyson. 2013a. The girdles of the oldest fossil turtle, *Proterochersis robusta*, and the age of the turtle crown. BMC Evolutionary Biology 13: 266. doi:10.1186/1471-2148-13-266.

Joyce, W. G., J. Sterli, and S. D. Chapman. 2014. The skeletal morphology of the solemydid turtle *Naomichelys speciosa* from the Early Cretaceous of Texas. Journal of Paleontology 88: 1257-1287.

Joyce, W. G., I. Werneburg, and T. R. Lyson. 2013c. The hooked element in the pes of turtles (Testudines): a global approach to exploring primary and secondary homology. Journal of Anatomy 223: 421-441.

Kammerer, C. F., R. J. Butler, S. Bandyopadhyay, and M. R. Stocker. 2015. Relationships of the Indian phytosaur *Parasuchus hislopi* Lydekker, 1885. Papers in Palaeontology 2015: 1-23.

Kear, B. P., and M. S. Y. Lee. 2006. A primitive protostegid from Australia and early sea turtle evolution. Biology Letters 2: 116-119.

Kearney, M. 2003. Systematics of the Amphisbaenia (Lepidosauria: Squamata) based on morphological evidence from Recent and fossil forms. Herpetological Monographs 17: 1-74.

Kellner, A. W. A. 2003. Pterosaur phylogeny and comments on the evolutionary history of the group. Pp. 105-137 in E. Buffetaut and J.-M. Mazin (eds.), Evolution and Palaeobiology of Pterosaurs. Geological Society, London, Special Publications 217.

Kellner, A. W. A. 2004. New information on the Tapejaridae (Pterosauria, Pterodactyloidea) and discussion of the relationships of this clade. Ameghiniana 41: 521-534.

Kellner, A. W. A., and D. A. Campos 2002. The function of the cranial crest and jaws of a unique pterosaur from the Early Cretaceous of Brazil. Science 297: 389-392.

Kellner, A. W. A., and W. Langston, Jr. 1996. Cranial remains of *Quetzalcoatlus* (Pterosauria, Azhdarchidae) from Late Cretaceous sediments of Big Bend National Park. Journal of Vertebrate Paleontology 16: 222-231.

Kellner, A. W. A., and Y. Tomida. 2000. Description of a new species of Anhangueridae (Pterodactyloidea) with comments on the pterosaur fauna from the Santana Formation (Aptian-Albian), northeastern Brazil. National Science Museum Monographs 17: 1-135.

Kellner, A. W. A., X. Wang, H. Tischlinger, D. A. Campos, D. W. E. Hone, and X. Meng. 2010. The soft tissue of *Jeholopterus* (Pterosauria, Anurognathidae, Batrachognathinae) and the structure of the pterosaur wing membrane. Proceedings of the Royal Society B 277: 321-329.

Ketchum, H. F., and R. B. J. Benson. 2010. Global interrelationships of Plesiosauria (Reptilia, Sauropterygia) and the pivotal role of taxon sampling in determining the outcome of phylogenetic analyses. Biological Reviews 85: 361-392.

Kirkland, J. I., R. Gaston, and D. Burge. 1993. A large dromaeosaur (Theropoda) from the Lower Cretaceous of eastern Utah. Hunteria 2(10): 1-16.

Kissel, R. A., D. W. Dilkes, and R. R. Reisz. 2002. *Captorhinus magnus*, a new captorhinid (Amniota: Eureptilia) from the Lower Permian of Oklahoma, with new evidence on the homology of the astragalus. Canadian Journal of Earth Sciences 39: 1363-1372.

Klein, N., and P. M. Sander. 2007. Bone histology and growth of the prosauropod dinosaur *Plateosaurus engelhardti* von Meyer, 1837 from the Norian bonebeds of Trossingen (Germany) and Frick (Switzerland). Special Papers in Palaeontology 77: 169-206.

Klembara, J. 2008. A new anguimorph lizard from the Lower Miocene of north-west Bohemia, Czech Republic. Palaeontology 51: 81-94.

Klembara, J. 2012. A new species of *Pseudopus* (Squamata, Anguidae) from the early Miocene of Northwest Bohemia (Czech Republic). Journal of Vertebrate Paleontology 32: 854-866.

Klembara, J., J. A. Clack, A. R. Milner, and M. Ruta. 2014. Cranial anatomy, ontogeny, and relationships of the Late Carboniferous tetrapod *Gephyrostegus bohemicus* Jaekel, 1902. Journal of Vertebrate Paleontology 34: 774-792.

Knauss, G. E., W. G. Joyce, T. R. Lyson, and D. Pearson. 2011. A new kinosternoid from the Late Cretaceous Hell Creek Formation of North Dakota and Montana and the origin of the *Dermatemys mawii* line. Paläontologische Zeitschrift 85: 125-142.

Knoll, F., K. Padian, and A. de Ricqlès. 2010. Ontogenetic change and adult body size of the early ornithischian dinosaur *Lesothosaurus diagnosticus*: implications for basal ornithischian taxonomy. Gondwana Research 17: 171-179.

Knutsen, E. M. 2012. A taxonomic review of the genus *Pliosaurus* (Owen, 1841a) Owen, 1841b. Norwegian Journal of Geology 92: 259-276.

Kobayashi, Y., and J. Lü. 2003. A new ornithomimid dinosaur with gregarious habits from the Late Cretaceous of China. Acta Palaeontologica Polonica 48: 235-259.

Krahl, A., N. Klein, and P. M. Sander. 2013. Evolutionary implications of the divergent long bone histologies of *Nothosaurus* and *Pistosaurus* (Sauropterygia, Triassic). BMC Evolutionary Biology 13: 123. doi:10.1186/1471-2148-13-123.

Kraus, F. 2015. Impacts from invasive reptiles and amphibians. Annual Review of Ecology, Evolution, and Systematics 46: 75-97.

Krause, D. W., S. E. Evans, and K.-q. Gao. 2003. First definitive record of Mesozoic lizards from Madagascar. Journal of Vertebrate Paleontology 23: 842-856.

Krause, D. W., and N. J. Kley (eds.). 2010. *Simosuchus clarki* (Crocodyliformes: Notosuchia) from the Late Cretaceous of Madagascar. Society of Vertebrate Paleontology Memoir 10: 1-236.

Krebs, B. 1962. Ein *Steneosaurus*-Rest aus dem Oberen Jura von Dielsdorf, Kt. Zürich. Schweizerische Paläontologische Abhandlungen 79: 1-28.

Krebs, B. 1963. Bau und Funktion des Tarsus eines Pseudosuchiers aus der Trias des Monte San Giorgio (Kanton Tessin, Schweiz). Paläontologische Zeitschrift 37: 88-95.

Krebs, B. 1974. Die Archosaurier. Naturwissenschaften 61: 17-24.

Ksepka, D. T., K.-q. Gao, and M. A. Norell. 2005. A new choristodere from the Cretaceous of Mongolia. American Museum Novitates 3468: 1-22.

Kubo, T., M. T. Mitchell, and D. M. Henderson. 2012. *Albertonectes vanderveldei*, a new elasmosaur (Reptilia, Sauropterygia) from the Upper Cretaceous of Alberta. Journal of Vertebrate Paleontology 32: 557-572.

Kuch, U., J. Müller, C. Mödden, and D. Mebs. 2006. Snake fangs from the Lower Miocene of Germany: evolutionary stability of perfect weapons. Naturwissenschaften 93: 84-87.

Kuhn-Schnyder, E. 1962. Ein weiterer Schädel von *Macrocnemus bassanii* Nopcsa aus der anisischen Stufe der Trias des Monte San Giorgio (Kt. Tessin, Schweiz). Paläontologische Zeitschrift 36: 110-133.

Kurzanov, S. M. 1987. [Avimimidae and the problem of the origin of birds.] Sovmestnaya Sovyetskogo-Mongol'skaya Paleontologicheskaya Ekspeditsiya, Trudy 31: 1-96. [Russian]

Kuzmin, I. T., P. P. Skutschas, E. A. Boitsova, and H.-D. Sues. 2019. Revision of the large crocodyliform *Kansajsuchus* from the Late Cretaceous of Asia. Zoological Journal of the Linnean Society 185: 335-387.

Lacovara, K., et al. 2014. A gigantic, exceptionally complete titanosaurian sauropod dinosaur from southern Patagonia, Argentina. Scientific Reports 4: 6196. doi:10.1038/srep06196. [16 co-authors]

LaDuke, T., D. W. Krause, J. D. Scanlon, and N. J. Kley. 2010. A Late Cretaceous (Maastrichtian) snake assemblage from the Maevarano Formation, Mahajanga Basin, Madagascar. Journal of Vertebrate Paleontology 30: 109-138.

Lamanna, M. C., H.-D. Sues, E. R. Schachner, and T. R. Lyson. 2014. A new large-bodied oviraptorosaurian theropod dinosaur from the latest Cretaceous of western North America. PLoS ONE 9(3): e92022. doi:10.1371/journal.pone0092022.

Lambe, L. M. 1920. The hadrosaur *Edmontosaurus* from the Upper Cretaceous of Alberta. Geological Survey of Canada Memoir 120: 1-79.

Landberg, T., J. D. Mailhot, and E. L. Brainerd. 2003. Lung ventilation during treadmill locomotion in a terrestrial turtle, *Terrapene carolina*. Journal of Experimental Biology 206: 3391-3404.

Langer, M. C. 2003. The pelvic and hind limb anatomy of the stem-sauropodomorph *Saturnalia tupiniquim* (Late Triassic, Brazil). PaleoBios 23(2): 1-30.

Langer, M. C., F. Abdala, M. Richter, and M. J. Benton. 1999. A sauropodomorph dinosaur from the Upper Triassic (Carnian) of southern Brazil. Comptes Rendus de l'Académie des Sciences Paris, Sciences de la Terre et des planètes 329: 511-517.

Langer, M. C., and M. J. Benton. 2006. Early dinosaurs: a phylogenetic study. Journal of Systematic Palaeontology 4: 309-358.

Langer, M. C., M. D. Ezcurra, J. S. Bittencourt, and F. E. Novas. 2010. The origin and early evolution of dinosaurs. Biological Reviews 85: 55-110.

Langer, M. C., S. J. Nesbitt, J. S. Bittencourt, and R. B. Irmis. 2013. Non-dinosaurian Dinosauromorpha. Pp. 157-186 in S. J. Nesbitt, J. B. Desojo, and R. B. Irmis (eds.), Anatomy, Phylogeny and Palaeobiology of Early Archosaurs and Their Kin. Geological Society, London, Special Publications 379.

Langston, W., Jr. 1965. Fossil crocodilians from Colombia and the Cenozoic history of the Crocodilia in South America. University of California Publications in Geological Sciences 52: 1-157.

Larsson, H. C. E., and H.-D. Sues. 2007. Cranial osteology and phylogenetic relationships of *Hamadasuchus rebouli* (Crocodyliformes: Mesoeucrocodylia) from the Cretaceous of Morocco. Zoological Journal of the Linnean Society 149: 533-567.

Laurenti, J. N. 1768. Specimen Medicum, exhibens synopsin Reptilium emendatum cum experimentis circa venena et antidota Reptilium Austriacorum. J. T. de Trattern, Vienna.

Laurin, M. 1991. The osteology of a Lower Permian eosuchian from Texas and a review of diapsid phylogeny. Zoological Journal of the Linnean Society 101: 59-95.

Laurin, M., and R. R. Reisz. 1995. A reevaluation of early amniote phylogeny. Zoological Journal of the Linnean Society 113: 165-223.

Lawson, R., J. B. Slowinski, and F. T. Burbrink. 2004. A molecular approach to discerning the phylogenetic placement of the enigmatic snake *Xenophidion schaeferi* among the Alethinophidia. Journal of Zoology, London 263: 285-294.

Lazzell, J. D., Jr. 1965. An *Anolis* (Sauria, Iguanidae) in amber. Journal of Paleontology 39: 379-382.

Le, M., C. Raxworthy, W. McCord, and L. Mertz. 2006. A molecular phylogeny of tortoises (Testudines: Testudinidae) based on mitochondrial and nuclear genes. Molecular Phylogenetics and Evolution 40: 517-531.

Leahey, L. G., R. E. Molnar, K. Carpenter, L. M. Witmer, and S. W. Salisbury. 2015. Cranial osteology of the ankylosaurian dinosaur formerly known as *Minmi* sp. (Ornithischia: Thyreophora) from the Lower Cretaceous Allaru Mudstone of Richmond, Queensland, Australia. PeerJ 3: e1475. doi: 10.7717/peerj.1475.

Leal, L. A., S. A. K. Azevedo, A. W. A. Kellner, and A. A. S. Da Rosa. 2004. A new early dinosaur (Sauropodomorpha) from the Caturrita Formation (Late Triassic), Paraná Basin, Brazil. Zootaxa 690: 1-24.

LeBlanc, A. R. H., K. S. Brink, T. M. Cullen, and R. R. Reisz. 2017. Evolutionary implications of tooth attachment versus tooth implantation: a case study using dinosaur, crocodilian, and mammal teeth. Journal of Vertebrate Paleontology 37: e1354006. doi:10.1080/02724634.2017.1354006.

LeBlanc, A. R. H., and R. R. Reisz. 2015. Patterns of tooth development and replacement in captorhinid reptiles: a comparative approach for understanding the origin of multiple tooth rows. Journal of Vertebrate Paleontology 35: e919928. doi:10.1080/02724634.2014.919928.

LeBlanc, A. R. H., R. R. Reisz, D. C. Evans, and A. M. Bailleul. 2016. Ontogeny reveals function and evolution of the hadrosaurid dinosaur dental battery. BMC Evolutionary Biology 16: 152. doi:10.1186/s12862-016-0721-1.

Lee, M. S. Y. 1995. Historical burden in systematics and the interrelationships of 'parareptiles'. Biological Reviews 70: 459-547.

Lee, M. S. Y. 1997a. Pareiasaur phylogeny and the origin of turtles. Zoological Journal of the Linnean Society 120: 197-280.

Lee, M. S. Y. 1997b. The phylogeny of varanoid lizards and the affinities of snakes. Philosophical Transactions of the Royal Society of London B 352: 53-91.

Lee, M. S. Y. 2005. Squamate phylogeny, taxon sampling and data congruence. Organisms Diversity & Evolution 5: 25-45.

Lee, M. S. Y. 2009. Hidden support from unpromising data sets strongly unites snakes with anguimorph 'lizards'. Journal of Evolutionary Biology 22: 1308-1316.

Lee, M. S. Y., and M. W. Caldwell. 2000. *Adriosaurus* and the affinities of mosasaurs, dolichosaurs, and snakes. Journal of Paleontology 74: 915-937.

Lee, M. S. Y., M. N. Hutchinson, T. H. Worthy, M. Archer, A. J. D. Tennyson, J. P. Worthy, and R. P. Scofield. 2009. Miocene skinks and geckos reveal long-term conservatism of New Zealand's lizard fauna. Biology Letters 5: 833-837.

Lee, M. S. Y., A. Palci, M. E. H. Jones, M. W. Caldwell, J. D. Holmes, and R. R. Reisz. 2016. Aquatic adaptations in the four limbs of the snake-like reptile *Tetrapodophis* from the Lower Cretaceous of Brazil. Cretaceous Research 66: 194-199.

Lee, M. S. Y., T. W. Reeder, J. B. Slowinski, and R. Lawson. 2004. Resolving reptile relationships: molecular and morphological markers. Pp. 451-467 in J. Cracraft and M. J. Donoghue (eds.), Assembling the Tree of Life. Oxford University Press, New York.

Lee, M. S. Y., and J. D. Scanlon. 2002. Snake phylogeny based on osteology, soft anatomy and ecology. Biological Reviews 77: 333-401.

Lee, Y., R. Barsbold, P. J. Currie, Y. Kobayashi, H. Lee, P. Godefroit, F. Escuillié, and T. Chinzorig. 2014. Resolving the long-standing enigmas of a giant ornithomimosaur *Deinocheirus mirificus*. Nature 515: 257-260.

Lehman, T. M., and H. N. Woodward. 2008. Modeling growth rates for sauropod dinosaurs. Paleobiology 34: 264-281.

Leidy, J. 1858. [Remarks concerning *Hadrosaurus*.] Proceedings of the Academy of Natural Sciences of Philadelphia 10: 215-218.

Li, C., N. C. Fraser, O. Rieppel, and X.-c. Wu. 2018. A Triassic stem turtle with an edentulous beak. Nature 560: 476-479.

Li, C., D.-y. Jiang, L. Cheng, X.-c. Wu, and O. Rieppel. 2014. A new species of *Largocephalosaurus* (Diapsida: Saurosphargidae), with implications for the morphological diversity and phylogeny of the group. Geological Magazine 151: 100-120.

Li, C., and O. Rieppel. 2002. A new cyamodontoid placodont from Triassic of Guizhou, China. Chinese Science Bulletin 47: 403-407.

Li, C., O. Rieppel, C. Long, and N. C. Fraser. 2016. The earliest herbivorous marine reptile and its remarkable

jaw apparatus. Science Advances 2: e1501659. doi:10.1126/sciadv.1501659.

Li, C., X.-c. Wu, Y.-n. Cheng, T. Sato, and L. Wang. 2006. An unusual archosaurian from the marine Triassic of China. Naturwissenschaften 93: 200-206.

Li, C., X.-c. Wu, O. Rieppel, L.-t. Wang, and L.-j. Zhao. 2008. An ancestral turtle from the Late Triassic of southwestern China. Nature 456: 497-501.

Li, C., X.-c. Wu, L.-j. Zhao, S. J. Nesbitt, M. R. Stocker, and L.-t. Wang. 2016. A new armored archosauriform (Diapsida: Archosauromorpha) from the marine Middle Triassic of China, with implications for the diverse life styles of archosauriforms prior to the diversification of Archosauria. The Science of Nature 103: 95. doi:10.1007/s00114-016-1418-4.

Li, C., X.-c. Wu, L.-j. Zhao, and L.-t. Wang. 2011. A new Triassic marine reptile from southwestern China. Journal of Vertebrate Paleontology 31: 303-312.

Li, D., M. A. Norell, K.-q. Gao, N. D. Smith, and P. J. Makovicky. 2010. A longirostrine tyrannosauroid from the Early Cretaceous of China. Proceedings of the Royal Society B 277: 183-190.

Li, L., W. G. Joyce, and J. Liu. 2015. The first soft-shelled turtle from the Jehol Biota of China. Journal of Vertebrate Paleontology 35: e909450. doi:10.1080/02724634.2014.909450.

Li, P.-p., K.-q. Gao, L.-h. Hou, and X. Xu. 2007. A gliding lizard from the Early Cretaceous of China. Proceedings of the National Academy of Sciences of the United States of America 104: 5507-5509.

Li, Q., K.-q. Gao, J. Vinther, M. D. Shawkey, J. A. Clarke, L. D'Alba, Q. Meng, D. E. G. Briggs, and R. O. Prum. 2010. Plumage color patterns of an extinct dinosaur. Science 327: 1369-1372.

Lin, K., and O. Rieppel. 1998. Functional morphology and ontogeny of *Keichousaurus hui* (Reptilia, Sauropterygia). Fieldiana Geology, n. s. 39: 1-35.

Lindgren, J., J. W. M. Jagt, and M. W. Caldwell. 2007. A fishy mosasaur: the axial skeleton of *Plotosaurus* (Reptilia, Squamata) reassessed. Lethaia 40: 153-160.

Lindgren, J., H. F. Kaddumi, and M. J. Polcyn. 2013. Soft tissue preservation in a fossil marine lizard with a bilobed tail fin. Nature Communications 4: 2423. doi:10.1038/ncomms3423.

Lindgren, J., P. Sjövall, R. M. Carney, P. Uvdal, J. A. Gren, G. Dyke, B. P. Schultz, M. D. Shawkey, K. R. Barnes, and M. J. Polcyn. 2014. Skin pigmentation provides evidence of convergent melanism in extinct marine reptiles. Nature 506: 484–488.

Linnaeus, C. 1758. Systema naturae per regna tria naturae, secundum classes, ordines, genera, species, cum characteribus, differentiis, synonymis, locis. Editio decima. Reformata. Tomus I. Laurentius Salvius, Stockholm.

Liu, J., C. L. Organ, M. J. Benton, M. C. Brandley, and J. C. Aitchison. 2017. Live birth in an archosauromorph reptile. Nature Communications 8: 14445. doi:10.1038/ncomms14445.

Liu, J., and O. Rieppel. 2005. Restudy of *Anshunsaurus huangguoshuensis* (Reptilia: Thalattosauria) from the Middle Triassic of Guizhou, China. American Museum Novitates 3488: 1–34.

Liu, J., L.-j. Zhao, C. Li, and T. He. 2013. Osteology of *Concavispina biseridens* (Reptilia, Thalattosauria) from the Xiaowa Formation (Carnian), Guanling, Guizhou, China. Journal of Paleontology 87: 341–350.

Liu, J., et al. 2014. A gigantic nothosaur (Reptilia: Sauropterygia) from the Middle Triassic of SW China and its implication for the Triassic biotic recovery. Scientific Reports 4: 7142. doi:10.1038/srep07142. [12 co-authors]

Liu, S., A. S. Smith, Y. Gu, J. Tan, C. K. Liu, and G. Turk. 2015. Computer simulations imply forelimb-dominated underwater flight in plesiosaurs. PLoS Computational Biology 11(12): e1004605. doi:10.1371/journal.pcbi.1004605/.

Liu, Y.-q., H.-w. Kuang, X.-j. Jiang, N. Peng, H. Xu, and H.-y. Sun. 2012. Timing of the earliest known feathered dinosaurs and transitional pterosaurs older than the Jehol Biota. Palaeogeography, Palaeoclimatology, Palaeoecology 323–325: 1–12.

Lloyd, G. T., K. E. Davis, D. Pisani, J. E. Tarver, M. Ruta, M. Sakamoto, D. W. E. Hone, R. Jennings, and M. J. Benton. 2008. Dinosaurs and the Cretaceous terrestrial revolution. Proceedings of the Royal Society B 275: 2483–2490.

Long, R. A., and P. A. Murry. 1995. Late Triassic (Carnian and Norian) tetrapods from the southwestern United States. New Mexico Museum of Natural History & Science Bulletin 4: 1–254.

Longrich, N., B.-A. S. Bhullar, and J. Gauthier. 2012a. A transitional snake from the Late Cretaceous period of North America. Nature 488: 205–208.

Longrich, N., and P. J. Currie. 2009. A microraptorine (Dinosauria-Dromaeosauridae) from the Late Cretaceous of North America. Proceedings of the National Academy of Sciences of the United States of America 106: 5002–5007.

Longrich, N., J. Vinther, Q. Meng, Q. Li, and A. P. Russell. 2012b. Primitive wing feather arrangement in *Archaeopteryx lithographica* and *Anchiornis huxleyi*. Current Biology 22: 2262–2267.

Longrich, N., J. Vinther, R. A. Pyron, D. Pisani, and J. A. Gauthier. 2015. Biogeography of worm lizards (Amphisbaenia) driven by end-Cretaceous mass extinction. Proceedings of the Royal Society B 282: 20143034. doi:10.1098/rspb.2014.3034.

Lortet, L. 1892. Les reptiles fossiles du Bassin du Rhône. Archives du Muséum d'Histoire Naturelle de Lyon 15: 1–139.

Losos, J. B. 2011. Lizards in an Evolutionary Tree: Ecology and Adaptive Radiation of Anoles. University of California Press, Berkeley.

Losos, J. B., D. M. Hillis, and H. W. Greene. 2012. Who speaks with a forked tongue? Science 338: 1428–1429.

Lovich, J.E., J. R. Ennen, M. Agha, J. W. Gibbons. 2018. Where have all the turtles gone, and why does it matter? BioScience 68: 771–781.

Løvtrup, S. 1985. On the classification of the taxon Tetrapoda. Systematic Zoology 34: 463–470.

Lu, B., W. Yang, Q. Dai, and J. Fu. 2013. Using genes as characters and a parsimony analysis to explore the phylogenetic position of turtles. PLoS ONE 8(11): e79348. doi:10.1371/journal.pone.0079348.

Lü, J., and Q. Ji. 2005. New azhdarchid pterosaur from the Early Cretaceous of Western Liaoning. Acta Geologica Sinica 79: 301–307.

Lü, J., S. Ji, C. Yuan, Y. Gao, Z. Sun, and Q. Ji. 2006. New pterodactyloid pterosaur from the Lower Cretaceous Yixian Formation of Western Liaoning. Pp. 195–203 in J. Lü, Y. Kobayashi, D. Huang, and Y. N. Lee (eds.), Papers from the 2005 Heyuan International Dinosaur Symposium. Geological Publishing House, Beijing.

Lü, J., D. M. Unwin, D. C. Deeming, X. Jin, Y. Liu, and Q. Ji. 2011. An egg-adult association, gender, and reproduction in pterosaurs. Science 331: 321–324.

Lü, J., D. M. Unwin, X. Jin, Y. Liu, and Q. Ji. 2010. Evidence for modular evolution in a long-tailed pterosaur with a pterodacyloid skull. Proceedings of the Royal Society B 277: 383–389.

Lü, J., D. M. Unwin, L. Xu, and X. Zhang. 2008. A new azhdarchoid pterosaur from the Lower Cretaceous of China and its implications for pterosaur phylogeny and evolution. Naturwissenschaften 95: 891–897.

Lü, J., L. Yi, S. L. Brusatte, L. Yang, H. Li, and L. Chen. 2014. A new clade of Asian Late Cretaceous long-snouted tyrannosaurids. Nature Communications 5: 3788. doi: 10.1038/ncomms4788.

Luan, X., C. Walker, S. Dangaria, Y. Ito, R. Druzinsky, K. Jarosius, H. Lesot, and O. Rieppel. 2009. The mosasaur tooth attachment apparatus as paradigm for the evolution of the gnathostome periodontium. Evolution & Development 11: 247-259.

Lucas, S. G. 2016. Dinosaurs: The Textbook. Sixth Edition. Columbia University Press, New York.

Lucas, S. G., J. A. Spielmann, L. F. Rinehart, A. B. Heckert, M. C. Herne, A. P. Hunt, J. R. Foster, and R. M. Sullivan. 2006. Taxonomic status of *Seismosaurus hallorum*, a Late Jurassic sauropod dinosaur from New Mexico. New Mexico Museum of Natural History & Science Bulletin 36: 149-161.

Lull, R. S. 1933. A revision of the Ceratopsia or horned dinosaurs. Memoirs of the Peabody Museum of Natural History 3(3): 1-135.

Lull, R. S., and N. E. Wright. 1942. Hadrosaurian dinosaurs of North America. Geological Society of America Special Paper 40: 1-242.

Lydekker, R. 1885. Indian Pre-Tertiary Vertebrata. The Reptilia and Amphibia of the Maleri and Denwa groups. Palaeontologia Indica, series 4, 1(5): 1-38.

Lyson, T. R., G. S. Bever, B.-A. S. Bhullar, W. G. Joyce, and J. A. Gauthier. 2010. Transitional fossils and the origin of turtles. Biology Letters 6: 830-833.

Lyson, T. R., G. S. Bever, T. M. Scheyer, A. Y. Hsiang, and J. A. Gauthier. 2013. Evolutionary origin of the turtle shell. Current Biology 23: 1113-1119.

Lyson, T. R., and W. G. Joyce. 2010. A new baenid turtle from the Late Cretaceous (Maastrichtian) Hell Creek Formation of North Dakota and a preliminary taxonomic revision of Cretaceous Baenidae. Journal of Vertebrate Paleontology 30: 394-402.

Lyson, T. R., B. S. Rubidge, T. M. Scheyer, K. de Queiroz, E. R. Schachner, R. M. H. Smith, J. Botha-Brink, and G. S. Bever. 2016. Fossorial origin of the turtle shell. Current Biology 26: 1887-1894.

Macartney, J. 1802. Preface and Table III. In Lectures on Comparative Anatomy. Translated from the French of G. Cuvier by William Ross, under the inspection of James Macartney. Vol. 1. On the Organs of Motion. Longman & Rees, London.

Madsen, J. H., Jr. 1976. *Allosaurus fragilis*: a revised osteology. Utah Geological and Mineral Survey Bulletin 109: 1-163.

Madsen, J. H., Jr., J. S. McIntosh, and D. S Berman. 1995. Skull and atlas-axis complex of the Upper Jurassic sauropod *Camarasaurus* Cope (Reptilia: Saurischia). Bulletin of Carnegie Museum of Natural History 31: 1-115.

Madsen, J. H., Jr., and S. P. Welles. 2000. *Ceratosaurus* (Dinosauria: Theropoda): a revised osteology. Utah Geological Survey, Miscellaneous Publication 00-2: 1-80.

Maidment, S. C. R., and P. M. Barrett. 2014. Osteological correlates for quadrupedality in ornithischian dinosaurs. Acta Palaeontologica Polonica 59: 53-70.

Maidment, S. C. R., C. Brassey, and P. M. Barrett. 2015. The postcranial skeleton of an exceptionally complete individual of the plated dinosaur *Stegosaurus stenops* (Dinosauria: Thyreophora) from the Upper Jurassic Morrison Formation of Wyoming, U.S.A. PLoS ONE 10(10): e0138352. doi:10.1371/journal.pone.0138352.

Maidment, S. C. R., D. B. Norman, P. M. Barrett, and P. Upchurch. 2008. Systematics and phylogeny of Stegosauria (Dinosauria: Ornithischia). Journal of Systematic Palaeontology 6: 367-407.

Maidment, S. C. R., and L. B. Porro. 2010. Homology of the palpebral and origin of supraorbital ossifications in ornithischian dinosaurs. Lethaia 43: 95-111.

Maisch, M. W. 1998. A new ichthyosaur genus from the Posidonia Shale (Lower Toarcian, Jurassic) of Holzmaden, SW-Germany with comments on the phylogeny of post-Triassic ichthyosaurs. Neues Jahrbuch für Geologie und Paläontologie Abhandlungen 209: 47-78.

Makádi, L., M. W. Caldwell, and A. Ösi. 2012. The first freshwater mosasauroid (Upper Cretaceous, Hungary) and a new clade of basal mosasauroids. PLoS ONE 7(12): e51781. doi:10.1371/journal.pone.0051781.

Makovicky, P. J. 2001. A *Montanoceratops cerorhynchus* (Dinosauria: Ceratopsia) braincase from the Horseshoe Canyon Formation of Alberta. Pp. 243-262 in D. H. Tanke and K. Carpenter (eds.), Mesozoic Vertebrate Life. Indiana University Press, Bloomington.

Makovicky, P. J., S. Apesteguía, and F. L. Agnolin. 2005. The earliest dromaeosaurid theropod from South America. Nature 437: 1007-1011.

Makovicky, P. J., and M. A. Norell. 2004. Troodontidae. Pp. 184-195 in D. B. Weishampel, P. Dodson, and H. Osmólska (eds.), The Dinosauria. Second Edition. University of California Press, Berkeley.

Makovicky, P. J., and M. A. Norell. 2006. *Yamaceratops dorngobiensis*, a new primitive ceratopsian (Dinosauria:

Ornithischia) from the Cretaceous of Mongolia. American Museum Novitates 3530: 1-42.

Malafaia, E., F. Ortega, F. Escaso, and B. Silva. 2015. New evidence of *Ceratosaurus* (Dinosauria: Theropoda) from the Late Jurassic of the Lusitanian Basin, Portugal. Historical Biology 27: 938-946.

Maleev, E. A. 1974. [Giant carnosaurs of the family Tyrannosauridae.] Pp. 132-191 in Fauna i Biostratigrafiya Mesozoya i Kainozoya Mongolii. Sovmestmaya Sovyetskogo-Mongol'skaya Paleontologicheskaya Ekspeditiiya Trudy 1. Izdatel'stvo "Nauka," Moscow. [Russian]

Mallison, H. 2010. The digital *Plateosaurus* II: an assessment of the range of motion of the limbs and vertebral column and of previous reconstructions using a digital skeletal mount. Acta Palaeontologica Polonica 55: 433-458.

Mallon, J. C., D. C. Evans, M. J. Ryan, and J. S. Anderson. 2013. Feeding height stratification among the herbivorous dinosaurs from the Dinosaur Park Formation (upper Campanian) of Alberta, Canada. BMC Ecology 13: 14. doi:10.1186/1472-6785-13-14.6.

Manning, P. L., et al. 2013. Synchroton-based chemical imaging reveals plumage patterns in a 150 million year old early bird. Journal of Analytical Atomic Spectrometry 28: 1024-1030. [11 co-authors]

Mantell, G. A. 1825. Notice on the Iguanodon, a newly discovered fossil reptile, from the sandstone of Tilgate forest, in Sussex. Philosophical Transactions of the Royal Society of London 115: 179-186.

Mantell, G. A. 1831. The geological Age of Reptiles. Edinburgh New Philosophical Journal 11: 181-185.

Marinho, T. S., and I. S. Carvalho. 2009. An armadillo-like sphagesaurid crocodyliform from the Late Cretaceous of Brazil. Journal of South American Earth Sciences 27: 36-41.

Markwick, P. J. 1998. Fossil crocodilians as indicators of Late Cretaceous and Cenozoic climates: implications for using palaeontological data in reconstructing palaeoclimate. Palaeogeography, Palaeoclimatology, Palaeoecology 137: 205-271.

Martill, D. M. 1986. The diet of *Metriorhynchus*, a Mesozoic marine crocodile. Neues Jahrbuch für Geologie and Paläontologie Monatshefte 1986: 621-625.

Martill, D. M. 1987. Prokaryote mats replacing soft tissues in Mesozoic marine reptiles. Modern Geology 11: 265-269.

Martill, D. M., H. Tischlinger, and N. R. Longrich. 2015. A four-legged snake from the Early Cretaceous of Gondwana. Science 349: 416-419.

Martill, D. M., and D. M. Unwin. 2012. The world's largest toothed pterosaur, NHMUK R481, an incomplete rostrum of *Coloborhynchus capito* (Seeley, 1870) from the Cambridge Greensand of England. Cretaceous Research 34: 1-9.

Martin, A. J. 2012. Life Traces of the Georgia Coast. Indiana University Press, Bloomington.

Martin, E. G., and C. Palmer. 2014. Air space proportion in pterosaur limb bones using computed tomography and its implications for previous estimates of pneumaticity. PLoS ONE 9(5): e97159. doi:10.1371/journal.pone.0097159.

Martin, J. E. 2010. A new species of *Diplocynodon* (Crocodylia, Alligatoroidea) from the Late Eocene of the Massif Central, France, and the evolution of the genus in the climatic context of the Late Palaeogene. Geological Magazine 147: 596-610.

Martin, J. E., and F. de Lapparent de Broin. 2016. A miniature notosuchian with multicuspid teeth from the Cretaceous of Morocco. Journal of Vertebrate Paleontology 36: e1211534. doi:10.1080/02724634.2016.1211534.

Martin, J. E., and M. Gross. 2011. Taxonomic clarification of *Diplocynodon* Pomel, 1847 (Crocodilia) from the Miocene of Styria, Austria. Neues Jahrbuch für Geologie und Paläontologie Abhandlungen 261: 177-193.

Martin, J. E., K. Lauprasert, E. Buffetaut, R. Liard, and V. Suteethorn. 2014b. A large pholidosaurid in the Phu Kradung Formation of north-eastern Thailand. Palaeontology 57: 757-769.

Martin, J. E., M. Rabi, Z. Cisiki-Sava, and S. Vasile. 2014a. Cranial morphology of *Theriosuchus sympiestodon* (Mesoeucrocodylia, Atoposauridae) and the widespread occurrence of *Theriosuchus* in the Late Cretaceous of Europe. Journal of Paleontology 88: 444-456.

Martínez, R. N., and O. Alcober. 2009. A basal sauropodomorph (Dinosauria: Saurischia) from the Ischigualasto Formation (Triassic, Carnian) and the early evolution of Sauropodomorpha. PLoS ONE 4(2): e4397. doi:10.1371/journal.pone.0004397.

Martínez, R. N., C. Apaldetti, C. E. Colombi, A. Praderio, E. Fernandez, P. S. Malnis, G. A. Correa, D. Abelin, and O. Alcober. 2013a. A new sphenodontian (Lepidosauria: Rhynchocephalia) from the Late Triassic of Argentina and the early origin of the herbivore opisthodontians. Proceedings of the Royal Society B 280: 20132057. doi:10.1098/rspb.2013.2057.

Martínez, R. N., C. Apaldetti, O. A. Alcober, C. E. Colombi, P. C. Sereno, E. Fernandez, P. S. Malnis, G. A. Correa, and D. Abelin. 2013b. Vertebrate succession in the Ischigualasto Formation. Society of Vertebrate Paleontology Memoir 12: 10-30.

Martínez, R. N., C. Apaldetti, and D. Abelin. 2013c. Basal sauropodomorphs from the Ischigualasto Formation. Society of Vertebrate Paleontology Memoir 12: 51-69.

Martínez, R. N., C. Apaldetti, G. A. Correa, and D. Abelín. 2016. A Norian lagerpetid dinosauromorph from the Quebrada del Barro Formation, northwestern Argentina. Ameghiniana 53: 1-13.

Martinez, R. N., P. C. Sereno, O. A. Alcober, C. E. Colombi, P. R. Renne, I. P. Montañez, and B. S. Currie. 2011. A basal dinosaur from the dawn of the dinosaur era in southwest Pangaea. Science 331: 206-209.

Martínez, R. D. F., M. C. Lamanna, F. E. Novas, R. C. Ridgely, G. A. Casal, J. E. Martínez, J. R. Vita, and L. M. Witmer. 2016. A basal lithostrotian titanosaur (Dinosauria: Sauropoda) with a complete skull: implications for the evolution and paleobiology of Titanosauria. PLoS ONE 11(4): e0151661. doi:10.1371/journal. pone.0151661.

Maryańska, T. 1977. Ankylosauridae (Dinosauria) from Mongolia. Palaeontologia Polonica 37: 85-151.

Maryańska, T., and H. Osmólska. 1974. Pachycephalosauria, a new suborder of ornithischian dinosaurs. Palaeontologia Polonica 30: 45-102.

Maryańska, T., and H. Osmólska. 1975. Protoceratopsidae (Dinosauria) of Asia. Palaeontologia Polonica 33: 133-181.

Maryańska, T., and H. Osmólska. 1981. Cranial anatomy of *Saurolophus angustirostris* with comments on the Asian Hadrosauridae (Dinosauria). Palaeontologia Polonica 42: 5-24.

Maryańska, T., H. Osmólska, and M. Wolsam. 2002. Avialian status for Oviraptorosauria. Acta Palaeontologica Polonica 47: 97-116.

Massare, J. A. 1987. Tooth morphology and prey preference of Mesozoic marine reptiles. Journal of Vertebrate Paleontology 7: 121-137.

Massare, J. A., and D. R. Lomax. 2018. A taxonomic reassessment of *Ichthyosaurus communis* and *I. intermedius* and a revised diagnosis for the genus. Journal of Systematic Palaeontology 16: 263-277.

Mateus, O., S. C. R. Maidment, and N. A. Christiansen. 2009. A new long-necked 'sauropod-mimic' stegosaur and the evolution of plated dinosaurs. Proceedings of the Royal Society B 276: 1815-1821.

Mateus, O., P. D. Mannion, and P. Upchurch. 2014. *Zby atlanticus*, a new turiasaurian sauropod (Dinosauria, Eusauropoda) from the Late Jurassic of Portugal. Journal of Vertebrate Paleontology 34: 618-634.

Mateus, O., A. Walen, and M. T. Antunes. 2006. The large theropod fauna of the Lourinhã Formation (Portugal) and its similarity to that of the Morrison Formation, with a description of a new species of *Allosaurus*. New Mexico Museum of Natural History & Science Bulletin 36: 123-129.

Matsumoto, R., E. Buffetaut, F. Escuillié, S. Hervet, and S. E. Evans. 2013. New material of the choristodere *Lazarussuchus* (Diapsida, Choristodera) from the Paleocene of France. Journal of Vertebrate Paleontology 33: 319-339.

Mayr, G. 2017. Avian Evolution: The Fossil Record of Birds and Its Paleobiological Siginificance. John Wiley & Sons Ltd, Chichester.

Mayr, G., D. S. Peters, G. Plodowski, and O. Vogel. 2002. Bristle-like integumentary structures at the tail of the horned dinosaur *Psittacosaurus*. Naturwissenschaften 89: 361-365.

Mazzetta, G. V., P. Christiansen, and R. A. Fariña. 2004. Giants and bizarres: body size of some southern South American Cretaceous dinosaurs. Historical Biology 16: 71-83.

McAliley, L. R., R. E. Willis, D. A. Ray, P. S. White, C. A. Brochu, and L. D. Densmore III. 2006. Are crocodiles really monophyletic?—Evidence for subdivisions from sequence and morphological data. Molecular Phylogenetics and Evolution 39: 16-32.

McCartney, J. A., N. J. Stevens, and P. M. O'Connor. 2014. The earliest colubroid-dominated snake fauna from Africa: perspectives from the late Oligocene Nsungwe Formation of southwestern Tanzania. PLoS ONE 9(3): e90415. doi:10.1371/journal.pone.0090415.

McDonald, A. T. 2011. The taxonomy of species assigned to *Camptosaurus* (Dinosauria: Ornithopoda). Zootaxa 2783: 52-68.

McDonald, A. T., J. Bird, J. I. Kirkland, and P. Dodson. 2012. Osteology of the basal hadrosauroid *Eolambia caroljonesa* (Dinosauria: Ornithopoda) from the Cedar Mountain Formation of Utah. PLoS ONE 7(10): e45712. doi:10.1371/journal.pone.0045712.

McDonald, A. T., J. I. Kirkland, D. D. DeBlieux, S. K. Madsen, J. Cavin, A. R. C. Milner, and L. Panzarin. 2010. New basal iguanodonts from the Cedar Mountain Formation of Utah and the evolution of thumb-spiked dinosaurs. PLoS ONE 5(11): e14075. doi:10.1371/journal. pone.0014075.

McDowell, S. B., and C. M. Bogert. 1954. The systematic position of *Lanthanotus* and the affinities of anguinomorphan lizards. Bulletin of the American Museum of Natural History 105: 1-142.

McGowan, C., and R. Motani. 2003. Ichthyopterygia. Pp. 1-175 in H.-D. Sues (ed.), Handbook of Paleoherpetology, Part 8. Verlag Dr. Friedrich Pfeil, Munich.

McGuire, J. A., and R. Dudley. 2011. The biology of gliding in flying lizards (genus *Draco*) and their fossil and extant analogs. Integrative and Comparative Biology 51: 983-990.

McHenry, C. R., A. G. Cook, and S. Wroe. 2005. Bottom-feeding plesiosaurs. Science 310: 75.

McIntosh, J. S., W. E. Miller, K. L. Stadtman, and D. D. Gillette. 1996. The osteology of *Camarasaurus lewisi* (Jensen, 1988). Brigham Young University Geology Studies 41: 73-115.

McPhee, B. W., J. N. Choiniere, A. M. Yates, and P. A. Vigletti. 2015. A second species of *Eucnemesaurus* Van Hoepen, 1920 (Dinosauria, Sauropodomorpha): new information on the diversity and evolution of the sauropodomorph fauna of South Africa's Lower Elliot Formation (latest Triassic). Journal of Vertebrate Paleontology 35: e980504. doi:10.1080/02724634.2015.980504.

McPhee, B. W., A. M. Yates, J. N. Choiniere, and F. Abdala. 2014. The complete anatomy and phylogenetic relationships of *Antetonitrus ingenipes* (Sauropodiformes, Dinosauria): implications for the origins of Sauropoda. Zoological Journal of the Linnean Society 171: 151-205.

Mead, J. I. 2013. Scolecophidia (Serpentes) of the Late Oligocene and Early Miocene, North America, and a fossil history overview. Geobios 46: 225-231.

Meng, Q.-j., J.-y. Liu, D. J. Varricchio, T. Huang, and C.-l. Gao. 2004. Parental care in an ornithischian dinosaur. Nature 431: 145-146.

Merriam, J. C. 1905. The Thalattosauria: a group of marine reptiles from the Triassic of California. Memoirs of the California Academy of Sciences 5: 1-52.

Merriam, J. C. 1908. Triassic Ichthyosauria with special reference to the American forms. Memoirs of the University of California 1: 1-196.

Mertens, R. 1942. Die Familie der Warane (Varanidae). Teil 1-3. Abhandlungen der Senckenbergischen naturforschenden Gesellschaft 462: 1-116; 465: 1-118; 466: 1-160.

Metzger, K. 2002. Cranial kinesis in lepidosaurs: skulls in motion. Pp. 15-46 in P. Aerts, K. D'Août, A. Herrel, and R. Van Damme (eds.), Topics in Functional and Ecological Vertebrate Morphology. Shaker Publishing, Maastricht.

Meylan, P. A. 1987. The phylogenetic relationships of soft-shelled turtles (Family Trionychidae). Bulletin of the American Museum of Natural History 186: 1-101.

Meylan, P. A., and E. S. Gaffney. 1989. The skeletal morphology of the Cretaceous cryptodiran turtle, *Adocus*, and the relationships of the Trionychoidea. American Museum Novitates 2941: 1-60.

Mickoleit, G. 2004. Phylogenetische Systematik der Wirbeltiere. Verlag Dr. Friedrich Pfeil, Munich.

Miralles, A., J. Martin, D. Markus, A. Herrel, S. B. Hedges, and N. Vidal. 2018. Molecular evidence for the paraphyly of Scolecophidia and its evolutionary implications. Journal of Evolutionary Biology doi :10.1111/jeb.13373

Mo, J.-y., X. Xu, and S. E. Evans. 2010. The evolution of the lepidosaurian lower temporal bar: new perspectives from the Late Cretaceous of South China. Proceedings of the Royal Society B 277: 331-336.

Moazen, M., N. Curtis, P. O'Higgins, S. E. Evans, and M. J. Fagan. 2009. Biomechanical assessment of evolutionary changes in the lepidosaurian skull. Proceedings of the National Academy of Sciences of the United States of America 106: 8373-8277.

Modesto, S. P. 1999. Observations on the structure of the Early Permian reptile *Stereosternum tumidum* Cope. Palaeontologia africana 35: 7-19.

Modesto, S. P. 2006. The cranial skeleton of the Early Permian aquatic reptile *Mesosaurus tenuidens*: implications for relationships and palaeobiology. Zoological Journal of the Linnean Society 146: 345-368.

Modesto, S. P., and J. S. Anderson. 2004. The phylogenetic definition of Reptilia. Systematic Biology 53: 815-821.

Modesto, S. P., and R. R. Reisz. 2002. An enigmatic new diapsid reptile from the Upper Permian of eastern Europe. Journal of Vertebrate Paleontology 22: 851-855.

Modesto, S. P., D. M. Scott, D. S. Berman, J. Müller, and R. R. Reisz. 2007. The skull and the palaeoecological significance of *Labidosaurus hamatus*, a captorhinid reptile from the Lower Permian of Texas. Zoological Journal of the Linnean Society 149: 237-262.

Modesto, S. P., D. M. Scott, M. J. MacDougall, H.-D. Sues, D. C. Evans, and R. R. Reisz. 2015. The oldest parareptile and the early diversification of reptiles. Proceedings of the Royal Society B 282: 20141912. doi:10.1098/rspb.2014.1912.

Modesto, S. P., D. M. Scott, and R. R. Reisz. 2009a. A new parareptile with temporal fenestration from the Middle Permian of South Africa. Canadian Journal of Earth Sciences 46: 9-20.

Modesto, S. P., D. M. Scott, and R. R. Reisz. 2009b. Arthropod remains in the oral cavities of fossil reptiles support inference of early insectivory. Biology Letters 5: 838-840.

Modesto, S. P., and H.-D. Sues. 2004. The skull of the Early Triassic archosauromorph reptile *Prolacerta broomi* and its phylogenetic significance. Zoological Journal of the Linnean Society 140: 335-351.

Modesto, S. P., H.-D. Sues, and R. J. Damiani. 2001. A new Triassic procolophonoid reptile and its implications for procolophonoid survivorship during the Permo-Triassic extinction event. Proceedings of the Royal Society of London B 268: 2047-2052.

Mohabey, D. M., J. J. Head, and J. A. Wilson. 2011. A new species of the snake *Madtsoia* from the Upper Cretaceous of India and its paleobiogeographic implications. Journal of Vertebrate Paleontology 31: 588-595.

Molnar, R. E., T. Worthy, and P. M. A. Willis. 2002. An extinct Pleistocene endemic mekosuchine crocodylian from Fiji. Journal of Vertebrate Paleontology 22: 612-628.

Montañez, I. P., N. J. Tabor, D. Niemeier, W. A. DiMichele, T. D. Frank, C. R. Fielding, J. L. Isbell, L. P. Birgenheier, and M. C. Rygel. 2007. CO_2-forced climate and vegetation instability during late Paleozoic deglaciation. Science 315: 87-91.

Montefeltro, F. C., M. C. Langer, and C. L. Schultz. 2010. Cranial anatomy of a new genus of hyperodapedontine rhynchosaur (Diapsida, Archosauromorpha) from the Upper Triassic of southern Brazil. Earth and Environmental Science Transactions of the Royal Society of Edinburgh 101: 27-52.

Montefeltro, F. C., H. C. E. Larsson, M. A. G. de França, and M. C. Langer. 2013. A new neosuchian with Asian affinities from the Jurassic of northeastern Brazil. Naturwissenschaften 100: 835-841.

Moody, R. T. J. 1974. The taxonomy and morphology of *Puppigerus camperi*, an Eocene sea turtle from northern Europe. Bulletin of the British Museum (Natural History), Geology 25: 155-186.

Moser, M. 2003. *Plateosaurus engelhardti* Meyer, 1837 (Dinosauria: Sauropodomorpha) aus dem Feuerletten (Mittelkeuper; Obertrias) von Bayern. Zitteliana B 24: 3-186.

Motani, R. 1997. Temporal and spatial distribution of tooth implantation in ichthyosaurs. Pp. 81-103 in J. M. Callaway and E. L. Nicholls (eds.), Ancient Marine Reptiles. Academic Press, New York.

Motani, R. 1999. Phylogeny of the Ichthyopterygia. Journal of Vertebrate Paleontology 19: 473-496.

Motani, R. 2002. Scaling effects in caudal fin kinematics and the speeds of ichthyosaurs. Nature 415: 309-312.

Motani, R. 2005. Evolution of fish-shaped reptiles (Reptilia: Ichthyopterygia) in their physical environments and constraints. Annual Review of Earth and Planetary Sciences 33: 395-420.

Motani, R., X.-h. Chen, D.-y. Jiang, L. Cheng, A. Tintori, and O. Rieppel. 2015b. Lunge feeding in early marine reptiles and fast evolution of marine tetrapod feeding guilds. Scientific Reports 5: 8900. doi:10.1038/srep08900.

Motani, R., C. Ji, T. Tomita, N. Kelley, E. Maxwell, D.-j. Jiang, and P. M. Sander. 2013. Absence of suction feeding ichthyosaurs and its implications for Triassic mesopelagic paleoecology. PLoS ONE 8(12): e66075. doi:10.1371/journal.pone.0066075.

Motani, R., D.-y. Jiang, G.-b. Chen, A. Tintori, O. Rieppel, C. Ji, and J.-d. Huang. 2015a. A basal ichthyosauriform with a short snout from the Lower Triassic of China. Nature 517: 485-488.

Motani, R., D.-j. Jiang, A. Tintori, O. Rieppel, and G.-b. Chen. 2014. Terrestrial origin of viviparity in Mesozoic marine reptiles indicated by Early Triassic embryonic fossils. PLoS ONE 9(2): e88640. doi:10.1371/journal.pone.0088640.

Müller, J. 2004. The relationships among diapsid reptiles and the influence of taxon selection. Pp. 379-408 in G. Arratia, M. V. H. Wilson, and R. Cloutier (eds.), Recent Advances in the Origin and Early Radiation of Vertebrates. Verlag Dr. Friedrich Pfeil, Munich.

Müller, J. 2005. The anatomy of *Askeptosaurus italicus* from the Middle Triassic of Monte San Giorgio and the interrelationships of thalattosaurs (Reptilia, Diapsida). Canadian Journal of Earth Sciences 42: 1347-1367.

Müller, J., C. A. Hipsley, J. J. Head, N. Kardjilov, A. Hilger, M. Wuttke, and R. R. Reisz. 2011. Eocene lizard from Germany reveals amphisbaenian origins. Nature 473: 364-367.

Müller, J., C. A. Hipsley, and J. A. Maisano. 2016. Skull osteology of the Eocene amphisbaenian *Spathorhynchus fossorium* (Reptilia, Squamata) suggests convergent evolution and reversals of fossorial adaptations in worm lizards. Journal of Anatomy 229: 615-630.

Müller, J., and R. R. Reisz. 2006. The phylogeny of early eureptiles: comparing parsimony and Bayesian approaches in the investigation of a basal fossil clade. Systematic Biology 55: 503-511.

Müller, J., S. Renesto, and S. E. Evans. 2005. The marine diapsid reptile *Endennasaurus* from the Upper Triassic of Italy. Palaeontology 48: 15-30.

Müller, J., and L. A. Tsuji. 2007. Impedance-matching hearing in Paleozoic reptiles: evidence of advanced sensory perception at an early stage of amniote evolution. PLoS ONE 2(9): e889. doi:10.1371/journal.pone.0000889.

Müller, R. T., M. C. Langer, M. Bronzati, C. P. Pacheco, S. F. Cabreira, and S. Dias-da-Silva. 2018. Early evolution of sauropodomorphs: anatomy and phylogenetic relationships of a remarkably well-preserved dinosaur from the Upper Triassic of southern Brazil. Zoological Journal of the Linnean Society 184: 1187-1248.

Munk, W., and H.-D. Sues. 1993. Gut contents of *Parasaurus* (Pareiasauria) and *Protorosaurus* (Archosauromorpha) from the Kupferschiefer (Upper Permian) of Hessen, Germany. Paläontologische Zeitschrift 67: 169-176.

Muscutt, L. E., G. Dyke, G. D. Weymouth, D. Naish, C. Palmer, and B. Ganapathisubramani. 2017. The four-flipper swimming method of plesiosaurs enabled efficient and effective locomotion. Proceedings of the Royal Society B 284: 20170951. doi:10.1098/rspb.2017.0951.

Myers, N. 1988. Tropical forests and their species: going, going . . . ? Pp. 28-35 in E. O. Wilson (ed.), Biodiversity. National Academy Press, Washington, DC.

Nabavizadeh, A. 2014. Hadrosauroid jaw mechanics and the functional significance of the predentary bone. Pp. 467-481 in D. A. Eberth and D. C. Evans (eds.), Hadrosaurs. Indiana University Press, Bloomington.

Nabavizadeh, A., and D. B. Weishampel. 2016. The predentary bone and its significance in the evolution of feeding mechanisms in ornithischian dinosaurs. The Anatomical Record 299: 1358-1388.

Nagashima, H., F. Sugahara, M. Takechi, R. Ericsson, Y. Kawashima-Ohya, Y. Narita, and S. Kuratani. 2009. Evolution of the turtle body plan by the folding and creation of new muscle connections. Science 325: 193-196.

Narváez, I., C. A. Brochu, F. Escaso, A. Pérez-García, and F. Ortega. 2015. New crocodyliforms from southwestern Europe and definition of a diverse clade of European Late Cretaceous basal eusuchians. PLoS ONE 10(11): e0140679. doi:10.1371/journal.pone.0140679.

Nascimento, P. M., and H. Zaher. 2010. A new species of *Baurusuchus* (Crocodyliformes, Mesoeucrocodylia) from the Upper Cretaceous of Brazil, with the first complete postcranial skeleton described for the family Baurusuchidae. Papéis Avulsos de Zoologia 50: 323-361.

Nee, S., and R. M. May. 1997. Extinction and the loss of evolutionary history. Science 278: 692-694.

Neenan, J. M., N. Klein, and T. M. Scheyer. 2013. European origin of placodont marine reptiles and the evolution of crushing dentition in Placodontia. Nature Communications 4: 1621. doi:10.1038/ncomms2633.

Neenan, J. M., C. Li, O. Rieppel, and T. M. Scheyer. 2015. The cranial anatomy of Chinese placodonts and the phylogeny of Placodontia (Diapsida: Sauropterygia). Zoological Journal of the Linnean Society 175: 415-428.

Nesbitt, S. J. 2005. Osteology of the Middle Triassic pseudosuchian archosaur *Arizonasaurus babbitti*. Historical Biology 17: 19-47.

Nesbitt, S. J. 2007. The anatomy of *Effigia okeeffeae* (Archosauria, Suchia), theropod-like convergence, and the distribution of related taxa. Bulletin of the American Museum of Natural History 302: 1-84.

Nesbitt, S. J. 2011. The early evolution of archosaurs: relationships and the origin of major clades. Bulletin of the American Museum of Natural History 352: 1-292.

Nesbitt, S. J., P. M. Barrett, S. Werning, C. A. Sidor, and A. J. Charig. 2012. The oldest dinosaur? A Middle Triassic dinosauriform from Tanzania. Biology Letters 9: 20120949. doi:10.1098/rsbl.2012.0949.

Nesbitt, S. J., S. L. Brusatte, J. B. Desojo, A. Liparini, M. A. G. De França, J. C. Weinbaum, and D. J. Gower. 2013. Rauisuchia. Pp. 241-274 in S. J. Nesbitt, J. B. Desojo, and R. B. Irmis (eds.), Anatomy, Phylogeny and Palaeobiology of Early Archosaurs and Their Kin. Geological Society, London, Special Publications 379.

Nesbitt, S. J., and R. J. Butler. 2013. Redescription of the archosaur *Parringtonia gracilis* from the Middle Triassic Manda beds of Tanzania, and the antiquity of Erpetosuchidae. Geological Magazine 150: 225-238.

Nesbitt, S. J., R. J. Butler, M. D. Ezcurra, A. J. Charig, and P. M. Barrett. 2018. The anatomy of *Teleocrater rhadinus*, an early avemetatarsalian from the lower portion of the Lifua Member of the Manda Beds (Middle Triassic). Memoirs of the Society of Vertebrate Paleontology 17: 142-177.

Nesbitt, S. J., and D. W. E. Hone. 2010. An extenal mandibular fenestra and other archosauriform character states in basal pterosaurs. Palaeodiversity 3: 225-233.

Nesbitt, S. J., R. B. Irmis, and W. G. Parker. 2007. A critical re-evaluation of the Late Triassic dinosaur taxa of North America. Journal of Systematic Palaeontology 5: 209-243.

Nesbitt, S. J., R. B. Irmis, W. G. Parker, N. D. Smith, A. H. Turner, and T. Rowe. 2009b. Hindlimb osteology and distribution of basal dinosauromorphs from the Late Triassic of North America. Journal of Vertebrate Paleontology 29: 498-516.

Nesbitt, S. J., J. Liu, and C. Li. 2011. A sail-backed suchian from the Heshanggou Formation (Early Triassic: Olenekian) of China. Earth and Environmental Science Transactions of the Royal Society of Edinburgh 101: 271-284.

Nesbitt, S. J., C. A. Sidor, R. B. Irmis, K. D. Angielczyk, R. M. H. Smith, and L. A. Tsuji. 2010. Ecologically distinct dinosaurian sister group shows early diversification of Ornithodira. Nature 464: 95-98.

Nesbitt, S. J., N. D. Smith, R. B. Irmis, A. H. Turner, A. Downs, and M. A. Norell. 2009c. A complete skeleton of a Late Triassic saurischian and the early evolution of dinosaurs. Science 326: 1530-1533.

Nesbitt, S. J., M. R. Stocker, B. J. Small, and A. Downs. 2009a. The osteology and relationships of *Vancleavea campi* (Reptilia: Archosauriformes). Zoological Journal of the Linnean Society 157: 814-864.

Nesbitt, S. J., A. H. Turner, M. Spaulding, J. L. Conrad, and M. A. Norell. 2009d. The theropod furcula. Journal of Morphology 270: 856-879.

Nesbitt, S. J., et al. 2017. The earliest bird-line archosaurs and the assembly of the dinosaur body plan. Nature 544: 484-487. [11 co-authors]

Nicholls, E. L. 1999. A reexamination of *Thalattosaurus* and *Nectosaurus* and the relationships of the Thalattosauria (Reptilia, Diapsida). PaleoBios 19: 1-29.

Nicholls, E. L., and M. Manabe. 2004. Giant ichthyosaurs of the Triassic—a new species of *Shonisaurus* from the Pardonet Formation (Norian: Late Triassic) of British Columbia. Journal of Vertebrate Paleontology 24: 838-849.

Nielsen, E. 1963. On the post-cranial skeleton of *Eosphargis breineri* Nielsen. Meddelelser fra Dansk Geologisk Forening 15: 281-313.

Noè, L. F., M. A. Taylor, and M. Gómez-Pérez. 2017. An integrated approach to understanding the role of a long neck in plesiosaurs. Acta Palaeontologica Polonica 62: 137-162.

Nopcsa, F. 1923. *Eidolosaurus* und *Pachyophis*. Zwei neue Neocom-Reptilien. Palaeontographica 65: 97-154.

Norell, M. A. 1989. Late Cenozoic lizards of the Anza Borrego Desert, California. Natural History Museum of Los Angeles County, Contributions in Science 414: 1-31.

Norell, M. A., and J. M. Clark. 1991. A reanalysis of *Bernissartia fagesii*, with comments on its phylogenetic position and its bearing on the origin and diagnosis of the Eusuchia. Bulletin de l'Institut Royal des Sciences Naturelles de Belgique, Sciences de la Terre 60: 115-128.

Norell, M. A., and K. de Queiroz. 1991. The earliest iguanine lizard (Reptilia: Squamata) and its bearing on iguanine phylogeny. American Museum Novitates 2997: 1-16.

Norell, M. A., and K.-q. Gao. 1997. Braincase and phylogenetic relationships of *Estesia mongoliensis* from the Late Cretaceous of the Gobi desert and the recognition of a new clade of lizards. American Museum Novitates 3211: 1-25.

Norell, M. A., K.-q. Gao, and J. L. Conrad. 2007. A new platynotan lizard (Diapsida: Squamata) from the Late Cretaceous Gobi Desert (Ömnögov), Mongolia. American Museum Novitates 3605: 1-22.

Norell, M. A., and P. J. Makovicky. 1997. Important features of the dromaeosaur skeleton: information from a new specimen. American Museum Novitates 3215: 1-28.

Norell, M. A., M. C. McKenna, and M. J. Novacek. 1992. *Estesia mongoliensis*, a new fossil varanoid from the Cretaceous Barun Goyot Formation of Mongolia. American Museum Novitates 3045: 1-24.

Norman, D. B. 1980. On the ornithischian dinosaur *Iguanodon bernissartensis* from the Lower Cretaceous of Bernissart (Belgium). Institut Royal des Sciences Naturelles de Belgique Mémoires 178: 1-103.

Norman, D. B. 1986. On the anatomy of *Iguanodon atherfieldensis* (Ornithischia: Ornithopoda). Bulletin de l'Institut Royal des Sciences Naturelles de Belgique: Sciences de la Terre 56: 281-372.

Norman, D. B. 2002. On Asian ornithopods (Dinosauria: Ornithischia). 4. Redescription of *Probactrosaurus gobiensis* Rozhdestvensky, 1966. Zoological Journal of the Linnean Society 136: 113-144.

Norman, D. B. 2012. Iguanodontian taxa (Dinosauria: Ornithischia) from the Lower Cretaceous of Britain and Belgium. Pp. 174-212 in P. Godefroit (ed.), Bernissart Dinosaurs and Early Cretaceous Terrestrial Ecosystems. Indiana University Press, Bloomington.

Norman, D. B. 2014. Iguanodonts from the Wealden of England: do they contribute to the discussion concerning hadrosaur origins? Pp. 10-43 in D. A. Eberth and

D. C. Evans (eds.), Hadrosaurs. Indiana University Press, Bloomington.

Norman, D. B. 2015. On the history, osteology, and systematic position of the Wealden (Hastings group) dinosaur *Hypselospinus fittoni* (Iguanodontia: Styracosterna). Zoological Journal of the Linnean Society 173: 92-189.

Norman, D. B., A. W. Crompton, R. J. Butler, L. B. Porro, and A. J. Charig. 2011. The Lower Jurassic ornithischian dinosaur *Heterodontosaurus tucki* Crompton & Charig, 1962: cranial anatomy, functional morphology, taxonomy, and relationships. Zoological Journal of the Linnean Society 163: 182-276.

Norman, D. B., and D. B. Weishampel. 1985. Ornithopod feeding mechanisms: their bearing on the evolution of herbivory. American Naturalist 126: 151-164.

Nosotti, S. 2007. *Tanystropheus longobardicus* (Reptilia, Protorosauria): re-interpretations of the anatomy based on new specimens from the Middle Triassic of Besano (Lombardy, northern Italy). Memorie della Società Italiana di Scienze Naturali e del Museo Civico di Storia Naturale di Milano 35(3): 1-88.

Nosotti, S., and O. Rieppel. 2003. *Eusaurosphargis dalassoi* n. gen. n. sp., a new, unusual diapsid reptile from the Middle Triassic of Besano (Lombardy, N Italy). Memorie della Società Italiana di Scienze Naturali e del Museo Civico di Storia Naturale di Milano 31(2): 1-33.

Novas, F. E. 1992. Phylogenetic relationships of the basal dinosaurs, the Herrerasauridae. Palaeontology 16: 51-62.

Novas, F. E. 1994. New information on the systematics and postcranial skeleton of *Herrerasaurus ischigualastensis* (Theropoda: Herrerasauridae) from the Ischigualasto Formation (Upper Triassic) of Argentina. Journal of Vertebrate Paleontology 13: 400-423.

Novas, F. E. 1997. Anatomy of *Patagonykus puertai* (Theropoda, Avialae, Alvarezsauridae), from the Late Cretaceous of Patagonia. Journal of Vertebrate Paleontology 17: 137-166.

Novas, F. E., A. V. Cambiaso, and A. Ambrioso. 2004. A new basal iguanodontian (Dinosauria, Ornithischia) from the Upper Cretaceous of Patagonia. Ameghiniana 41: 75-82.

Novas, F. E., M. D. Ezcurra, F. L. Agnolin, D. Pol, and R. Ortíz. 2012. New Patagonian Cretaceous theropod sheds light about the early radiation of Coelurosauria. Revista del Museo Argentino de Ciencias Naturales, n. s. 14: 57-81.

Novas, F. E., D. Pol, J. I. Canale, J. D. Porfiri, and J. O. Calvo. 2009. A bizarre Cretaceous theropod dinosaur from Patagonia and the evolution of Gondwanan dromaeosaurids. Proceedings of the Royal Society B 276: 1101-1107.

Novas, F. E., L. Salgado, M. Suárez, F. L. Agnolín, M. D. Ezcurra, N. R. Chimento, R. de la Cruz, M. P. Isasi, A. O. Vargas, and D. Rubilar-Rogers. 2015. An enigmatic plant-eating theropod from the Late Jurassic Period of Chile. Nature 522: 331-334.

Nydam, R. L., and R. L. Cifelli. 2005. New data on the dentition of the scincomorphan lizard *Polyglyphanodon sternbergi*. Acta Palaeontologica Polonica 50: 73-78.

Nydam, R., J. G. Eaton, and J. Sankey. 2007. New taxa of transversely-toothed lizards (Squamata: Scincomorpha) and new information on the evolutionary history of "teiids." Journal of Paleontology 81: 538-549.

Nydam, R. L., and B. M. Fitzpatrick. 2009. The occurrence of *Contogenys*-like lizards in the Late Cretaceous and Early Tertiary of the Western Interior of the U.S.A. Journal of Vertebrate Paleontology 29: 677-701.

Oaks, J. R. 2011. A time-calibrated species tree of Crocodylia reveals a recent radiation of the true crocodiles. Evolution 65: 3285-3297.

O'Connor, P. M., J.W. Sertich, N. J. Stevens, E. M. Roberts, M. D. Gottfried, T. L. Hieronymus, Z. A. Jinnah, R. Ridgely, S. E. Ngasala, and J. Temba. 2010. The evolution of mammal-like crocodyliforms in the Cretaceous Period in Gondwana. Nature 466: 748-751.

Oelofson, B. W., and D. C. Araújo. 1987. *Mesosaurus tenuidens* and *Stereosternum tumidum* from the Permian Gondwana of both southern Africa and South America. South African Journal of Science 83: 370-372.

Oftedal, O. T. 2002. The origin of lactation as a water source for parchment-shelled eggs. Journal of Mammary Gland Biology and Neoplasia 7: 253-266.

O'Keefe, F. R. 2001a. A cladistic analysis and taxonomic revision of the Plesiosauria (Reptilia: Plesiosauria). Acta Zoologica Fennica 213: 1-63.

O'Keefe, F. R. 2001b. Ecomorphology of plesiosaur flipper geometry. Journal of Evolutionary Biology 14: 987-991.

O'Keefe, F. R. 2002. The evolution of plesiosaur and pliosaur morphotypes in the Plesiosauria (Reptilia: Sauropterygia). Paleobiology 28: 101-112.

O'Keefe, F. R. 2004. Preliminary description and phylogenetic position of a new genus and species of plesiosaur (Reptilia: Sauropterygia) from the Toarcian of Holzmaden, Germany. Journal of Paleontology 78: 973-988.

O'Keefe, F. R., and L. M. Chiappe. 2011. Viviparity and K-selected life history in a Mesozoic marine plesiosaur (Reptilia, Sauropterygia). Science 333: 870-873.

O'Keefe, F. R., R. A. Otero, S. Soto-Acuña, J. P. O'Gorman, S. J. Godfrey, and S. Chatterjee. 2017. Cranial anatomy of *Morturneria seymourensis* from Antarctica, and the evolution of filter feeding in plesiosaurs of the Austral Late Cretaceous. Journal of Vertebrate Paleontology 37: e1347570. doi:10.1080/02724634.2017.1347570.

O'Keefe, F. R., C. A. Sidor, H. C. E. Larsson, A. Maga, and O. Ide. 2006. Evolution and homology of the astragalus in early amniotes: new fossils, new perspectives. Journal of Morphology 267: 415-425.

Olsen, P. E. 1979. A new aquatic eosuchian from the Newark Supergroup (Late Triassic-Early Jurassic) of North Carolina and Virginia. Postilla 176: 1-14.

Olson, E. C. 1947. The family Diadectidae and its bearing on the classification of reptiles. Fieldiana Geology 11: 1-53.

Organ, C. L. 2006. Thoracic epaxial muscles in living archosaurs and ornithopod dinosaurs. The Anatomical Record 288A: 782-793.

Ortega, F., F. Escaso, and J. L. Sanz. 2010. A bizarre, humped Carcharodontosauria (Theropoda) from the Lower Cretaceous of Spain. Nature 467: 203-206.

Ortega, F., Z. Gasparini, A. D. Buscalioni, and J. O. Calvo. 2000. A new species of *Araripesuchus* (Crocodylomorpha, Mesoeucrocodylia) from the Lower Cretaceous of Patagonia. Journal of Vertebrate Paleontology 20: 57-76.

Osborn, H. F. 1903. The reptilian subclasses Diapsida and Synapsida and the early history of the Diaptosauria. Memoirs of the American Museum of Natural History 1: 449-507.

Osborn, H. F., and C. C. Mook. 1921. *Camarasaurus, Amphicoelias* and other sauropods of Cope. Memoirs of the American Museum of Natural History 3(3): 247-387.

Ösi, A. 2014. The evolution of jaw mechanism and dental function in heterodont crocodyliforms. Historical Biology 26: 279-414.

Ösi, A., R. J. Butler, and D. B. Weishampel. 2010. A Late Cretaceous ceratopsian dinosaur from Europe with Asian affinities. Nature 465: 466-468.

Ösi, A., D. B. Weishampel, and C. M. Jianu. 2005. First evidence of azhdarchid pterosaurs from Hungary. Acta Palaeontologica Polonica 50: 777-787.

Osmólska, H., S. Hua, and E. Buffetaut. 1997. *Gobiosuchus kielanae* (Protosuchia) from the Late Cretaceous of Mongolia: anatomy and relationships. Acta Palaeontologica Polonica 42: 257-289.

Osmólska, H., E. Roniewicz, and R. Barsbold. 1972. A new dinosaur, *Gallimimus bullatus* n. gen., n. sp. from the Upper Cretaceous of Mongolia. Palaeontologia Polonica 27: 103-143.

Ostrom, J. H. 1961. Cranial morphology of the hadrosaurian dinosaurs of North America. Bulletin of the American Museum of Natural History 122: 33-186.

Ostrom, J. H. 1964. A functional analysis of the jaw mechanics in the dinosaur *Triceratops*. Postilla 88: 1-35.

Ostrom, J. H. 1969. Osteology of *Deinonychus antirrhopus*, an unusual theropod from the Lower Cretaceous of Montana. Bulletin of the Peabody Museum of Natural History, Yale University 30: 1-165.

Ostrom, J. H. 1970. Stratigraphy and paleontology of the Cloverly Formation (Lower Cretaceous) of the Bighorn Basin area, Wyoming and Montana. Bulletin of the Peabody Museum of Natural History, Yale University 35: 1-234.

Ostrom, J. H. 1973. The ancestry of birds. Nature 242: 136.

Ostrom, J. H. 1976a. *Archaeopteryx* and the origin of birds. Biological Journal of the Linnean Society 8: 91-182.

Ostrom, J. H. 1976b. On a new specimen of the Lower Cretaceous theropod dinosaur *Deinonychus antirrhopus*. Breviora 439: 1-21.

Ostrom, J. H. 1978. The osteology of *Compsognathus longipes* Wagner. Zitteliana 4: 73-118.

Otero, A., E. Krupandan, D. Pol, A. Chinsamy, and J. N. Choiniere. 2015. A new basal sauropodiform from South Africa and the phylogenetic relationships of basal sauropodomorphs. Zoological Journal of the Linnean Society 174: 589-634.

Otero, A., and D. Pol. 2013. Postcranial anatomy and phylogenetic relationships of *Mussaurus patagonicus* (Dinosauria, Sauropodomorpha). Journal of Vertebrate Paleontology 33: 1138-1168.

Owen, R. 1840. Report on British fossil reptiles. Part I. Report of the British Association for Advancement of Science, 9th Meeting, Birmingham 1839: 43-126.

Owen, R. 1840-1845. Odontography; or, a Treatise on the Comparative Anatomy of the Teeth; Their Physiological Relations, Mode of Development, and Microscopic Structure, in the Vertebrate Animals. Hippolyte Bailliere, London.

Owen, R. 1842. Report on British fossil reptiles. Part II. Report of the British Association for Advancement of Science, 11th Meeting, Plymouth 1841: 60-204.

Owen, R. 1860. Palaeontology, or a Systematic Summary of Extinct Animals and Their Geological Relations. Adam and Charles Black, Edinburgh.

Owen, R. 1863. A monograph of the fossil Reptilia of the Liassic formations. 2. *Scelidosaurus harrisoni* Owen of the Lower Lias. Palaeontographical Society Monographs 14: 1-26.

Packard, M. J., and R. S. Seymour. 1997. Evolution of the amniote egg. Pp. 265-290 in S. S. Sumida and K. L. M. Martin (eds.), Amniote Origins: Completing the Transition to Land. Academic Press, San Diego.

Padian, K. 1983a. A functional analysis of flying and walking in pterosaurs. Paleobiology 9: 218-239.

Padian, K. 1983b. Description and reconstruction of new material of *Dimorphodon macronyx* (Buckland) (Pterosauria: Rhamphorhynchoidea) in the Yale Peabody Museum. Postilla 189: 1-44.

Padian, K. 1984. The origin of pterosaurs. Pp. 163-168 in W.-E. Reif and F. Westphal (eds.), Third Symposium on Mesozoic Terrestrial Ecosystems, Short Papers. Attempto Verlag, Tübingen.

Padian, K. 1985. The origins and aerodynamics of flight in extinct vertebrates. Palaeontology 28: 413-433.

Padian, K. 2008. The Early Jurassic pterosaur *Dorygnathus banthensis* (Theodori, 1830) and the Early Jurassic pterosaur *Campylognathoides* Strand, 1928. Special Papers in Palaeontology 80: 1-107.

Padian, K. 2013. The problem of dinosaur origins: integrating three approaches to the rise of Dinosauria. Earth and Environmental Science Transactions of the Royal Society of Edinburgh 103: 423-442.

Padian, K. 2017. Structure and evolution of the ankle bones in pterosaurs and other ornithodirans. Journal of Vertebrate Paleontology 37: e1364651. doi: 10.1080/02724634.2017.1364651.

Padian, K., A. J. de Ricqlès, and J. R. Horner. 2001. Dinosaurian growth rates and bird origins. Nature 412: 405-408.

Padian, K., and J. R. Horner. 2011. The evolution of 'bizarre structures' in dinosaurs: function, sexual selection, social selection, or species recognition? Journal of Zoology, London 283: 3-17.

Padian, K., and C. L. May. 1993. The earliest dinosaurs. Pp. 379-380 in S. G. Lucas and M. Morales (eds.), The Nonmarine Triassic. New Mexico Museum of Natural History and Science Bulletin 3.

Palci, A., M. W. Caldwell, and A. M. Albino. 2013. Emended diagnosis and phylogenetic relationships of the Upper Cretaceous fossil snake *Najash rionegrina* Apesteguía and Zaher, 2006. Journal of Vertebrate Paleontology 33: 131-140.

Palmer, C. 2011. Flight in slow motion: aerodynamics of the pterosaur wing. Proceedings of the Royal Society B 278: 1881-1885.

Palmer, C. 2014. The aerodynamics of gliding flight and its application to the arboreal flight of the Chinese feathered dinosaur *Microraptor*. Biological Journal of the Linnean Society of London 113: 828-835.

Palmer, C., and G. J. Dyke. 2010. Biomechanics of the unique pterosaur pteroid. Proceedings of the Royal Society B 277: 1121-1127.

Paolillo, A., and O. J. Linares. 2007. Nuevos cocodrilos Sebecosuchia del Cenozoico suramericano (Mesosuchia: Crocodylia). Paleobiología Neotropical 3: 1-23.

Parham, J. F., C. R. Feldman, and J. L. Boone. 2006. The complete mitochondrial genome of the enigmatic bigheaded turtle (*Platysternon*): description of unusual genomic features and the reconciliation of phylogenetic hypotheses based on mitochondrial and nuclear DNA. BMC Evolutionary Biology 6: 11. doi:10.1186/1471-2148-6-11.

Parham, J. F., and N. D. Pyenson. 2010. New sea turtle from the Miocene of Peru and the iterative evolution of feeding ecomorphologies since the Cretaceous. Journal of Paleontology 84: 231-242.

Parker, W. G. 2007. Reassessment of the aetosaur *"Desmatosuchus" chamaensis* with a reanalysis of the phylogeny of the Aetosauria (Archosauria: Pseudosuchia). Journal of Systematic Palaeontology 5: 41-68.

Parker, W. G. 2008. Description of new material of the aetosaur *Desmatosuchus spurensis* (Archosauria: Suchia) from the Chinle Formation of Arizona and a revision of the genus *Desmatosuchus*. PaleoBios 28: 1-40.

Parker, W. G. 2016. Revised phylogenetic analysis of the Aetosauria (Archosauria: Pseudosuchia): assessing the effects of incongruent morphological character sets. PeerJ 4: e1583. doi:10.7717/peerj.1583.

Parker, W. G., R. B. Irmis, S. J. Nesbitt, J. W. Martz, and L. S. Browne. 2005. The Late Triassic pseudosuchian *Revueltosaurus callenderi* and its implications for the diversity of early ornithischian dinosaurs. Proceedings of the Royal Society of London B 272: 963-969.

Parmley, D., and J. A. Holman 1995. Hemphillian (late Miocene) snakes from Nebraska, with comments on Arikareean through Blancan snakes of midcontinental North America. Journal of Vertebrate Paleontology 15: 79-95.

Parrish, J. M. 1987. The origin of crocodilian locomotion. Paleobiology 13: 396-414.

Parrish, J. M. 1993. Phylogeny of Crocodylotarsi, with reference to archosaurian and crurotarsan monophyly. Journal of Vertebrate Paleontology 13: 287-308.

Paton, R. L., T. R. Smithson, and J. A. Clack. 1999. An amniote-like skeleton from the Early Carboniferous of Scotland. Nature 398: 508-513.

Patterson, C. 1981. Significance of fossils in determining evolutionary relationships. Annual Review of Ecology and Systemstics 12: 195-223.

Peecook, B. R., C. A. Sidor, S. J. Nesbitt, R. M. H. Smith, J. S. Steyer, and K. D. Angielczyk. 2013. A new silesaurid from the upper Ntawere Formation of Zambia (Middle Triassic) demonstrates the rapid diversification of Silesauridae (Avemetatarsalia, Dinosauriformes). Journal of Vertebrate Paleontology 33: 1127-1137.

Pérez-García, A., and V. Codrea. 2018. New insights on the anatomy and systematics of *Kallokibotion* Nopcsa, 1923, the enigmatic uppermost Cretaceous basal turtle (stem Testudines) from Transylvania. Zoological Journal of the Linnean Society 182: 419-443.

Pérez-García, A., J. M. Gasulla, and F. Ortega. 2014. *Eodortoka morellana* gen. et sp. nov., the first pan-pleurodiran turtle (Dortokidae) defined in the Lower Cretaceous of Europe. Cretaceous Research 48: 130-138.

Pérez-García, A., R. Royo-Torres, and A. Cobos. 2015. A new European Late Jurassic pleurosternid (Testudines, Paracryptodira) and a new hypothesis of paracryptodiran phylogeny. Journal of Systematic Palaeontology 13: 351-369.

Pérez-García, A., and E. Vlachos. 2014. New generic proposal for the European Neogene large testudinids (Cryptodira) and the first phylogenetic hypothesis for the medium and large representatives of the European Cenozoic record. Zoological Journal of the Linnean Society 172: 653-719.

Pérez-Moreno, B. P., J. L. Sanz, A. D. Buscalioni, J. J. Moratalla, F. Ortega, and D. Rasskin-Gutman. 1994. A unique multitoothed ornithomimosaur from the Lower Cretaceous of Spain. Nature 370: 363-367.

Perle, A., T. Maryańska, and H. Osmólska. 1982. *Goyocephale lattimorei* gen. et sp. n., a new flat-headed pachycephalosaur (Ornithischia, Dinosauria) from the Upper Cretaceous of Mongolia. Acta Palaeontologica Polonica 27: 115-127.

Perle, A., M. A. Norell, and J. M. Clark. 1993. Flightless bird from the Cretaceous of Mongolia. Nature 362: 623-626.

Peterson, J. E., C. Dischler, and N. R. Longrich. 2013. Distributions of cranial pathologies provide evidence for head-butting in dome-headed dinosaurs (Pachycephalosauridae). PLoS ONE 8(7): e68620. doi:10.1371/journal.pone.0068620.

Peyer, B. 1931a. Die Triasfauna der Tessiner Kalkalpen. II. *Tanystropheus longobardicus* Bass. sp. Abhandlungen der Schweizerischen Paläontologischen Gesellschaft 50: 7-110.

Peyer, B. 1931b. Die Triasfauna der Tessiner Kalkalpen. III. Placodontia. Abhandlungen der Schweizerischen Paläontologischen Gesellschaft 51: 1-25.

Peyer, B. 1936. Die Triasfauna der Tessiner Kalkalpen. X. *Clarazia schinzi* nov. gen. nov. spec. Abhandlungen der Schweizerischen Paläontologischen Gesellschaft 57: 1-26.

Peyer, B. 1937. Die Triasfauna der Tessiner Kalkalpen. XII. *Macrocnemus bassanii* Nopcsa. Abhandlungen der Schweizerischen Paläontologischen Gesellschaft 59: 1-140.

Peyer, B. 1944. Die Reptilien vom Monte San Giorgio. 1924-1944. Neujahrsblatt auf das Jahr 1944. Naturforschende Gesellschaft, Zürich.

Peyer, B. 1955. Die Triasfauna der Tessiner Kalkalpen. XVIII. *Helveticosaurus zollingeri* n. g. n. sp. Schweizerische Paläontologische Abhandlungen 72: 1-50.

Peyer, B. 1968. Comparative Odontology. The University of Chicago Press, Chicago.

Peyer, K. 2006. A reconsideration of *Compsognathus* from the upper Tithonian of Canjuers, southeastern France. Journal of Vertebrate Paleontology 26: 879-896.

Peyer, K., J. G. Carter, H.-D. Sues, S. E. Novak, and P. E. Olsen. 2008. A new suchian archosaur from the Upper Triassic of North Carolina. Journal of Vertebrate Paleontology 28: 363-381.

Philippe, H., H. Brinkman, D. V. Lavrov, D. T. J. Littlewood, M. Manuel, G. Wörheide, and D. Baurain. 2011. Resolving difficult phylogenetic questions: why more sequences are not enough. PLoS Biology 9(3): e1000602. doi:10.1371/journal.pbio.1000602.

Pianka, E. R., and D. R. King. 2004. Varanoid Lizards of the World. Indiana University Press, Bloomington.

Pianka, E. R., and L. J. Vitt. 2003. Lizards: Windows to the Evolution of Diversity. University of California Press, Berkeley.

Pierce, S. E., K. D. Angielczyk, and E. J. Rayfield. 2009a. Shape and mechanics in thalattosuchian (Crocodylomorpha) skulls: implications for feeding behaviour and niche partitioning. Journal of Anatomy 215: 555-576.

Pierce, S. E., K. D. Angielczyk, and E. J. Rayfield. 2009b. Morphospace occupation in thalattosuchian crocodylomorphs: skull shape variation, species delineation and temporal patterns. Palaeontology 52: 1057-1097.

Pimm, S., C. N. Jenkins, R. Abell, T. M. Brooks, J. L. Gittleman, L. N. Joppa, P. H. Raven, C. M. Roberts, and J. O. Sexton. 2014. The biodiversity of species and their rates of extinction, distribution, and protection. Science 344: 1246752. doi:10.1126/science.1246752.

Piñeiro, G., J. Ferigolo, M. Meneghel, and M. Laurin. 2012a. The oldest known amniotic embryos suggest viviparity in mesosaurs. Historical Biology 24: 620-630.

Piñeiro, G., J. Ferigolo, A. Ramos, and M. Laurin. 2012b. Cranial morphology of the Early Permian mesosaurid *Mesosaurus tenuidens* and the evolution of lower temporal fenestration reassessed. Comptes Rendus Palevol 11: 379-391.

Pinheiro, F. L., M. A. G. França, M. B. Lacerda, R. J. Butler, and C. L. Schultz. 2016. An exceptional fossil skull from South America and the origins of the archosauriform radiation. Scientific Reports 6: 22817. doi:10.1038/srep22817.

Pinna, G. 1984. Osteologia di *Drepanosaurus unguicaudatus*, Lepidosauro triassico del Sottordine Lacertilia. Memoria della Società Italiana di Scienze Naturali e del Museo Civico di Storia Naturale di Milano 24: 7-28.

Pinna, G., and S. Nosotti. 1989. Anatomia, morfologia funzionale e paleoecologia del rettile placodonte *Psephoderma alpinum* Meyer, 1858. Memoria della Società Italiana di Scienze Naturali e del Museo Civico di Storia Naturale di Milano 25: 17-49.

Poe, S. 1996. Data set incongruence and the phylogeny of crocodilians. Systematic Biology 45: 393-414.

Pol, D. 2003. New remains of *Sphagesaurus huenei* (Crocodylomorpha: Mesoeucrocodylia) from the Late Cretaceous of Brazil. Journal of Vertebrate Paleontology 23: 817-831.

Pol, D., and Z. Gasparini. 2009. Skull anatomy of *Dakosaurus andiniensis* (Thalattosuchia: Crocodylomorpha) and the phylogenetic position of Thalattosuchia. Journal of Systematic Palaeontology 7: 163-197.

Pol, D., S.-a. Ji, J. M. Clark, and L. M. Chiappe. 2004. Basal crocodyliforms from the Lower Cretaceous Tugulu Group (Xinjiang, China), and the phylogenetic position of *Edentosuchus*. Cretaceous Research 25: 603-622.

Pol, D., J. M. Leardi, A. Lecuona, and M. Krause. 2012. Postcranial anatomy of *Sebecus icaeorhinus* (Crocodyliformes, Sebecidae) from the Eocene of Patagonia. Journal of Vertebrate Paleontology 32: 328-354.

Pol, D., P. M. Nascimento, A. B. Carvalho, C. Riccomini, R. A. Pires-Domingues, and H. Zaher. 2014. A new notosuchian from the Late Cretaceous of Brazil and the phylogeny of advanced neosuchians. PLoS ONE 9(4): e93105. doi:10.1371/journal.pone.0093105.

Pol, D., and M. A. Norell. 2004. A new crocodyliform from Zos Canyon, Mongolia. American Museum Novitates 3445: 1-36.

Pol, D., and J. E. Powell. 2007. Skull anatomy of *Mussaurus patagonicus* (Dinosauria: Sauropodomorpha) from the Late Triassic of Patagonia. Historical Biology 19: 125-144.

Pol, D., and O. W. M. Rauhut. 2012. A Middle Jurassic abelisaurid from Patagonia and the early diversification of theropod dinosaurs. Proceedings of the Royal Society B 279: 3170-3175.

Pol, D., O. W. M. Rauhut, A. Lecuona, J. M. Leardi, X. Xu, and J. M. Clark. 2013. A new fossil from the Jurassic of Patagonia reveals the early basicranial evolution and the origins of Crocodyliformes. Biological Reviews 88: 862-872.

Pol, D., A. H. Turner, and M. A. Norell. 2009. Morphology of the Late Cretaceous crocodylomorph *Shamosuchus djadochtensis* and a discussion of neosuchian phylogeny as related to the origin of Eusuchia. Bulletin of the American Museum of Natural History 324: 1-103.

Polcyn, M. J., and G. L. Bell, Jr. 2005. *Russellosaurus coheni* n. gen., n. sp., a 92 million-year-old mosasaur from Texas (USA), and the definition of the parafamily Russellosaurina. Netherlands Journal of Geosciences 84: 321-333.

Poropat, S. F., L. Kool, P. Vickers-Rich, and T. H. Rich. 2016. Oldest meiolaniid turtle remains from Australia: evidence from the Eocene Kerosene Creek Member of the Rundle Formation, Queensland. Alcheringa 41: 231-239.

Porro, L. B., L. M. Witmer, and P. M. Barrett. 2015. Digital preparation and osteology of the skull of *Lesothosaurus diagnosticus* (Ornithischia: Dinosauria). PeerJ 3: e1494. doi:10.7717/peerj.1494.

Pregill, G. K., J. A. Gauthier, and H. W. Greene. 1986. The evolution of helodermatid squamates, with description of a new taxon and an overview of Varanoidea. Transactions of the San Diego Society of Natural History 21: 167-202.

Presch, W. 1983. The lizard family Teiidae: is it a monophyletic group? Zoological Journal of the Linnean Society 77: 189-197.

Price, L. I. 1964. Sôbre o crânio de um grande crocodilídeo extinto do alto Rio Juruá, Estado do Acre. Anais da Academia Brasileira de Ciências 36: 59-66.

Prieto-Márquez, A. 2010a. Global phylogeny of Hadrosauridae (Dinosauria: Ornithopoda) using parsimony and Bayesian methods. Zoological Journal of the Linnean Society 159: 435-502.

Prieto-Márquez, A. 2010b. Global historical biogeography of hadrosaurid dinosaurs. Zoological Journal of the Linnean Society 159: 503-525.

Prieto-Márquez, A., G. M. Erickson, and J. A. Ebersole. 2016. A primitive hadrosaurid from southeastern North America and the origin and early evolution of 'duck-billed' dinosaurs. Journal of Vertebrate Paleontology 36: e1054495. doi:10.1080/02724634.2015.1054495.

Prieto-Márquez, A., and M. A. Norell. 2011. Redescription of a nearly complete skull of *Plateosaurus* (Dinosauria: Sauropodomorpha) from the Late Triassic of Trossingen (Germany). American Museum Novitates 3727: 1-58.

Prieto-Márquez, A., and and J. R. Wagner. 2014. Soft-tissue structures of the nasal vestibular region of saurolophine hadrosaurids (Dinosauria, Ornithopoda) revealed in a 'mummified' specimen of *Edmontosaurus annectens*. Pp. 591-599 in D. Eberth and D. C. Evans (eds.), Hadrosaurs. Indiana University Press, Bloomington.

Pritchard, A. C., J. A. McCartney, D. W. Krause, and N. J. Kley. 2014. New snakes from the Upper Cretaceous (Maastrichtian) Maevarano Formation, Mahajanga Basin, Madagascar. Journal of Vertebrate Paleontology 34: 1080-1093.

Pritchard, A. C., and S. J. Nesbitt. 2017. A bird-like skull in a Triassic diapsid reptile increases heterogeneity of the morphological and phylogenetic radiation of Diapsida. Royal Society Open Science 4: 170499. doi:10.1098/rsos.170499.

Pritchard, A. C., A. H. Turner, E. R. Allen, and M. A. Norell. 2013. Osteology of a North American goniopholidid (*Eutretauranosuchus delfsi*) and palate evolution in Neosuchia. American Museum Novitates 3783: 1-56.

Pritchard, A. C., A. H. Turner, R. B. Irmis, S. J. Nesbitt, and N. D. Smith. 2016. Extreme modification of the tetrapod forelimb in a Triassic diapsid reptile. Current Biology 26: P2779-P2786.

Pritchard, A. C., A. H. Turner, S. J. Nesbitt, R. B. Irmis, and N. D. Smith. 2015. Late Triassic tanystropheids (Reptilia, Archosauromorpha) from northern New Mexico (Petrified Forest Member, Chinle Formation) and the biogeography, functional morphology, and evolution of Tanystropheidae. Journal of Vertebrate Paleontology 35: e911186. doi:10.1080/02724634.2014.911186.

Pu, H., Y. Kobayashi, J. Lü, L. Xu, Y. Wu, H. Chang, J. Zhang, and S. Jia. 2013. An unusual basal therizinosaur dinosaur with an ornithischian dental arrangement from northeastern China. PLoS ONE 8(5): e63423. doi:10.1371/journal.pone.0063423.

Puértolas-Pascual, E., J. I. Canudo, and M. Moreno-Azanza. 2014. The eusuchian crocodylomorph *Allodaposuchus subjuniperus* sp. nov., a new species from the latest Cretaceous (upper Maastrichtian) of Spain. Historical Biology 26: 91-109.

Pyron, R. A., F. T. Burbrink, and J. J. Wiens. 2013. A phylogeny and revised classification of Squamata, including 4161 species of lizards and snakes. BMC Evolutionary Biology 13: 93. doi:10.1186/1471-2148-13-93.

Rabi, M., V. B. Sukhanov, V. N. Egorova, I. Danilov, and W. G. Joyce. 2014. Osteology, relationships, and ecology of *Annemys* (Testudines, Eucryptodira) from the Late Jurassic of Shar Teg, Mongolia, and phylogenetic definitions for Xinjiangchelyidae, Sinemydidae, and Macrobaenidae. Journal of Vertebrate Paleontology 34: 327-352.

Rage, J.-C. 1977. An erycine snake (Boidae) of the genus *Calamagras* from the French lower Eocene, with comments on the phylogeny of the Erycinae. Herpetologica 33: 459-463.

Rage, J.-C. 1982. La phylogénie des Lépidosauriens: une approche cladistique. Comptes Rendus de l'Académie des Sciences Paris, sér. II, 294: 563-566.

Rage, J.-C. 1984. Serpentes. Pp. 1-80 in P. Wellnhofer (ed.), Handbuch der Paläoherpetologie, Teil 11. Gustav Fischer Verlag, Stuttgart.

Rage, J.-C. 2001. Fossil snakes from the Palaeocene of São José de Itaboraí, Brazil. Part II. Boidae. Palaeovertebrata 30: 111-150.

Rage, J.-C. 2013. Mesozoic and Cenozoic squamates of Europe. Palaeobiodiversity and Palaeoenvironments 93: 517-534.

Rage, J.-C., and M. L. Augé. 2010. Squamate reptiles from the middle Eocene of Lissieu (France). A landmark in the middle Eocene of Europe. Geobios 43: 253-268.

Rage, J.-C., S. Bajpai, J. G. M. Thewissen, and B. N. Tiwari. 2003. Early Eocene snakes from Kutch, Western India, with a review of the Palaeophiidae. Geodiversitas 25: 695-716.

Rage, J.-C., and F. Escuillié. 2000. Un nouveau serpent bipède du Cénomanien (Crétacé). Implications phylétiques. Comptes Rendus de l'Académie des Sciences Paris, Sciences de la Terre et des planètes 330: 513-520.

Rage, J.-C., A. Folie, R. S. Rana, H. Singh, K. D. Rose, and T. Smith. 2008. A diverse snake fauna from the Early Eocene of Vastan Lignite Mine, Gujarat, India. Acta Palaeontologica Polonica 53: 391-403.

Rage, J.-C., G. Métais, A. Bartolini, I. A. Brohi, R. A. Lashari, L. Marivaux, D. Merle, and S. H. Solangi. 2014. First report of the giant snake *Gigantophis* (Madtsoiidae) from the Paleocene of Pakistan: paleobiogeographic implications. Geobios 47: 147-153.

Rage, J.-C., and Z. Szyndlar. 1994. Latest Oligocene-early Miocene in Europe: Dark Period for booid snakes. Comptes Rendus Palevol 4: 428-435.

Rage, J.-C., R. Vuillo, and D. Néraudeau. 2016. The mid-Cretaceous snake *Simoliophis rochebrunei* Sauvage, 1880 (Squamata: Ophidia) from its type area (Charentes, southwestern France): redescription, distribution, and palaeoecology. Cretaceous Research 58: 234-253.

Rana, R. S., M. Augé, A. Folie, K. D. Rose, K. Kumar, L. Singh, A. Sahni, and T. Smith. 2013. High diversity of acrodontan lizards in the Early Eocene Vastan Lignite Mine of India. Geologica Belgica 16: 290-301.

Rauhut, O. W. M. 2003. The interrelationships and evolution of basal theropod dinosaurs. Special Papers in Palaeontology 69: 1-213.

Rauhut, O. W. M., and M. T. Carrano. 2016. The theropod dinosaur *Elaphrosaurus bambergi* Janensch, 1920, from the Late Jurassic of Tendaguru, Tanzania. Zoological Journal of the Linnean Society 178: 546-610.

Rauhut, O. W. M., C. Foth, H. Tischlinger, and M. A. Norell. 2012b. Exceptionally preserved juvenile megalosauroid theropod dinosaur with filamentous integument from the Late Jurassic of Germany. Proceedings of the National Academy of Sciences of the United States of America 109: 11746-11751.

Rauhut, O. W. M., A. M. Heyng, A. López-Arbarello, and A. Hecker. 2012a. A new rhynchocephalian from the Late Jurassic of Germany with a dentition that is unique amongst tetrapods. PLoS ONE 7(10): e46839. doi:10.1371/journal.pone.0046839.

Rauhut, O. W. M., T. R. Hübner, and K.-P. Lanser. 2016. A new megalosaurid theropod dinosaur from the late Middle Jurassic (Callovian) of north-western Germany: implications for theropod evolution and faunal turnover in the Jurassic. Palaeontologia Electronica 19.2.26A: 1-65. doi:10.26879/654.

Rauhut, O. W. M., A. C. Milner, and S. Moore-Fay. 2010. Cranial osteology and phylogenetic position of the theropod dinosaur *Proceratosaurus bradleyi* (Woodward, 1910) from the Middle Jurassic of England. Zoological Journal of the Linnean Society 158: 155-195.

Rauhut, O. W. M., K. Remes, R. Fechner, G. Cladera, and P. Puerta. 2005. Discovery of a short-necked sauropod dinosaur from the Late Jurassic period of Patagonia. Nature 435: 670-672.

Rauhut, O. W. M., and X. Xu. 2005. The small theropod dinosaurs *Tugulusaurus* and *Phaedrolosaurus* from the Early Cretaceous of Xinjiang, China. Journal of Vertebrate Paleontology 25: 107-118.

Rayfield, E. J., A. C. Milner, V. B. Xuan, and P. G. Young. 2007. Functional morphology of spinosaur 'crocodile-mimic' dinosaurs. Journal of Vertebrate Paleontology 27: 892-901.

Reeder, T. W., T. M. Townsend, D. G. Mulcahy, B. P. Noonan, P. L. Wood Jr., J. W. Sites, Jr., and J. J. Wiens. 2015. Integrated analyses resolve conflicts over squamate reptile phylogeny and reveal unexpected placements for fossil taxa. PLoS ONE 10(3): e0118199. doi:10.1371/journal.pone.0118199.

Reisz, R. R. 1981. A diapsid reptile from the Pennsylvanian of Kansas. Special Publications, Museum of Natural History, The University of Kansas 7: 1-74.

Reisz, R. R. 1997. The origin and early evolutionary history of amniotes. Trends in Ecology and Evolution 12: 218-222.

Reisz, R. R. 2007. The cranial anatomy of basal diadectomorphs and the origin of amniotes. Pp. 228-252 in J. S. Anderson and H.-D. Sues (eds.), Major Transitions in Vertebrate Evolution. Indiana University Press, Bloomington.

Reisz, R. R., D. S. Berman, and D. Scott. 1984. The anatomy and relationships of the Lower Permian reptile *Araeoscelis*. Journal of Vertebrate Paleontology 4: 57-67.

Reisz, R. R., D. C. Evans, H.-D. Sues, and D. Scott. 2010. Embryonic skeletal anatomy of the sauropodomorph dinosaur *Massospondylus* from the Lower Jurassic of South Africa. Journal of Vertebrate Paleontology 30: 1653-1665.

Reisz, R. R., and J. Fröbisch. 2014. The oldest caseid synapsid from the Late Pennsylvanian of Kansas, and the evolution of herbivory in terrestrial vertebrates. PLoS ONE 9(4): e94518. doi:10.1371/journal.pone.0094518.

Reisz, R. R., and M. Laurin. 1991. *Owenetta* and the origin of turtles. Nature 349: 324-326.

Reisz, R. R., M. J. MacDougall, and S. P. Modesto. 2014. A new species of the parareptile genus *Delorhynchus*, based on articulated skeletal remains from Richards Spur, Lower Permian of Oklahoma. Journal of Vertebrate Paleontology 34: 1033-1043.

Reisz, R. R., S. P. Modesto, and D. M. Scott. 2011. A new Early Permian reptile and its significance in early diapsid evolution. Proceedings of the Royal Society B 278: 3731-3737.

Reisz, R.R., J. Müller, L. Tsuji, and D. Scott. 2007. The cranial osteology of *Belebey vegrandis* (Parareptilia: Bolosauridae), from the Middle Permian of Russia, and its bearing on reptilian evolution. Zoological Journal of the Linnean Society 151: 191-214.

Reisz, R. R., and D. Scott. 2002. *Owenetta kitchingorum*, sp. nov., a small parareptile (Procolophonia: Owenettidae) from the Lower Triassic of South Africa. Journal of Vertebrate Paleontology 22: 244-256.

Reisz, R. R., and H.-D. Sues. 2000. Herbivory in late Paleozoic and Triassic tetrapods. Pp. 9-41 in H.-D. Sues (ed.), Evolution of Herbivory in Terrestrial Vertebrates: Perspectives from the Fossil Record. Cambridge University Press, New York.

Remes, K., F. Ortega, I. Fierro, U. Joger, R. Kosma, J. M. Marín Ferrer, O. A. Ide, and A. Maga. 2009. A new basal sauropod dinosaur from the Middle Jurassic of Niger and the early evolution of Sauropoda. PLoS ONE 4(9): e6924. doi:10.1371/journal.pone.0006924.

Renesto, S., and M. Bernardi. 2014. Redescription and phylogenetic relationships of *Megachirella wachtleri* Renesto et Posenato, 2003 (Reptilia, Diapsida). Paläontologische Zeitschrift 88: 197-210.

Renesto, S., and F. M. Dalla Vecchia. 2000. The unusual dentition and feeding habits of the prolacertiform reptile *Langobardisaurus* (Late Triassic, northern Italy). Journal of Vertebrate Paleontology 20: 622-627.

Renesto, S., J. A. Spielmann, S. G. Lucas, and G. T. Spagnoli. 2010. The taxonomy and paleobiology of the Late Triassic (Carnian-Norian: Adamanian-Apachean) drepanosaurs (Diapsida: Archosauromorpha: Drepanosauromorpha). New Mexico Museum of Natural History & Science Bulletin 46: 1-81.

Reynolds, S. H. 1913. The Vertebrate Skeleton. Cambridge University Press, Cambridge, UK.

Reynoso, V.-H. 1996. A Middle Jurassic *Sphenodon*-like sphenodontian (Diapsida: Lepidosauria) from Huizachal Canyon, Tamaulipas, Mexico. Journal of Vertebrate Paleontology 16: 210-221.

Reynoso, V.-H. 1998. *Huehuecuetzpalli mixtecus* gen. et sp. nov., a basal squamate (Reptilia) from the Early Cretaceous of Tepexi de Rodríguez, Central México. Philosophical Transactions of the Royal Society of London B 353: 477-500.

Reynoso, V.-H. 2000. An unusual aquatic sphenodontian (Reptilia: Diapsida) from the Tlayua Formation (Albian), central Mexico. Journal of Paleontology 74: 133-148.

Reynoso, V.-H. 2005. Possible evidence of a venom apparatus in a Middle Jurassic sphenodontian from the Huizachal red beds of Tamaulipas, México. Journal of Vertebrate Paleontology 25: 646-654.

Reynoso, V.-H., and G. Callison. 2000. A new scincomorph lizard from the Early Cretaceous of Puebla, México. Zoological Journal of the Linnean Society 130: 183-212.

Reynoso, V.-H., and J. M. Clark. 1998. A dwarf sphenodontian from the Jurassic La Boca Formation of Tamaulipas, México. Journal of Vertebrate Paleontology 18: 333-339.

Rich, T. H., P. Vickers-Rich, O. Giménez, R. Cúneo, P. Puerta, and R. Vacca. 1999. A new sauropod dinosaur from Chubut Province, Argentina. Pp. 61-84 in Y. Tomida, T. H. Rich, and P. Vickers-Rich (eds.), Proceedings of the Second Gondwanan Dinosaur Symposium. National Science Museum Monographs 15. National Science Museum, Tokyo.

Richter, A. 1994. Lacertilia aus der Unteren Kreide von Uña und Galve (Spanien) und Anoual (Marokko). Berliner Geowissenschaftliche Abhandlungen (E) 14: 1-147.

Rieppel, O. 1977. Studies on the skull of the Henophidia (Reptilia: Serpentes). Journal of Zoology, London 181: 145-173.

Rieppel, O. 1980a. The phylogeny of anguimorph lizards. Denkschriften der Schweizerischen Naturforschenden Gesellschaft 94: 1-86.

Rieppel, O. 1980b. The evolution of the ophidian feeding system. Zoologisches Jahrbuch, Abteilung Anatomie und Ontogenie der Tiere 103: 551-564.

Rieppel, O. 1987a. *Clarazia* and *Hescheleria*: a re-investigation of two problematic reptiles from the Middle Triassic of Monte San Giorgio, Switzerland. Palaeontographica A 195: 101-129.

Rieppel, O. 1987b. The phylogenetic relationships within the Chameleonidae, with comments on some aspects of cladistic analysis. Zoological Journal of the Linnean Society 89: 41-62.

Rieppel, O. 1988. A review of the origin of snakes. Pp. 37-130 in M. K. Hecht, B. Wallace, and G. T. Prance (eds.), Evolutionary Biology. Volume 22. Plenum, New York.

Rieppel, O. 1989a. A new pachypleurosaur (Reptilia: Sauropterygia) from the Middle Triassic of Monte San Giorgio, Switzerland. Philosophical Transactions of the Royal Society of London B 323: 1-73.

Rieppel, O. 1989b. The hind limb of *Macrocnemus bassanii* (Nopcsa) (Reptilia, Diapsida): development and functional anatomy. Journal of Vertebrate Paleontology 9: 373-387.

Rieppel, O. 1989c. *Helveticosaurus zollingeri* Peyer (Reptilia, Diapsida): skeletal paedomorphosis, functional anatomy and systematic affinities. Palaeontographica A 208:123-152.

Rieppel, O. 1993. Euryapsid relationships: a preliminary analysis. Neues Jahrbuch für Geologie und Paläontologie Abhandlungen 188: 241-264.

Rieppel, O. 1994. Osteology of *Simosaurus gaillardoti* and the relationships of stem-group Sauropterygia. Fieldiana Geology, n. s. 28: 1-85.

Rieppel, O. 1995. The genus *Placodus*: systematics, morphology, paleobiogeography and paleobiology. Fieldiana Geology, n. s. 31: 1-44.

Rieppel, O. 1998. *Corosaurus alcovensis* Case and the phylogenetic interrelationships of Triassic stem-group Sauropterygia. Zoological Journal of the Linnean Society 124: 1-41.

Rieppel, O. 2000a. Turtles as diapsid reptiles. Zoologica Scripta 29: 199-212.

Rieppel, O. 2000b. Sauropterygia I: Placodontia, Pachypleurosauria, Nothosauroidea, Pistosauroidea. Pp. 1-134 in P. Wellnhofer (ed.), Handbuch der Paläoherpetologie, Teil 12A. Verlag Dr. Friedrich Pfeil, Munich.

Rieppel, O. 2000c. *Paraplacodus* and the phylogeny of the Placodontia (Reptilia: Sauropterygia). Zoological Journal of the Linnean Society 130: 635-659.

Rieppel, O. 2001. The cranial anatomy of *Placochelys placodonta* Jaekel, 1902, and a review of the Cyamodontoidea (Reptilia, Placodontia). Fieldiana Geology, n. s. 45: 1-104.

Rieppel, O. 2002. Feeding mechanics in Triassic stem-group sauropterygians: the anatomy of a successful invasion of Mesozoic seas. Zoological Journal of the Linnean Society 135: 33-63.

Rieppel, O., and M. deBraga. 1996. Turtles as diapsid reptiles. Nature 384: 453-455.

Rieppel, O., J. Gauthier, and J. Maisano. 2008. Comparative morphology of the dermal palate in squamate reptiles, with comments on phylogenetic implications. Zoological Journal of the Linnean Society 152: 131-152.

Rieppel, O., and L. Grande. 2007. The anatomy of the fossil varanid lizard *Saniwa ensidens* Leidy, 1870, based on a newly discovered complete skeleton. Journal of Paleontology 81: 643-665.

Rieppel, O., and R. W. Gronowski. 1981. The loss of the lower temporal arcade in diapsid reptiles. Zoological Journal of the Linnean Society 72: 203-217.

Rieppel, O., and J. J. Head. 2004. New specimens of the fossil snake genus *Eupodophis* Rage & Escuillié, from Cenomanian (Late Cretaceous) of Lebanon. Memorie della Società Italiana di Scienze Naturali e del Museo Civico di Storia Naturale di Milano 32(2): 1-26.

Rieppel, O., N. J. Kley, and J. A. Maisano. 2009. Morphology of the skull of the white-nosed blindsnake, *Liotyphlops albirostris* (Scolecophidia: Anomalepididae). Journal of Morphology 270: 536-557.

Rieppel, O., A. G. Kluge, and H. Zaher. 2002. Testing the phylogenetic relationships of the Pleistocene snake *Wonambi naracoortensis* Smith. Journal of Vertebrate Paleontology 22: 812-829.

Rieppel, O., C. Li, and N. C. Fraser. 2008. The skeletal anatomy of the Triassic protorosaur *Dinocephalosaurus orientalis* Li, from the Middle Triassic of Guizhou Province, China. Journal of Vertebrate Paleontology 28: 95-110.

Rieppel, O., and R. R. Reisz. 1999. The origin and early evolution of turtles. Annual Review of Ecology and Systematics 30: 1-22.

Rieppel, O., P. M. Sander, and G. W. Storrs. 2002. The skull of the pistosaur *Augustasaurus* from the Middle Triassic of northwestern Nevada. Journal of Vertebrate Paleontology 22: 577-592.

Rieppel, O., and R. Wild. 1996. A revision of the genus *Nothosaurus* (Reptilia: Sauropterygia) from the Germanic Triassic, with comments on the status of *Conchiosaurus clavatus*. Fieldiana Geology, n. s. 34: 1-82.

Rieppel, O., and H. Zaher. 2000a. The braincases of mosasaurs and *Varanus*, and the relationships of snakes. Zoological Journal of the Linnean Society 129: 489-514.

Rieppel, O., and H. Zaher. 2000b. The intramandibular joint in squamates and the phylogenetic relationships of the fossil snake *Pachyrhachis problematicus* Haas. Fieldiana Geology 43: 1-69.

Rieppel, O., H. Zaher, E. Tchernov, and M. J. Polcyn. 2003. The anatomy and relationships of *Haasiophis terrasanctus*, a fossil snake with well-developed hind limbs from the Mid-Cretaceous of the Middle East. Journal of Paleontology 77: 536-558.

Riff, D., and O. A. Aguilera. 2008. The world's largest gharials *Gryposuchus*: description of *G. croizati* n. sp. (Crocodylia, Gavialidae) from the Upper Miocene Urumaco Formation, Venezuela. Paläontologische Zeitschrift 82: 178-195.

Riggs, E. S. 1904. Structure and relationships of opisthocoelian dinosaurs. Part II: The Brachiosauridae. Field Columbian Museum Publications, Geological Series 2: 229-247.

Robinson, J. A. 1975. The locomotion of plesiosaurs. Neues Jahrbuch für Geologie und Paläontologie Abhandlungen 149: 286-332.

Robinson, J. A. 1977. Intracorporal force transmission in plesiosaurs. Neues Jahrbuch für Geologie und Paläontologie Abhandlungen 153: 86-128.

Robinson, P. L. 1962. Gliding lizards from the Upper Keuper of Great Britain. Proceedings of the Geological Society of London 1601: 137-146.

Robinson, P. L. 1967. The evolution of the Lacertilia. Pp. 395-407 in Problèmes Actuels de Paléontologie (Évolution des Vertébrés). Colloques Internationaux du Centre National de la Recherche Scientifique, no. 163. Éditions du Centre National de la Recherche Scientifique, Paris.

Robinson, P. L. 1973. A problematic reptile from the British Upper Trias. Journal of the Geological Society of London 129: 457-479.

Robinson, P. L. 1975. The function of the hooked fifth metatarsal in lepidosaurian reptiles. Pp. 461-483 in Problèmes Actuels de Paléontologie—Évolution des Vertébrés. Colloques Internationaux du Centre National de la Recherche Scientifique, no. 218. Éditions du Centre National de la Recherche Scientifique, Paris.

Rogers, J. V. II. 2003. *Pachycheilosuchus trinquei*, a new procoelous crocodyliform from the Lower Cretaceous (Albian) Glen Rose Formation of Texas. Journal of Vertebrate Paleontology 23: 128-145.

Romano, P. S. R., V. Gallo, R. R. C. Ramos, and L. Antonioli. 2014. *Atolchelys lepida*, a new side-necked turtle from the Early Cretaceous of Brazil and the age of crown Pleurodira. Biology Letters 10: 20140290. doi:10.1098/rsbl.2014.0290.

Romer, A. S. 1927. The pelvic musculature of ornithischian dinosaurs. Acta Zoologica (Stockholm) 8: 225-275.

Romer, A. S. 1956. Osteology of the Reptiles. The University of Chicago Press, Chicago.

Romer, A. S. 1966. Vertebrate Paleontology. Third Edition. The University of Chicago Press, Chicago.

Romer, A. S. 1971. Unorthodoxies in reptilian phylogeny. Evolution 25: 103-112.

Romer, A. S. 1972. The Chañares (Argentina) Triassic reptile fauna. XIII. An early ornithosuchid pseudosuchian, *Gracilisuchus stipanicicorum*, gen. et sp. nov. Breviora 389: 1-24.

Romer, A. S., and L. I. Price. 1939. The oldest vertebrate egg. American Journal of Science 237: 826-829.

Ross, J. P. (ed.). 1998. Crocodiles. Status Survey and Conservation Action Plan. Second Edition. IUCN/SSC Crocodile Specialist Group. IUCN, Gland, Switzerland.

Rossmann, T. 2000. Skelettanatomische Beschreibung von *Pristichampsus rollinatii* (Gray) (Crocodilia, Eusuchia) aus dem Paläogen von Europa, Nordamerika und Ostasien. Courier Forschungsinstitut Senckenberg 221: 1-107.

Rowe, T., C. A. Brochu, and K. Kishi (eds.). 1999. Cranial morphology of *Alligator mississippiensis* and phylogeny of Alligatoroidea. Society of Vertebrate Paleontology Memoir 6: 1-100.

Royo-Torres, R., A. Cobos, and L. Alcalá. 2006. A giant European dinosaur and a new sauropod clade. Science 314: 1925-1927.

Royo-Torres, R., P. Upchurch, J. I. Kirkland, D. D. De Blieux, J. R. Foster, A. Cobos, and L. Alcalá. 2017. Descendants of the Jurassic turiasaurs from Iberia found refuge in the Early Cretaceous of western USA. Scientific Reports 7: 14311. doi:10.1038/s41598-017-14677-2.

Ruddiman, W. F. 2014. The Anthropocene. Annual Review of Earth and Planetary Sciences 41: 45-68.

Russell, A. P., and A. M. Bauer 2008. The appendicular locomotor apparatus in *Sphenodon* and normal-limbed squamates. Pp. 1-465 in C. Gans, A. S. Gaunt, and K. Adler (eds.), Biology of the Reptilia. Volume 21: The Skull and Appendicular Locomotor Apparatus of Lepidosauria. Society for the Study of Amphibians and Reptiles, Ithaca, New York.

Russell, A. P., L. D. Dijkstra, and G. L. Powell. 2001. Structural characteristics of the patagium of *Ptychozoon kuhli* (Reptilia: Gekkonidae) in relation to parachuting locomotion. Journal of Morphology 247: 252-263.

Russell, D. A. 1967. Systematics and morphology of American mosasaurs (Reptilia, Sauria). Bulletin of the Peabody Museum of Natural History, Yale University 23: 1-237.

Russell, D. A., and T. P. Chamney. 1967. Notes on the biostratigraphy of dinosaurian and microfossil faunas in the Edmonton Formation (Cretaceous), Alberta. National Museum of Canada Natural History Papers 35: 1-22.

Russell-Sigogneau, D., and D. E. Russell. 1978. Étude ostéologique du reptile *Simoedosaurus* (Choristodera). Annales de Paléontologie (Vertébrés) 64: 1-84.

Ruta, M., M. I. Coates, and D. L. J. Quicke. 2003. Early tetrapod relationships revisited. Biological Reviews 78: 251-345.

Ryan, M. J., R. Holmes, and A. P. Russell. 2007. A revision of the late Campanian centrosaurine ceratopsid genus *Styracosaurus* from the Western Interior of North

America. Journal of Vertebrate Paleontology 27: 944-962.

Ryan, M. J., A. P. Russell, D. A. Eberth, and P. J. Currie. 2001. The taphonomy of a *Centrosaurus* (Ornithischia: Ceratopsidae) bone bed from the Dinosaur Park Formation (Upper Campanian), Alberta, Canada, with comments on cranial ontogeny. Palaios 16: 482-506.

Rybczynski, N., and M. K. Vickaryous. 2001. Evidence of complex jaw movement in the Late Cretaceous ankylosaurid, *Euoplocephalus tutus*. Pp. 299-317 in K. Carpenter (ed.), The Armored Dinosaurs. Indiana University Press, Bloomington.

Sadleir, R., P. M. Barrett, and H. P. Powell. 2008. The anatomy and systematics of *Eustreptospondylus oxoniensis*, a theropod dinosaur from the Middle Jurassic of Oxfordshire, England. Palaeontographical Society Publication 627: 1-82.

Sahney, S., M. J. Benton, and H. Falcon-Lang. 2010. Rainforest collapse triggered Caboniferous tetrapod diversification in Euramerica. Geology 38:1079-1082.

Säilä, L. K. 2010a. The phylogenetic position of *Nyctiphruretus acudens*, a parareptile from the Permian of Russia. Journal of Iberian Geology 36: 123-143.

Säilä, L. K. 2010b. Osteology of *Leptopleuron lacertinum* Owen, a procolophonoid parareptile from the Upper Triassic of Scotland, with remarks on ontogeny, ecology and affinities. Earth and Environmental Science Transactions of the Royal Society of Edinburgh 101: 1-25.

Salgado, L., and J. F. Bonaparte. 1991. Un nuevo sauropodo Dicraeosauridae, *Amargasaurus cazaui* gen. et sp. nov., de la Provincia del Neuquén, Argentina. Ameghiniana 28: 333-346.

Salgado, L., R. A. Coria, and J. O. Calvo. 1997. Evolution of titanosaurid sauropods. I: Phylogenetic analysis based on the postcranial evidence. Ameghiniana 34: 3-32.

Salgado, L., A. Garrido, S. Cocca, and J. R. Cocca. 2004. Lower Cretaceous rebbachisaurids from Cerro Aguada del León (Lohan Cura Formation), Neuquén Province, northwestern Patagonia, Argentina. Journal of Vertebrate Paleontology 24: 903-912.

Salisbury, S. W., E. Frey, D. M. Martill, and M.-C. Buchy. 2003. A new crocodilian from the Lower Cretaceous Crato Formation of north-eastern Brazil. Palaeontographica A 270: 3-47.

Salisbury, S. W., R. E. Molnar, E. Frey, and P. M. A. Willis. 2006. The origin of modern crocodyliforms: new evidence from the Cretaceous of Australia. Proceedings of the Royal Society B 273: 2439-2448.

Salisbury, S. W., and P. M. A. Willis. 1996. A new crocodylian from the Early Eocene of southeastern Queensland and a preliminary investigation of the phylogenetic relationships of crocodyloids. Alcheringa 20: 179-226.

Sampson, S. D. 1995. Two new horned dinosaurs from the Upper Cretaceous Two Medicine Formation of Montana, with a phylogenetic analysis of the Centrosaurinae (Ornithischia: Ceratopsidae). Journal of Vertebrate Paleontology 15: 743-760.

Sampson, S. D., and D. W. Krause (eds.). 2007. *Majungasaurus crenatissimus* (Theropoda: Abelisauridae) from the Late Cretaceous of Madagascar. Society of Vertebrate Paleontology Memoir 8: 1-184.

Sampson, S. D., and M. A. Loewen. 2010. Unraveling a radiation: a review of the diversity, stratigraphic distribution, biogeography, and evolution of horned dinosaurs (Ornithischia: Ceratopsidae). Pp. 405-427 in M. J. Ryan, B. J. Chinnery-Allgeier, and D. A. Eberth (eds.), New Perspectives on Horned Dinosaurs. Indiana University Press, Bloomington.

Sampson, S. D., M. A. Loewen, A. A. Farke, E. M. Roberts, C. A. Forster, J. A. Smith, and A. L. Titus. 2010. New horned dinosaurs from Utah provide evidence for intracontinental dinosaur endemism. PLoS ONE 5(9): e12292. doi:10.1371/journal.pone.0012292.

Sampson, S. D., M. J. Ryan, and D. H. Tanke. 1997. Craniofacial ontogeny in centrosaurine dinosaurs (Ornithischia: Ceratopsidae): taxonomic and behavioral implications. Zoological Journal of the Linnean Society 121: 293-337.

Sander, P. M. 1989. The pachypleurosaurids (Reptilia: Nothosauria) from the Middle Triassic of Monte San Giorgio (Switzerland) with the description of a new species. Philosophical Transactions of the Royal Society of London B 325: 561-666.

Sander, P. M. 2000. Ichthyosauria: their diversity, distribution and phylogeny. Paläontologische Zeitschrift 74: 1-35.

Sander, P. M., X. Chen, L. Cheng, and X. Wang. 2011. Short-snouted toothless ichthyosaur from China suggests Late Triassic diversification of suction feeding ichthyosaurs. PLoS ONE 6(5): e19480. doi:10.1371/journal.pone.0019480.

Sander, P. M., O. Mateus, T. Laven, and N. Knötschke. 2006. Bone histology indicates insular dwarfism in a new Late Jurassic sauropod dinosaur. Nature 441: 739-741.

Sander, P. M., O. Rieppel, and H. Bucher. 1997. A new pistosaurid (Reptilia, Sauropterygia) from the Middle Triassic of Nevada and its implications for the origin of the plesiosaurs. Journal of Vertebrate Paleontology 17: 526-533.

Sander, P. M., et al. 2011. Biology of the sauropod dinosaurs: the evolution of gigantism. Biological Reviews 86: 117-155. [16 co-authors]

Sanders, K. L., M. S. Y. Lee, R. Leijs, R. Foster, and J. S. Keogh. 2008. Molecular phylogeny and divergence dates for Australasian elapids and sea snakes (Hydrophiinae): evidence from seven genes for rapid evolutionary radiations. Journal of Evolutionary Biology 21: 682-695.

Sanders, K. L., M. S. Y. Lee, Mumpuni, T. Bertozzi, and A. R. Rasmussen. 2011. Multilocus phylogeny and recent rapid radiation of the viviparous sea snakes (Elapidae: Hydrophiinae). Molecular Phylogenetics and Evolution 66: 575-591.

Sanders, K. L., Mumpuni, A. Hamidy, J. J. Head, and D. J. Gower. 2010. Phylogeny and divergence times of filesnakes (*Acrochordus*): inferences from morphology, fossils and three molecular loci. Molecular Phylogenetics and Evolution 56: 857-867.

Sasaki, K., D. Lesbarrères, C. T. Beaulieu, G. Watson, and J. Litzgus. 2016. Effects of a mining-altered environment on individual fitness of amphibians and reptiles. Ecosphere 7(6): e01360. doi:10.1002/ecs2.1360.

Sato, T., and X.-c. Wu. 2008. A new Jurassic pliosaur from Melville Island, Canadian Arctic Archipelago. Canadian Journal of Earth Sciences 45: 303-320.

Sato, T., L.-j. Zhao, X.-c. Wu, and C. Li. 2014. A new specimen of the Triassic pistosauroid *Yunguisaurus*, with implications for the origin of Plesiosauria (Reptilia, Sauropterygia). Palaeontology 57: 55-76.

Savitzky, A. H. 1980. The role of venom-delivery strategies in snake evolution. Evolution 34: 1194-1204.

Scanferla, A., K. T. Smith, and S. F. K. Schaal. 2016. Revision of the cranial anatomy and phylogenetic relationships of the Eocene minute boas *Messelophis variatus* and *Messelophis ermannorum* (Serpentes, Booidea). Zoological Journal of the Linnean Society 176: 182-206.

Scanferla, A., H. Zaher, F. E. Novas, C. de Muizon, and R. Céspedes. 2013. A new snake skull from the Paleocene of Bolivia sheds light on the evolution of macrostomatans. PLoS ONE 8(3): e57583. doi:10.1371/journal.pone.0057583.

Scanlon, J. D. 2001. *Montypythonoides*: the Miocene snake *Morelia riversleighensis* (Smith and Plane, 1985) and the geographical origin of pythons. Memoirs of the Association of Australasian Palaeontologists 25: 1-35.

Scanlon, J. D. 2003. The basicranial morphology of madtsoiid snakes (Squamata, Ophidia) and the earliest Alethinophidia (Serpentes). Journal of Vertebrate Paleontology 23: 971-976.

Scanlon, J. D. 2005. Cranial morphology of the Plio-Pleistocene giant madtsoiid snake *Wonambi naracoortensis*. Acta Palaeontologica Polonica 50: 139-180.

Scanlon, J. D. 2006. Skull of the large non-macrostomatan snake *Yurlunggur* from the Australian Oligo-Miocene. Nature 439: 839-842.

Scanlon, J. D., and M. S. Y. Lee. 2011. The major clades of living snakes: morphological evolution, molecular phylogeny, and divergence dates. Pp. 55-95 in R. D. Aldridge and D. M. Sever (eds.), Reproductive Biology and Phylogeny of Snakes. CRC Press, Boca Raton.

Scanlon, J. D., M. S. Y. Lee, and M. Archer. 2003. Mid-Tertiary elapid snakes (Squamata, Colubroidea) from Riversleigh, northern Australia: early steps in a continent-wide adaptive radiation. Geobios 36: 573-601.

Scarpetta, S. 2018. The earliest known occurrence of *Elgaria* (Squamata: Anguidae) and a minimum age for crown Gerrhonotinae: fossils from the Split Rock Formation, Wyoming, USA. Palaeontologia Electronica 21.1.1FC: 1-9. doi:10.26879/837.

Schachner, E. R., R. L. Cieri, J. P. Butler, and C. G. Farmer. 2013b. Unidirectional pulmonary airflow patterns in the savannah monitor lizard. Nature 506: 367-370.

Schachner, E. R., J. R. Hutchinson, and C. G. Farmer. 2013a. Pulmonary anatomy in the Nile crocodile and the evolution of unidirectional airflow in Archosauria. PeerJ 1: e60. doi:10.7717/peerj.60.

Schaumberg, G., D. M. Unwin, and S. Brandt. 2007. New information on the anatomy of the Late Permian gliding reptile *Coelurosauravus*. Paläontologische Zeitschrift 81: 160-173.

Scheyer, T. M. 2010. New interpretation of the postcranial skeleton and overall body shape of *Cyamodus hildegardis* Peyer, 1931 (Reptilia, Sauropterygia). Palaeontologia Electronica 13(2): 15A.

Scheyer, T. M., O. A. Aguilera, M. Delfino, D. C. Fortier, A. A. Carlini, R. Sánchez, J. D. Carrillo-Briceño, L. Quiroz, and M. R. Sánchez-Villagra. 2013. Crocodylian diversity peak and extinction in the late Cenozoic of the northern Neotropics. Nature Communications 4: 1907. doi:10.1038/ncomms2940.

Scheyer, T. M., and M. Delfino. 2016. The late Miocene caimanine fauna (Crocodylia: Alligatoroidea) of the

Urumaco Formation, Venezuela. Palaeontologia Electronica 19.3.48A: 1-57.

Scheyer, T. M., J. M. Neenan, T. Bodogan, H. Furrer, C. Obrist, and M. Plamondon. 2017. A new, exceptionally preserved juvenile specimen of *Eusaurosphargis dalsassoi* (Diapsida) and implications for Mesozoic marine diapsid phylogeny. Scientific Reports 7: 4406. doi:10.1038/s41598-017-04514-x.

Scheyer, T. M., C. Romano, J. Jenks, and H. Bucher. 2014. Early Triassic marine biotic recovery: the predators' perspective. PLoS ONE 9(3): e88987. doi:10.1371/journal.pone.0088987.

Scheyer, T. M., and P. M. Sander. 2002. Histology of ankylosaur osteoderms: implications for systematics and function. Journal of Vertebrate Paleontology 24: 874-893.

Scheyer, T. M., P. M. Sander, W. G. Joyce, W. Böhme, and U. Witzel. 2007. A plywood structure in the shell of fossil and living soft-shelled turtles (Trionychidae) and its evolutionary implications. Organisms Diversity & Evolution 7: 136-144.

Schoch, R. R. 2007. Osteology of the small archosaur *Aetosaurus* from the Upper Triassic of Germany. Neues Jahrbuch für Geologie und Paläontologie Abhandlungen 246: 1-35.

Schoch, R. R. 2015. Reptilien des Lettenkeupers. Pp. 231-264 in H. Hagdorn, R. Schoch, and G. Schweigert (eds.), Der Lettenkeuper—Ein Fenster in die Zeit vor den Dinosauriern. Palaeodiversity Sonderband, Stuttgart and Ingelfingen.

Schoch, R. R., and H.-D. Sues. 2014. A new archosauriform reptile from the Middle Triassic (Ladinian) of Germany. Journal of Systematic Palaeontology 12: 113-131.

Schoch, R. R., and H.-D. Sues. 2015. A Middle Triassic stem-turtle and the evolution of the turtle body plan. Nature 523: 584-587.

Schoch, R. R., and H.-D. Sues. 2018a. Osteology of the Middle Triassic stem-turtle *Pappochelys rosinae* and the early evolution of the turtle skeleton. Journal of Systematic Palaeontology 16: 927-965.

Schoch, R. R., and H.-D. Sues. 2018b. A new lepidosauromorph reptile from the Middle Triassic (Ladinian) of Germany and its phylogenetic relationships. Journal of Vertebrate Paleontology 38: e1444619. doi:10.1080/02724634.2018.1444619.

Schott, R. K., D. C. Evans, M. B. Goodwin, J. R. Horner, C. M. Brown, and N. R. Longrich. 2011. Cranial ontogeny in *Stegoceras validum* (Dinosauria: Pachycephalosauria): a quantitative model of pachycephalosaur dome growth and variation. PLoS ONE 6(6): e21092. doi:10.1371/journal.pone.0021092.

Schulte, J. A. II, J. P. Valladares, and A. Larson. 2003. Phylogenetic relationships within Iguanidae inferred using molecular and morphological data and a phylogenetic taxonomy of iguanian lizards. Herpetologica 59: 399-419.

Schultze, H.-P. 1969. Die Faltenzähne der rhipidistiiden Crossopterygier, der Tetrapoden und der Actinopterygier-Gattung *Lepisosteus*; nebst einer Beschreibung der Zahnstruktur von *Onychodus* (struniiformer Crossopterygier). Palaeontographica Italica 65: 63-137.

Schumacher, G.-H. 1973. The head muscles and hyolaryngeal skeleton of turtles and crocodilians. Pp. 101-199 in C. Gans and T. S. Parsons (eds.), Biology of the Reptilia. Volume 4: Morphology D. Academic Press, London.

Schwarz, D., M. Raddatz, and O. Wings. 2017. *Knoetschkesuchus langenbergensis* gen. nov. sp. nov., a new atoposaurid crocodyliform from the Upper Jurassic Langenberg Quarry (Lower Saxony, northwestern Germany), and its relationships to *Theriosuchus*. PLoS ONE 12(2): e0160617. doi:10.1371/journal.pone.0160617.

Schweitzer, M., J. A. Watt, R. Avci, L. Knapp, L. Chiappe, M. Norell, and M. Marshall. 1999. Beta-keratin specific immunological reactivity in feather-like structures of the Cretaceous alvarezsaurid, *Shuvuuia deserti*. Journal of Experimental Zoology (Molecular and Developmental Evolution) 285: 146-157.

Schwenk, K. 1994. Why snakes have forked tongues. Science 263: 1573-1577.

Schwenk, K. 2000. Feeding in lepidosaurs. Pp. 175-291 in K. Schwenk (ed.), Feeding: Form, Function, and Evolution in Tetrapod Vertebrates. Academic Press, San Diego.

Seeley, H. G. 1887. On the classification of the fossil animals commonly named Dinosauria. Proceedings of the Royal Society of London 43: 165-171.

Sengupta, S., M. D. Ezcurra, and S. Bandyopadhyay. 2017. A new horned and long-necked herbivorous stem-archosaur from the Middle Triassic of India. Scientific Reports 7: 8366. doi:10.1038/s41598-017-08658-8.

Senn, D. G., and R. G. Northcutt. 1973. The forebrain and midbrain of some squamates and their bearing on the origin of snakes. Journal of Morphology 140: 135-152.

Sennikov, A. G. 2011. New tanystropheids (Reptilia: Archosauromorpha) from the Triassic of Europe. Paleontological Journal 45: 90-104.

Senter, P. 2004. Phylogeny of Drepanosauridae (Reptilia: Diapsida). Journal of Systematic Palaeontology 2: 257-268.

Senter, P. 2007. Analysis of forelimb function in basal ceratopsians. Journal of Zoology, London 273: 305-314.

Sereno, P. C. 1986. Phylogeny of the bird-hipped dinosaurs (Order Ornithischia). National Geographic Research 2: 234-256.

Sereno, P. C. 1991a. Basal archosaurs: phylogenetic relationships and functional implications. Society of Vertebrate Paleontology Memoir 2: 1-53.

Sereno, P. C. 1991b. *Lesothosaurus*, "fabrosaurids," and the early evolution of Ornithischia. Journal of Vertebrate Paleontology 11: 168-197.

Sereno, P. C. 1994. The pectoral girdle and forelimb of the basal theropod *Herrerasaurus ischigualastensis*. Journal of Vertebrate Paleontology 13: 425-450.

Sereno, P. C. 1997. The origin and evolution of dinosaurs. Annual Review of Earth and Planetary Sciences 25: 435-489.

Sereno, P. C. 1998. A rationale for phylogenetic definitions, with application to the higher-level taxonomy of Dinosauria. Neues Jahrbuch für Geologie und Paläontologie Abhandlungen 210: 41-83.

Sereno, P. C. 1999. The evolution of dinosaurs. Science 284: 2137-2146.

Sereno, P. C. 2000. The fossil record, systematics and evolution of pachycephalosaurs and ceratopsians from Asia. Pp. 480-516 in M. J. Benton, M. A. Shishkin, D. M. Unwin, and E. N. Kurochkin (eds.), The Age of Dinosaurs in Russia and Mongolia. Cambridge University Press, Cambridge, UK.

Sereno, P. C. 2005. Taxonsearch: Database for Suprageneric Taxa & Phylogenetic Definitions. The University of Chicago, Chicago. www.taxonsearch.org.

Sereno, P. C. 2007. The phylogenetic relationships of early dinosaurs: a comparative report. Historical Biology 19: 145-155.

Sereno, P. C. 2010. Taxonomy, cranial morphology, and relationships of parrot-beaked dinosaurs (Ceratopsia: *Psittacosaurus*). Pp. 21-58 in M. J. Ryan, B. J. Chinnery-Allgeier, and D. A. Eberth (eds.), New Perspectives on Horned Dinosaurs. Indiana University Press, Bloomington.

Sereno, P. C. 2012. Taxonomy, morphology, masticatory function and phylogeny of heterodontosaurid dinosaurs. ZooKeys 226: 1-225.

Sereno, P. C. 2017. Early Cretaceous ornithomimosaurs (Dinosauria: Coelurosauria) from Africa. Ameghiniana 54: 576-616.

Sereno, P. C., and A. B. Arcucci. 1989. The monophyly of crurotarsal archosaurs and the origin of bird and crocodile ankle joints. Neues Jahrbuch für Geologie und Paläontologie Abhandlungen 180: 21-52.

Sereno, P. C., and A. B. Arcucci. 1994a. Dinosaurian precursors from the Middle Triassic of Argentina: *Lagerpeton chanarensis*. Journal of Vertebrate Paleontology 13: 385-399.

Sereno, P. C., and A. B. Arcucci. 1994b. Dinosaurian precursors from the Middle Triassic of Argentina: *Marasuchus lilloensis*, gen. nov. Journal of Vertebrate Paleontology 14: 53-73.

Sereno, P. C., and Z.-m. Dong. 1992. The skull of the basal stegosaur *Huayangosaurus taibaii* and a cladistic diagnosis of Stegosauria. Journal of Vertebrate Paleontology 12: 318-343.

Sereno, P. C., D. B. Dutheuil, M. Iarochene, H. C. E. Larsson, G. H. Lyon, P. M. Magwene, C. A. Sidor, D. J. Varricchio, and J. A. Wilson. 1996. Predatory dinosaurs from the Sahara and Late Cretaceous faunal differentiation. Science 272: 986-991.

Sereno, P. C., C. A. Forster, R. R. Rogers, and A. M. Monetta. 1993. Primitive dinosaur skeleton from Argentina and the early evolution of Dinosauria. Nature 361: 64-66.

Sereno, P. C., and H. C. E. Larsson. 2009. Cretaceous crocodyliforms from the Sahara. ZooKeys 28: 1-143.

Sereno, P. C., H. C. E. Larsson, C. A. Sidor, and B. Gado. 2001. The giant crocodyliform *Sarcosuchus* from the Cretaceous of Africa. Science 294: 1516-1519.

Sereno, P. C., R. N. Martínez, and O. A. Alcober. 2013. Osteology of *Eoraptor lunensis* (Dinosauria, Sauropodomorpha). Society of Vertebrate Paleontology Memoir 12: 83-179.

Sereno, P. C., and F. E. Novas. 1994. The skull and neck of the basal theropod *Herrerasaurus ischigualastensis*. Journal of Vertebrate Paleontology 13: 451-476.

Sereno, P. C., J. A. Wilson, L. M. Witmer, J. A. Whitlock, A. Maga, O. Ide, and T. B. Rowe. 2007. Structural extremes in a Cretaceous dinosaur. PLoS ONE 2(11): e1230. doi:10.1371/journal.pone.0001230.

Sereno, P. C., X. Zhao, and L. Tan. 2010. A new psittacosaur from Inner Mongolia and the parrot-like structure and function of the psittacosaur skull. Proceedings of the Royal Society B 277: 199-209.

Sereno, P. C., et al. 1999. Cretaceous sauropods from the Sahara and the uneven rate of skeletal evolution among dinosaurs. Science 286: 1342-1347. [11 co-authors]

Seymour, R. S. 1982. Physiological adaptations to aquatic life. Pp. 1-51 in C. Gans and F. H. Pough (eds.), Biology of the Reptilia, Vol. 13: Physiology D. Academic Press, London.

Sherratt, E., M. del Rosario Castañeda, R. J. Garwood, D. L. Mahler, T. J. Sanger, A. Herrel, K. de Queiroz, and J. B. Losos. 2015. Amber fossils demonstrate deep-time stability of Caribbean lizard communities. Proceedings of the National Academy of Sciences of the United States of America 112: 9961-9966.

Shi, H., J. F. Parham, Z. Fan, M. Hong, and F. Yin. 2008. Evidence for the massive scale of turtle farming in China. Oryx 42: 147-150.

Sill, W. D. 1967. *Proterochampsa barrionuevoi* and the early evolution of the Crocodilia. Bulletin of the Museum of Comparative Zoology, Harvard University 135: 415-446.

Silva, R. R., J. Ferigolo, P. Bajdek, and G. Piñeiro. 2017. The feeding habits of Mesosauridae. Frontiers in Earth Science 5: 23. doi:10.3389/feart.2017.00023.

Simões, T. R., M. W. Caldwell, R. L. Nydam, and P. Jiménez-Huidobro. 2017. Osteology, phylogeny, and functional morphology of two Jurassic lizard species and the early evolution of scansoriality in geckoes. Zoological Journal of the Linnean Society 180: 216-241.

Simões, T. R., M. W. Caldwell, M. Talanda, M. Bernardi, A. Palci, O. Vernygora, F. Bernardini, L. Mancini, and R. L. Nydam. 2018. The origin of squamates revealed by a Middle Triassic lizard from the Italian Alps. Nature 557: 706-709.

Simões, T. R., E. Wilner, M. W. Caldwell, L. C. Weinschütz, and A. W. A. Kellner. 2015. A stem acrodontan lizard in the Cretaceous of Brazil revises early lizard evolution in Gondwana. Nature Communications 6: 8149. doi:10.1038/ncomms9149.

Sinervo, B., et al. 2010. Erosion of lizard diversity by climate change and altered thermal niches. Science 328: 894-899. [25 co-authors]

Slowinski, J. B., and R. Lawson 2002. Snake phylogeny: evidence from nuclear and mitochondrial genes. Molecular Phylogenetics and Evolution 24: 194-202.

Smith, A. S., and G. J. Dyke. 2008. The skull of the giant predatory pliosaur *Rhomaleosaurus cramptoni*: implications for plesiosaur phylogenetics. Naturwissenschaften 95: 975-980.

Smith, A. S., and P. Vincent. 2010. A new genus of pliosaur (Reptilia: Sauropterygia) from the Lower Jurassic of Holzmaden, Germany. Palaeontology 53: 1049-1063.

Smith, H. F., J. H. Hutchinson, K. E. B. Townsend, B. Adrian, and D. Jager. 2017. Morphological variation, phylogenetic relationships, and geographic distribution of the Baenidae (Testudines), based on new specimens from the Uinta Formation (Uinta Basin), Utah (USA). PLoS ONE 12(7): e0180574. doi:10.1371/journal.pone 0180574.

Smith, K. K. 1980. Mechanical significance of streptostyly in lizards. Nature 283: 778-779.

Smith, K. K. 1982. An electromyographic study of the function of the jaw adducting muscles in *Varanus exanthematicus* (Varanidae). Journal of Morphology 173: 137-158.

Smith, K. T. 2009a. Eocene lizards of the clade *Geiseltaliellus* from Messel and Geiseltal, Germany, and the early radiation of Iguanidae (Reptilia: Squamata). Bulletin of the Peabody Museum of Natural History, Yale University 50: 219-306.

Smith, K. T. 2009b. A new lizard assemblage from the earliest Eocene (Zone Wa0) of the Bighorn Basin, Wyoming, USA: biogeography during the warmest interval of the Cenozoic. Journal of Systematic Palaeontology 7: 299-358.

Smith, K. T. 2011a. The evolution of mid-latitude faunas during the Eocene: late Eocene lizards from the Medicine Pole Hills reconsidered. Bulletin of the Peabody Museum of Natural History, Yale University 52: 3-105.

Smith, K. T. 2011b. On the phylogenetic affinity of the extinct acrodontan lizard *Tinosaurus*. Pp. 9-28 in K.-L. Schuchmann (ed.), Tropical Vertebrates in a Changing World. (Bonner Zoologische Monographien 57.) Zoologisches Forschungsmuseum Alexander Koenig, Bonn.

Smith, K. T. 2013. New constraints on the evolution of the snake clades Ungaliophiinae, Loxocemidae and Colubridae (Serpentes), with comments on the fossil history of erycine boids in North America. Zoologischer Anzeiger 252: 157-182.

Smith, K. T. 2017. First crocodile-tailed lizard (Squamata: *Pan-Shinisaurus*) from the Paleogene of Europe. Journal of Vertebrate Paleontology 37: e1313743. doi:10.1080/027246 34.2017.1313743.

Smith, K. T., and J. A. Gauthier. 2013. Early Eocene lizards of the Wasatch Formation near Bitter Creek, Wyoming: diversity and paleoenvironment during an interval of global warming. Bulletin of the Peabody Museum of Natural History, Yale University 54: 135-230.

Smith, K. T., S. F. K. Schaal, and J. Habersetzer (eds.). 2018. Messel. An Ancient Greenhouse Ecosystem. Senckenberg Gesellschaft für Naturforschung, Frankfurt am Main.

Smith, K. T., and M. Wuttke. 2012. From tree to shining sea: taphonomy of the arboreal lizard *Geiseltaliellus maarius* from Messel, Germany. Palaeobiodiversity and Palaeoenvironments 92: 45-65.

Smith, N. D., P. J. Makovicky, W. R. Hammer, and P. J. Currie. 2007. Osteology of *Cryolophosaurus ellioti* (Dinosauria: Theropoda) from the Early Jurassic of Antarctica and implications for early theropod evolution. Zoological Journal of the Linnean Society 151: 377-421.

Smith, N. D., and D. Pol. 2007. Anatomy of a basal sauropodomorph from the Early Jurassic Hanson Formation of Antarctica. Acta Palaeontologica Polonica 52: 657-674.

Smithson, T. R., R. L. Carroll, R. L. Panchen, and S. M. Andrews. 1994. *Westlothiana lizziae* from the Viséan of East Kirkton, West Lothian, Scotland, and the amniote stem. Transactions of the Royal Society of Edinburgh: Earth Sciences 84: 383-412.

Smithwick, F. M., R. Nicholls, I. C. Cuthill, and J. Vinther. 2017. Countershading and stripes in the theropod dinosaur *Sinosauropteryx* reveal heterogeneous habitats in the Early Cretaceous Jehol Biota. Current Biology 27: P3337-3343.

Snively, E., and J. M. Theodor. 2011. Common functional correlates of head-strike behavior in the pachycephalosaur *Stegoceras validum* (Ornithischia, Dinosauria) and combative artiodactyls. PLoS ONE 6(6): e21422. doi:10.1371/journal.pone.0021422.

Soto, M., D. Pol, and D. Perea. 2011. A new specimen of *Uruguaysuchus aznarezi* (Crocodyliformes: Notosuchia) from the middle Cretaceous of Uruguay and its phylogenetic relationships. Zoological Journal of the Linnean Society 163 (Supplement 1): S173-S198.

Spener, C. M. 1710. Disquisitio de crocodilo in lapide scissili expresso aliisque Lithozois. Miscellanea Berolinensia ad incrementum scientiarum, ex scriptis Societati Regiae Scientiarum exhibitis edita 1710: 99-118.

Spielmann, J. A., S. G. Lucas, A. B. Heckert, L. F. Rinehart, and H. R. Richards III. 2009. Redescription of *Spinosuchus caseanus* (Archosauromorpha: Trilophosauridae) from the Upper Triassic of North America. Palaeodiversity 2: 283-313.

Spielmann, J. A., S. G. Lucas, L. F. Rinehart, and A. B. Heckert. 2008. The Late Triassic archosauromorph *Trilophosaurus*. New Mexico Museum of Natural History & Science Bulletin 43: 1-177.

Spinks, P. Q., R. C. Thomson, M. Gidis, and H. B. Shaffer. 2004. Multilocus phylogeny of the New-World mud turtles (Kinosternidae) supports the traditional classification of the group. Molecular Phylogenetics and Evolution 76: 254-260.

Spinks, P. Q., R. C. Thomson, E. McCartney-Melstad, and H. B. Shaffer. 2016. Phylogeny and temporal diversification of the New World pond turtles (Emydidae). Molecular Phylogenetics and Evolution 103: 85-97.

Stein, K., C. Palmer, P. G. Gill, and M. J. Benton. 2008. The aerodynamics of the British Late Triassic Kuehneosauridae. Palaeontology 51: 967-981.

Sterli, J. 2010. Phylogenetic relationships among extinct and extant turtles: the position of Pleurodira and the effects of the fossils on rooting crown-group turtles. Contributions to Zoology 79: 93-106.

Sterli, J. 2015. A review of the fossil record of Gondwanan turtles of the clade *Meiolaniformes*. Bulletin of the Peabody Museum of Natural History, Yale University 56: 21-45.

Sterli, J., and M. S. de la Fuente. 2011. Redescription and evolutionary remarks on the Patagonian horned turtle *Niolamia argentina* Ameghino, 1899 (Testudinata, Meiolaniidae). Journal of Vertebrate Paleontology 31: 1210-1229.

Sterli, J., and M. S. de la Fuente. 2013. New evidence from the Palaeocene of Patagonia (Argentina) on the evolution and palaeobiogeography of Meiolaniformes (Testudinata, new taxon name). Journal of Systematic Palaeontology 11: 835-852.

Sterli, J., M. S. de la Fuente, and G. W. Rougier. 2007. Anatomy and relationships of *Palaeochersis talampayensis*, a Late Triassic turtle from Argentina. Palaeontographica A 281: 1-61.

Sternberg, C. M. 1951. Complete skeleton of *Leptoceratops gracilis* Brown from the Upper Edmonton Member on Red Deer River, Alberta. National Museum of Canada, Annual Report 123: 225-255.

Stevens, K. 2006. Binocular vision in theropod dinosaurs. Journal of Vertebrate Paleontology 26: 321-330.

Stevens, M. S. 1977. Further study of Castolon Local Fauna (Early Miocene), Big Bend National Park, Texas. Texas Memorial Museum, The Pearce-Sellards Series 28: 1-69.

Stewart, J. R. 1997. Morphology and evolution of the egg of oviparous amniotes. Pp. 291-326 in S. S. Sumida and K. L. M. Martin (eds.), Amniote Origins: Completing the Transition to Land. Academic Press, San Diego.

Stocker, M. R., and R. J. Butler. 2013. Phytosauria. Pp. 91-117 in S. J. Nesbitt, J. B. Desojo, and R. B. Irmis (eds.), Anatomy, Phylogeny and Palaeobiology of Early Archosaurs and Their Kin. Geological Society, London, Special Publications 379.

Stocker, M. R., S. J. Nesbitt, K. E. Criswell, W. G. Parker, L. M. Witmer, T. B. Rowe, R. Ridgely, and M. A. Brown. 2016. A dome-headed stem archosaur exemplifies convergence among dinosaurs and their distant relatives. Current Biology 26: 2674-2680.

Stocker, M. R., L.-J. Zhao, S. J. Nesbitt, X.-c. Wu, and C. Li. 2017. A short-snouted, Middle Triassic phytosaur and its implications for the morphological evolution and biogeography of Phytosauria. Scientific Reports 7: 46028. doi:10.1038/srep46028.

Storrs, G. W. 1991. Anatomy and relationships of *Corosaurus alcovensis* (Diapsida: Sauropterygia) from the Triassic Alcova Limestone of Wyoming. Bulletin of the Peabody Museum of Natural History, Yale University 44: 1-151.

Storrs, G. W. 1997. Morphological and taxonomic clarification of the genus *Plesiosaurus*. Pp. 145-190 in J. M. Callaway and E. L. Nicholls (eds.), Ancient Marine Reptiles. Academic Press, New York.

Storrs, G. W. 2003. Late Miocene-Early Pliocene crocodilian fauna of Lothagam, southwest Turkana Basin, Kenya. Pp. 137-159 in M. G. Leakey and J. M. Harris (eds.), Lothagam: The Dawn of Humanity in Eastern Africa. Columbia University Press, New York.

Storrs, G. W., and M. B. Efimov. 2000. Mesozoic crocodyliforms of north-central Eurasia. Pp. 402-419 in M. J. Benton, M. A. Shishkin, D. M. Unwin, and E. N. Kurochkin (eds.), The Age of Dinosaurs in Russia and Mongolia. Cambridge University Press, Cambridge, UK.

Storrs, G. W., D. J. Gower, and N. F. Large. 1996. The diapsid reptile, *Pachystropheus rhaeticus*, a probable choristodere from the Rhaetian of Europe. Palaeontology 39: 323-349.

Stromer, E. 1915. Ergebnisse der Forschungsreisen Prof. E. Stromers in den Wüsten Ägyptens. II. Wirbeltier-Reste der Baharîje-Stufe (unterstes Cenoman). 3. Das Original des Theropoden *Spinosaurus aegyptiacus* nov. gen., nov. spec. Abhandlungen der Königlich Bayerischen Akademie der Wissenschaften, mathematisch-physikalische Klasse 28(3): 1-32.

Stromer, E. 1925. Ergebnisse der Forschungsreisen Prof. E. Stromers in den Wüsten Ägyptens. II. Wirbeltier-Reste der Baharîje-Stufe (unterstes Cenoman). 7. *Stomatosuchus inermis* Stromer, ein schwach bezahnter Krokodilier und 8. Ein Skelettrest des Pristiden *Onchopristis numidus* Haug sp. Abhandlungen der Bayerischen Akademie der Wissenschaften, mathematisch-naturwissenschaftliche Abteilung 30(6): 1-22.

Stromer, E. 1931. Ergebnisse der Forschungsreisen Prof. E. Stromers in den Wüsten Ägyptens. II. Wirbeltier-Reste der Baharîje-Stufe (unterstes Cenoman). 10. Ein Skelett-Rest von *Carcharodontosaurus* gen. nov. Abhandlungen der Bayerischen Akademie der Wissenschaften, mathematisch-naturwissenschaftliche Abteilung, Neue Folge 9: 1-23.

Sues, H.-D. 1987a. Postcranial skeleton of *Pistosaurus* and interrelationships of the Sauropterygia (Diapsida). Zoological Journal of the Linnean Society 90: 109-131.

Sues, H.-D. 1987b. On the skull of *Placodus gigas* and the relationships of the Placodontia. Journal of Vertebrate Paleontology 7: 138-144.

Sues, H.-D. 2003. An unusual new archosauromorph reptile from the Upper Triassic Wolfville Formation of Nova Scotia. Canadian Journal of Earth Sciences 40: 635-649.

Sues, H.-D., and A. Averianov. 2009. *Turanoceratops tardabilis*—the first ceratopsid dinosaur from Asia. Naturwissenschaften 96: 645-652.

Sues, H.-D., and N. C. Fraser. 2010. Triassic Life on Land: The Great Transition. Columbia University Press, New York.

Sues, H.-D., E. Frey, D. M. Martill, and D. M. Scott. 2002. *Irritator challengeri*, a spinosaurid (Dinosauria: Theropoda) from the Lower Cretaceous of Brazil. Journal of Vertebrate Paleontology 22: 535-547.

Sues, H.-D., and P. M. Galton. 1987. Anatomy and classification of the North American Pachycephalosauria (Dinosauria: Ornithischia). Palaeontographica A 198: 1-40.

Sues, H.-D., S. J. Nesbitt, D. S Berman, and A. C. Henrici. 2011. A late-surviving basal theropod dinosaur from the latest Triassic of North America. Proceedings of the Royal Society B 278: 3459-3464.

Sues, H.-D., P. E. Olsen, J. G. Carter, and D. M. Scott. 2003. A new crocodylomorph reptile from the Upper Triassic of North Carolina. Journal of Vertebrate Paleontology 23: 329-343.

Sues, H.-D., P. E. Olsen, D. M. Scott, and P. S. Spencer. 2000. Cranial osteology of *Hypsognathus fenneri*, a latest Triassic procolophonid reptile from the Newark Supergroup of eastern North America. Journal of Vertebrate Paleontology 20: 275-284.

Sues, H.-D., and R. R. Reisz. 1998. Origins and early evolution of herbivory in tetrapods. Trends in Ecology and Evolution 13: 141-145.

Sues, H.-D., and R. R. Reisz. 2008. Anatomy and phylogenetic relationships of *Sclerosaurus armatus* Meyer in Fischer, 1857 (Amniota: Parareptilia) from the Buntsandstein

(Triassic) of Europe. Journal of Vertebrate Paleontology 28: 1031-1042.

Sukhanov, V. B. 2006. An archaic turtle, *Heckerochelys romani* gen. et sp. nov., from the Middle Jurassic of Moscow Region, Russia. Pp. 112-118 in I. G. Danilov and J. F. Parham (eds.), Fossil Turtle Research. Russian Journal of Herpetology 13, Supplement.

Sulimski, A. 1975. Macrocephalosauridae and Polyglyphanodontidae (Sauria) from the Late Cretaceous of Mongolia. Palaeontologia Polonica 33: 25-102.

Sullivan, C. 2010. The role of the calcaneal 'heel' as a propulsive lever in basal archosaurs and extant monitor lizards. Journal of Vertebrate Paleontology 30: 1422-1432.

Sullivan, C. 2015. Evolution of hind limb posture in Triassic archosauriforms. Pp. 107-123 in K. P. Dial, N. Shubin, and E. L. Brainerd (eds.), Great Transformations in Vertebrate Evolution. The University of Chicago Press, Chicago.

Sullivan, R. M. 1979. Revision of the Paleogene genus *Glyptosaurus* (Reptilia, Anguidae). Bulletin of the American Museum of Natural History 163: 1-72.

Sullivan, R. M. 2006. A taxonomic review of the Pachycephalosauridae (Dinosauria: Ornithischia). New Mexico Museum of Natural History & Science Bulletin 35: 347-366.

Sullivan, R. M., and R. Estes. 1997. A reassessment of the fossil Tupinambinae. Pp. 100-112 in R. F. Kay, R. H. Madden, R. L. Cifelli, and J. J. Flynn (eds.), Vertebrate Paleontology in the Neotropics: The Miocene Fauna of La Venta, Colombia. Smithsonian Institution Press, Washington, DC.

Sullivan, R. M., T. Keller, and J. Habersetzer. 1999. Middle Eocene (Geiseltalian) anguid lizards from Geiseltal and Messel, Germany. I. *Ophisauriscus quadrupes* Kuhn 1940. Courier Forschungsinstitut Senckenberg 216: 97-129.

Sullivan, R. M., and T. E. Williamson. 1999. A new skull of *Parasaurolophus* (Dinosauria: Hadrosauridae) from the Kirtland Formation of New Mexico and a revision of the genus. New Mexico Museum of Natural History & Science Bulletin 15: 1-52.

Sumida, S. S. 1997. Locomotor features of taxa spanning the origin of amniotes. Pp. 353-398 in S. S. Sumida and K. L. M. Martin (eds.), Amniote Origins: Completing the Transition to Land. Academic Press, San Diego.

Sumida, S. S., and R. E. Lombard. 1991. The atlas-axis complex in the late Paleozoic genus *Diadectes* and the characteristics of the atlas-axis complex across the amphibian to amniote transition. Journal of Paleontology 65: 973-983.

Szczygielski, T., and T. Sulej. 2016. Revision of the Triassic European turtles *Proterochersis* and *Murrhardtia* (Reptilia, Testudinata, Proterochersidae), with description of new taxa from Poland and Germany. Zoological Journal of the Linnean Society 177: 395-427.

Szyndlar, Z. 1984. Fossil snakes from Poland. Acta Zoologica Cracoviensia 28: 1-156.

Szyndlar, Z. 1994. Oligocene snakes of southern Germany. Journal of Vertebrate Paleontology 14: 24-37.

Szyndlar, Z., and H. H. Schleich. 1994. Two species of the genus *Eryx* (Serpentes: Boidae: Erycinae) from the Spanish Neogene with comments on the past distribution of the genus in Europe. Amphibia-Reptilia 15: 233-248.

Talanda, M. 2016. Cretaceous roots of the amphisbaenian lizards. Zoologica Scripta 45: 1-8.

Taquet, P. 1976. Géologie et paléontologie du gisement de Gadoufaoua (Aptien du Niger). (Cahiers de Paléontologie.) Éditions du Centre National de la Recherche Scientifique, Paris.

Tatarinov, L. P. 2006. [Sketch of Reptile Evolution.] GEOS, Moscow. [Russian]

Tatarinov, L. P. 2009. [Sketch of Reptile Evolution: Archosaurians and Theromorphs.] GEOS, Moscow. [Russian]

Taylor, M. A. 1987. A reinterpretation of ichthyosaur swimming and buoyancy. Palaeontology 30: 531-535.

Taylor, M. P. 2009. A re-evaluation of *Brachiosaurus altithorax* Riggs 1903 (Dinosauria, Sauropoda) and its generic separation from *Giraffatitan brancai* (Janensch 1914). Journal of Vertebrate Paleontology 29: 787-806.

Tchernov, E., O. Rieppel, H. Zaher, M. J. Polcyn, and L. L. Jacobs. 2000. A fossil snake with limbs. Science 287: 2010-2012.

Thewissen, J. G. M., and M. A. Taylor. 2007. Aquatic adaptations in the limbs of amniotes. Pp. 310-322 in B. K. Hall (ed.), Fins into Limbs: Evolution, Development, and Transformation. The University of Chicago Press, Chicago.

Thomas, D. A. 2015. The cranial anatomy of *Tenontosaurus tilletti* Ostrom, 1970 (Dinosauria, Ornithopoda). Palaeontologia Electronica 18.2.37A: 1-99. doi:10.26879/450.

Thompson, R. S., J. C. Parish, S. C. R. Maidment, and P. M. Barrett. 2012. Phylogeny of the ankylosaurian dinosaurs (Ornithischia: Thyreophora). Journal of Systematic Palaeontology 10: 301-312.

Throckmorton, G. S., J. A. Hopson, and P. Parks. 1981. A redescription of *Toxolophosaurus cloudi* Olson, a Lower

Cretaceous herbivorous sphenodontid reptile. Journal of Paleontology 55: 586-597.

Thulborn, R. A. 1980. The ankle joints of archosaurs. Alcheringa 4: 241-261.

Tong, H., and P. A. Meylan. 2013. Morphology and relationships of *Brachyopsemys tingitana* gen. et sp. nov. from the Early Paleocene of Morocco and recognition of the new eucryptodiran turtle family: Sandowniidae. Pp. 187-212 in D. B. Brinkman, P. A. Holroyd, and J. D. Gardner (eds.), Morphology and Evolution of Turtles. Springer, Dordrecht.

Townsend, T. M., A. Larson, E. Louis and J. R. Macey. 2004. Molecular phylogenetics of Squamata: the position of snakes, amphisbaenians, and dibamids, and the root of the squamate tree. Systematic Biology 53: 735-757.

Trotteyn, M. J., A. B. Arcucci, and T. Raugust. 2013. Proterochampsia: an endemic archosauriform clade from South America. Pp. 59-90 in S. J. Nesbitt, J. B. Desojo, and R. B. Irmis (eds.), Anatomy, Phylogeny and Palaeobiology of Early Archosaurs and Their Kin. Geological Society, London, Special Publications 379.

Tschopp, E., O. Mateus, and R. B. J. Benson. 2015. A specimen-level phylogenetic analysis and taxonomic revision of Diplodocidae (Dinosauria, Sauropoda). PeerJ 3: e857. doi:10.7717/peerj.857.

Tsuji, L. A. 2006. Cranial anatomy and phylogenetic relationships of the Permian parareptile *Macroleter poezicus*. Journal of Vertebrate Paleontology 26: 849-865.

Tsuji, L. A. 2013. Anatomy, cranial ontogeny and phylogenetic relationships of the pareiasaur *Deltavjatia rossicus* from the Late Permian of central Russia. Earth and Environmental Science Transactions of the Royal Society of Edinburgh 104: 81-122.

Tsuji, L. A., and J. Müller. 2009. Assembling the history of Parareptilia: phylogeny, diversification, and a new definition of the clade. Fossil Record 12: 71-81.

Tsuji, L. A., J. Müller, and R. R. Reisz. 2010. *Microleter mckinzieorum* gen. et sp. nov. from the Lower Permian of Oklahoma: the basalmost parareptile from Laurasia. Journal of Systematic Palaeontology 8: 245-255.

Tsuji, L. A., J. Müller, and R. R. Reisz. 2012. Anatomy of *Emeroleter levis* and the phylogeny of nycteroleter parareptiles. Journal of Vertebrate Paleontology 32: 45-67.

Turner, A. H. 2015. A review of *Shamosuchus* and *Paralligator* (Crocodyliformes, Neosuchia) from the Cretaceous of Asia. PLoS ONE 10(2): e0118116. doi:10.1371/journal.pone.0118116.

Turner, A. H., and G. A. Buckley. 2008. *Mahajangasuchus insignis* (Crocodyliformes: Mesoeucrocodylia) cranial anatomy and new data on the origin of the eusuchian-style palate. Journal of Vertebrate Paleontology 28: 382-408.

Turner, A. H., P. J. Makovicky, and M. A. Norell. 2012. A review of dromaeosaurid systematics and paravian phylogeny. Bulletin of the American Museum of Natural History 371: 1-206.

Turner, A. H., and A. C. Pritchard. 2015. The monophyly of Susisuchidae (Crocodyliformes) and its phylogenetic placement in Neosuchia. PeerJ 3: e759. doi:10.7717/peerj.759.

Turner, A. H., and J. W. Sertich. 2010. Phylogenetic history of *Simosuchus clarki* (Crocodyliformes: Notosuchia) from the Late Cretaceous of Madagascar. Society of Vertebrate Paleontology Memoir 10: 177-236.

Turner, M. L., L. A. Tsuji, O. Ide, and C. A. Sidor. 2015. The vertebrate fauna of the upper Permian of Niger—IX. The appendicular skeleton of *Bunostegos akokanensis* (Parareptilia: Pareiasauria). Journal of Vertebrate Paleontology 35: e994746. doi:10.1080/02724634.2014.994746.

Tykoski, R., T. B. Rowe, R. A. Ketcham, and M. W. Colbert. 2002. *Calsoyasuchus valliceps*, a new crocodyliform from the Early Jurassic Kayenta Formation of Arizona. Journal of Vertebrate Paleontology 22: 593-611.

Tzika, A. C., R. Helaers, G. Schramm, and M. C. Milinkovitch. 2011. Reptilian-transcriptome v1.0, a glimpse in the brain transcriptome of five divergent Sauropsida lineages and the phylogenetic position of turtles. EvoDevo 2: 19. doi:10.1186/2041-9139-2-19.

Underwood, G. 1957. *Lanthanotus* and the anguinomorphan lizards: a critical review. Copeia 1957: 20-30.

Underwood, G. 1967. A Contribution to the Classification of Snakes. British Museum (Natural History), London.

Unwin, D. M. 2001. An overview of the pterosaur assemblage from the Cambridge Greensand (Cretaceous) of eastern England. Geowissenschaftliche Mitteilungen, Museum für Naturkunde Berlin 4: 189-221.

Unwin, D. M. 2003. On the phylogeny and evolutionary history of pterosaurs. Pp. 139-190 in E. Buffetaut and J.-M. Mazin (eds.), Evolution and Palaeobiology of Pterosaurs. Geological Society, London, Special Publications 217.

Unwin, D. M. 2006. The Pterosaurs from Deep Time. Pi Press, New York.

Unwin, D. M., and J. Lü. 1997. On *Zhejiangopterus* and the relationships of pterodactyloid pterosaurs. Historical Biology 12: 199-210.

Upchurch, P. 1998. The phylogenetic relationships of sauropod dinosaurs. Zoological Journal of the Linnean Society 124: 43-103.

Upchurch, P., P. M. Barrett, and P. M. Galton. 2007. A phylogenetic analysis of basal sauropodomorph relationships: implications for the origin of sauropod dinosaurs. Special Papers in Palaeontology 77: 57-90.

Upchurch, P., and J. Martin. 2002. The Rutland *Cetiosaurus*: the anatomy and relationships of a Middle Jurassic British sauropod dinosaur. Palaeontology 45: 1049-1074.

Valenzuela, N. 2004. Temperature-dependent sex determination. Pp. 229-252 in D. C. Deeming (ed.), Reptilian Incubation Environment, Evolution and Behaviour. Nottingham University Press, Nottingham.

van der Reest, A., A. P. Wolfe, and P. J. Currie. 2016. A densely feathered ornithomimid (Dinosauria: Theropoda) from the Upper Cretaceous Dinosaur Park Formation, Alberta, Canada. Cretaceous Research 58: 108-117.

Varricchio, D. J., A. J. Martin, and Y. Katsura. 2007. First trace and body fossil evidence of a burrowing, denning dinosaur. Proceedings of the Royal Society B 274: 1361-1368.

Varricchio, D. J., P. C. Sereno, X. Zhao, L. Tan, J. A. Wilson, and G. H. Lyon. 2008. Mud-trapped herd captures evidence of distinctive dinosaur sociality. Acta Palaeontologica Polonica 53: 567-578.

Vasile, Ş., Z. Csiki-Sava, and M. Venczel. 2013. A new madtsoiid snake from the Upper Cretaceous of the Haţeg Basin, western Romania. Journal of Vertebrate Paleontology 33: 1100-1119.

Vaughn, P. P. 1955. The Permian reptile *Araeoscelis* restudied. Bulletin of the Museum of Comparative Zoology, Harvard College 113: 305-467.

Vélez-Juarbe, J., C. A. Brochu, and H. Santos. 2007. A gharial from the Oligocene of Puerto Rico: transoceanic dispersal in the history of a non-marine reptile. Proceedings of the Royal Society B 274: 1245-1254.

Versluys, J. 1912. Das Streptostylie-Problem und die Bewegungen im Schädel bei Sauropsiden. Zoologische Jahrbücher, Supplement 15, 2: 545-716.

Vickaryous, M. K., and B. K. Hall. 2006. Homology of the reptilian coracoid and a reappraisal of the evolution and development of the amniote pectoral apparatus. Journal of Anatomy 208: 263-285.

Vickaryous, M. K., T. Maryańska, and D. B. Weishampel. 2004. Ankylosauria. Pp. 362-393 in D. B. Weishampel,

P. Dodson, and H. Osmólska (eds.), The Dinosauria. Second Edition. University of California Press, Berkeley.

Vidal, N., and S. B. Hedges. 2005. The phylogeny of squamate reptiles (lizards, snakes, and amphisbaenians) inferred from nine nuclear protein-coding genes. Comptes Rendus Biologies 328: 1000-1008.

Vidal, N., and S. B. Hedges. 2009. The molecular evolutionary tree of lizards, snakes, and amphisbaenians. Comptes Rendus Biologies 332: 129-139.

Vidal, N., J. Marin, M. Morini, S. Donnellan, W. R. Branch, R. Thomas, M. Vences, A. Wynn, C. Cruaud, and S. B. Hedges. 2009. Blindsnake evolutionary tree reveals long history on Gondwana. Biology Letters 6: 558-561.

Vinther, J., R. Nicholls, S. Lautenschlager, M. Pittman, T. G. Kaye, E. Rayfield, G. Mayr, and I. C. Cuthill. 2016. 3D camouflage in an ornithischian dinosaur. Current Biology 26: 2456-2462.

Vitt, L. J., and J. P. Caldwell. 2009. Herpetology: An Introductory Biology of Amphibians and Reptiles. Third Edition. Academic Press, San Diego.

Völker, H. 1913. Über das Stamm-, Gliedmaßen- und Hautskelet von *Dermochelys coriacea* L. Zoologische Jahrbücher, Abteilung für Anatomie und Ontogenie der Tiere 33: 431-552.

Vremir, M., M. Witton, D. Naish, G. Dyke, S. L. Brusatte, M. Norell, and R. Totoianu. 2015. A medium-sized robust-necked azhdarchid pterosaur (Pterodactyloidea: Azhdarchidae) from the Maastrichtian of Pui (Haţeg Basin, Transylvania, Romania). American Museum Novitates 3827: 1-16.

Wagner, G. P., and J. A. Gauthier. 1999. 1, 2, 3 + 2, 3, 4: a solution to the problem of the homology of the digits in the avian hand. Proceedings of the National Academy of Sciences of the United States of America 96: 5111-5116.

Waldman, M., and S. E. Evans. 1994. Lepidosauromorph reptiles from the Middle Jurassic of Skye. Zoological Journal of the Linnean Society 112: 135-150.

Walker, A. D. 1961. Triassic reptiles from the Elgin area: *Stagonolepis*, *Dasygnathus* and their allies. Philosophical Transactions of the Royal Society of London B 244: 103-204.

Walker, A. D. 1964. Triassic reptiles from the Elgin area: *Ornithosuchus* and the origin of carnosaurs. Philosophical Transactions of the Royal Society of London B 248: 53-134.

Walker, A. D. 1990. A revision of *Sphenosuchus acutus* Haughton, a crocodylomorph reptile from the Elliot Formation (Late Triassic or Early Jurassic) of South

Africa. Philosophical Transactions of the Royal Society of London B 330: 1-120.

Wang, J., Y. Ye, R. Pei, Y. Tian, C. Feng, D. Zheng, and S.-C. Chang. 2018. Age of Jurassic basal sauropods in Sichuan, China: a reappraisal of basal sauropod evolution. Geological Society of America Bulletin 130: 1493-1500.

Wang, S., J. Stiegler, R. Amiot, X. Wang, G.-h. Du, J. M. Clark, and X. Xu. 2017. Extreme ontogenetic changes in a ceratosaurian theropod. Current Biology 27: 144-148.

Wang, X., A. W. A. Kellner, S. Jiang, and X. Meng. 2009. An unusual long-tailed pterosaur with elongated neck from western Liaoning of China. Anais da Academia Brasileira de Ciências 81: 793-812.

Wang, X., and J. Lü. 2001. Discovery of a pterodactyloid pterosaur from the Yixian Formation of western Liaoning, China. Chinese Science Bulletin 46: 1112-1117.

Wang, X., Z. Zhou, F. Zhang, and X. Xu. 2002. A nearly complete articulated rhamphorhynchid pterosaur with exeptionally well-preserved wing membranes and "hairs" from Inner Mongolia, northeast China. Chinese Science Bulletin 47: 226-230.

Wang, X., et al. 2014. Sexually dimorphic tridimensionally preserved pterosaurs and their eggs from China. Current Biology 24: 1323-1330. [11 co-authors]

Wang, X., et al. 2017. Egg accumulation with 3D embryos provides insight into the life history of a pterosaur. Science 338: 1197-1201. [16 co-authors]

Wang, Y.-m., H.-l. You, and T. Wang. 2017. A new basal sauropodiform dinosaur from the Lower Jurassic of Yunnan Province, China. Scientific Reports 7: 41881. doi:10.1038/srep41881.

Wang, Z., et al. 2013. The draft genomes of soft-shell turtle and green sea turtle yield insights into the development and evolution of the turtle-specific body plan. Nature Genetics 45: 701-706. [33 co-authors]

Watson, D. M. S. 1914. *Eunotosaurus africanus* Seeley, and the ancestry of the Chelonia. Proceedings of the Zoological Society of London 1914: 1011-1020.

Watson, D. M. S. 1924. The elasmosaurid shoulder-girdle and fore-limb. Proceedings of the Zoological Society of London 28: 85-95.

Weber, S. 2004. *Ornatocephalus metzleri* gen. et spec. nov. (Lacertilia, Scincoida)—taxonomy and paleobiology of a basal scincoid lizard from the Messel Formation (Middle Eocene: basal Lutetian, Geiseltalium). Abhandlungen der Senckenbergischen Naturforschenden Gesellschaft 561: 1-159.

Wedel, M. J. 2005. Postcranial skeletal pneumaticity in sauropods and implications for mass estimates. Pp. 201-228 in K. A. Curry Rogers and J. A. Wilson (eds.), The Sauropods: Evolution and Paleobiology. University of California Press, Berkeley.

Wedel, M. J. 2009. Evidence for bird-like air sacs in saurischian dinosaurs. Journal of Experimental Zoology 311A: 611-628.

Wedel, M. J., R. L. Cifelli, and R. K. Sanders. 2000. Osteology, paleobiology, and relationships of the sauropod dinosaur *Sauroposeidon*. Acta Palaeontologica Polonica 45: 343-388.

Weinbaum, J. 2011. The skull of *Postosuchus kirkpatricki* (Archosauria: Paracrocodyliformes) from the Upper Triassic of the United States. PaleoBios 30: 18-44.

Weishampel, D. B. 1981. Acoustic analyses of potential vocalization in lambeosaurine dinosaurs (Reptilia: Ornithischia). Paleobiology 7: 252-261.

Weishampel, D. B. 1984. Evolution of jaw mechanisms in ornithischian dinosaurs. Advances in Anatomy, Embryology and Cell Biology 87: 1-110.

Weishampel, D. B., P. Dodson, and H. Osmólska (eds.). 2004. The Dinosauria. Second Edition. University of California Press, Berkeley.

Weishampel, D. B., C.-M. Jianu, Z. Csiki, and D. B. Norman. 2003. Osteology and relationships of *Zalmoxes* (N.G.), an unusual euornithopod dinosaur from the latest Cretaceous of Romania. Journal of Systematic Palaeontology 1: 65-123.

Weishampel, D. B., D. B. Norman, and D. Grigorescu. 1993. *Telmatosaurus transsylvanicus* from the Late Cretaceous of Romania: the most basal hadrosaurid dinosaur. Palaeontology 36: 361-385.

Welles, S. P. 1943. Elasmosaurid plesiosaurs with description of new material from California and Colorado. Memoirs of the University of California 13: 125-254.

Welles, S. P. 1952. A review of the North American Cretaceous elasmosaurs. University of California Publications in Geological Sciences 29: 47-143.

Welles, S. P. 1984. *Dilophosaurus wetherilli* (Dinosauria, Theropoda). Osteology and comparisons. Palaeontographica A 185: 85-180.

Wellik, D. M. 2007. *Hox* patterning of the vertebrate axial skeleton. Developmental Dynamics 236: 2454-2463.

Wellnhofer, P. 1970. Die Pterodactyloidea (Pterosauria) der Oberjura-Plattenkalke Süddeutschlands. Abhandlungen der Bayerischen Akademie der Wissenschaften, mathematisch-naturwissenschaftliche Klasse, Neue Folge, 141: 1-133.

Wellnhofer, P. 1975. Die Rhamphorhynchoidea der Oberjura-Plattenkalke Süddeutschlands. I-III. Palaeontographica A 148: 1-33 and 132-186; 149: 1-30.

Wellnhofer, P. 1978. Pterosauria. Pp. 1-82 in P. Wellnhofer (ed.), Handbuch der Paläoherpetologie, Part 19. Gustav Fischer Verlag, Stuttgart.

Wellnhofer, P. 1987. New crested pterosaurs from the Lower Cretaceous of Brazil. Mitteilungen der Bayerischen Staatssammlung für Paläontologie und historische Geologie 27: 175-186.

Wellnhofer, P. 1988. Terrestrial locomotion in pterosaurs. Historical Biology 1: 3-16.

Wellnhofer, P. 1991a. The Illustrated Encyclopedia of Pterosaurs. Crescent Books, New York.

Wellnhofer, P. 1991b. Weitere Pterosaurierfunde aus der Santana-Formation (Apt) der Chapada do Araripe, Brasilien. Palaeontographica A 215: 43-101.

Wellnhofer, P. 2009. *Archaeopteryx*: The Icon of Evolution. Verlag Dr. Friedrich Pfeil, Munich.

Wellnhofer, P., and A. W. A. Kellner. 1991. The skull of *Tapejara wellnhoferi* Kellner (Reptilia, Pterosauria) from the Lower Cretaceous Santana Formation of the Araripe Basin, northeastern Brazil. Mitteilungen der Bayerischen Staatssammlung für Paläontologie und historische Geologie 31: 89-106.

Werneburg, I. 2012. Temporal bone arrangements in turtles: an overview. Journal of Experimental Biology (Molecular and Developmental Evolution) 318: 235-249.

Westphal, F. 1962. Die Krokodilier des deutschen und englischen Oberen Lias. Palaeontographica A 118: 23-118.

Westphal, F. 1975. Bauprinzipien im Panzer der Placodonten (Reptilia triadica). Paläontologische Zeitschrift 49: 97-125.

Whiteside, D. I. 1986. The head skeleton of the Rhaetian sphenodontid *Diphydontosaurus avonis* gen. et sp. nov. and the modernizing of a living fossil. Philosophical Transactions of the Royal Society of London B 312: 379-430.

Whiteside, D. I., and C. J. Duffin. 2017. Late Triassic terrestrial microvertebrates from Charles Moore's '*Microlestes*' quarry, Holwell, Somerset, UK. Zoological Journal of the Linnean Society 179: 677-705.

Whiteside, D. I., C. J. Duffin, and H. Furrer. 2017. The Late Triassic lepidosaur fauna from Hallau, north-eastern Switzerland, and a new 'basal' rhynchocephalian

Deltadectes elvetica gen. et sp. nov. Neues Jahrbuch für Geologie und Paläontologie Abhandlungen 285: 53-74.

Whiting, E. T., and A. K. Hastings, 2015. First fossil *Alligator* from the late Eocene of Nebraska and the Late Paleogene record of alligators in the Great Plains. Journal of Herpetology 49: 560-569.

Whiting, E. T., D. W. Steadman, and K. A. Vliet. 2016. Cranial polymorphism and systematics of Miocene and living *Alligator* in North America. Journal of Herpetology 50: 306-315.

Whitlock, J. A. 2011. A phylogenetic analysis of Diplodocoidea (Saurischia: Sauropoda). Zoological Journal of the Linnean Society 161: 872-915.

Wiens, J. J., C. R. Hutter, D. G. Mulcahy, B. P. Noonan, T. M. Townsend, J. W. Sites, Jr., and T. W. Reeder. 2012. Resolving the phylogeny of lizards and snakes (Squamata) with extensive sampling of genes and species. Biological Letters 8: 1043-1046.

Wiens, J. J., C. A. Kuczynski, T. Townsend, T. W. Reeder, D. G. Mulcahy, and J. W. Sites, Jr. 2010. Combining phylogenomics and fossils in higher-level squamate reptile phylogeny: molecular data change the placement of fossil taxa. Systematic Biology 59: 674-688.

Wilberg, E. W. 2015. A new metriorhynchoid (Crocodylomorpha, Thalattosuchia) from the Middle Jurassic of Oregon and the evolutionary timing of marine adaptations in thalattosuchian crocodylomorphs. Journal of Vertebrate Paleontology 35: e902846. doi:10.1080/027246 34.2014.902846.

Wild, R. 1973. Die Triasfauna der Tessiner Kalkalpen. XXIII. *Tanystropheus longobardicus* (Bassani) (Neue Ergebnisse). Schweizerische Paläontologische Abhandlungen 95: 1-162.

Wild, R. 1978. Die Flugsaurier (Reptilia, Pterosauria) aus der Oberen Trias von Cene bei Bergamo, Italien. Bolletino della Società Paleontologica Italiana 17: 176-256.

Williams, E. E. 1950. Variation and selection in the cervical central articulations of living turtles. Bulletin of the American Museum of Natural History 94: 505-562.

Williston, S. W. 1917. The phylogeny and classification of reptiles. Journal of Geology 25: 411-421.

Wilson, J. A. 2002. Sauropod dinosaur phylogeny: critique and cladistic analysis. Zoological Journal of the Linnean Society 136: 217-276.

Wilson, J. A. 2005. Redescription of the Mongolian sauropod *Nemegtosaurus mongoliensis* Nowinski (Dinosauria: Saurischia) and comments on Late Cretaceous sauropod diversity. Journal of Systematic Palaeontology 3: 283-318.

Wilson, J. A., and R. Allain. 2015. Osteology of *Rebbachisaurus garasbae* Lavocat, 1954, a diplodocoid (Dinosauria, Sauropoda) from the early Late Cretaceous-aged Kem Kem beds of southeastern Morocco. Journal of Vertebrate Paleontology 35: e1000701. doi:10.1080/02724634.2014.1000701.

Wilson, J. A., D. M. Mohabey, S. E. Peters, and J. J. Head. 2010. Predation upon hatchling dinosaurs by a new snake from the Late Cretaceous of India. PLoS Biology 8(3): e1000322. doi:10.1371/journal.pbio.1000322.

Wilson, J. A., and P. C. Sereno. 1998. Early evolution and higher-level phylogeny of sauropod dinosaurs. Society of Vertebrate Paleontology Memoir 5: 1-68.

Wilson, J. A., P. C. Sereno, S. Srivastava, D. K. Bhatt, A. Khosla, and A. Sahni. 2003. A new abelisaurid (Dinosauria, Theropoda) from the Lameta Formation (Cretaceous, Maastrichtian) of India. Contributions from the Museum of Paleontology, The University of Michigan 31: 1-42.

Wilson, J. A., and P. Upchurch. 2009. Redescription and reassessment of the phylogenetic affinities of *Euhelopus zdanskyi* (Dinosauria: Sauropoda) from the Early Cretaceous of China. Journal of Systematic Palaeontology 7: 199-239.

Wiman, C. 1929. Die Kreide-Dinosaurier aus Shantung. Palaeontologia Sinica, Ser. C, 6(1): 1-67.

Wiman, C. 1931. *Parasaurolophus tubicen* n. sp. aus der Kreide in New Mexico. Nova Acta Regiae Societatis Scientiarum Upsaliensis, Ser. 4, 7: 1-11.

Winkler, D. A., P. A. Murry, and L. L. Jacobs. 1997. A new species of *Tenontosaurus* (Dinosauria: Ornithopoda) from the Early Cretaceous of Texas. Journal of Vertebrate Paleontology 17: 330-348.

Wintrich, T., S. Hayashi, A. Houssaye, Y. Nakajima, and P. M. Sander. 2017. A Triassic plesiosaurian skeleton and bone histology inform on the evolution of a unique body plan. Science Advances 3: e1701144. doi:10.1126/sciadv.1701144.

Witmer, L. M. 1997. The evolution of the antorbital cavity of archosaurs: a study in soft-tissue reconstruction in the fossil record with an analysis of the function of pneumaticity. Society of Vertebrate Paleontology Memoir 3: 1-73.

Witmer, L. M., S. Chatterjee, J. Franzosa, and T. Rowe. 2003. Neuroanatomy of flying reptiles and implications for flight, posture and behaviour. Nature 425: 950-953.

Witmer, L. M., and R. C. Ridgely. 2008. The paranasal sinuses of predatory and armored dinosaurs (Archosauria: Theropoda and Ankylosauria) and their contribution to cephalic structure. The Anatomical Record 291: 1362-1388.

Witton, M. P. 2012. New insights into the skull of *Istiodactylus latidens* (Ornithocheiroidea, Pterodactyloidea). PLoS ONE 7(3): e33170. doi:10.1371/journal.pone.0033170.

Witton, M. P. 2013. Pterosaurs. Princeton University Press, Princeton, New Jersey.

Witton, M. P., and M. B. Habib. 2010. On size and flight diversity of giant pterosaurs, the use of birds as pterosaur analogues and comments on pterosaur flightlessness. PLoS ONE 5(11): e13982. doi:10.1371/journal.pone.0013982.

Witton, M. P., and D. Naish. 2008. A reappraisal of azhdarchid pterosaur functional morphology and paleoecology. PLoS ONE 3(5): e2271. doi:10.1371/journal.pone.0002271.

Wolfe, D. G., and J. I. Kirkland. 1998. *Zuniceratops christopheri* n. gen. & n. sp., a ceratopsian dinosaur from the Moreno Hill Formation (Cretaceous, Turonian) of west-central New Mexico. New Mexico Museum of Natural History & Science Bulletin 14: 303-317.

Woltering, J. M. 2012. From lizard to snake: behind the evolution of an extreme body plan. Current Genomics 13: 289-299.

Wood, R. C. 1976. *Stupendemys geographicus*, the world's largest turtle. Breviora 436: 1-31.

Wood, R. C., J. Johnson-Gove, E. S. Gaffney, and K. F. Maley. 1996. Evolution and phylogeny of leatherback turtles (Dermochelyidae), with descriptions of new fossil taxa. Chelonian Conservation and Biology 2: 266-286.

Wu, W., and P. Godefroit. 2012. Anatomy and relationships of *Bolong yixianensis*, an Early Cretaceous iguanodontoid dinosaur from western Liaoning. Pp. 293-333 in P. Godefroit (ed.), Bernissart Dinosaurs and Early Cretaceous Terrestrial Ecosystems. Indiana University Press, Bloomington.

Wu, X.-c., Z. Cheng, and A. P. Russell. 2001c. Cranial anatomy of a new crocodyliform (Archosauria: Crocodylomorpha) from the Lower Cretaceous of Song-Liao Plain, northeastern China. Canadian Journal of Earth Sciences 38: 1653-1663.

Wu, X.-c., D. B. Brinkman, A. P. Russell, Z.-m. Dong, P. J. Currie, L.-h. Hou, and G.-h. Cui. 1993. Oldest known amphisbaenian from the Upper Cretaceous of Chinese Inner Mongolia. Nature 366: 57-59.

Wu, X.-c., D. B. Brinkman, and A. P. Russell. 1996. *Sineoamphisbaena hexatabularis*, an amphisbaenian (Diapsida: Squamata) from the Upper Cretaceous redbeds at Bayan

Mandahu (Inner Mongolia, People's Republic of China), and comments on the phylogenetic relationships of the Amphisbaenia. Canadian Journal of Earth Sciences 33: 541-577.

Wu, X.-c., J. Li, and X. Li. 1994. Phylogenetic relationship of *Hsisosuchus*. Vertebrata PalAsiatica 32: 166-180.

Wu, X.-c., A. P. Russell, and D. B. Brinkman. 2001a. A review of *Leidyosuchus canadensis* Lambe, 1907 (Archosauria: Crocodylia) and an assessment of cranial variation based upon new material. Canadian Journal of Earth Sciences 38: 1665-1687.

Wu, X.-c., A. P. Russell, and S. L. Cumbaa. 2001b. *Terminonaris* (Archosauria: Crocodyliformes): new material from Saskatchewan, Canada, and comments on its phylogenetic relationships. Journal of Vertebrate Paleontology 21: 492-514.

Wu, X.-c., and H.-D. Sues. 1996. Anatomy and phylogenetic relationships of *Chimaerasuchus paradoxus*, an unusual crocodyliform reptile from the Lower Cretaceous of Hubei, China. Journal of Vertebrate Paleontology 16: 688-702.

Wu, X.-c., H.-D. Sues, and Z.-m. Dong. 1997. *Sichuanosuchus shuhanensis*, a new ?Early Cretaceous protosuchian (Archosauria: Crocodyliformes) from Sichuan (China) and the monophyly of Protosuchia. Journal of Vertebrate Paleontology 17: 89-103.

Xu, X., J. M. Clark, C. A. Forster, M. A. Norell, G. M. Erickson, D. A. Eberth, C. Jia, and Q. Zhao. 2006a. A basal tyrannosauroid from the Late Jurassic of China. Nature 439: 715-718.

Xu, X., et al. 2009. A Jurassic ceratosaur from China helps clarify avian digital homologies. Nature 459: 940-944. [14 co-authors]

Xu, X., P. Currie, M. Pittman, L. Xing, Q. Meng, J. Lü, D. Hu, and C. Yu. 2017. Mosaic evolution in an asymmetrically feathered troodontid dinosaur with transitional features. Nature Communications 8: 14972. doi:10.1038/ncomms14972.

Xu, X., C. A. Forster, J. M. Clark, and J. Mo. 2006b. A basal ceratopsian with transitional features from the Late Jurassic of northwestern China. Proceedings of the Royal Society B 273: 2135-2140.

Xu, X., P. J. Makovicky, X.-l. Wang, M. A. Norell, and H.-l. You. 2002. A ceratopsian dinosaur from China and the early evolution of Ceratopsia. Nature 416: 314-317.

Xu, X., M. A. Norell, X. Kuang, X. Wang, Q. Zhao, and C. Jia. 2004. Basal tyrannosauroids from China and evidence of protofeathers in tyrannosauroids. Nature 431: 680-684.

Xu, X., and D. Pol. 2014. *Archaeopteryx*, paravian phylogenetic analyses, and the use of probability-based methods for palaeontological datasets. Journal of Systematic Palaeontology 12: 323-334.

Xu, X., P. Upchurch, Q. Ma, M. Pittman, J. Choiniere, C. Sullivan, D. W. E. Hone, Q. Tan, L. Tan, D. Xiao, and F. Han. 2013. Osteology of the Late Cretaceous alvarezsauroid *Linhenykus monodactylus* from China and comments on alvarezsauroid biogeography. Acta Palaeontologica Polonica 58: 25-46.

Xu, X., K. Wang, K. Zhang, Q. Ma, L. Xing, C. Sullivan, D. Hu, S. Cheng, and S. Wang. 2012. A gigantic feathered dinosaur from the Lower Cretaceous of China. Nature 484: 92-95.

Xu, X., K. Wang, X. Zhao, C. Sullivan, and S. Chen. 2010b. A new leptoceratopsid (Ornithischia: Ceratopsia) from the Upper Cretaceous of Shandong, China, and its implications for neoceratopsian evolution. PLoS ONE 5(11): e13835. doi:10.1371/journal.pone.0013835.

Xu, X., X.-l. Wang, and X.-c.Wu. 1999. A dromaeosaurid dinosaur with a filamentous integument from the Yixian Formation of China. Nature 401: 262-266.

Xu, X., X. Zheng, C. Sullivan, X. Wang, L. Xing, Y. Wang, X. Zhang, J. K. O'Connor, F. Zhang, and Y. Pan. 2015. A bizarre Jurassic maniraptoran theropod with preserved evidence of membranous wings. Nature 521: 70-73.

Xu, X., X. Zheng, and H. You. 2010a. Exceptional dinosaur fossils show ontogenetic development of early feathers. Nature 464: 1338-1341.

Xu, X., Z. Zhou, R. Dudley, S. Mackem, C.-m. Chuong, G. M. Erickson, and D. J. Varricchio. 2014. An integrative approach to understanding bird origins. Science 346: 1253293. doi:10.1126/science.1253293.

Xu, X., Z. Zhou, X. Wang, X. Kuang, F. Zhang, and X. Du. 2003. Four-winged dinosaurs from China. Nature 421: 335-340.

Yates, A. M. 2003a. A new species of the primitive dinosaur *Thecodontosaurus* (Saurischia: Sauropodomorpha) and its implications for the systematics of early dinosaurs. Journal of Systematic Palaeontology 1: 1-42.

Yates, A. M. 2003b. The species taxonomy of the sauropodomorph dinosaurs from the Löwenstein Formation (Norian, Late Triassic) of Germany. Palaeontology 46: 317-337.

Yates, A. M. 2004. *Anchisaurus polyzelus* (Hitchcock): the smallest known sauropod dinosaur and the evolution of gigantism among sauropodomorph dinosaurs. Postilla 230: 1-58.

Yates, A. M. 2007a. Solving a dinosaurian puzzle: the identity of *Aliwalia rex* Galton. Historical Biology 19: 93-123.

Yates, A. M. 2007b. The first complete skull of the Triassic dinosaur *Melanorosaurus* Haughton (Sauropodomorpha: Anchisauria). Special Papers in Palaeontology 77: 9-55.

Yates, A. M. 2010. A revision of the problematic sauropodomorph dinosaurs from Manchester, Connecticut and the status of *Anchisaurus* Marsh. Palaeontology 53: 739-752.

Yates, A. M., M. F. Bonnan, J. Neveling, A. Chinsamy, and M. G. Blackbeard. 2010. A new transitional sauropodomorph dinosaur from the Early Jurassic of South Africa and the evolution of sauropod feeding and quadrupedalism. Proceedings of the Royal Society B 277: 787-794.

Yates, A. M., and J. W. Kitching. 2003. The earliest known sauropod dinosaur and the first steps towards sauropod locomotion. Proceedings of the Royal Society of London B 270: 1753-1758.

Young, C.-c. 1941. A complete osteology of *Lufengosaurus huenei* Young (gen. et sp. nov.) from Lufeng, Yunnan, China. Palaeontologia Sinica, New Series, C 7: 1-53.

Young, C.-c. 1951. The Lufeng saurischian fauna in China. Palaeontologia Sinica, New Series, C 13: 1-96.

Young, C.-c., and X.-j. Zhao 1972. [*Mamenchisaurus* from Ho Chuan.] Institute of Vertebrate Paleontology and Paleoanthropology, Monograph Series A, no. 8. Science Press, Beijing. [Chinese]

Young, M. T., and M. B. de Andrade. 2009. What is *Geosaurus*? Redescription of *Geosaurus giganteus* (Thalattosuchia: Metriorhynchidae) from the Upper Jurassic of Bayern, Germany. Zoological Journal of the Linnean Society 157: 551-585.

Young, M. T., S. L. Brusatte, M. B. de Andrade, J. B. Desojo, B. L. Beatty, L. Steel, M. S. Fernández, M. Sakamoto, J. I. Ruiz-Omeñaca, and R. R. Schoch. 2012. The cranial osteology and feeding ecology of the metriorhynchid crocodylomorph genera *Dakosaurus* and *Plesiosuchus* from the Late Jurassic of Europe. PLoS ONE 7(9): e44985. doi:10.1371/journal.pone.0044985.

Young, M. T., S. L. Brusatte, M. Ruta, and M. B. de Andrade. 2010. The evolution of Metriorhynchoidea (Mesoeucrocodylia, Thalattosuchia): an integrated approach using geometric morphometrics, analysis of disparity, and biomechanics. Zoological Journal of the Linnean Society 158: 801-859.

Young, M. T., S. Hua, L. Steel, D. Foffa, S. L. Brusatte, S. Thüring, O. Mateus, J. I. Ruiz-Omeñaca, P. Havlik, Y. Lepage, and M. B. de Andrade. 2014. Revision of the Late Jurassic teleosaurid genus *Machimosaurus* (Crocodylomor-pha, Thalattosuchia). Royal Society Open Science 1: 140222. doi:10.1098/rsos.140222.

Young, M. T., J. P. Tennant, S. L. Brusatte, T. J. Challands, N. C. Fraser, N. D. L. Clark, and D. A. Ross. 2016. The first definitive Middle Jurassic atoposaurid (Crocodylomorpha, Neosuchia), and a discussion on the genus *Theriosuchus*. Zoological Journal of the Linnean Society 176: 443-462.

Zaher, H., S. Apesteguía, and C. A. Scanferla. 2009. The anatomy of the Upper Cretaceous snake *Najash rionegrina* Apesteguía & Zaher, 2006, and the evolution of limblessness in snakes. Zoological Journal of the Linnean Society 156: 801-826.

Zaher, H., and C. A. Scanferla. 2012. The skull of the Upper Cretaceous snake *Dinilysia patagonica* Smith-Woodward, 1901, and its phylogenetic position revisited. Zoological Journal of the Linnean Society 164: 194-238.

Zanazzi, A., M. J. Kohn, B. J. MacFadden, and D. O. Terry Jr. 2007. Large temperature drop across the Eocene-Oligocene transition in central North America. Nature 445: 639-642.

Zangerl, R. 1944. Contribution to the osteology of the skull of the Amphisbaenidae. American Midland Naturalist 31: 417-454.

Zangerl, R. 1953. The vertebrate fauna of the Selma Formation of Alabama. Part III. The turtles of the family Protostegidae. Part IV. The turtles of the family Toxochelyidae. Fieldiana Geology Memoirs 3: 57-277.

Zangerl, R. 1969. The turtle shell. Pp. 311-339 in C. Gans, A. d'A. Bellairs, and T. S. Parsons (eds.), Biology of the Reptilia. Volume 1: Morphology A. Academic Press, London.

Zanno, L. E. 2010a. Osteology of *Falcarius utahensis* (Dinosauria: Theropoda): characterizing the anatomy of basal therizinosaurs. Zoological Journal of the Linnean Society 158: 196-230.

Zanno, L. E. 2010b. A taxonomic and phylogenetic re-evaluation of Therizinosauria (Dinosauria: Maniraptora). Journal of Systematic Palaeontology 8: 503-543.

Zanno, L. E., S. Drymala, S. J. Nesbitt, and V. P. Schneider. 2015. Early crocodylomorph increases top tier predator diversity during rise of dinosaurs. Scientific Reports 5: 9276. doi:10.1038/srep09276.

Zanno, L. E., and P. J. Makovicky. 2011. Herbivorous ecomorphology and specialization patterns in theropod dinosaur evolution. Proceedings of the National Academy of Sciences of the United States of America 108: 232-237.

Zanno, L. E., and P. J. Makovicky. 2013. Neovenatorid theropods are apex predators in the Late Cretaceous of North America. Nature Communications 4: 2827. doi:10.1038/ncomms3827.

Zardoya, R., and A. Meyer. 1998. Complete mitochondrial genome indicates diapsid affinities of turtles. Proceedings of the National Academy of Sciences of the United States of America 95: 14226-14231.

Zeletnisky, D. K., F. Therrien, G. M. Erickson, C. L. DeBuhr, Y. Kobayashi, D. A. Eberth, and F. Hadfield. 2012. Feathered non-avian dinosaurs from North America provide insight into wing origins. Science 338: 510-514.

Zeletnisky, D. K., F. Therrien, and Y. Kobayashi. 2009. Olfactory acuity in theropods: palaeobiological and evolutionary implications. Proceedings of the Royal Society B 276: 667-673.

Zhang, F., S. L. Kearns, P. J. Orr, M. J. Benton, Z. Zhou, D. Johnson, X. Xu, and X. Wang. 2010. Fossilized melanosomes and the colour of Cretaceous dinosaurs and birds. Nature 463: 1075-1078.

Zhang, F., Z. Zhou, X. Xu, and X. Wang. 2002. A juvenile coelurosaurian theropod from China indicates arboreal habits. Naturwissenschaften 89: 394-398.

Zhang, F., Z. Zhou, X. Xu, X. Wang and C. Sullivan. 2008. A bizarre Jurassic maniraptoran from China with elongate ribbon-like feathers. Nature 455: 1105-1108.

Zhang, Y. 1988. [The Middle Jurassic dinosaur fauna from Dashanpu, Zigong, Sichuan. Sauropod dinosaur (I) *Shunosaurus*.] Sichuan Publishing House of Science and Technology, Chengdu. [Chinese]

Zhao, Q., M. J. Benton, X. Xu, and P. M. Sander. 2014. Juvenile-only clusters and behaviour of the Early Cretaceous dinosaur *Psittacosaurus*. Acta Palaeontologica Polonica 59: 827-833.

Zhao, X., Z. Cheng, and X. Xu. 1999. The earliest ceratopsian from the Tuchengzi Formation of Liaoning, China. Journal of Vertebrate Paleontology 19: 681-691.

Zhao, X., Z. Cheng, X. Xu, and P. J. Makovicky. 2006. A new ceratopsian from the Upper Jurassic Houcheng Formation of Hebei, China. Acta Geologica Sinica (English Edition) 80: 467-473.

Zhao, X., and P. J. Currie. 1994. A large crested theropod from the Jurassic of Xinjiang, People's Republic of China. Canadian Journal of Earth Sciences 30: 2027-2036.

Zheng, X.-t., H.-l. You, X. Xu, and Z.-m. Dong. 2009. An Early Cretaceous heterodontosaurid dinosaur with filamentous integumentary structures. Nature 458: 333-336.

Index